Oocyte Growth and Maturation

Oocyte Growth and Maturation

Edited by
T. A. Dettlaff
and
S. G. Vassetzky
N. K. Kol'tsov Institute of Developmental Biology
Academy of Sciences of the USSR
Moscow, USSR

Translation Edited by
Frank Billett
University of Southampton
Southampton, England

Consultants Bureau • New York and London

Library of Congress Cataloging in Publication Data

Oocyte growth and maturation / edited by T. A. Dettlaff and S. G. Vassetzky: translation edited by Frank S. Billett.

p. cm.

Translated from the Russian.

Includes bibliographies and index.

ISBN-13: 978-1-4684-0684-9 e-ISBN-13: 978-1-4684-0682-5

DOI: 10.1007/978-1-4684-0682-5

1. Oogenesis. I. Detlaf, Tat'iana Antonovna. II. Vassetzky, S. G. III. Billett, Frank S.

QL965.065 1988

591.3′2—dc19 88-3966

CIP

This translation is published under an agreement with the Copyright Agency of the USSR (VAAP).

Softcover reprint of the hardcover 1st edition 1988
A Division of Plenum Publishing Corporation
233 Spring Street, New York, N.Y. 10013

LIST OF CONTRIBUTORS

From the N. K. Kol'tsov Institute of Developmental Biology, Academy of Sciences of the USSR, Moscow:

T. B. Aizenshtadt
T. A. Dettlaff
L. S. Milman
N. D. Ozernyuk
M. N. Skoblina
S. G. Vassetzky
L. P. Yermolaeva
Yu. G. Yurowitzky

From the Institute of Cytology, Academy of Sciences of the USSR, Leningrad:

M. N. Gruzova

PREFACE

Studies of oogenesis occupy an important place in current investigations in developmental biology. Today no one has any doubt whatsoever that oogenesis is not just the prelude to development, but is development itself, and a very essential part of it. These words of an eminent Soviet scientist, B. L. Astaurov, taken by T. B. Aizenshtadt as an epigraph to her chapter in this book, make a good epigraph for the entire book. It is now clear that during oogenesis not only vast reserves of ribosomes and mitochondria, of yolk, carbohydrates, and lipids, and of enzymes for protein and nucleic acid synthesis and for carbohydrate and fat metabolism (which ensures the energy supply and metabolism of the oocyte and the developing embryo) are formed, but also long-lived mRNA and proteins are synthesized, which determine both the completion of oocyte maturation and the initial stages of embryonic development.

In the last 15-20 years, the use of molecular biology methods, electron microscopy, autoradiography, and microsurgical methods of experimental embryology in studying the pre-embryonic development of animals has greatly increased our knowledge of oogenesis. This has led to the need to systematize the data obtained, to reinterpret old ideas, and to review the results obtained by new research trends which have emerged in the last few years and which are of general biological interest. Such a task was undertaken in the book *Sovremennye Problemy Oogeneza* (Current Problems of Oogenesis), published in 1977 (in Russian).

Since then, a great mass of new information has emerged; therefore, the new edition of this book, now published in English, necessitated an extensive revision and enlargement of the original

Chapters 1-4, and a new chapter on maturation of mammalian oocytes has been included. Some other material has been omitted. In particular, the present edition does not contain a review of numerous and very important experiments in which oocytes are used as a model system for studying macromolecular synthesis – replication, transcription, and translation – with the use of heterogeneous mRNA, DNA, and individual genes. These experiments were periodically reviewed by J. B. Gurdon et al. (see, for example, Gurdon and Melton, 1981; references in Chapter 3) and our readers are referred to these reviews.

In this book, special attention is paid to the processes occurring in the cytoplasm of the oocyte during the period of its growth and maturation; changes in the chromosomal apparatus are dealt with relatively briefly.

Like all the monographs in the series *Problems of Developmental Biology*, which are published by Nauka, it is the purpose of this book to attract the attention of biologists to the most important and challenging problems of developmental biology, and to facilitate their further study. To this end, many reviews not only put forward new facts and make generalizations, but also suggest subjects for discussion and questions demanding further research. Methods of research are described as well. Currently important data are being obtained on the causal relationship between phenomena occurring in the oocyte at different levels of its organization, on the hormonal control of oogenesis, on the origin and functions of the organelles in the growing oocyte, and on the functional organization of the cytoplasm, as well as on the involvement of the nucleus and cytoplasm in oocyte maturation. The cytoplasmic control of nuclear behavior and the regulation of gene activity in oogenesis are also being intensively studied, as well as the molecular mechanisms of processes common to any dividing cell, such as the periodic disintegration of the nuclear envelope, the condensation of the chromosomes, and the emergence of factors ensuring the capacity for DNA synthesis.

The study of oogenesis has important practical implications, among them: 1) improvement of the methods of hormonal stimulation of maturation and ovulation in animals bred in captivity, and determination of the optimal conditions for these processes; 2) elimination of the causes of phenotypic defects in oogenesis (disturbances of the normal disjunction of the homologous chromosomes, leading to the appearance of genetic diseases and to decreased egg fertilizability, abnormal cleavage, and gastrulation resulting in serious structural abnormalities and embryonic death); 3) experimental influences on the oocyte during the period of meiotic divisions for the purpose of artificial sex control of animals; and many others.

Hopefully, the English edition of this book will facilitate scientific contacts between Soviet and foreign scientists who are working in the same general area and researching many of the problems discussed in it.

The major part of the investigations of the Soviet authors reviewed in this book was published in the journal *Ontogenez*, which is translated into English under the title *Soviet Journal of Developmental Biology*, where, in some cases, the transliteration of the names of Soviet authors using the British–American Standard system has been preserved.

T. A. Dettlaff
S. G. Vassetzky

FOREWORD TO THE ENGLISH EDITION

Although attempts to understand the role of the egg in the formation of the embryo date back to the time of Aristotle, the epigenetic view that the intrinsic structure of the egg itself plays a key role in animal development is relatively recent, the modern view owing much to the interaction of experimental embryology and genetics around the beginning of this century. Since the 1920s few have disagreed with the central theme, formulated by T. H. Morgan, that the interaction of the genetic information of the nucleus with the heterogeneous cytoplasm of the egg provides the basis for the precisely ordered transformation of the egg into embryo; for many years now, advances in molecular biology, experimental analysis, and developmental genetics have confirmed the essential correctness of the epigenetic view. Much recent work has focused attention on the structure of the egg itself and, in particular, the way in which this structure is established during oogenesis; the present volume concentrates on these essential themes.

Soviet developmental biologists have always recognized oogenesis as an integral part of animal development studies and have made many significant contributions to this area, ranging from work on the origin of the germ cell line to the transformation of oocyte to egg to maturation. Although much has been published in English, a large proportion of the original work remains in Russian in Soviet publications and is therefore inaccessible to many readers. More generally such language difficulties prevent access to important reviews and developmental textbooks which have been published in the Soviet Union in recent years. Outstanding among this Soviet work is the comprehensive series *Problems of Developmental Biology* published by Nauka Press. This consists of a number of multi-author volumes covering major aspects of developmental biology, each volume consisting of a number of chapters dealing with particular aspects of a particular topic. The series as a whole provides a comprehensive insight into the work of Soviet developmental biologists over the past two decades or so. The decision by Plenum Press, in association with Nauka Press, to publish the series in English will ensure that this important work reaches a much wider readership. Appropriately the first volume to be translated is *Oocyte growth and Maturation*.

This translation of *Oocyte Growth and Maturation* is an updated version of the original Russian text, each author revising his contribution during the translation and editing of the new volume. The original translation into English was made by the Soviet authors themselves in collaboration with their editors, Prof. T. A. Dettlaff and Dr. S. G. Vassetzky and other colleagues. In editing the English version I have tried to achieve a necessary uniformity of presentation without detriment to the individual style of the authors. Thus I hope that much of the Russian flavor of the contributions is retained and thereby the essential meaning of the authors themselves is preserved, as well as the intentions of the original editors.

Frank Billett
University of Southampton
August 1987

CONTENTS

Chapter 1
OOCYTE GROWTH AND VITELLOGENESIS
T. B. Aizenshtadt

Chapter 4
MEIOTIC DIVISIONS
S. G. Vassetzky

Chapter 5
MATURATION OF MAMMALIAN OOCYTES IN VITRO
M. N. Skoblina

Chapter 6
SPECIAL FEATURES OF CONTROLLING RESPIRATION DURING OOGENESIS
N. D. Ozernyuk

Chapter 7
CONTROL OF CARBOHYDRATE METABOLISM AT DIFFERENT STAGES OF OOGENESIS
L. S. Milman, Yu. G. Yurowitzky, and L. P. Yermolaeva

Chapter 1

OOCYTE GROWTH AND VITELLOGENESIS

T. B. Aizenshtadt

"...The formation of the ovum is not just the prelude to development, but is development itself, and a very responsible part of it at that – when the very foundations of the promorphological organization of the egg are laid, the very architectural plan of the future individual..."

B. L. Astaurov

In the last few years, due to the advances made in biochemistry, electron microscopy, and cytochemistry, a functional approach to the solution of a number of problems regarding the biology of development has become possible. One of these problems is the development of the organization of the mature egg. It is impossible to solve this problem without a knowledge of the mechanisms involved in the enormous growth of oocytes, which is linked with the accumulation in them of a large quantity of ribosomes and yolk intended for the future embryo.

The size of eggs in animals varies within a wide range, from a few microns to several centimeters, and is dependent to a considerable extent on the animal's type of development. But even in placental mammals, where the embryo receives all the substances it needs from the maternal organism from the earliest stages of development, the eggs are large in size (80-100 μm) compared with the somatic cells.

The mechanisms of growth are undoubtedly similar in all cells. However, whereas the growth of somatic cells is frequently accompanied by their polyploidization, this does not occur in female germ cells.

This chapter examines the peculiar features of the synthesis of ribosomal RNA (rRNA) in the oogenesis of various animals. It

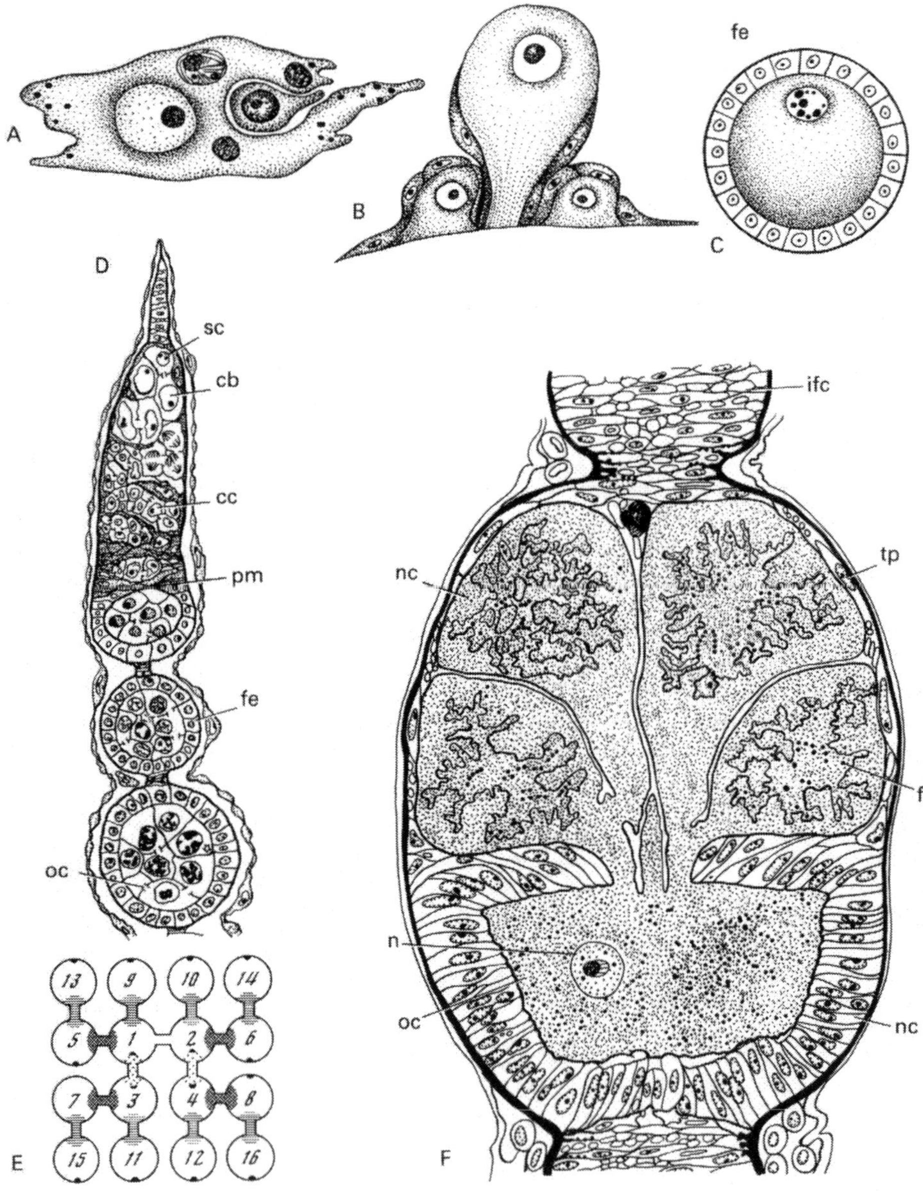
A
B
fe
C
D
sc
cb
cc
pm
fe
oc
13
9
10
14
5
1
2
6
7
3
4
8
15
11
12
16
E
ifc
nc
tp
fe
n
oc
nc
F

also considers data on the origin of yolk proteins, the mechanisms by which they enter the oocytes, and their subsequent accumulation in the oocytes.

This review makes no pretense to being complete. The flood of new publications dealing with the development of oocytes and vitellogenesis is so great that it is impossible to consider all the relevant literature in one article. The author's main aim is that of acquainting those interested in oogenesis with these new data, which have been obtained in the last 10 to 15 years with the help of modern research methods and which have, to a considerable extent, changed our ideas on the sources of yolk proteins and on the origins and functions of organelles of the growing oocyte.

1.1. SOURCES OF RIBOSOMAL RNA IN DIFFERENT TYPES OF OOGENESIS

The diverse patterns of oogenesis (Fig. 1A-E) make it difficult to generalize the data on the sources of ribonucleic acids. Here we will consider the sources of the rRNA, which is synthesized in an enormous quantity during oogenesis, contributing to the composition of the ribosomes, most of which remains functionless during the development of the oocyte. In the oocytes of *Xenopus laevis*, 95% of the ribosomes do not take part in protein synthesis and are only activated after fertilization (Davidson, 1969). These ribosomes are not stable structures, and even in

---

Fig. 1. Diagram of various types of oogenesis in animals. A. Diffuse oogenesis: the amoeboid oocyte exhibits phagocytic activity – found in sponges, some coelenterates, and a number of worms. B. Solitary type of oogenesis: follicle cells are either absent or exist temporarily for a short time during oogenesis – found in coelenterates, worms, molluscs, and echinoderms. C. Follicle type of oogenesis: follicle cells form an epithelial layer (or several layers) around the oocyte – found in several molluscs, arthropods, and in all vertebrates, including man. D. Nutrimentary type of oogenesis: the oocyte is associated with nurse cells by cytoplasmic bridges and is, together with them, surrounded by follicular epithelium (Koch et al., 1967) – found in insects and annelids. E. Spatial relationship between cystocytes in the germarium of *Drosophila* after four divisions (Koch et al., 1967): 1, 2) pro-oocytes (cystocytes each have four cytoplasmic bridges); 3-16) pro-trophocytes. F. Egg chamber of *Cecropia* before the beginning of vitellogenesis (King and Aggaral, 1965). cb) cystoblast; cc) cystocyte; fe) follicle epithelium; ifc) interfollicle cells; n) nucleus; nc) nurse cells; oc) oocyte; pm) pre-follicle mesoderm; sc) stem cell; tp) tunica propria.

large oocytes they are constantly being renewed (Leonard and LaMarca, 1975). The synthesis of a large quantity of rRNA also presupposes an increase in the amount of DNA coding for it. However, the stability of the genetic material of the germ cell is a necessary prerequisite for the preservation of the species. There are only two main ways by which the oocytes are supplied with a large quantity of rRNA: by accessory cells and by synthesis in the nucleus of the germ cell. In the latter case, in many animals an increase occurs in the quantity of ribosomal DNA (rDNA) coding the synthesis of rRNA.

1.1.1. Origin and Functional Significance of the Nurse Cells of the Ovary

Since rRNA does not exist in a free form in the cell, it must be transported from one cell to another as part of the ribosomes or their subunits, which presupposes the existence of cytoplasmic bridges. As a rule, cytoplasmic bridges form as a result of incomplete cell division during oogonial mitoses when one germ cell and several nurse cells are formed. This type of oogenesis, which has been called "nutrimentary," is found in many insects and a number of worms (see Fig. 1D, F).

Cytoplasmic bridges (ring canals, fusoms) between the germ cells are a unique phenomenon in cytology. It is thought (Gondos, 1973) that their formation is linked with the disturbances in the processes of cell division which hinder further mitosis and result in the degeneration of part of the female germ cells. It may be surmised that the emergence of nurse cells during evolution became possible through the production of excess numbers of gametes, on the one hand, and to the appearance of the mechanism of incomplete cytoplasmic division and degeneration of part of the cells, on the other. Both phenomena are found in vertebrates possessing the follicular type of oogenesis, such as fish (Satch, 1974), birds (Skalko et al., 1972), and mammals, including man (Franchi and Mandl, 1962; Gondos, 1973), where cytoplasmic bridges between the early oocytes frequently occur and atresia of some germ cells is found.

The ovaries of insects possessing nurse cells are called "meroistic." They are of two types: telotrophic, when the nurse cells are concentrated in the apical end of the ovariole and are united with the growing oocytes by trophic cords, and polytrophic, when each oocyte is associated with a certain number of nurse cells by cytoplasmic bridges and, together with them, is surrounded by follicle cells (Ramamurty, 1964b; King and Aggarwal, 1965; Telfer, 1975).

The development of nurse cells has been studied extensively in *Drosophila* (Koch and King, 1966; Koch et al., 1967; King, 1974, 1975), in which the oogonial divisions take place in the ovarioles of the larva. There are one or two oogonia, which are also called "stem cells," in each germarium. The stem cell, which is located in the apical end of the germarium, divides mitotically, while the long axis of the spindle is situated parallel to the long axis of the ovariole. The upper cell increases its volume to that of the original stem cell and itself becomes a stem cell, while the lower one becomes a cystoblast, a cell "programmed" for incomplete cytoplasmic division. At present, it is believed that this property of the cystoblast is determined by ecdysone secreted by the prothoracic gland of the nymph and is transferred to the daughter cells, the cystocytes, the division of which takes place in the imago, when there are no prothoracic glands. Ecdysone does not, however, prevent the complete division of the stem cell. The prefollicle cells surrounding the stem cell apparently perform a protective function (King, 1975).

Four divisions of the cystoblasts in *Drosophila* (the number of these divisions varies in different insects) give rise to sixteen small cystocytes connected with one another by cytoplasmic bridges (see Fig. 1E). Graphic reconstruction using microscopic sections reveals the spatial relationship of the cystocytes in each group of cells (Koch et al., 1967). Two cells out of the sixteen have four ring-shaped canals, and both enter into meiotic prophase. However, only one of these two cells (pro-oocytes) becomes an oocyte; it is the one which comes into contact with the follicle cells in the connecting zone between the chambers.

It is thought that in some way the formation of the cells with four canals delays the mitotic division of the remaining fourteen cells, which become differentiated as nurse cells (see King, 1975). What happens if the mitoses of the cystocytes continue is seen in *Drosophila* females homozygous for the recessive gene *fes*. In such cases, "swellings" consisting of 100-1000 cystocytes are formed instead of normal egg chambers in the ovaries. In such accumulations very few cells have four ring-shaped canals. Consequently, some cells are eliminated from the accumulation because they divide completely. When the temperature is lowered to 18°C, the division of the cystocytes in *fes/fes* females ceases; such cells may then be transformed into nurse cells. In 2% of such cases oocytes capable of vitellogenesis appear, but this applies only to those cells that have four ring-shaped canals. Thus, the study of *fes/fes* females shows the important part played by the number of mitotic divisions and the number of cytoplasmic bridges in the differentiation of cystocytes into germ and nurse cells. It is not quite clear how the lowering of temperature affects the cystocytes of mutant females. King suggested that, under these conditions, the growth of the cystocytes between the mitoses is delayed, and they are not able to reach that "critical mass" when the division of cells is still possible. It is, moreover, possible that a product of the *fes* gene is thermolabile.

The polarity of cytoplasmic bridges has been demonstrated by the microinjection of fluorescein-labeled rabbit globulin into the oocytes and nurse cells of *Cecropia* and the simultaneous measurement of the potential with the help of a microelectrode (Woodruff and Telfer, 1973). These elegant experiments showed that the protein can cross the bridge from the nurse cells to the oocyte, but not vice versa. The difference in potential between the nurse cells and the oocytes was 10 mV. The polarity of the bridges can be reversed through 180° by applying a current of $5 \cdot 10^{-8}$ A; under these conditions, the labeled globulin begins to pass from the oocyte to the nurse cells.

Apparently, the polarity of the cytoplasmic bridges plays an important part in the movement of structures, such as ribosomes or 80S-monosomes (see below), from one cell to another. Moreover, the transfer of ribonucleoproteins into the oocyte may result from a concentration gradient since the nurse cells produce much more rRNA than the oocyte.

However, this does not mean that the cytoplasm flows from the nurse cells into the oocyte and transfers the yolk material. Such assertions are groundless; they were frequently found in the older literature, and even more recently (Raven, 1961). The fact that mitochondria are found in the trophic cord in the telotrophic ovarioles of some insects does not mean that they move in the direction of the oocytes (Huebner and Anderson, 1972).

The nurse cells in the polytrophic ovarioles of insects grow rapidly at the beginning of oogenesis and become polyploid. Later they grow more slowly than the oocyte, shrinking before the end of vitellogenesis and sometimes at the beginning of it (Urbani, 1969). In the telotrophic ovarioles in *Homoptera* the trophic cord degenerates before the beginning of vitellogenesis (Matsuzaki, 1975). In some cases, the remains of the nurse cells are, as it were, drawn into the oocyte, but they cannot then play a substantial part in its growth.

Initially, the transfer of RNA from nurse cells to oocyte was discovered by the use of autoradiography in the polytrophic ovarioles of *Musca domestica* and *Calliphora erythrocephala* (Bier, 1963a, b, 1964) and in *Panorpa communis* (Ramamurty, 1963, 1964a). Autoradiography was also used to show the movement of RNA from the nurse chamber to the oocytes in the telotrophic ovarioles of insects (Macgregor and Stebbings, 1970). Electron microscopy has revealed a high concentration of ribosome-like particles in the trophic cord of telotrophic ovarioles (Matsuzaki, 1965). Biochemical investigation of the isolated trophic cords of telotrophic ovarioles of the bug *Dysdercus intermedius* has shown that rRNA passes from trophocytes into the oocyte as 80S-monosomes, stable ribonucleoprotein particles, which form a depot of 28S- and 18S-rRNA (Winter, 1974). In the ovarian follicle of the giant silkmoth *Antheraea polyphemus*, the cytoplasm of the highly polyploid nurse cells accumulates inactive 39S-monosomes (ribosomal precursors), which are subsequently transported to the ooplasm (Hughes and Berry, 1970). In *Cecropia*, in just a few days the amount of RNA in the oocyte increases fivefold (from 0.5 to 2.5 μg) but, despite active synthesis of RNA, the RNA content of the nurse cell remains at 0.5 μg.

Over a long period of oogenesis there is no nucleolus in the nucleus of the oocyte in most insects with polytrophic and telotrophic ovarioles. The chromosomes form a karyosphere, indicating that in the "oocyte—trophocyte" system the nuclei of the trophocytes are responsible for rRNA synthesis (see Chapter 2).

Nurse cells of oogonial origin have been described in several worms. In the coelomic fluid in the polychaete *Diopatra cuprea* strings of uniform cells are found. The centrally situated cell in a string rapidly increases in size and becomes an oocyte, to which two cords of nurse cells are attached. These are connected with the oocyte and with each other by cytoplasmic bridges, through which the ribosomes enter the oocyte (Anderson and Huebner, 1968). In the snail leech *Glossiphonia complanata* there are paired ovaries, each of which is a long cord of syncytially connected pyriform cells (Plate I, 1, 1A), the elongated ends of which form a cytoplasmic shaft, the rachis (Aizenshtadt, 1964). A small number of these cells form oocytes, and the rest become nurse cells which provide the oocytes with ribosomes through the rachis (Aizenshtadt et al., 1967). In the leech there also is no nucleolus in the nucleus of the oocyte; a karyosphere is formed (Gruzova and Zaichikova, 1967).

Analysis of RNA synthesis in insects with different types of ovarioles shows that, in the case of meroistic ovaries, the extent to which the nucleus of the oocyte participates in RNA synthesis varies (Bier, 1963a, 1970; Ramamurty, 1963). *Panorpa* occupies an intermediate position between the insects with panoistic ovaries, in which the entire synthesis of RNA is concentrated in the nucleus of the oocyte, and the insects with highly developed meroistic ovaries, for example *Diptera*, where almost all the RNA is synthesized in the nuclei of the nurse cells during vitellogenesis.

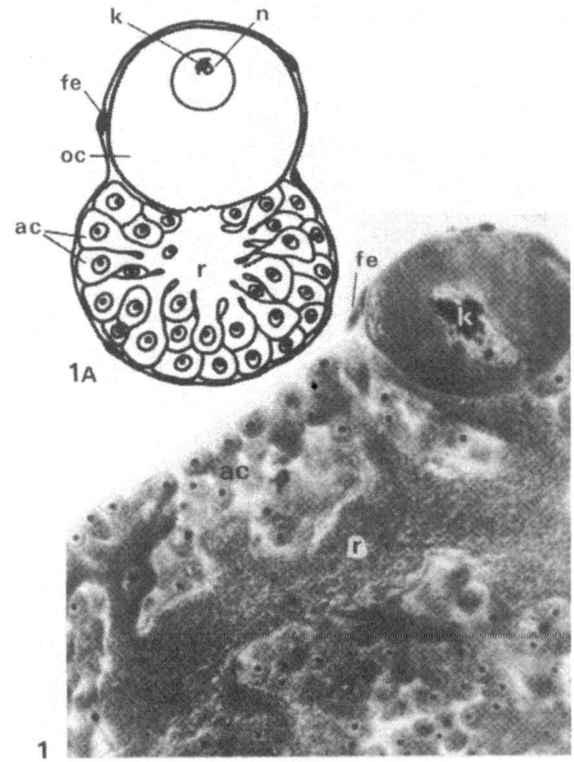

1A

1

3

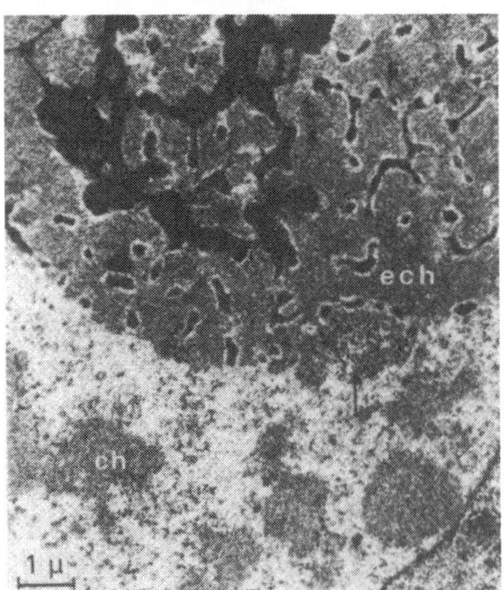

2

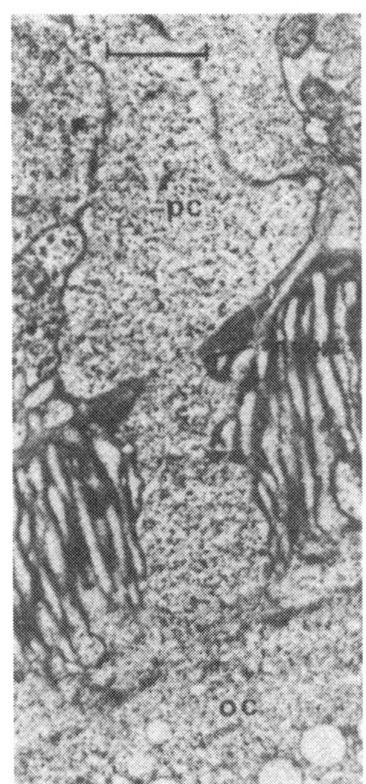

4

In the annelid *Ophryotrocha puerilis*, the oocyte is associated with a single polyploid nurse cell and both synthesize RNA (Ruthmann, 1964). Autoradiography on the telotrophic ovarioles of milkweed bug *Oncopeltus fasciatus* demonstrates that RNA synthesis occurs both in the nuclei of the nurse cells and in the nucleus of the oocyte (Zinsmeister and Davenport, 1971). The participation of the oocyte nucleus in rRNA synthesis has also been revealed in the telotrophic ovarioles of the beetle *Bruchidius obtectus* (Büning, 1972).

It has long been known that in many insects the nurse cells reach a high level of ploidy (Painter and Reindorp, 1939). Later, cytophotometry established that a gradient exists in the distribution of ploidy in the nurse cells in *Drosophila* (Jacob and Sirlin, 1959). The DNA of cells closest to the oocyte undergoes eight replications, and the DNA of the rest of the cells, seven. At later stages in oogenesis the nurse cells lose parts of the DNA. The cells located near the oocyte and in contact with the follicle cells have larger nuclei (Brown and King, 1964). It has been calculated that during oogenesis in *Drosophila* the volume of the nuclei in nurse cells increases 6000 times, and the volume of the nucleus in the oocyte, 900 times (Koch and King, 1966). Although the degree of ploidy in the nurse cells of the telotrophic ovarioles of beetles is not very great, a high rate of RNA synthesis has been observed in their nuclei (Büning, 1972).

In the snail leech *Glossiphonia complanata*, in the ovary of which there are about 2000 nurse cells to every growing oocyte, the former are not subject to polyploidy (Aizenshtadt and Marshak, 1969).

Thus it is clear that an alternative way of creating a large number of ribosomes for the future embryo is to transfer the function of rRNA synthesis from the nucleus of the oocyte to the nuclei of the nurse cells, which have, in some cases, reached high classes of ploidy.

Plate I. 1. The ovary of a snail leech, *Glossiphonia complanata* (longitudinal section). Magnification ×150. Diagram of a microscopic section. 2. The nucleus of a beetle oocyte with an extrachromosomal body of DNA (Kato, 1968). 3. Pyriform follicle cell in the ovary of lizard. 4. The cytoplasmic bridge between this cell and the oocyte (Neaves, 1971). ac) accessory cells; ch) chromosomes; ech) extrachromosomal chromatin; fe) follicle epithelium; k) karyosphere; oc) oocyte; pc) pyriform cells; r) rachis; n) nucleus.

1.1.2. Participation of Follicle Cells in the Synthesis of rRNA for the Oocyte

The idea that nucleic acids are transferred from the follicle cells to the oocyte is long-standing (Raven, 1961). It was based on the discovery that in follicle cells there was polarity in the distribution of the pyroninophilic material, with the higher concentration in the areas near the oocyte. However, we believe that these data do not indicate the transfer of nucleic acids from the follicle cells to the oocyte, but rather the concentration in these areas of granular endoplasmic reticulum, which is involved in the synthesis of the glycoproteins secreted by the follicle cells in the perioocyte space (see below). The transfer of ribonucleoproteins can only be carried out across the cytoplasmic bridges.

The development of electron microscopy made it clear that structures described in the old literature as cytoplasmic bridges between the follicle cells and the oocyte, as, for instance, in birds and mammals, were in reality extensions of the follicle cells stretching toward the surface of the oocyte. However, despite the fact that the tips of the extensions may become enclosed in the surface layer of the oocyte, the cytoplasm of the latter is always separated from the cytoplasm of the follicle cell by two plasma membranes.

Somewhat surprising was the appearance of electron microscopic observations pointing unquestionably to the existence of cytoplasmic bridges between the oocyte and the follicle cells in lizards (Hubert, 1971; Neaves, 1971; Bou-Resli, 1974; Olmo and Taddei, 1974). Not all the follicle cells were united with the oocyte by bridges, but only the so-called pyriform cells, which differed from the rest in their large size (Plate I, 3, 4). These cells degenerate at relatively early stages in oogenesis and do not transfer yolk proteins to the oocyte (see below) or synthesize secondary egg membranes (Neaves, 1971; Hubert, 1970). Neaves believes that the cytoplasmic bridges in the follicle of lizard are the result of the degeneration of the plasma membranes between the pyriform cells and the oocyte. They are formed when the follicle is multilayer and does not have a typical structure, i.e., a thickened plasma membrane (Filosa and Taddei, 1976). Unlike the small follicle cells, these pyriform cells are strictly diploid and do not become polyploid (Olmo and Taddei, 1974).

Although it has been suggested that these cells are of oogonial origin (Hubert, 1970), i.e., correspond to the trophocytes of insects, this point of view is, at present, rejected by the majority of investigators in the field. During oogenesis the follicular epithelium of the lizard is modified and becomes polymorphic;

autoradiography indicates that the large pyriform cells develop from small follicle cells (Filosa et al., 1979).

The nuclei of pyriform cells possess large nucleoli, and active synthesis of rRNA obviously takes place. The nucleolus (Taddei, 1972) of the oocyte disappears during the functioning of the cells, whose number reaches 10,000. It may be assumed that ribosomes of their precursors pass from the pyriform cells into the oocyte, but additional research is needed to confirm this.

In 1959, enigmatic structures resembling mitochondria and the vacuoles of the Golgi apparatus were described in the oocytes of birds and were at first thought to be due to the poor electron-microscopy technique. Later, in 1964, the author who discovered them (Press, 1964) showed that these structures were the expanded end sections of the follicle cell outgrowths which had penetrated into the ooplasm (Fig. 2). Press called them "transosomes."

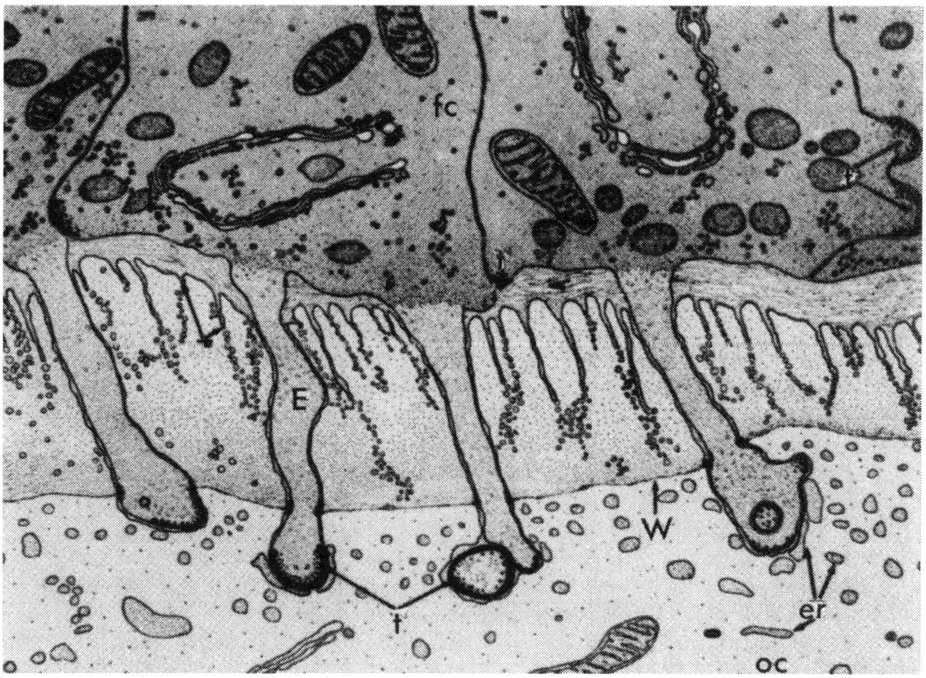

Fig. 2. Transosomes in the follicle of a hen (after Press, 1964). er) endoplasmic reticulum; fc) follicle cell; oc) oocyte; t) transosomes. E) follicle cell extension; W) terminal web.

Another term, "lining bodies" (Bellairs, 1965) was also proposed almost at the same time. The transosomes are covered by two membranes. A dense material borders upon the internal membrane, which is apparently the plasma membrane of the follicle cell. Evenly spaced ribosome-like granules are attached to this material. The formation of transosomes is foreseen inside the follicle cell where, in certain slightly thickened sections of the plasma membrane, dense granules are formed. Subsequently this section of the cell protrudes and begins to grow toward the oocyte. Inside the oocyte the transosomes become separated, and they have been found inside forming yolk granules (Schjeide et al., 1974). The concentration of transosomes in microscopic sections of previtellogenic oocytes indicates that they are present in enormous numbers in these cells. In recent years, the ribosome nature of the dense granules inside transosomes has been demonstrated (Paulson and Rosenberg, 1972, 1974). This relates to the fact that a typical nucleolus is absent in the oocytes of some birds, and a karyosphere is formed (Gaginskaya and Gruzova, 1969). This method of transferring ribosomes from cells of somatic origin to the growing oocyte only exists in birds and apparently in some reptiles (Rahil and Narbaitz, 1973).

Thus, although in most animals the follicle cells do not contribute rRNA to the oocyte and the future embryo, some reptiles and birds are an exception to this rule.

1.1.3. Synthesis of rRNA in the Oocyte Nucleus. The Amplification of rDNA

This topic is only dealt with briefly, as it is covered in more detail in Chapter 2 of this book (Section 2.5). Although the existence of an extrachromosomal heterochromatic body in the oocytes of insects has long been known (see Kato, 1968), the amplification of rDNA cistrons in the oocytes of various animals is a relatively recent discovery, dependent on improved cytological methods, especially the hybridization of RNA—DNA on microscopic sections. Insects and amphibians are best studied in this respect (Bier, 1970). Amplification of the genes of the nucleolar organizer has been discovered in insects with various types of ovarioles. In water beetles *Dytiscidae* extrachromosomal chromatin (see Plate I, 2), also known as "elimination chromatin," appears in the nucleus of an oogonium before it divides, and is subsequently found only in the nucleus of the daughter cell destined to become an oocyte (Kato, 1968). This also occurs in the beetle family *Gyrinidae* (Hoser, 1975). In some *Tipulidae* this chromatin is also found in the nucleus of one of the nurse cells (Coggins, 1973). Extrachromosomal chromatin actively incorporates ^{3}H-uridine during oogenesis and is characterized by a low histone content. In beetles the DNA content is more than 40 C. The selective multiplication of the nucleolar organizer occurs in oocytes of house cricket *Acheta domesticus*, which develop without nurse cells (Allen and Cave, 1968; Kunz, 1969). In this species small globules of DNA appear in contact with the nucleoli in the interphase nuclei of the oogonia. The number of these globules decreases and their size increases as the oogonia divide. In oocytes there is only one globule containing amplified rDNA. It reaches a maximum size at middle pachytene, and during late pachytene to early diplotene an accumulation of granules 150 Å in diameter surround it. The emission of the granular material from the nucleus to the cytoplasm can be seen in electron micrographs (Allen and Cave, 1968).

In *Xenopus laevis* oocytes the main synthesis of rDNA takes place at the pachytene stage; however, amplification commences before this in oogonia (Gall, 1969; Kalt, 1972). At the diplotene stage the number of the nucleolar structures themselves increases, and at the previtellogenic stage oocytes about 400 μm in diameter each possess, on the average, 1500 nucleoli. At the lampbrush chromosome stage more than 30% of the proteins being synthesized are ribosome proteins, and it is suggested that the genes, which code for the synthesis of these proteins, are also amplified (Hallberg and Smith, 1975). In *Xenopus* this synthesis appears to be coordinated with the synthesis of 28S- and 18S-RNA, but this needs to be confirmed.

A crescent-shaped DNA-containing body which actively incorporates ^{3}H-thymidine appears adjacent to the nuclear envelope in the pachytene oocytes of some teleosts and in the sturgeon (follicular type of oogenesis) (Chmilevsky, 1970; Steopoe et al., 1970). Numerous nucleoli are formed in association with these bodies. Cytophotometry demonstrates that the DNA content in the diplotene oocytes of sturgeon is 50 times greater than the haploid amount and 31 times greater in the oocytes of sterlet (Shmantsar, 1976). Recently, the amplification of rDNA in the early oocytes of man has been described (Stahl and Luciani, 1975). The lists of animals with oocytes in which amplification of rDNA is found are given in Brown and Dawid (1969) and Gall (1969).

Although it would seem that amplification of ribosome genes should occur in oocytes of the solitary and follicular type of oogenesis, and be absent in oocytes of the nutrient type, no such principle emerges. For example, the method of RNA–DNA hybridization demonstrates that there is no amplification of rDNA in the oocytes of the milkweed bug *Oncopeltus fasciatus*, which has telotrophic ovarioles (Cave, 1975), but in the oocytes of beetles with polytrophic ovarioles, amplification of the genes of the nucleolar organizer does occur (Kato, 1968; Urbani, 1969; Matuszewski and Kloc, 1976). However, in the beetles, the nurse cells degenerate very early (see above). Extrachromosomal globules of DNA have

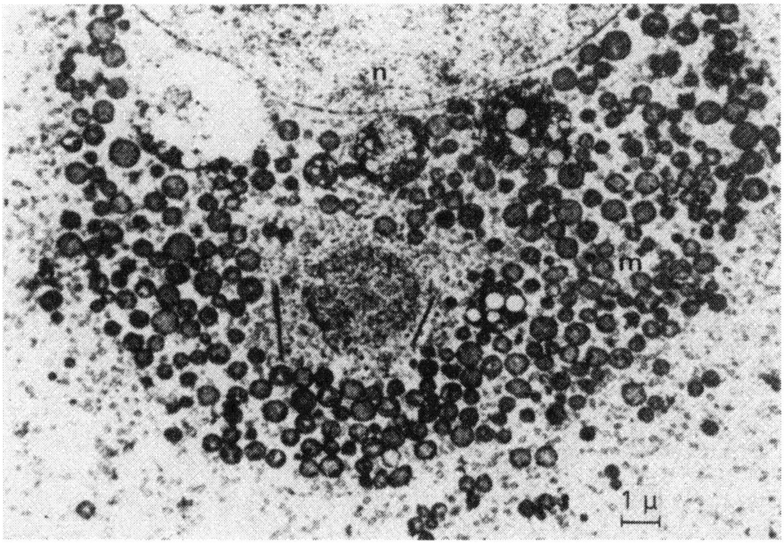
n
m
1 μ

1

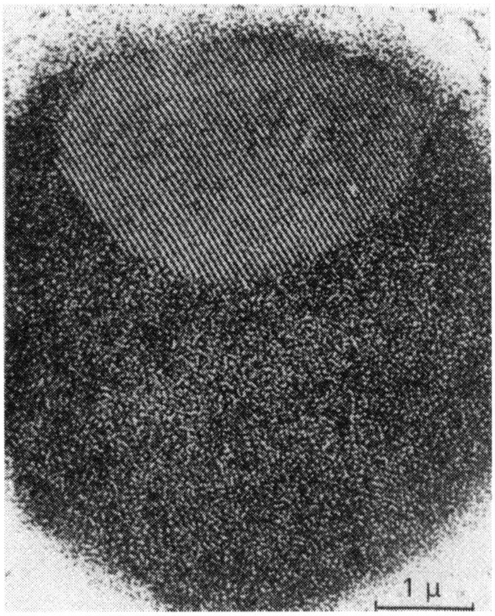
1 μ

2

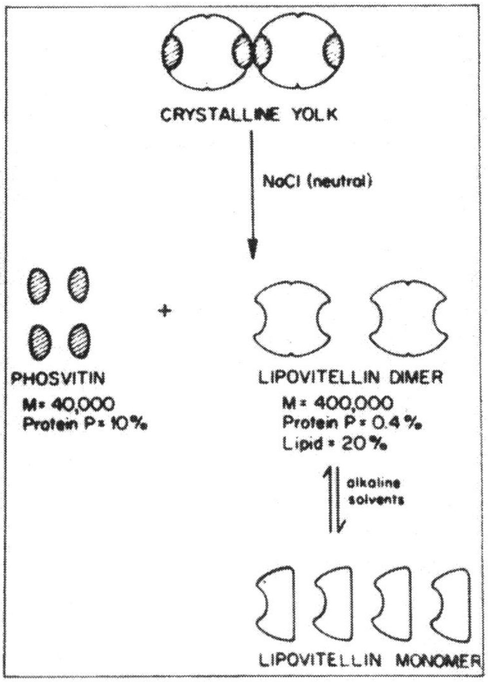
CRYSTALLINE YOLK
NaCl (neutral)
+
PHOSVITIN
M= 40,000
Protein P= 10%
LIPOVITELLIN DIMER
M= 400,000
Protein P= 0.4%
Lipid = 20%
alkaline
solvents
LIPOVITELLIN MONOMER

3

also been observed in the oocytes of golden-eyed fly *Chrysopa perla*, which has polytrophic ovarioles (Gruzova et al., 1972). Amplification of rDNA is not found in starfish species (Vincent et al., 1969), whose oocytes do not receive rRNA from other cells, but in the echiuroid *Urechis caupo*, whose oocytes also develop according to the solitary type, rDNA amplification occurs. The process is similar in echiuroids and amphibians. In both cases, the rate of tRNA and rRNA synthesis increases at later stages in oogenesis. However, there are differences. In frogs, only tRNA and 5S-rRNA are synthesized in early oocytes and 28S- and 18S-rRNA later on at the stage of lampbrush chromosomes (Mairy and Denis, 1970), while in echiuroids rRNA synthesis takes place at all stages of oogenesis (Davis and Wilt, 1972).

1.2. VITELLOGENESIS

Vitellogenesis has attracted the attention of embryologists and cytologists for many decades; however, many of the ideas on the synthesis and accumulation of yolk in oocytes, which were based on morphological research, have been proven to be wrong. For a long time, for example, the so-called yolk nucleus or Balbiani's body was considered to be of great importance in the process of vitellogenesis. It was first described in the early oocytes of spiders, and then in other animals back in the middle of the last century, and was believed to be the site in which yolk inclusions were formed which then spread throughout the cytoplasm.

Electron microscopy reveals that in the vicinity of the nucleus is an accumulation of organelles and inclusions. Such accumulations have now been discovered in the early oocytes of many animals and man (Plate II, 1) (Rebhun, 1961; Hertig and Adams, 1967; Hertig, 1968; Holland, 1976). In different animals the yolk nucleus differs in size; the special features of its structure, as seen by light-microscope studies of the oocytes, are determined by the predominance of one type of organelle or another. For instance, the striation of the yolk nucleus in spiders, molluscs, and insects is determined by a large number of concentrically located cisternae of endoplasmic reticulum, the dictyosomes of the Golgi apparatus, or the annulate lamellae (Kessel

---

Plate II. 1. Balbiani's body (yolk nucleus) in an early human oocyte (Hertig and Adams, 1967). 2. Platelet of primordial yolk in the frog oocyte (Massover, 1971b). 3. Molecular model of the crystalline yolk in amphibians (Wallace and Dumont, 1968). m) mitochondria; n) nucleus.

and Beams, 1969). In some cases where the Golgi complexes are formed in the zone of the yolk nucleus and endogenous synthesis of the yolk prevails, the first yolk granules may appear in the Golgi region, i.e., inside the yolk nucleus; this suggests that the older ideas are not completely unfounded. The part played by the Golgi apparatus in vitellogenesis is discussed in detail below. The appearance of the yolk nucleus in early oocytes is evidently connected with the reproduction of cell structures, which are almost absent in the dividing oogonia in many animals (see also Guraya, 1975). This explains the closeness of the yolk nucleus to the nuclear envelope of the oocyte, since in many germ and embryonic cells such structures, like the endoplasmic reticulum, the Golgi complexes, and the annulate lamellae, appear to be formed from the perinuclear cisterna (Kessel, 1968a; Stang-Voss, 1970). In this region, reproduction of mitochondria evidently takes place, and in the oocytes of some animals they make up most of the yolk nucleus (Anteunis et al., 1964; Greenfield, 1966).

The yolk is a complex of macromolecules; its division into protein, carbohydrate, and fatty yolk (Raven, 1961) has become obsolete. All yolk granules, which are composed of proteins, carbohydrate-protein complexes, and lipoproteins, are at present commonly called protein yolk, in contrast to the so-called fat yolk, the lipid droplets, which mainly contain phospholipids and triglycerides. The lipid droplets appear in the cytoplasm, without visible connection with any of the cell structures. Questions connected with the formation of the fat yolk are not dealt with, since the process does not differ from the formation of the lipid inclusions in the somatic cells. Moreover, the very concept of fat yolk may not be valid.

1.2.1. Endogenous Yolk. The Origin of Yolk Proteins

In some animals, the growth rate of the oocytes during vitellogenesis is quite striking. Thus, the oocyte in *Drosophila* increases its volume 90,000 times in 3 days (King, 1975), that of the cockroach 2,500,000 times in a few days (Anderson, 1969b), and that of the crab 20,000 times in a few weeks (Wolin et al., 1973). In the snail leech the oocytes increase in volume more than 1000 times in 2 days (Aizenshtadt et al., 1964).

Nevertheless, the widely spread opinion that the nucleocytoplasmic ratios in the developing oocytes are disturbed is apparently incorrect. The ratio of the volume of the nucleus to the volume of the active cytoplasm, the so-called functional volume of the egg, does not change during the development of the oocyte. The volume of the active cytoplasm not occupied by the yolk is determined quantitatively by the influx of various substances, for example, metabolites. The comparatively small volume of this cytoplasm rapidly becomes saturated with the necessary low-molecular-weight compounds by simple diffusion (Melton and Smorul, 1974).

In recent times it has been shown that the yolk proteins in most animals are of double origin; the greater part of them is synthesized outside the gonad and enters the oocyte via micropinocytosis, and the lesser part is synthesized by the oocyte itself.

Although a similarity between yolk proteins and the proteins found in chicken blood serum had already been discovered in 1932 (Jukes and Kay, 1932), and immunological methods had demonstrated that yolk proteins in insects are similar to hemolymph proteins (Telfer, 1954), it is only in the last 10-15 years that the exogenous origin of the bulk of the yolk proteins in different animals has become established beyond doubt: in insects (Bier, 1962, 1963a; Anderson, 1964, 1969b; Droller and Roth, 1966; Cummings and King, 1970), in amphibians (Follett et al., 1968; Rudack and Wallace, 1968; Dehn and Wallace, 1973; Sportnitz and Kress, 1973; Follett and Redshaw, 1974), in fishes (Chmilevsky, 1972; Heesen and Engels, 1973), and in birds (Heald and MacLachlan, 1965; Williams, 1977).

Nowadays it is usual to divide macromolecules into two groups: autosynthetic, which are synthesized by the cell itself, and heterosynthetic, which are synthesized outside the cell but then used by it. Phosvitin and lipovitellin belong to the second category. It is precisely these proteins which are most rapidly assimilated by embryonic cells. Embryo transplants develop very well in the presence of lipovitellin alone. This protein is the main source of amino acids essential for developing vertebrate embryos. Phosvitin is also rapidly absorbed by the embryonic cells (Schjeide et al., 1970).

In insects, the bulk of the yolk proteins is synthesized by the cells of the fat body and subsequently taken up by the hemolymph (Engelman, 1970; Tanaka and Ishizaki, 1974). In the cockroach, the hemocytes circulating in the hemolymph (Buhlmann, 1974) may take part in the synthesis of yolk proteins. Vitellogenin, the main fraction of the exogenous yolk protein, has been isolated and purified in a number of insects (see Oie et al., 1975).

In *Brachydanio rerio*, a glycoprotein fraction absent in the liver or the blood has been isolated from the ovaries of 4-month-old immature females. This fraction was identified as endogenous yolk material. However, in 6-month-old mature females, a protein fraction appears in the ovaries which is similar to that found in the blood and liver. This fraction was identified as exogenous yolk material, synthesized in the liver (Heesen and Engels, 1973).

In amphibians, the yolk protein is synthesized in the liver of the "vitellogenic" females as a serum lipophosphoprotein known as "vitellogenin" (Wallace, 1963; Wallace and Dumont, 1968).

The vitellogenins (lipovitellins) of different animals differ in their chemical and physical properties. For example, in vertebrates they contain a rather large amount (16-20%) of lipids (Wallace, 1965). In ticks (*Ornithodorus moubata* and *Dermacentor andersoni)*, two fractions of lipovitellin are found; these are hemoglycolipoproteins (Diehl, 1970; Boctor and Kamel, 1976). This protein is immunochemically similar to components in the hemolymph of the females, but not to the blood of the host. In the cockroach *Lucophaea maderae*, two of the vitellogenins also have a carbohydrate component (Dejmal and Brookes, 1972). Experiments using various fractions of labeled serum proteins showed that vitellogenin is taken up by the ovary 50 times more rapidly than other blood proteins. For instance, when oocytes of *Xenopus laevis* are incubated with labeled vitellogenin, the rate of uptake of this protein is 0.70 μg/oocyte/h (Wallace et al., 1972).

Having entered the oocytes, the vitellogenin breaks up into lipovitellin (mw 400,000) and phosvitin (mw 40,000) which, together, form the crystalline yolk (Plate II, 3). The crystalline structure of yolk in the amphibian oocytes was first described over 20 years ago (Karasaki and Komoda, 1958; Ward, 1962; Karasaki, 1963a). The transformation of vitellogenin occurs in the spaces limited by the membrane of pinocytotic vesicles, and in granules of "primordial yolk" (see Plate II, 2) formed by the fusion of the pinocytotic vesicles. One molecule of phosvitin is bound with each monomer of the lipovitellin dimer (Wallace and Dumont, 1968; Leonard et al., 1972). When vitellogenin is injected into ooplasm, it is degraded (Dehn and Wallace, 1973). The formation of crystalline yolk is a very rapid process. For instance, in the honeybee (*Apis mellifica*) it takes slightly more than 2 h from the beginning of active pinocytosis to the appearance of large yolk inclusions (Engels, 1973). In amphibians the volume of yolk platelets increases to 27,000 times its original volume in a short time (Massover, 1971b).

The detailed mechanism of crystalline yolk formation is still not understood. Despite the similarity in the chemical composition (phosphoproteins and lipoproteins in a ratio of 2:1), the yolk in teleosts has a different crystalline lattice than that of amphibians (Yamamoto and Oota, 1967). However, the crystalline structure of the yolk in the lamprey (*Petromyzon marinas*) is similar to that in amphibians (Karasaki, 1967).

1.2.2. Hormonal Control of Vitellogenesis

This section is not aimed at an exhaustive review of data on the hormonal control of vitellogenesis. Attention is focused chiefly on insects and amphibians, where the problem has been studied most extensively. Of particular interest is vitellogenin synthesis in the liver of vertebrates. This system allows one to study various aspects of cell biology during a reversible induction of a stable protein achieved by means of a well-defined stimulus, namely estradiol. In studies of transcription, such model systems as the polytene chromosomes of *Diptera*, which produce puffs in response to ecdysone, or the avian oviduct cells, which synthesize albumin when exposed to an estrogen, are inferior to the liver of egg-laying vertebrates since liver cells are capable of different responses to different hormones. Modifications of vitellogenin following its synthesis, namely phosphorylation, enrichment with lipids, or cleavage inside the oocyte, appear to be a model system for the study of post-translational modification of proteins.

In insects there is no single scheme of hormonal control of vitellogenesis. All aspects of oocyte maturation in the mosquito are controlled by ecdysone, while in the cockroach and the locust this part is played by the juvenile hormone (JH) produced by the *corpora allata* and the brain neurohormone secreted by the *corpora cardiaca*, respectively.

The medial neurosecretory cells of the brain and *corpora cardiaca* of the locust control vitellogenin synthesis in the fat body, while the JH confers on the oocyte the ability to absorb the protein from the hemolymph (Highnam et al., 1963; Handler and Postlethwait, 1977). At later stages of vitellogenesis a neurohormonal factor induces follicular cells of the ovary to secrete ecdysone, which is accumulated by mature oocytes and partially appears in the hemolymph. The role of ecdysone accumulated in the locust oocytes is unknown.

The imaginal moult of an anautogenous mosquito, *Aedes aegypti*, whose vitellogenesis starts only after feeding with blood, stimulates the development of ovaries where oogonia begin to differentiate into oocytes and trophocytes. However, in starving mosquitos the oocyte growth is blocked at the previtellogenic stage. To reach this stage, the mosquito needs JH (Leah, 1963; Hagedorn, 1974). After feeding with blood, neurosecretory cells of the brain produce a hormone which, by analogy with vertebrate hormones, is sometimes called the folliculotropic hormone. Follicle cells exposed to it secrete ecdysone which, after turning into 20-hydroxyecdysone, induces the synthesis of vitellogenin in the fat body. Vitellogenin enters the oocytes by way of micropinocytosis, while mature oocytes produce a factor which inhibits development of the next batch of eggs until the second feeding. All the above is outlined schematically in Fig. 3 (Hanaoka and Hagedorn, 1980).

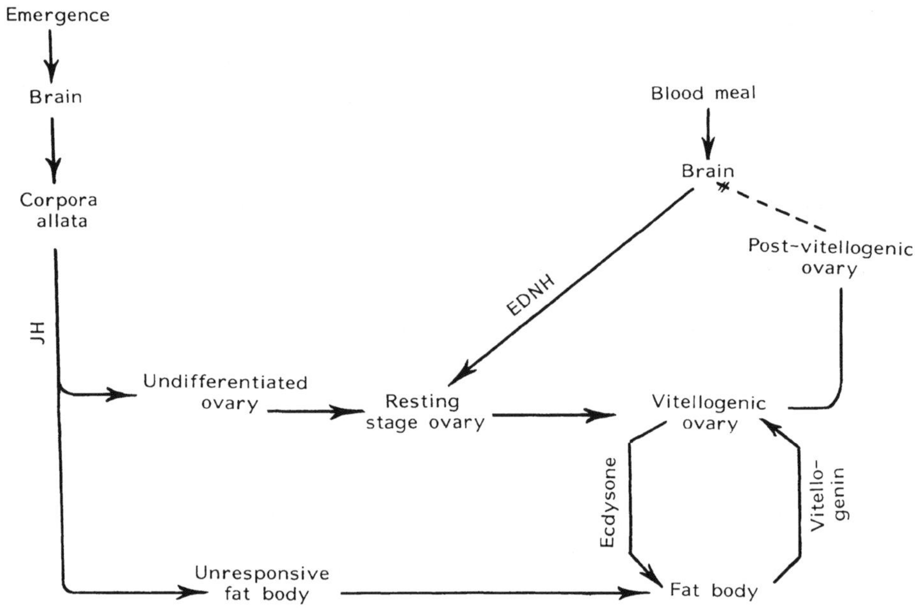

Fig. 3. A simplified scheme of endocrine regulation of egg development in *Aedes aegypti* (Hanaoka and Hagedorn, 1980).

The autogenous mosquito, *Aedes atropalpus*, whose vitellogenesis is independent of feeding, has a similar scheme of vitellogenesis control (Masler et al., 1980), although the ecdysone titer is higher: a female contains 323 pg of ecdysone compared with 275 pg in *Aedes aegypti* (Hagedorn et al., 1975). The role played by JH in mosquito vitellogenin synthesis remains obscure. Although yolk protein synthesis by the fat body is stimulated by ecdysone, in vitro the fat body must first be treated with JH and then with ecdysone for vitellogenin to be induced (Masler et al., 1980). The synthesis of two different vitellogenins under the influence of the two hormones has not been ruled out (Handler and Postlethwait, 1978).

In the cockroach, JH is responsible for all aspects of oocyte maturation (Bell, 1969; Brookes, 1969; Wilhelm and Luscher, 1974), and JH can stimulate egg development even in starving cockroach females (Larsen and Bodenstein, 1959). However, removal of the head of *Leucophaea maderae* 1 day after imaginal moult stops vitellogenin synthesis in the fat body, and injection of JH, though restoring the synthesis of this protein, is insufficient to make the headless females complete their oocyte growth (Koeppe et al., 1980). At the end of each ovarian cycle the ovaries of cockroach secrete ecdysterone to suppress the function of the *corpora allata*

(Stay et al., 1980). It may be supposed that JH is responsible for the removal of vitellogenin from the fat body, since after the vitellogenesis is completed and the ecdysone titer has risen, there is a high concentration of vitellogenin in the fat body but not in the hemolymph (Tanaka and Ishizaki, 1974; Stay and Tobe, 1981).

In *Drosophila* the processes that trigger vitellogenin synthesis are synchronized with the imaginal moult and require functioning of the brain. It is often held (see Tedesco et al., 1981) that a brain factor, as yet uncharacterized, controls vitellogenin synthesis in *Drosophila* together with ecdysone, while JH is responsible for the uptake of this protein by the oocytes. Vitellogenin synthesis in males can be stimulated by an injection of 20-hydroxyecdysone (Postlethwait et al., 1980a). In isolated female abdomens vitellogenin synthesis is induced only jointly by JH and 20-hydroxyecdysone. It has been suggested, therefore, that JH stimulates ecdysone synthesis in intact fly ovaries and ecdysone triggers vitellogenin synthesis in the fat body (Postlethwait et al., 1980b). A different hypothesis was proposed by Srdić et al. (1979), who induced vitellogenesis in the immature ovaries of *Drosophila mercatorum* by culturing them in the pupae of the same species, which are known to lack JH, but have a high titer of ecdysone. According to this hypothesis, in *Drosophila* JH is engaged in yolk protein synthesis control in the fat body while ecdysone performs the same function inside the ovaries. The authors suggested that vitellogenin could be synthesized endogenously, with the participation of the fat body.

Thus, the involvement of JH and ecdysone in the synthesis and secretion of vitellogenin of *Drosophila* is not completely clear, although it is beyond doubt that JH controls uptake of vitellogenin by oocytes (Postlethwait and Handler, 1978; Tedesko et al., 1981). Vitellogenin is present in the hemolymph of homozygous ap^4 females of *Drosophila*, but the flies fail to carry out vitellogenesis in their oocytes due to defective *corpora allata*. Following transplantation of ovaries of the mutants to normal females, or applying JH to the abdomens of ap^4/ap^4 females, the oocytes begin to take up vitellogenin. The oocytes show microvilli and numerous endocytotic pits and vesicles at their surface. The authors cited above proposed that JH exerts its effect directly on the oocyte surface.

There is variation in the number of vitellogenin polypeptides and their molecular size among some species of *Drosophila*, and transplantation of ovaries between such species does not lead to the uptake of host vitellogenin by the graft oocyte. However, the application of JH analog ZR 515 can "force" the oocytes to take up vitellogenin (Bownes, 1980).

In beetles and bugs vitellogenesis is controlled mainly by JH (Loof and Lagasse, 1970; Laverdure, 1972). The ability of the *corpora allata* to trigger vitellogenesis depends on how normally it functions, i.e., on the preservation of its nervous connections with the brain (Lender, 1965). The bug *Rhodnius prolixus* responds to JH by widening the spaces between the follicle epithelium cells through which vitellogenin passes from hemolymph into the oocyte (Davey, 1981).

The hormonal control of vitellogenesis in insects has been studied for many years, yet a number of questions remain. Vitellogenesis in these animals is triggered by external stimuli such as light, temperature, and feeding, but how such stimuli induce secretion of a hormone is not understood. In some insects, e.g., in the bean weevil *Acanthoscelides obtectus*, vitellogenesis is initiated without any external stimuli (Biemont, 1979).

In crustaceans, vitellogenesis is associated with moulting and requires ecdysteroid for the normal course of events. Some workers (Charniaux-Cotton, 1973; Meusy, 1980) believe that the onset of vitellogenesis in crustaceans is triggered by a short, moult-related stimulus, the source of which resides somewhere outside the eyestalks. In their opinion, the crustacean ovaries secrete a hormone similar to estradiol of vertebrates which induces vitellogenin synthesis. It is still unclear, however, what organ synthesizes this protein. Most of the workers are inclined to think that this is the hepatopancreas and/or fat body, although there is as yet no direct evidence for this.

In polychaetes, vitellogenesis is controlled by a brain endocrine center which, in its turn, is under the influence of environmental factors such as light and temperature. The gonad may also produce substances that affect this center (the feedback) (see Durchon, 1975). The removal of brain ganglia of nemertines stimulates gametogenesis (Rue and Bierne, 1980), and hence the brain of these animals produces factors that inhibit reproduction. It is hardly worthwhile to look for endocrine organs in these animals that induce vitellogenin synthesis, since yolk proteins seem to be synthesized continually in the oocyte, and only a factor that can inhibit vitellogenesis at certain time periods is required.

In cephalopods the yolk material synthesized by follicle cells is controlled by factors known as gonadotropins secreted by the optic glands (Wells and Wells, 1959; O'Dor and Wells, 1975; Buckley, 1976). The secretion of the optic glands suppresses protein synthesis in muscles, thereby increasing amino acid content in the hemolymph (O'Dor and Wells, 1978). This explains why the rapid growth of follicles of the ovary is accompanied by a decrease in the weight of other organs.

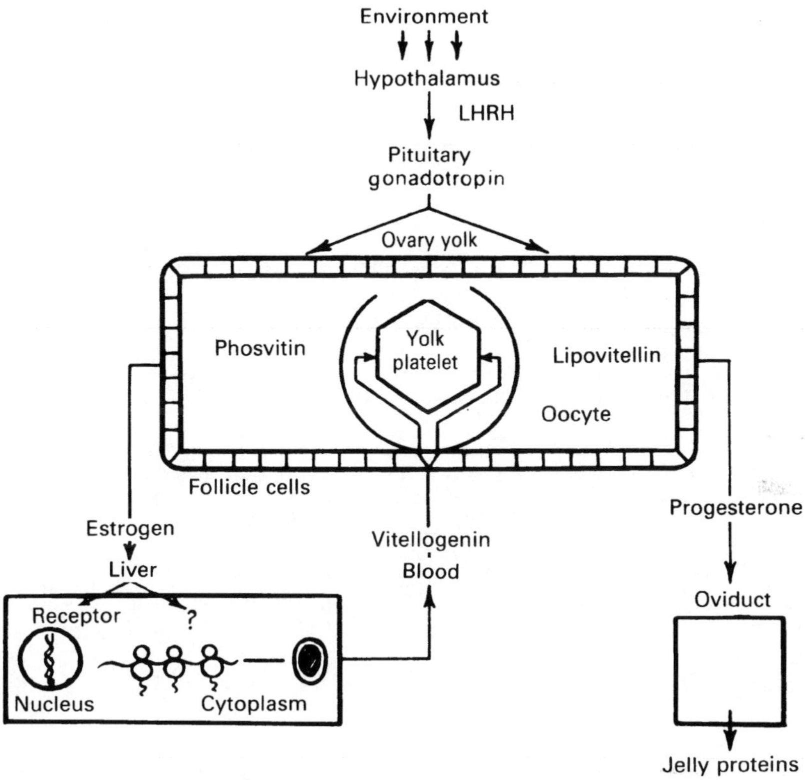

Fig. 4. Physiology of vitellogenesis in *Xenopus laevis* (Tata, 1976).

In *Cyclostomata* and fishes, vitellogenin is synthesized under the influence of estrogens (Sundararaj and Nath, 1981; Yu et al., 1981). Hormones such as progesterone, testosterone, or cortisone do not show any effect on vitellogenin synthesis.

In amphibians, the gonadotropins of the hypophysis, acting on the steroidogenic tissue of the ovary, induce the synthesis of estrogens, which are responsible for the synthesis of vitellogenin by the liver. The vitellogenin is transported by the blood to the vitellogenic follicles, and is taken up by the oocytes by way of micropinocytosis (Fig. 4). This established general sequence of the processes still requires more research.

Some 12-24 h after the introduction of 17β-estradiol into nonvitellogenic frogs, active synthesis of vitellogenin begins in the liver and continues for 3 weeks (Dolphin et al., 1971). This occurs not only in females, but in males as well (Wallace and

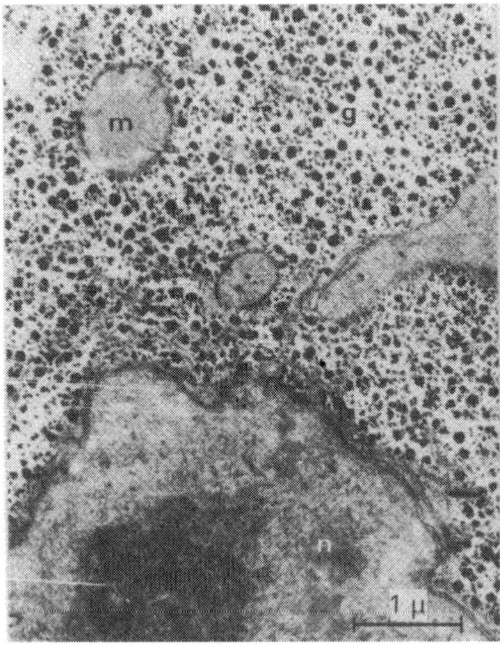

1

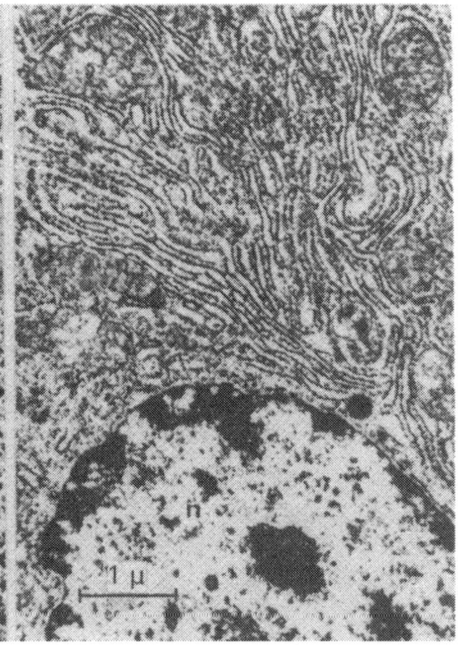

2

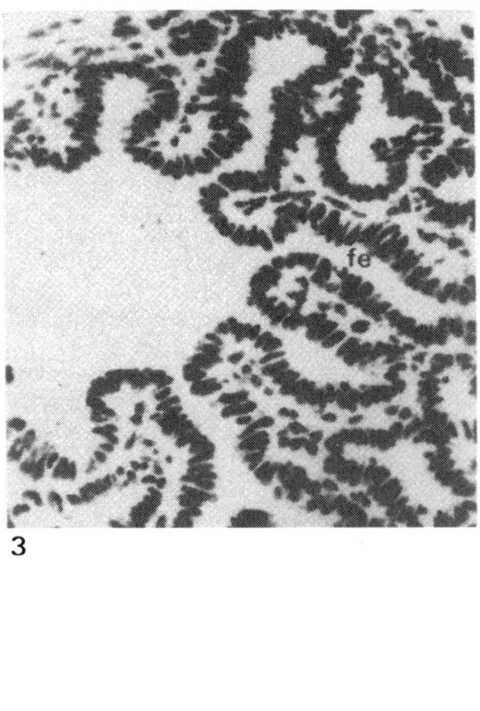

3

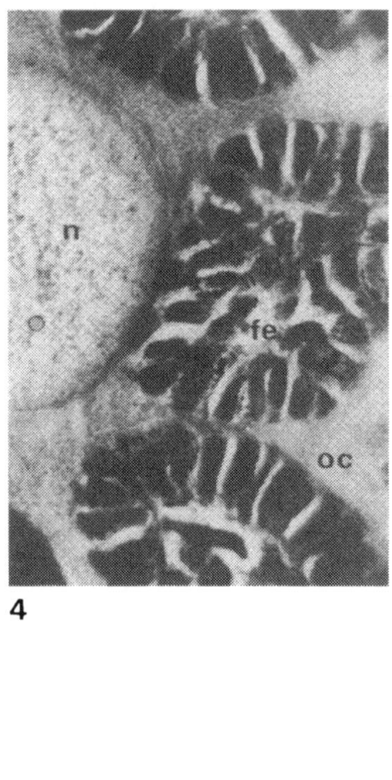

4

Jared, 1968). Vitellogenin enters the blood 2 h after the commencement of synthesis. In liver cultivated in vitro, vitellogenin is released into the incubation medium and may be isolated from it. The amount of vitellogenin synthesized is directly proportional to the dose of estradiol (Wangh and Knowland, 1975). The results suggest that the activation of genes coding for vitellogenin depends only on the presence of the hormone, and that cell division is not required.

It is interesting to note that in contrast to other hormones, such as growth and thyroid hormones, estrogens do not increase the rate of protein synthesis in general, but induce only the synthesis of vitellogenin. Estrogens produce marked changes in the parenchymal cells of the liver (Plate III, 1). Just 5 days after the introduction of 17β-estradiol into the females of *Xenopus laevis* the nucleus of their cells becomes rounded, the nucleolus is activated (its granular component increases), and an extensive granular endoplasmic reticulum develops in the cytoplasm (see Plate III, 2) (Follett et al., 1968). As the synthesis of vitellogenin increases, the concentration of serum albumin in the blood falls sharply (the main product of the synthesizing activity of the liver).

Although data indicate that both gonadotropin and estrogen are needed to stimulate the synthesis of vitellogenin by the liver, it is gonadotropin which is essential for the transfer of the protein from the circulation to the gonad (Wallace and Jared, 1969). The hormone regulation of vitellogenesis is complex (Holland and Dumont, 1975). It appears that in *Xenopus laevis* the pinocytotic activity of oocytes and, consequently, the uptake of vitellogenin by them, may be suppressed by the additional injection of estrogen and, conversely, may be stimulated by the introduction of chorionic gonadotropin. In vitro, however, estrogen stimulates the absorption of vitellogenin by the oocytes. Holland and Dumont suggest that in vivo estrogen suppresses the secretion of gonadotropins, and thus reduces the level of pinocytosis in the oocytes.

When introduced into the females of *Xenopus laevis*, FSH causes a drop in the concentration of vitellogenin in the blood and a

---

Plate III. 1. Cells in the liver of *Xenopus laevis* before the introduction of 17β-estradiol. 2. Five days after the introduction of this hormone (Follett et al. 1968). 3. A follicle of the cuttlefish at the beginning of vitellogenesis (Feulgen reaction) (Aizenshtadt and Marshak, 1969). Magnification ×112.5. 4. The oocyte nucleus of the cuttlefish and the follicle epithelium folds in contact with it. Magnification ×250. g) glycogen; fe) follicle epithelium; m) mitochondria; n) nucleus; oc) oocyte.

sixfold increase in the concentration of vitellogenin in the ovary. From these observations Follett et al. (1968) suggest that FSH not only stimulates pinocytosis, but also changes the chemical properties of vitellogenin, rendering it insoluble. These authors suppose that the estrogen acts in some way on the fat stores of the organism releasing lipids, which are then used for the synthesis of vitellogenin.

It still remains a mystery why vitellogenin "settles" in the ovary and becomes included in the vitellogenic follicles. It may be suggested that, under the influence of the gonadotropins, the follicle cells secrete some kind of substances which increase the permeability of the capillary walls (see Clavert, 1958).

Unsolved problems also include the interaction of estrogen with the cell surface, its transfer into the nucleus, and the subsequent activation of the gene coding for vitellogenin mRNA. Identification of the intermediate products of the supposed precursors of this mRNA is under investigation (Ryffel et al., 1980). Eleven hours after the administration of estradiol to *Xenopus* the RNA content of the liver increases eightfold (Tata, 1978), but the rRNA content only begins to increase after this time. Thus, vitellogenin mRNA is synthesized first. It has been calculated that 12 days after estradiol administration each liver cell of *Xenopus* contains 40,000 vitellogenin mRNA molecules (Hayward and Shapiro, 1981).

Steroid hormones interact with the target cell in two steps. The hormone first binds to a cytoplasmic receptor and then the hormone—receptor complex is transported into the nucleus, where it interacts with an acceptor site of the chromatin. The translocation mechanism of this complex remains completely obscure.

Vitellogenin synthesis can be induced in *Xenopus* liver at the end of metamorphosis (stages 57-62, Nieuwkoop and Faber, 1956), i.e., long before this protein is needed (May and Knowland, 1981). Even before metamorphosis begins, the liver cells possess estrogen receptors. However, the level of these receptors cannot be increased by estrogen treatment until the liver becomes inducible to vitellogenin synthesis, i.e., after stage 57.

Each nucleus of *Ambystoma mexicanum* liver cells contains 1300 receptor sites for estradiol (May et al., 1981), i.e., 13 times more than the nuclei of *Xenopus* liver. However, as the latter are about 12 times smaller, the concentration of receptor sites on the nucleus of each species appears to be nearly the same.

1.2.3. Participation of the Follicular Epithelium in Vitellogenesis

At present, it may be considered an established fact that

in most animals the follicle cells do not take a direct part in the synthesis of yolk proteins. However, a number of authors consider it possible that yolk material is synthesized by the follicle cells and secreted into the perioocyte space in such animals as insects (Anderson and Telfer, 1969; Loof, 1971) and fish (Hirose, 1971). Bast and Telfer (1976) have shown that the follicle cells in the ovary of the *Cecropia* moth secrete polypeptides into the interstitial space, one of which is included in the composition of the yolk granules. However, the part played by this polypeptide is unclear. The authors suggest that it either stimulates micropinocytosis in the oocyte (see above) or binds the vitellogenin with the plasma membrane of the oocyte. Secretion of yolk material by the follicle cells has been described in the ovary of the body louse (Ries and van Well, 1934). However, this work concerned the late stages of oogenesis, when the formation of the secondary egg envelope begins; therefore, the material secreted may relate to the formation of the chorion.

At the moment, the only example of the direct participation of the follicular epithelium in yolk synthesis is that observed in the ovaries of cephalopods. During vitellogenesis in the cuttlefish, the follicle cells secrete a material of a glycoprotein nature into the oocyte, which has no envelope (Yung Ko Ching, 1930; Aizenshtadt, 1969, 1971). These observations were confirmed in the isolated follicles of squid using ^{3}H-leucine-autoradiography (Selman and Wallace, 1978). Before vitellogenesis begins, the folds in the follicular epithelium, whose cells are reproduced mitotically, penetrate the oocyte, reaching as far as the nucleus (see Plate III, 3, 4). As the oocyte develops, the increasing mass of PAS-positive material secreted by the follicle cells flattens these folds and, when the oocyte's growth is completed, the follicular epithelium is seen as a single layer of cells on the surface of the oocyte. Evidently, the folds of the follicular epithelium fulfill the need for a rapid increase in the surface area of the single-layered epithelium when the vitellogenesis begins and the actively secreting follicle cells cease to divide. A similar point of view was expressed earlier by Gabaeva et al. (1971), who discussed in detail the functional significance of the folds in the follicular epithelium in various groups of animals.

In some invertebrates the follicle cells are of oogonial origin and may take part in the resorption of the material, which is contained in the cavity of the gonad. They may phagocytize and digest the dying oocytes and other cells. The low-molecular-weight compounds secreted by them then enter the oocytes by diffusion through the plasma membrane. This is the case in the gonads in a number of molluscs (Recourt, 1961; Bottke, 1972) and echinoderms (Verhey and Moyer, 1967). However, in some molluscs, for example in chiton, the follicle cells apparently do not take part in supplying the oocyte with material (Selwood, 1968, 1970). It should be borne in mind that in some of these animals the follicle cells exist only for a short period of oogenesis and after their resorption the oocytes actually grow according to the solitary type (see Fig. 1B).

There is a major interest in the part played by the follicular epithelium in the transport of the exogenous yolk proteins, since it is precisely this property of the follicle cells that is important for vitellogenesis in most animals. In the classical work of Telfer (1960), it was shown that the oocytes of the *Cecropia* moth are capable of taking up all the proteins circulating in the hemolymph, even heterologous ones; however, the oocytes mainly accumulate proteins synthesized by the fat body of the female. When the ovaries of the moth are transplanted to males, the oocytes only develop if the males are injected with blood from vitellogenic females (Stay, 1965).

The proteins in the hemolymph of *Cecropia* males, when introduced into the females, accumulate in the interstitial space between the follicle cells, and the proteins from the hemolymph of females readily pass through the follicular epithelium and accumulate in the peripheral layer of the ooplasm (Melius and Telfer, 1969). Telfer et al. stressed the selective nature of the uptake of the yolk proteins into the follicles of insects, and attempted to explain its mechanism. In vitro experiments suggested that vitellogenin, passing through the follicular epithelium via interstitial spaces, becomes mixed with a secretory product of the follicle cells, which either by itself or combined with vitellogenin induced the micropinocytosis in the oocytes (Hausman et al., 1971).

Electron-microscope studies of vitellogenic follicles, using such markers of pinocytotic activity as peroxidase, ferritin, the dextran complex with iron (Myofer), and others, have considerably extended our ideas of the role of the follicular epithelium in vitellogenesis (Ramamurty, 1964a; Anderson and Spielman, 1971). The basal membrane of the follicle in insects appears to be a rough mechanical filter through which only molecules, no more than 110 Å in size, pass, precisely such molecules as serum proteins, peroxidase, cytochrome C, hemoglobin, and ferrin. Particles 300-500 Å in diameter, for example certain dyes, do not pass through this basal membrane (Anderson and Spielman, 1971). Proteins and polysaccharides pass through the basal membrane fairly quickly, and particles of the thorotrast type, slowly. Apparently the follicular epithelium is able to distinguish between the macromolecules entering the perioocyte zone.

The passage of macromolecules through the follicular epithelium has also been studied in vertebrate animals. Expansion of the interstitial spaces in the follicular epithelium of newt was initially described by Hope et al. (1963), and it was suggested that large protein-type molecules passed between the follicle cells. Subsequently, using electron microscopy, Wartenberg (1964) showed that Myofer, when injected into newt females, passed through the basal membrane of vitellogenic follicles and between the follicle cells, became incorporated in pinocytotic vesicles of the oocyte, and united with the developing yolk granules.

The introduction of peroxidase (1-5 mg, intraperitoneally) into the nonvitellogenic females of lizard imitates the process of vitellogenesis: the peroxidase is found inside the follicle in 90 min and actively enters the oocyte by micropinocytosis (Limatola et al., 1972).

Expansion of the spaces between the follicle cells of vitellogenic follicles does not occur in all animals. For instance, in mosquitos the width of the gap between these cells is only 200 Å, i.e., the same as before the beginning of vitellogenesis (Anderson and Spielman, 1971). Although Roth and Porter (1964) described a drastic expansion in the interstitial spaces in vitellogenic follicles in the same species of mosquito, this could be explained by the poor electron-microscope technique of the time.

In newts, vitellogenesis begins in oocytes 1.1-1.2 mm in diameter, precisely at the time when gaps appear between the follicle cells (Kessel and Panje, 1968).

Thus, when there is a follicular epithelium, the macromolecules pass between the follicle cells, but this is not always accompanied by the expansion of the interstitial spaces. Apparently in some still unknown way, the follicular epithelium participates in the selection of the macromolecules entering the perioocyte space. Once in the perioocyte space, vitellogenin enters the oocyte by micropinocytosis. This process is described below.

1.2.4. Pinocytotic Activity of Oocytes

Pinocytosis was discovered by Lewis in 1931, using light microscopy. However, detailed study only became possible with improved cytological methods such as electron microscopy and autoradiography. Such methods led to the discovery of micropinocytosis.

The pinocytotic activity of the cell, a facet of its endocytotic activity, is the ability to absorb substances dissolved in water from the surrounding medium. It is assumed that the process of pinocytosis is preceded by the concentration of molecules in a certain area of the plasma membrane. This area forms a pit in the cell surface and then, as a result of closing the edges of the plasma membrane, a vesicle 100-200 nm in diameter is formed in the cell which contains an extracellular substance. Frequently, the pinocytotic vesicles are bordered by a dense material (Fig. 5), and it is as part of these coated vesicles that the vitellogenin and other proteins enter the oocytes.

It is thought (see Zelenin, 1962) that the ability of the cell surface to adsorb molecules of different polarity, particularly proteins and inorganic salts, is the result of pinocytosis. Although at present the mechanisms of inducing pinocytosis have not been fully explained, some inductors of pinocytosis are known. Thus, it is known that proteins, inorganic salts, ferrous compounds, and amino acids are stimulants of pinocytosis (see Stockem and Wohlfarth-Bottermann, 1969). Unfortunately, in the review cited no mention is made of oocytes. However, the data available in the literature on the pinocytotic activity of vitellogenic oocytes throws some light on certain aspects of this important physiological process.

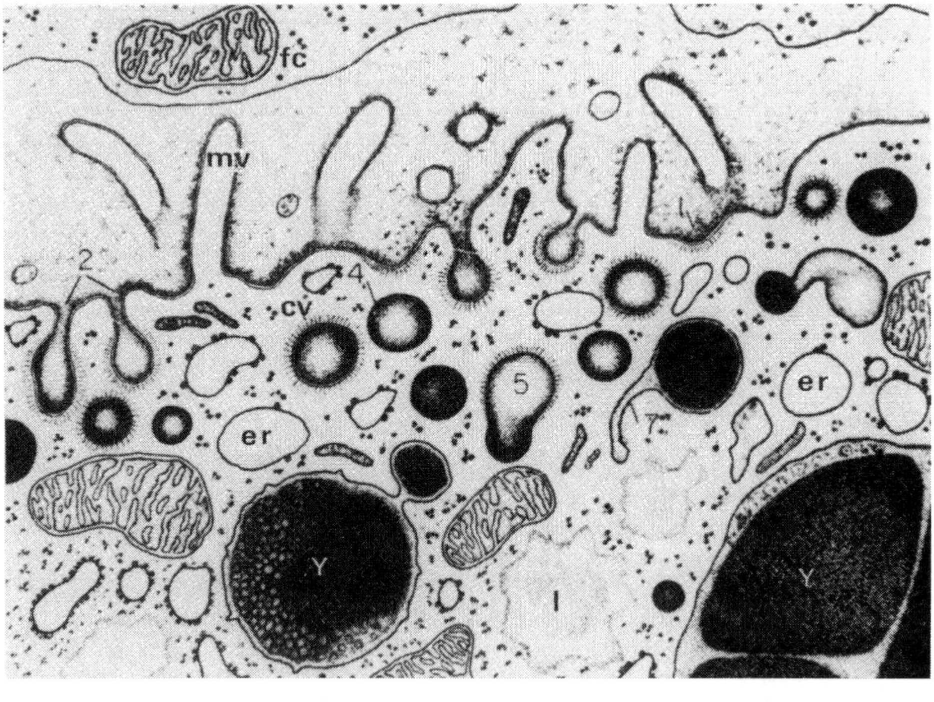

Fig. 5. The surface of the mosquito oocyte during the period of active pinocytosis and the formation of yolk granules (after Roth and Porter, 1964). cv) coated vesicles; er) endoplasmic reticulum; fc) follicle cell; l) lipid droplet; mv) microvilli; Y) yolk.

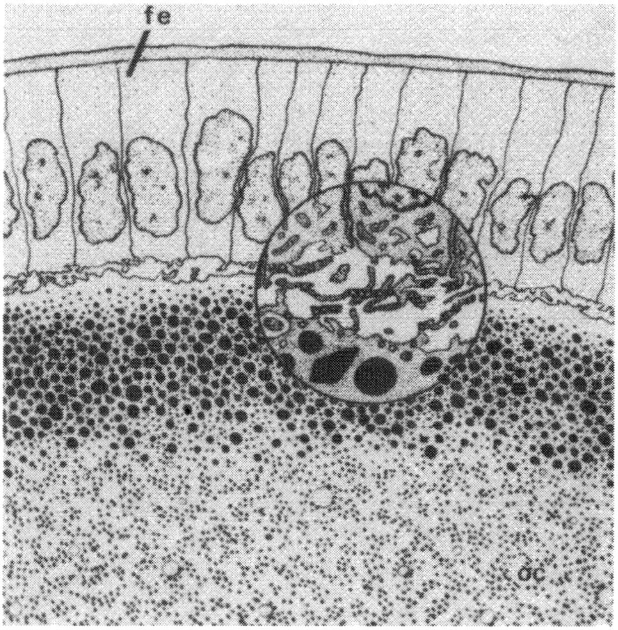

Fig. 6. Zone of contact between the oocyte and the follicle epithelium in *Cecropia* at the beginning of vitellogenesis (King and Aggaral, 1965). fe) follicle epithelium; oc) oocyte.

Although pinocytosis is inherent in practically all animal cells, some of them, such as endothelial cells, macrophages, vitellogenic oocytes, and also the protozoans, are characterized by very high activity. Thus, vitellogenic oocytes of amphibians incorporate 15 times more serum protein in the same interval of time than the previtellogenic oocyte (Jared and Wallace, 1969). In a fed mosquito numerous hollows and vesicles about 140 nm in size (Plate IV, 1; see Fig. 5) appear on the surface of the oocyte. The number of these structures per unit surface is approximately 15 times that in nonvitellogenic oocytes (Roth and Porter, 1964).

When studying the uptake of vitellogenin by the oocytes of amphibians in vitro, Wallace et al. (1970, 1973) came to the conclusion that the ionic strength of the medium plays an important part in this process. Moreover, it appeared that pinocytosis cannot generally take place without calcium ions. The transport to the oocyte of the substances making up the coated vesicles requires energy, which is supplied by glycolytic processes (Roth

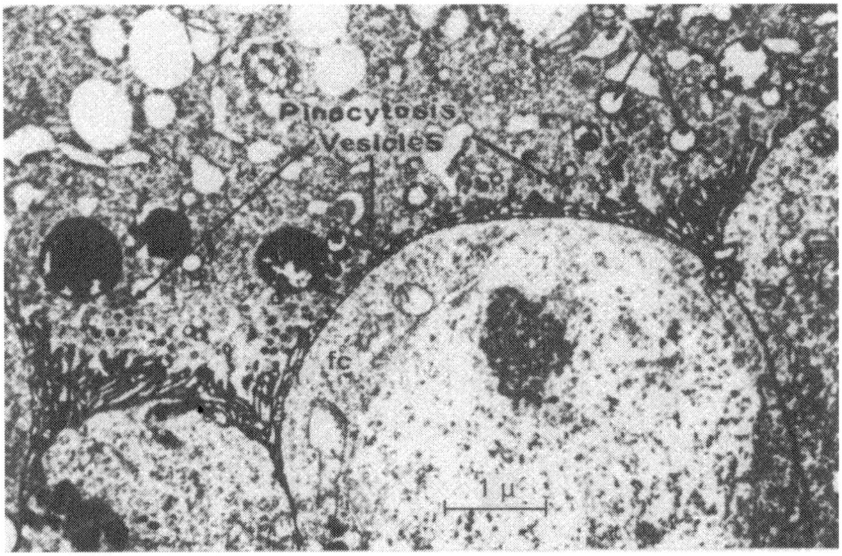

1

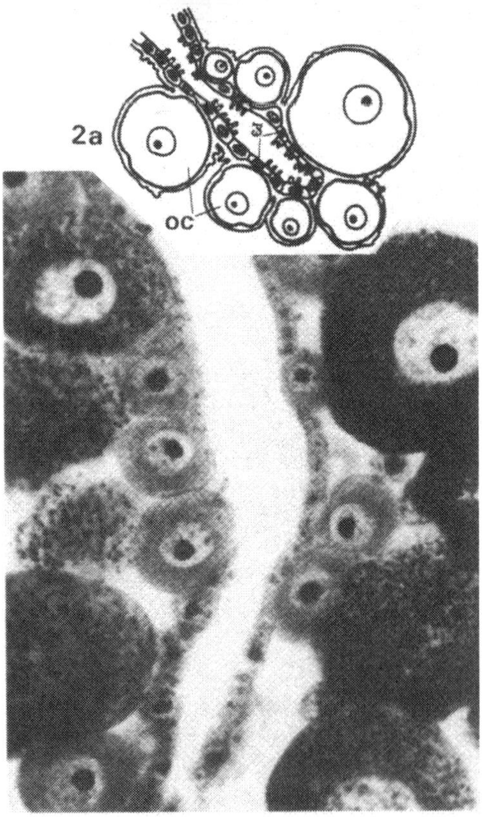

2

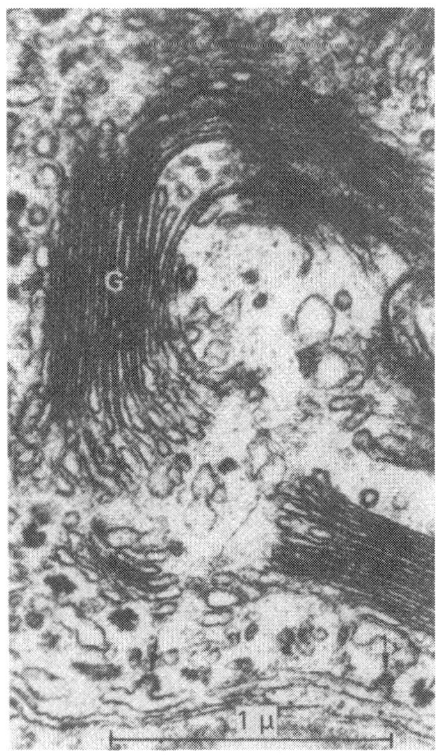

3

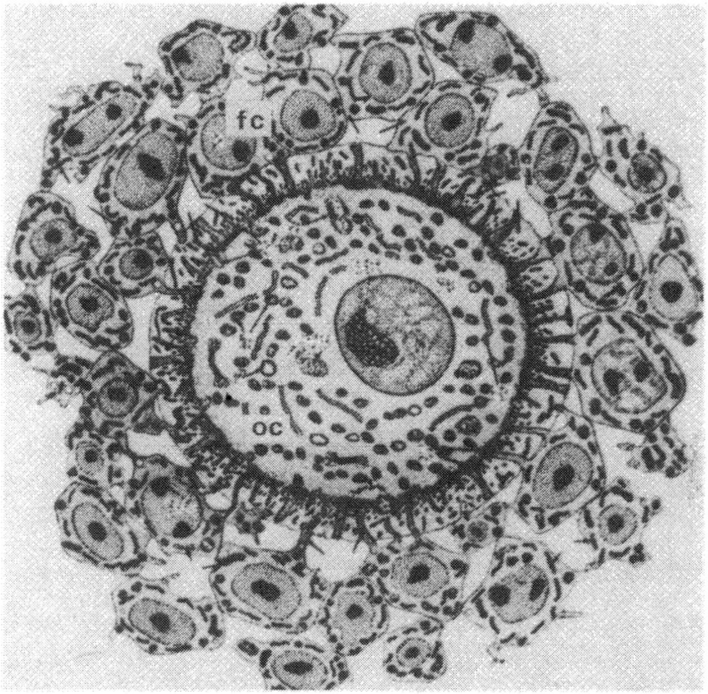

Fig. 7. Follicle of guinea pig (after Anderson and Beams, 1960). fc) follicle cells; oc) oocyte.

and Jackson, 1972). Inhibitors of glycolysis, 2-desoxyglucose and iodoacetic acid, suppress the uptake of the proteins by the oocytes by 95%.

The thin mucopolysaccharide layer (glycocalyx) on the surface of the plasma membrane of the oocyte plays an important part in the process of pinocytosis. In animals with follicular and nutrimentary oogenesis at an early stage of oogenesis, the oocyte closely adjoins the plasma membrane of the follicle cells. As the oocyte develops, a gap filled with PAS-positive material appears between it and the follicle cells (Figs. 6 and 7) (Bier, 1964;

Plate IV. 1. The surface layer of an oocyte in a mosquito. Incubation with peroxidase for 3 h after feeding with blood (Anderson and Spielman, 1971). 2. Cross section of the egg sac of *Priapulus caudatus*. Diagram of microscopic section. 3. The extensive Golgi apparatus in an ophiuroid oocyte (Kessel, 1968d). fc) follicle cells; G) Golgi complex; oc) oocyte.

Hope, 1965). Apparently both the follicle cells and the oocytes take part in secreting this material. In mammals this secretion is preceded by the appearance of numerous small Golgi vesicles on the surface layers of the follicle cells and oocyte cytoplasm. As the space between the oocyte and follicle cells expands, microvilli are formed on the surface of the oocyte and active pinocytosis commences. This is in agreement with observations on protozoans, in which the glycocalyx, a mucopolysaccharide layer 200-300 Å in thickness on the surface of the cell, also plays an important part in micropinocytosis (Stockem and Wohlfarth-Botterman, 1969).

Finally, it has been shown that one of the main conditions governing the pinocytotic activity of oocytes is the high concentration of proteins in the perioocyte space; this is also in agreement with observations on protozoans. Isolated oocytes from *Cecropia* lose their ability to take up hemolymph proteins since the concentration of these proteins in the incubation medium is much lower than in the perioocyte space of the intact follicle (Anderson, 1971).

In addition to inductors of pinocytosis, there are also factors that cause it to cease; their nature is still unclear. For example, during the development of the oocyte in *Cecropia* there is the so-called terminal phase, when proteins from the hemolymph cease to enter the oocyte. Twenty-four hours before the chorion material begins to be secreted by the follicle cells, certain factors appear to slow down pinocytosis (Telfer and Anderson, 1968). The cessation of vitellogenin synthesis or its entrance into the oocyte is regulated by hormones. Schuetz et al. (1974) discovered that corticosterone and progesterone in doses of 0.01-1.0 μg/ml suppress the inclusion of vitellogenin in the oocytes of amphibians in vitro and induce maturation. Similar doses of estrone neither suppressed pinocytosis nor stimulated maturation.

The vitellogenin that enters the oocytes of amphibians in the pinocytotic vesicles is washed out during preparation of the material for electron microscopy. The vesicles look empty, but when the protein becomes insoluble, dense contents appear in these vesicles. Similar observations have been made in insects (Roth and Porter, 1964).

Inside the oocyte the pinocytotic vesicles move and fuse with one another, forming precursor yolk inclusions. The latter become joined with new pinocytotic vesicles and Golgi vesicles (see below) and form definitive yolk inclusions. In *Xenopus* oocytes, vinblastine, which disturbs the polymerization of cytoplasmic microtubules and depolymerizes existing microtubules, hinders the movement of the pinocytotic vesicles in the cell (Dumont and Wallace, 1972). In oocytes treated with vinblastine, the process of combining the pinocytotic vesicles with the yolk platelets ceases, leading to the accumulation of the primordial yolk inclusions. Low temperatures have a similar effect, also causing depolymerization of microtubules.

The role of the follicular epithelium in the induction of pinocytotic activity in oocytes has been investigated mainly on insects and amphibians. Initially it appeared that isolated oocytes of *Cecropia* lost their ability to absorb the hemolymph proteins in vitro and form yolk platelets (Anderson, 1971); the same appeared to be true of amphibians (Wallace et al., 1970). However, later work showed that vitellogenin uptake by amphibian oocytes in vitro did not depend on the presence of follicle cells. Although oocytes treated with versene to remove the follicle cells ceased to take up vitellogenin, the addition of calcium to the medium restored this ability (Wallace et al., 1973).

The injection of animals with various dyes (Ramamurty, 1964a) or metal–colloid complexes impermeable to electrons (Anderson and Spielman, 1971) showed that oocytes incorporated all these substances provided the latter entered the perioocyte space. Numerous observations showed that oocytes, from which the follicular epithelium had been removed, had no selectivity in absorbing the various proteins. For example, the oocytes of chickens were capable of absorbing heterologous proteins such as bovine albumin, bovine globulin, and the serum protein of crustaceans (Knight and Schechtman, 1954). Contrast this with the selective uptake of vitellogenin by the amphibian oocytes (Wallace et al., 1970) and the selective uptake of phosvitin by chicken oocytes (Cutting and Roth, 1973). Obviously, data from research in vivo should not be compared with those obtained in experiments where heterologous proteins or different dyes were injected into animals, or where yolk proteins were introduced into a medium with isolated oocytes. Thus, the selective uptake of vitellogenin by amphibian oocytes in vivo may be a consequence of the high concentration of this protein in the perioocyte space. However, the exact mechanism of this selectivity remains to be determined. Observations by Wallace and Jared (1976) on the uptake of vitellogenin and bovine albumin by the oocytes of *Xenopus laevis* suggest that vitellogenin binds with specific receptor sites on the plasma membrane of the oocyte, and is then taken up by micropinocytosis. Roth et al. (1977) assert that in the oocytes of birds the receptor sections are located on the outer surface of coated alveoli, which then form coated vesicles.

Opresko et al. (1980) have shown that the intracellular stability and compartmentation of protein captured by endocytosis mediated by a specific receptor is different from the fate of proteins entering the cell via unspecific endocytosis. Endosomes containing vitellogenin-occupied receptors fuse together, and

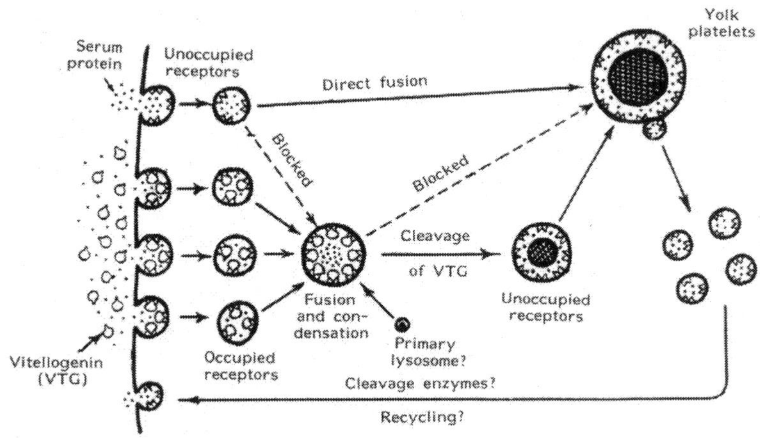

Fig. 8. Model of post-endocytotic compartmentation in oocytes (Opresko et al., 1980). Incoming endosomes containing occupied VTG (vitellogenin) receptors fuse and undergo subsequent condensation and processing of their contents. When the VTG is cleared from its receptor, fusion with the yolk platelets occurs. Endosomes lacking occupied VTG receptors cannot fuse with those containing occupied receptors, but instead fuse directly with the yolk platelets.

their contents are subject to condensation and processing. After vitellogenin leaves the receptor zones, the endosomes fuse with yolk platelets. Endosomes that contain no vitellogenin-occupied receptors do not fuse with endosomes that possess such receptors, although they can fuse with the yolk platelets directly (Fig. 8).

Our knowledge of the uptake of proteins into oocytes may be summarized as follows. The follicular epithelium can select heterosynthetic macromolecules, but the mechanism of this process is still not clear. Oocytes can take up various proteins, including heterologous ones, if they are present in the necessary concentration either in the perioocyte or in the incubation medium. The primary uptake of vitellogenin by the oocytes may be explained by its high concentration in the vitellogenic follicles, and also by the existence of receptor sites for this protein on the plasma membrane of the oocyte. The follicular epithelium ensures the high concentration of vitellogenin in the perioocyte space and participates in synthesizing the glycocalyx, which plays an important part in the pinocytotic activity of the oocyte.

1.2.5. Endogenous Yolk (General Comments)

Although a large part of the yolk in the oocytes of many

animals is of exogenous origin, endogenous synthesis, i.e., by the oocyte itself, is known.

Endogenous yolk is known in the oocytes of turbellarians (Boyer, 1972), crustaceans (Beams and Kessel, 1963), and fishes (Droller and Roth, 1966; Anderson, 1968; Ulrich, 1969). In the viviparous teleosts *Xiphophorus helleri* no similarity is in general found between the blood proteins and the yolk proteins, and it is suggested that all the yolk is endogenic in origin (Wegmann and Gotting, 1971). However, the vitellogenic oocytes of this fish are capable of the active uptake of Myofer from an incubation medium; they therefore may take up blood proteins during a certain period in their development.

Some yolk proteins may also be synthesized by the oocytes of those animals in which exogenous yolk is predominant, for example insects (Cummings and King, 1970). Initially this was assumed to be true of amphibians (Rudack and Wallace, 1968), but subsequently (Wallace et al., 1972) it was concluded that practically all the yolk in amphibians is exogenous.

Unfortunately, the extent to which the oocyte participates in the synthesis of yolk proteins cannot be determined precisely. The protein synthesis in oocytes revealed by autoradiography does not allow us to determine which proteins are synthesized; special biochemical methods should be employed.

Many electron-microscope studies describe the contact between the Golgi vesicles and the yolk platelets formed during the fusion of the pinocytotic vesicles, and the conclusion has been drawn that endogenous yolk material unites with the exogenous material. These pictures, however, should not be interpreted unambiguously. Thus, in the teleost *Brachydanio rerio* it has been shown (Korfsmeier, 1966) that glycoproteins are synthesized in vitellogenic oocytes, and then form a peripheral protective layer of the definitive yolk platelet. Using amphibians, Wallace et al. (1972) came to the same conclusion, considering that the Golgi vesicles in the oocytes, which fuse with the primordial yolk platelets, contain polysaccharides, which form part of the peripheral layer of the definitive yolk platelet.

In highly organized animals, such as insects or vertebrates, the ovaries possess accessory cells, and the oogenesis is of the alimentary type. However, in some molluscs (Bedford, 1966), worms (Eckelbarger, 1975; Fischer, 1975), coelenterates (Kessel, 1968a), and echinoderms (Kessel, 1968d), the oogenesis is mainly of the solitary type, since the follicle cells in the ovaries of these animals only exist for a short time or are generally absent (see Plate IV, 2, 2a). Although in this case there are no barriers between the oocytes and the surrounding medium (for instance the coelomic fluid), active pinocytosis does not take place in them. The oocytes of these animals differ in their ultrastructure from the oocytes that accumulate heterosynthetic molecules; they possess well-developed granular endoplasmic reticulum and Golgi regions (Bedford, 1966; Kessel, 1968a, 1968d; Eckelbarger, 1975). Some authors (Yamamoto et al., 1973) noted an inverse correlation between the degree of development of the endoplasmic reticulum and pinocytosis. In oocytes, which actively take up protein from outside, the endoplasmic reticulum is poorly represented; this is characteristic of vertebrates.

Before we deal in detail with the participation of the cell organelles in the synthesis and accumulation of yolk, we must dwell on some questions regarding the origin and functions of a number of cell organelles, first and foremost the Golgi apparatus, which plays an important part in vitellogenesis.

1.2.6. Origin and Functional Role of the Golgi Apparatus

In recent times, electron microscopy has played an important part in the development of modern cytology; however, we believe that this method has led to an incorrect idea of the cell as a system of compartments and canals sealed off from the matrix of the cytoplasm by the membranes. Many textbooks, reviews, and original research papers show diagrams indicating a direct link between the perinuclear space (the space between the two membranes of the nuclear envelope) and the extracellular environment through cavities in the branching endoplasmic reticulum. These views are in conflict with observations on living cells, in which movement of the cytoplasm and even rotation of the nucleus occurs.

From time to time attempts are made to examine the ultrastructure of cells from the point of view of colloid chemistry (Alov et al., 1969; Schmick, 1971) but, unfortunately, these works are few and their authors, like the author of this article, are not specialists in physical chemistry. Nevertheless, we believe the time has come to attract attention yet again to the problem of the supermolecular structure of the cell and to attempt to link, if only in general form, the pictures observed in the electron microscope and those that may occur in the living cell. Some aspects of this problem are considered below in light of data obtained in the last few years, mainly on the secretory cells of animals.

The accumulations of organic macromolecules, which form a number of cytoplasmic structures, exhibit certain physical and chemical properties, one of which is the ability to form a layer of protein and lipid polar molecules on the boundary with the dispersion medium, for example with water. The enzymes included in this layer play an important part in the chemical transformations of substances. The oriented lipid molecules allow good osmium impregnation, and thus the boundary layer is well preserved during the preparation of material for electron microscopy, while other components of the cell are washed away upon dehydration, especially after osmium fixation alone. In fixed preparations of the cell in electron microscopy one can clearly see vesicles, vacuoles, and cavities of various shapes formed by closed membranes, and one is involuntarily influenced by this lifeless scheme. Such terms as cisternae, cavities, vesicles, canals, etc., are in common usage. However, although this terminology is useful at the present time, it should be recognized that these concepts are merely conventional. More recently the use of complex research methods has advanced our understanding of the origin and functional role of cell organelles. As an example, we have chosen the formation of the Golgi apparatus.

In recent years the Golgi apparatus has been fruitfully studied by a number of techniques (Neutra and Leblond, 1966, 1969; Beams and Kessel, 1968; Favard, 1969; Morré et al., 1971). The main results were obtained in studies of secretory cells. Accumulation of substances in these cells, and their subsequent removal from them, are impossible without the participation of the Golgi apparatus: this allows the dynamics of the organelle to be studied. The use of autoradiography at the level of the electron microscope showed that the Golgi system is not only the site of "packing" of the secretory granules, but is also the place where various substances are synthesized, primarily complex carbohydrates (Neutra and Leblond, 1966). The protein component of the glycoprotein is synthesized in the endoplasmic reticulum and transported to the Golgi complex as small granules (Fig. 9a). Morphologically, this appears as the formation of numerous smooth-surfaced vesicles from the cisternae of the endoplasmic membrane; these unite with the external profiles of the flat Golgi sacs (dictyosomes) (Morré et al., 1971).

It is commonly believed that, unlike nonsecreted proteins, all secreted ones are glycoproteins, and that this is somehow linked with their ability to leave the cell. Secreted enzymes, some hormones, blood plasma proteins, antibodies, and glycocalyx do contain carbohydrates. Biochemical methods (see Neutra and Leblond, 1969) reveal the incorporation of labeled glucosamine into the glycoproteins not only in a fraction of the smooth-surfaced membranes, but also in the fraction rich in ribosomes. In the cells of the thyroid gland, labeled galactose is incorporated into the Golgi apparatus, and mannose into the area of small vesicles separating from the endoplasmic reticulum. An analysis of the lateral chains of molecules of glycoproteins from this

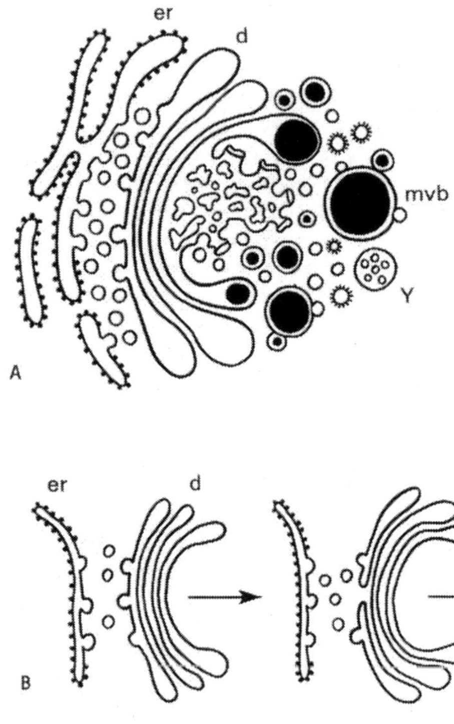

Fig. 9. Formation of Golgi complexes. A) Synthesis of the endogenous yolk; B) division of Golgi complexes (after Morre et al., 1971). d) dictyosome; er) endoplasmic reticulum; mvb) multivesicular body; Y) yolk.

organ showed that mannose is located close to the protein, while galactose is found at the end of the lateral chains. This suggests that a molecule of glycoprotein begins to form immediately after the formation of its protein skeleton when it is in close contact with the endoplasmic reticulum, and that the major part of the molecule completes its development in the Golgi apparatus (see also Leblond and Bennett, 1974). In general, the Golgi apparatus cannot be formed without the synthesis of certain proteins in the endoplasmic reticulum (or the perinuclear space). This is confirmed by the observation that during suppression of protein synthesis in the amoeba by emetine the Golgi complexes are reduced (Flickinger, 1971). The functional link between the nuclear envelope and the endoplasmic reticulum, on the one hand, and the plasma membrane on the other, necessarily involves the Golgi vesicles, since only their membrane is able to fuse with the

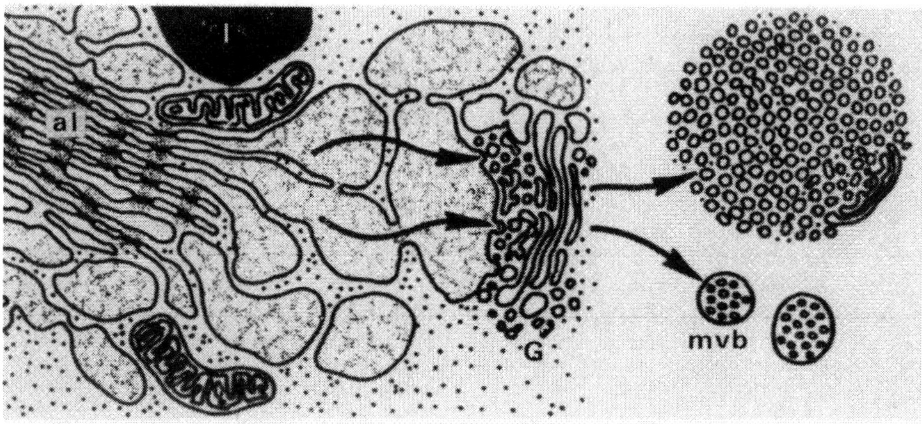

Fig. 10. The interrelation between annulate lamellae and Golgi apparatus in maturing sturgeon (*Acipenser stellatus*) oocytes (after Aizenshtadt and Dettlaff, 1972). al) annulate lamellae; G) Golgi complex; l) lipid droplet; mvb) multivesicular body.

plasma membrane (Morré et al., 1971; Frey-Wyssling, 1973). There is no direct contact between the perinuclear space or the cavity of the endoplasmic reticulum and the extracellular space.

Thus, the Golgi complexes, consisting of dictyosomes with small anastomosing tubules, vesicles, vacuoles, and multivesicular bodies, are a polar organelle. This organelle is constantly renewed and constantly expended in the formation of the final product of the cell's activity as vesicles, vacuoles, and granules. It has been calculated that the time necessary for the formation of a mucin granule from a Golgi sac is 2-4 min (Droz, 1966). Obviously, the Golgi apparatus is a transitory organelle, which is changing its volume all the time and may in some cases disappear temporarily. In the cells of plants, lower Metazoa, and in the oocytes and spermatocytes of a number of animals, active budding of the outer membrane of the nuclear envelope occurs, and Golgi complexes are formed directly from the nuclear envelope (Srivastava, 1965; Kessel, 1968c; Stang-Voss, 1970; Korotkova and Aizenshtadt, 1976; Frey-Wyssling, 1973). The formation of the Golgi complexes by budding of the end extensions of the annulate lamellae, which in their turn originated from the outer membrane of the nuclear envelope (Fig. 10), has been described in the maturing oocytes of amphibians and sturgeon (Ward, 1965; Aizenshtadt and Detlaff, 1972).

The elements of the Golgi complex, in particular the flat sacs located parallel to one another (dictysomes), may appear to be discrete on ultrathin sections, but are, in fact, linked with one another. This is demonstrated quite clearly on thick sections with the help of a high-voltage electron microscope (Carasso et al., 1971). It is obvious that the Golgi cisternae form a common mobile system; they may fuse, separate, and be perforated.

We have concentrated on the formation of the Golgi apparatus. However, there is a great deal of literature on the formation of the endoplasmic reticulum and annulate lamellae from the perinuclear cisterna. Morphologically, this takes the form of budding, i.e., the separation of vesicles from the outer membrane of the nuclear envelope. One can imagine that the system of the endoplasmic reticulum is a mobile compartment, constantly changing its form, surrounded by ribosomes, and actively participating in protein synthesis. In an electron microscope areas of the cell where synthesized protein molecules begin to unite with the carbohydrates resemble the budding cisternae of granular endoplasmic reticulum (see Fig. 9a).

Therefore, it is misleading to talk about the transport of the substances from one organelle to another through cavities or canals; changes in the chemical composition of substances and the synthesis of new ones is morphologically expressed in the emergence of new ultrastructure. In particular, the synthesis of glycoproteins is expressed morphologically as the separation of the small vesicles from the cisternae of the endoplasmic reticulum and the formation of the Golgi dictyosomes from them. The physical and chemical properties of the substances being synthesized determine the configuration of the membranes that may be seen in fixated preparations. This alone may explain the striking similarity between the ultrastructures in different cells.

It has been suggested that Golgi complexes may multiply (Morré et al., 1971). In actual fact, one can imagine that, at a certain stage in the synthesis of complex carbohydrates when the granules separating from the endoplasmic reticulum form new dictyosomes and the synthesis of the final product lags behind, the elongated flat sacs, which are increasing in size, are interlaced as in Fig. 9b. However, the question of the multiplication of the existing Golgi complexes in the cells, which are not dividing, remains open.

1.2.7. Formation of the Yolk Platelets during Endogenous Synthesis of the Yolk Proteins

There is great similarity between the processes of secretion and the endogenous synthesis of yolk proteins (see Fig. 9a). Although the yolk (unlike the secretion of the somatic cells) is not released from the cell, it is also, as a rule, a glycoprotein.

Cases are known where secretory granules in the somatic cells are crystallized and closely resemble yolk inclusions (Kurosumi, 1961).

The formation of yolk platelets in the Golgi apparatus has been described in echinoderms (Kessel, 1966a, 1968d), worms (Dhainaut, 1967; Dumont, 1967; Norrevang, 1968; Sawada et al., 1968), molluscs (Beams and Sekhon, 1966; Anderson, 1969a), coelenterates (Kessel, 1968a), tunicates (Kessel, 1966b), and spiders (Osaki, 1972).

Budding of endoplasmic reticulum cisternae and the formation of numerous Golgi complexes with yolk platelets occurs in the vitellogenic oocytes of *Ophiuroidea* (see Plate IV, 3) (Kessel, 1968d) and *Xiphosura* (Dumont and Anderson, 1967).

The Golgi apparatus also takes an active part in the synthesis and formation of yolk platelets in those cases when only part of the yolk is synthesized in the oocyte as, for example, in the rabbit (Fujitani, 1961), fishes (Wegmann and Gotting, 1971), and *Drosophila* (Mahowald, 1972).

Thus, both the secretory granules of somatic cells and the yolk platelets of oocytes pass, as it were, through the Golgi apparatus. However, observations exist which appear to contradict this. First Beams and Kessel (1963) discovered dense, disc-like granules 60-100 nm in diameter in the cisternae of the rough endoplasmic reticulum (Plate V, 1) in the oocytes of three species of crustaceans, which have a follicular type of oogenesis. During its passage from the nucleus to the surface of the oocyte, the endoplasmic reticulum loses ribosomes, and the accumulations of the disc-shaped granules become yolk platelets with a fine-granular structure (see Plate V, 2). Beams and Kessel suggest that yolk is entirely of endogenous origin, and have emphasized that the Golgi apparatus hardly takes any part in the synthesis of the yolk and its accumulation in the form of platelets. Further, autoradiographic and biochemical investigations confirmed this view (Kessel, 1968b; Ganion and Kessel, 1972; Lui and O'Connor, 1976). At first sight, the formation of yolk in crustaceans resembles the accumulation of secretory proteins in the endoplasmic reticulum in the cells of the pancreas of starving guinea pigs (Jamieson and Palade, 1971); in this case, however, the Golgi apparatus was activated and the secretory substance "passed" through it before leaving the cell.

Concerning the participation of cell organelles in vitellogenesis, one should always make a distinction between the synthesis of yolk material and its formation into platelets. The fact that yolk protein is synthesized in the endoplasmic reticulum in the oocytes of crustaceans is not surprising, but the fact that

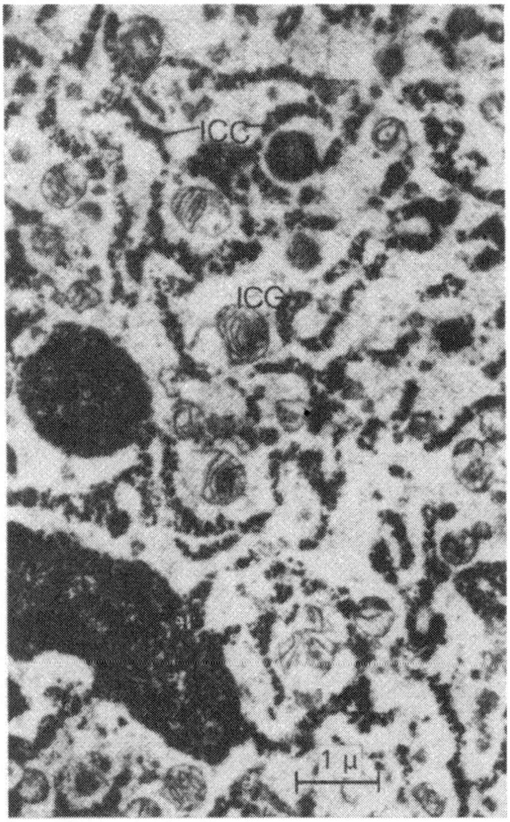

1

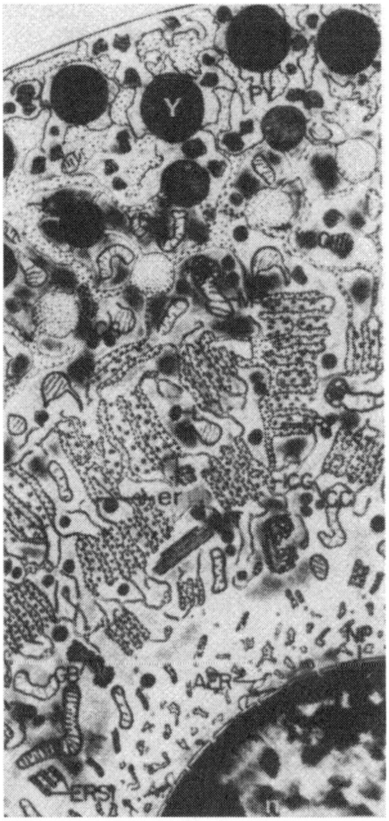

2

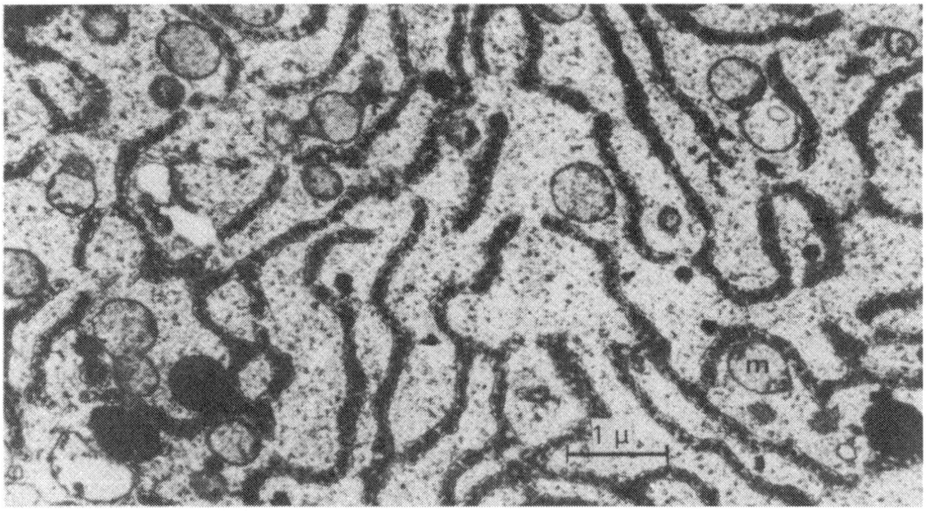

3

this protein is transformed into definitive yolk platelets within the endoplasmic reticulum as well, makes the oocytes of crustaceans an exception to the general rule. However, in the crab *Libinia emarginata*, for example, the Golgi apparatus does take an active part in the synthesis of endogenous yolk (Hinsch and Cone, 1969), and in this case yolk is also taken up from without by means of micropinocytosis. Lipovitellin, synthesized outside the ovary, has been isolated from crustaceans (Kerr, 1969), and a specific protein has been found in the vitellogenic female crustacean; in females without vitellogenic oocytes, and in the males, the protein is absent (Wolin et al., 1973). When labeled with fluorescein, the protein enters the oocytes and becomes incorporated into yolk platelets. These authors suggest that the hepatopancreas and the cells of the hemolymph are involved in the synthesis of vitellogenin in crustaceans. Although the methods used by Wolin et al. have been criticized and doubt has been cast on their observations (Lui and O'Connor, 1976), other evidence shows that there is a glycoprotein fraction in the hemolymph of female crustaceans which is identical to one of the yolk components (Croisille et al., 1974). Initially the significance of endogenous yolk proteins in the growth of the oocyte in crustaceans seems to have been overestimated by Beams and Kessel, and the possibility of exogenous yolk in these animals was in fact considered by Ganion and Kessel (1972).

Granular material has also been observed in the expanded endoplasmic reticulum of the vitellogenic oocytes of trout (see Plate V, 3) (Beams and Kessel, 1973), and in the oöcytes of freshwater mussels (Beams and Sekhon, 1966); however, the formation of the yolk platelets in these animals occurs in the Golgi apparatus.

Although our knowledge of the cytology of developing oocytes is incomplete, there is no doubt that in all animals the Golgi apparatus participates both in the synthesis of endogenous yolk and in the incorporation of substances that join up with the exogenous yolk platelets, for example carbohydrates (see above). In some cases, for instance in the annelid *Enchytraeus albidus* small pinocytotic vesicles, containing lipoproteins of exogenous origin (Dumont, 1969), join with the large platelets of endogenous yolk which are synthesized in the endoplasmic reticulum and the Golgi apparatus.

Plate V. 1. A part of the cytoplasm of a vitellogenic crayfish oocyte. 2. A diagram of a vitellogenic crustacean oocyte (Beams and Kessel, 1963). 3. A section of the cytoplasm of a vitellogenic trout oocyte (Beams and Kessel, 1973). er) endoplasmic reticulum; m) mitochondria; n) nucleus; Y) yolk; AER) agranular reticulum in perinuclear zone; ERSI) forming stacks of granular endoplasmic reticulum; PY) mature proteinaceous yolk body; ICC) granule-filled branching and communicating cistern; ICG) intracisternal granules.

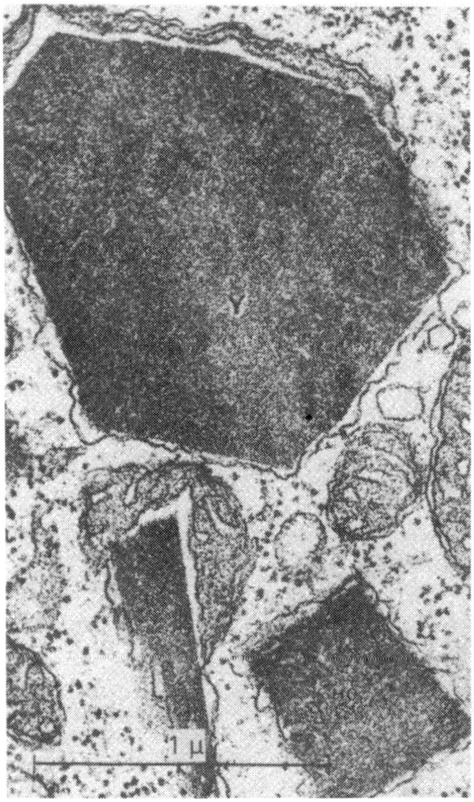

1

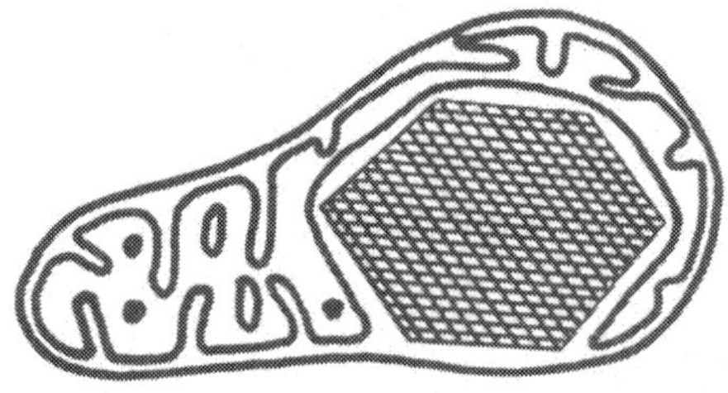

1a

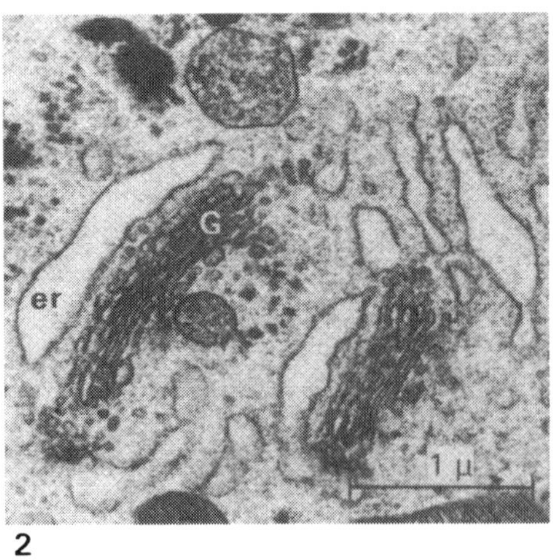

2

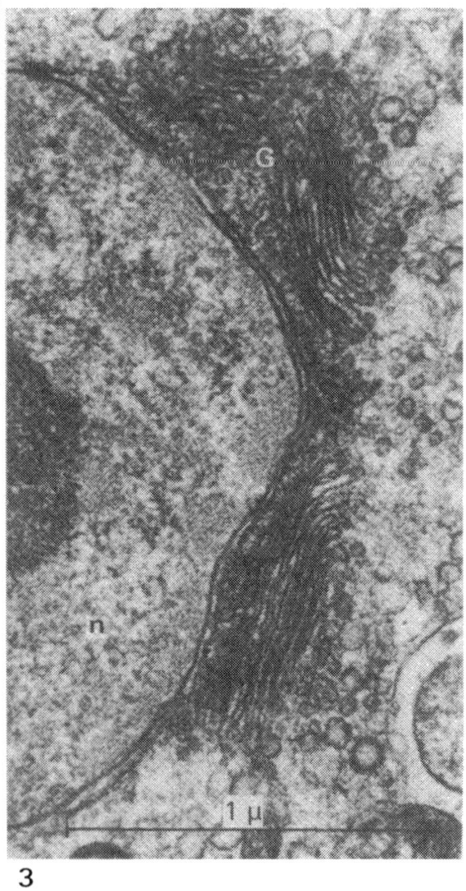

3

Various "lysosomal" hydrolases occur in the yolk platelets of a number of animals (Bluemink, 1969; Pasteels, 1973). For instance, acid nitrophenyl phosphatase and α-L-fucosidase (Schuel et al., 1975) have been discovered in the yolk platelets of sea urchin. The conclusion was drawn (Bluemink, 1969) that the yolk platelets are complex liposome-like organelles of oocytes, and their ability to digest themselves and other organelles fusing with them, as is the case, in particular, with the mollusc *Planorbis* (Favard and Carasso, 1958), allows us to regard the complex yolk inclusions as autophagous vacuoles in this case. Reger (1970) came to this conclusion when he discovered the tremendous similarity between yolk platelets in the oocytes of spider and autophagous vacuoles.

Morphological research alone cannot determine which component of the yolk is synthesized by the oocyte itself when a large part of the yolk material enters the oocyte by way of pinocytosis. Polysaccharides, endogenous yolk proteins, and hydrolases may be found in the small Golgi vesicles, which unite with the pinocytotic vesicles and the platelets of precursor yolk. However, although hydrolases, together with their substrates, are found inside the yolk platelets, activation of these enzymes only occurs after fertilization. The mechanism of this activation is not known, but it is believed (Schuel et al., 1975) that an important part is played in this process by the increase in the concentration of free calcium ions, which are released during fertilization from their binding sites on the plasma membrane of the oocyte or on the membranes surrounding the cell granules.

Initially, electron-microscope observations suggested that yolk platelets formed inside mitochondria (Lanzavecchia, 1961, 1965; Ward, 1962), supporting observations of previous authors suggesting the possibility of transforming mitochondria into yolk. At present, however, it would be more correct to talk not about transformation of mitochondria into yolk, but about the accumulation of yolk proteins in mitochondria. The possibility that these proteins are synthesized by the mitochondria themselves is denied by most researchers. The accumulation of yolk proteins inside the mitochondria is known only in some animals, in particular am-

---

Plate VI. 1. The crystalline yolk inside mitochondria (Massover, 1971c). 1a. A hexagonal crystalloid inside the mitochondria in the frog oocyte (scheme). 2. The formation of Golgi complexes by budding of the endoplasmic reticulum cisternae in early *Hydra* oocytes (Aizenshtadt and Marshak, 1974). 3. Formation of Golgi complexes by budding of perinuclear cisternae in sponge cells (Korotkova and Aizenshtadt, 1976). G) Golgi complex; er) endoplasmic reticulum; n) nucleus; Y) yolk.

phibians (see Massover, 1971a), fish (Yamamoto and Oota, 1967), and spiders (Reger, 1970). The crystalline yolk appears between the outer and inner membranes within the cristae (Plate VI, 1, 1a). As the yolk platelet develops, the mitochondrial structures on its surface decrease in size, and then disappear. Having reached a certain size, a granule becomes indistinguishable in its structure from yolk platelets formed by the fusion of pinocytotic vesicles. It is still not absolutely clear which part of the yolk granule is of mitochondrial origin. Kress and Spornitz (1972) suggest that about 20% of the mitochondria in *Rana* are converted into yolk; they also mention the formation of crystalline yolk in the matrix of the mitochondria. However, most authors believe that the yolk proteins only pass through the outer membrane of mitochondria and not the inner one; therefore, they cannot enter the matrix (Kessel, 1971; Massover, 1971a, b). Evidently, the mitochondria are a kind of phosvitin and lipovitellin depot, but it is still an open question as to where the intramitochondrial yolk is synthesized. Although it is known that most of the phosvitin and lipovitellin in amphibians results from the breakdown of vitellogenin synthesized in the liver, one cannot exclude the possibility of the synthesis of intramitochondrial yolk by the oocyte itself, all the more so since the transformation of exogenous vitellogenin begins and ends inside the pinocytotic vesicles in the peripheral ooplasm, and it is difficult to imagine how it could enter the mitochondria.

Some electron-microscope observations also suggest the formation of crystalline yolk platelets within mitochondria in the oocytes of the spider and the fusion of small Golgi vesicles to these mitochondria (Reger, 1970). However, these morphological observations do not seem to justify Reger's conclusion that endogenous yolk is synthesized in the mitochondria.

An interesting idea has been proposed by Karasaki (1963b), that yolk crystals inside mitochondria represent secondary yolk, which is indicative of a dynamic equilibrium between the formation and partial utilization of yolk which takes place during oogenesis.

We have seen that the growing oocytes in animals which store up huge quantities of ribosomes and yolk platelets in their cytoplasm have many adaptations which are, as a rule, absent from somatic cells. Oocytes are linked with the other frequently polyploid cells by means of cytoplasmic bridges, and may amplify genes coding for rRNA syntheses; they also possess the ability not only to synthesize yolk proteins but also to take them up in large quantities by means of micropinocytosis.

However, the properties of the oocytes, such as the link with the accessory cells supplying them with rRNA and the ability to take up and accumulate heterosynthetic macromolecules, did not suddenly appear during evolution. We have already noted the low level of pinocytotic activity in oocytes of lower invertebrates with the solitary type of oogenesis. In some cases there is no ovary as such; the oocytes are dispersed usually in the coelom. This condition is defined as "diffuse oogenesis." The growth of these oocytes, characteristic of lower metazoans and of some tunicates, is dealt with in the following section.

1.3. PHAGOCYTIC ACTIVITY OF THE OOCYTES DURING DIFFUSE OOGENESIS

In diffuse oogenesis, oocytes possess amoeboid mobility and the capacity for phagocytosis (Fig. 1A). Phagocytosis also occurs in the oocytes of higher animals under certain conditions. For instance, during the in vitro culture of human follicles, migration of the follicle cells across the *zona pellucida* is observed, and they are subjected to phagocytosis by the oocyte (Zamboni et al., 1972).

In many lower vertebrates, oogenesis has either not been studied at all or needs to be reinvestigated using modern methods. In many cases, the oogonial stage in the development of the germ cells has not been found, and the origin of germ cells themselves is still not known.

During diffuse oogenesis the development of the oocytes does not take place in gonads, but occurs in various parts of the animal's body. This particularly applies to sponges, coelenterates, turbellarians, and some nemertines.

In such animals as sponges and coelenterates, germ cells appear and develop throughout the life of the adult organism. We still cannot say whether sponges have a single source for all types of germ cells. Transformation of choanocytes into germ cells has been observed (Tuzet, 1962; Korotkova and Aizenshtadt, 1976). In such cases, disintegration of the flagellate chambers occurs and choanocytes are released into the mesohyl. The cells become rounded, lose their flagella and microvilli, and, in the case of differentiation into oocytes, they begin to actively phagocytize the fragments of the cells in the mesohyl. The phagocytizing oocytes of sea sponge *Halisarca dujardini* have pseudopodia, and their cytoplasm is filled with smooth-surfaced vesicles containing a dense material. In the Baikal sponge, oocytes at the beginning of vitellogenesis also acquire amoeboid mobility and phagocytize parenchymal cells (Gureeva, 1972). Electron-microscope studies of the sea sponge suggest that the extensive Golgi apparatus in the oocytes participates in the formation of the granules containing hydrolytic enzymes, which are necessary for

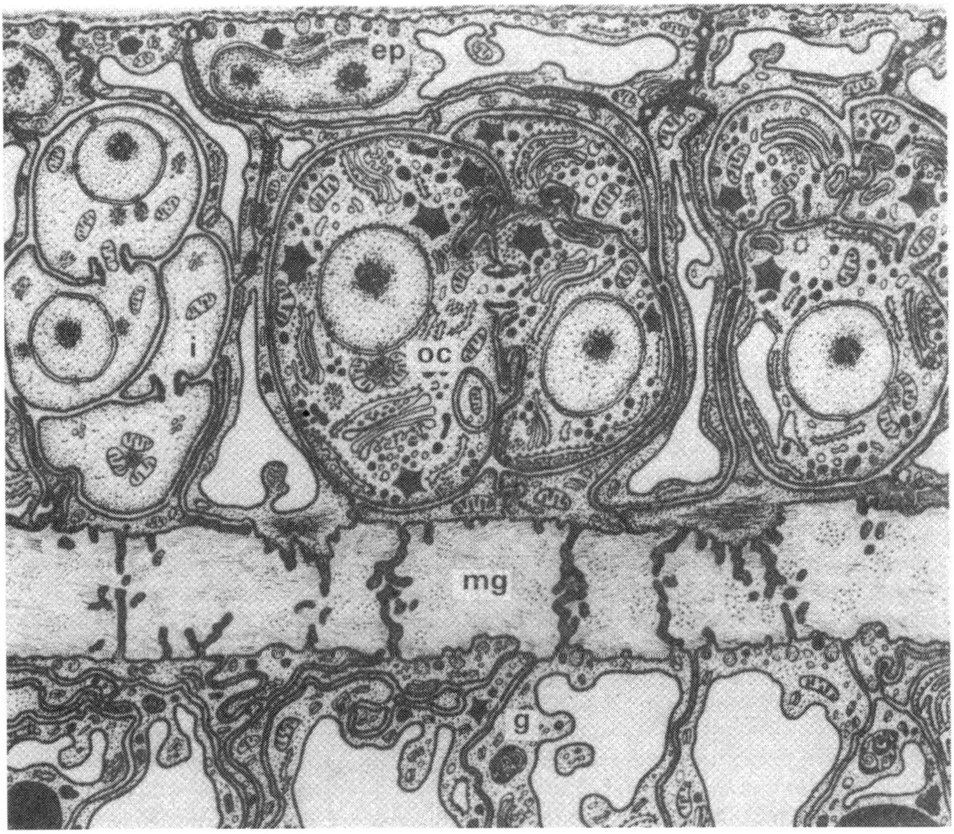

Fig. 11. Schematic cross section of the germ zone of *Hydra*, showing interstitial cells and early oocytes. ep) epiderm; g) gastroderm; i) interstitial cell; mg) mesoglea; oc) oocyte.

the digestion of phagocytized materials (Aizenshtadt and Korotkova, 1976). A characteristic feature of the cells in sponges is the poor development of the rough endoplasmic reticulum and the active participation of ribosomes associated with the nuclear envelope in the synthesis of proteins, apparently including hydrolases. Morphologically, this is reflected in the separation of a large number of smooth-surfaced vesicles from the outer membrane of the nuclear envelope and the formation of extensive Golgi complexes near the nucleus (see Plate VI, 3).

The fully grown oocyte of the sponge is filled with large inclusions known as yolk vacuoles; these resemble the heterophagous vacuoles of somatic cells. In fact, it is not plausible to say whether this is real yolk that will be used by the embryo, or whether it is the remains of the structures (residual bodies) digested in the process of oogenesis. For instance, in some molluscs and spiders the oocytes contain crystalline yolk in cytoplasmic structures resembling phagocytic vesicles.

The female germ cells of *Hydra* develop from interstitial cells (i-cells) which proliferate where buds are formed during asexual reproduction (Kanaev, 1952). Each i-cell divides mitotically and forms an accumulation of syncytially bound cells connected by cytoplasmic bridges (Fig. 11), and several of these accumulations form an epidermal thickening which can be seen with the naked eye. The central cell in each accumulation increases in size and begins to phagocytize the germ cells surrounding it (Zihler, 1972; Aizenshtadt, 1974). It is not known what determines the development of only one oocyte in each group of uniform cells, but the future oocyte is always at the center of the accumulation, and makes early contact with the mesoglea separating the epithelial muscle cells (Aizenshtadt and Marshak, 1974). The oocyte pseudopodia penetrate between the neighboring cells and remains of the cytoplasm of the latter are found inside the growing oocyte. Membranes surrounding these cell fragments become vesiculated, and they become incorporated into the cytoplasm of the growing oocyte, which increases in size as it begins to engulf the cells surrounding it (Fig. 12). These cells may be preserved for quite a long time in the oocyte cytoplasm, although some of them degenerate. In the older literature they were called "pseudocells," and, more recently, the term "Shrumpfzellen" (shrunken cells) (Zihler, 1972) has been used since there is only a thin layer of cytoplasm around their nuclei. Apparently this can be explained by the fact that part of the cytoplasm of these cells is incorporated into the ooplasm even before the beginning of its phagocytic activity. We suggest that the term "endocytes" is the appropriate name for these cells.

In the i-cells of *Hydra* which form groups of 6 to 10 cells in the epidermal layer, there are hardly any membrane structures. They are filled with ribosomes only and contain a few mitochondria and reduced Golgi complexes. However, when sexual reproduction commences and the mitotic division of the i-cells takes place, a rough endoplasmic reticulum appears in the latter from the cisternae of which smooth-surfaced vesicles separate and form Golgi complexes (see Plate VI, 2). Numerous osmiophilic granules, formed in the Golgi area, are transported into the pseudopodia of the growing oocyte.

Thus, in the oocytes of lower metazoans, endoplasmic reticulum is formed during the early stages of oogenesis, and granules form in the Golgi apparatus; this is, on the whole, not characteristic of the oocytes of higher animals, especially vertebrates. Apparently, this is connected with the synthesis of enzymes needed to digest phagocytized structures.

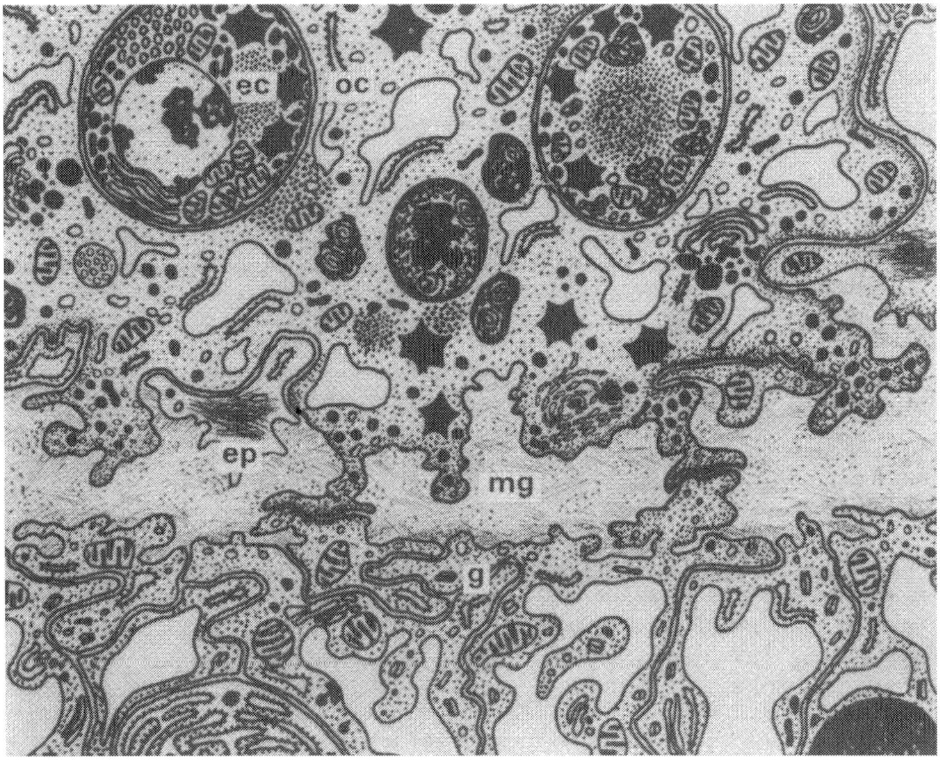

Fig. 12. Schematic section of a late oocyte of *Hydra* containing endocytes in the cytoplasm. The pseudopodia of the oocyte extend into the mesoglea and are in contact with the gastroderm. ec) endocyte; ep) epiderm; g) gastroderm; mg) mesoglea; oc) oocyte.

Besides typical cases of phagocytosis of other cells by the growing oocytes, the migration of cells inside the oocyte is described, where they may exist for a long time without being subjected to digestion. This happens in ascidians. In these animals the cells inside the oocyte have been called test cells or envelope cells, due to their similarity to real test cells. The latter arise from the mobile mesenchymal cells of the blood, which are capable of taking vanadium up from seawater and including it in the hemovanadin pigment. The cells containing this pigment are called vanadocytes. When vanadocytes penetrate the cuticle of the animal, they settle there, become test cells, and synthesize cellulose.

The origin of the test cells inside the oocyte has still not been explained. A number of authors (Kessel and Kemp, 1962;

Vitaioli and Materazzi, 1972) believe that these are follicle cells penetrating into the oocyte through a pedicle, which joins it to the germinative epithelium. Some observations suggest that both oocytes and follicle cells in the colonial forms of ascidians originate from hemoblasts, and that test cells are cells of the inner follicular layer which migrate into the oocyte (Mukai and Watanabe, 1976).

However, the follicle cells of ascidians do not contain vanadium, which is present in the test cells of the ovary, and therefore, a more correct view seems to be that the test cells of the ovary and the test cells themselves are of the same origin. The former become introduced in the ovary and penetrate into the oocyte's cytoplasm, while the latter settle in the cuticle. Apparently, the function of the test cells in the ovary is to transfer hemovanadin from the maternal organism through the oocyte to the larva, where it is accumulated in the test cells and is used during formation of the test during metamorphosis (Kalk, 1963).

It is believed that the test cells take part in the synthesis of the material of the secondary egg envelope (Vitaioli and Materazzi, 1972) or in the synthesis of the yolk material (Mancuso, 1964; Mansueto, 1964; Guraya, 1968). There is, however, still no evidence for this.

1.4. CONCLUDING REMARKS

Despite the fact that oocytes have been studied for many years, our knowledge about the mechanisms of their growth and their interrelations with other cells is far from complete.

In most animals, the follicle cells do not take part in the synthesis of yolk material or rRNA for the oocyte. Exceptions to this are squids, in which the follicle cells secrete yolk material of a glycoprotein nature into the oocyte, which has no envelope, and some reptiles and birds, which have special adaptations for the transfer of ribosomes from the follicle cells to the oocyte.

A simplified presentation of the functional relationships between the growing oocyte and accessory cells characteristic of different types of alimentary oogenesis is shown in Fig. 13. The nutrimental type is characterized by the supply of rRNA mainly from trophocyte nuclei, while in the follicular type rRNA comes from oocyte nuclei. At the same time, in the case of alimentary oogenesis, oocytes are capable of accumulating heterosynthetic molecules.

It has been suggested (Bier et al., 1969; Paschma and Ogor-

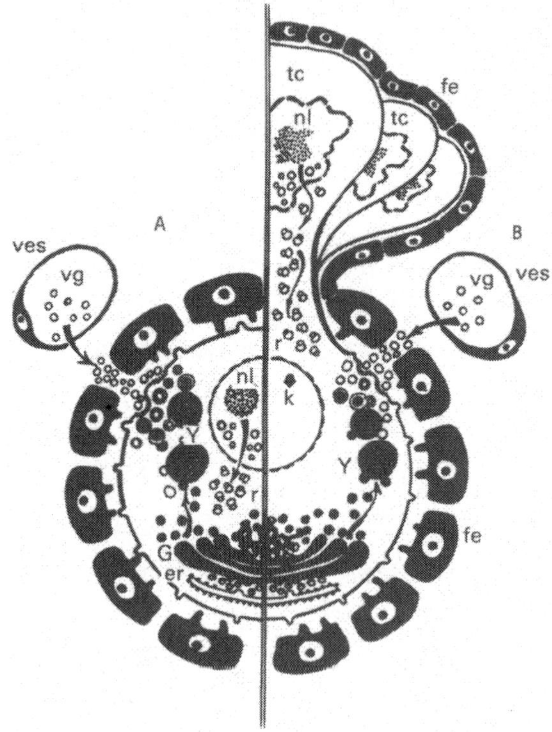

Fig. 13. Functional relationships between the growing oocyte and accessory cells in follicular (A) and nutrimentary types (B) of oogenesis. er) endoplasmic reticulum; fe) follicle epithelium; G) Golgi apparatus; k) karyosphere; nl) nucleolus; r) ribosomes; tc) trophocytes; vg) vitellogenin; ves) blood vessel; Y) yolk.

zalek, 1972) that the evolution of oogenesis led to the exclusion of the oocyte nucleus from metabolic processes and, from this point of view, the most advanced type of oogenesis, characteristic of some insects and worms, is associated with nurse cells; here the oocyte nucleus synthesizes either very little rRNA or none at all. Insects with panoistic ovaries, in which the oocyte nucleus actively synthesizes rRNA, belong to the most primitive orders. However, this view disregards such adaptations as amplification of rDNA in the nucleus of the oocytes and its ability to take up exogenic proteins and form yolk from them. Examining oogenesis from an evolutionary point of view, it should be noted, above all, that the autosynthetic yolk accumulation is peculiar to the more primitive groups of animals, and the heterosynthetic yolk accumulation is characteristic of more advanced ones (see also Boyer, 1972). The growth of the oocytes in lower metazoans is often accompanied by fusion of cells or phagocytosis; this necessarily excludes the complex interrelations with other cells and tissues that are characteristic of the oocytes of higher metazoans.

An important step in evolution was the appearance of follicular epithelium as a result of the adaptation of the somatic elements of the gonad to carry out auxiliary functions in oogenesis. Thus, follicle cells produce steroid which initiate the synthesis of vitellogenin in the liver; they select and accumulate heterosynthetic macromolecules in the perioocyte space; and they synthesize the material of the interstitial space and the secondary egg envelopes. Follicle cells also synthesize steroid hormones (vertebrates) and 1-methyladenine (starfish), which induce the maturation of the oocytes (see Chapter 3).

In this chapter we have not dealt with all aspects of the functions and the evolution of the follicular epithelium (see Gabaeva, 1970, 1980). We have concentrated on the role of the follicle cells in the heterosynthetic type of yolk accumulation. When the follicle cells are of oogonial origin as, for example, in some animals with the solitary type of oogenesis, they only exist for a short period of oogenesis and do not provide the conditions for active pinocytosis by the oocytes. In these cases, endogenous yolk synthesis predominates, and an extensive endoplasmic reticulum and Golgi apparatus develop in the oocytes.

When oocytes grow inside follicles, they exhibit active pinocytosis and for the most part exogenous protein is accumulated in them. In such oocytes endoplasmic reticulum and Golgi complexes are poorly developed.

The study of the pinocytotic activity of vitellogenic oocytes is of general interest in biology. The precise way in which macromolecules are selected at the cell surface and enter into the cells of the embryo or adult organism is still not known. Special attention is being paid to the role of the glycoprotein layer on the cell surface and in the interstitial spaces. The follicles and isolated oocytes are a useful model in this research area.

In spite of the advances made in the study of oogenesis, a number of questions regarding the growth of oocytes and the development of the mature egg organization remain unresolved. This is especially true of the sequence of events in the hormonal control of vitellogenesis in different animals, in particular the reception of estrogen and its transfer into the liver cell nucleus, as well as subsequent activation of the gene coding for vitellogenin mRNA. Mechanisms of pinocytosis and the conversion of vitellogenin inside the cell are also not known. Little light has been thrown on the factors that determine the growth of a single egg from a group of syncytially bound oogonia which takes place in a number of animal groups.

REFERENCES

Aizenshtadt, T. B., 1964, "Cytological studies in oogenesis. I. Morphology of the gonad of *Glossiphonia complanata* L. examined by light and electron microscope, Tsitologiya, *6*, 19-24.

Aizenshtadt, T. B., 1969, "Cytomorphological research into the follicles in the ovary of cuttlefish. I. The fine structure of the cells of the follicle epithelium," Tsitologiya, *11*, 401-409.

Aizenshtadt, T. B., 1971, "Cytomorphological research into the follicles in the ovary of cuttlefish. II. The secretory function of the follicle epithelium," Tsitologiya, *13*, 956-964.

Aizenshtadt, T. B., 1974, "Investigation of oogenesis in *Hydra*. I. Ultrastructure of interstitial cells at early stages of their transformation into oocytes," Ontogenez, *5*, 9-18.

Aizenshtadt, T. B., and Dettlaff, T. A., 1972, "Ultrastructure of stellate sturgeon oocytes during maturation. I. Annulate lamellae and Golgi complex," Ontogenez, *3*, 220-229.

Aizenshtadt, T. B., and Korotkova, G. P., 1976, "A study of oogenesis in the marine sponge *Halisarca dujardini*. II. Phagocytic activity of the oocytes and vitellogenesis," Tsitologiya, *18*, 818-822.

Aizenshtadt, T. B., and Marshak, T. L., 1969, "Cytophotometric research on DNA in the nuclei of the accessory cells in the ovaries of some invertebrates," Arkh. Anat., Gistol. Embryol., *56*, 15-21.

Aizenshtadt, T. B., and Marshak, T. L., 1974, "Investigation of oogenesis in *Hydra*. II. Cytophotometric determination of DNA content in nuclei of germ and somatic cells," Ontogenez, *5*, 394-401.

Aizenshtadt, T. B., Brodsky, V. Ya., and Ivanova, S. N., 1964, "Cytological studies on oogenesis. II. A cytochemical examination of the oocyte growth in *Glossiphonia complanata* L. by cytophotometry and interference microscopy," Tsitologiya, *6*, 77-81.

Aizenshtadt, T. B., Brodsky, V. Ya., and Gazaryan, K. G., 1967, "An autoradiographic study of the RNA and protein synthesis in gonads of animals with different types of oogenesis," Tsitologiya, *9*, 397-406.

Allen, E. R., and Cave, M. D., 1968, "Formation, transport, and storage of ribonucleic acid-containing structures in oocytes of *Acheta domesticus* (Orthoptera)," Z. Zellforsch., *92*, 477-486.
Alov, I. A., Braude, A. I., and Aspiz, M. E., 1969, The Bases of the Functional Morphology of a Cell [in Russian], Meditsina, Moscow.
Anderson, E., 1964, "Oocyte differentiation and vitellogenesis in the roach *Periplaneta americana*," J. Cell Biol., *20*, 131-135.
Anderson, E., 1968, "Cortical alveoli formation and vitellogenesis during oocyte differentiation in the pipefish *Synganthus fuscus*, and the killifish *Fundulus heteroclitus*," J. Morphol., *125*, 23-59.
Anderson, E., 1969a, "Oocyte-follicle cell differentiation in two species of amphineurans (Mollusca), *Mopalia mucosa* and *Chaetopleura apiculata*," J. Morphol., *129*, 89-125.
Anderson, E., 1969b, "Oogenesis in the cockroach *Periplaneta americana*, with special reference to the specialization of the oolemma and fate of coated vesicles," J. Microsc., *8*, 721-738.
Anderson, E., and Beams, H. W., 1960, "Cytological observations on the fine structure of the guinea pig ovary with special reference to the oogonium, primary oocyte, and associated follicle cells," J. Ultrastruct. Res., *3*, 432-446.
Anderson, E., and Huebner, E., 1968, "Development of the oocyte and its accessory cells of the polychaete *Diopatra cuprea* (Bosc)," J. Morphol.,*126*, 163-198.
Anderson, L. M., 1971, "Protein synthesis and uptake by isolated *Cecropia* oocytes," J. Cell Sci., *8*, 735-750.
Anderson, L. M., and Telfer, W. H., 1969, "A follicle cell contribution to the yolk spheres of moth oocytes," Tissue Cell, *1*, 633-644.
Anderson, W. A., and Spielman, A., 1971, "Permeability of the ovarian follicle of *Aedes aegypti* mosquitoes," J. Cell Biol., *50*, 201-221.
Anteunis, A., Fautrez-Firlefyn, N., Fautrez, J., and Lagasse, A., 1964, "L'ultrastructure du noyau vitellin de l'oeuf d'*Artemia salina*," Exp. Cell Res., *35*, 239-247.
Astaurov, B. L., 1948, "The significance of experiments on merogonia and androgenesis for the theory of development and heredity," Usp. Sovrem. Biol., *25*, 48-88.
Bast, R. E., and Telfer, W. H., 1976, "Follicle cell protein synthesis and its contribution to the yolk of the *Cecropia* moth oocyte," Dev. Biol., *52*, 83-97.
Beams, H. W., and Kessel, R. G., 1963, "Electron microscope studies on developing crayfish oocytes with special reference to the origin of yolk," J. Cell Biol., *18*, 621-649.
Beams, H. W., and Kessel, R. G., 1968, "The Golgi apparatus: structure and functions," Int. Rev. Cytol., *23*, 209-276.

Beams, H. W., and Kessel, R. G., 1973, "Oocyte structure and early vitellogenesis in the trout *Salmo gairdneri*," Am. J. Anat., 136 105-120.

Beams, H. W., and Sekhon, S. S., 1966, "Electron microscope studies on the oocyte of the freshwater mussel *Anodonta* with special reference to the stalk and mechanism of yolk deposition," J. Morphol., *119*, 477-501.

Bedford, L., 1966, "The electron microscopy and cytochemistry of oogenesis and the cytochemistry of embryonic development of the prosobranch gastropod *Bembicium nanum* L.," J. Embryol. Exp. Morphol., *15*, 15-37.

Bell, W. J., 1969, "The dual role of juvenile hormone in the control of yolk formation in *Periplaneta americana*," J. Insect Physiol., *15*, 1279-1290.

Bellairs, R., 1965, "The relationship between oocyte and follicle in the hen's ovary as shown by electron microscopy," J. Embryol. Exp. Morphol., *13*, 215-233.

Biemont, J. C., 1979, "Vitellogenesis in *Acanthoscelides obtectus* (Coleoptera, Bruchidae). I. Oocyte development and vitellogenesis in a European strain," Int. J. Invert. Reprod., *1*, 221-232.

Bier, K., 1962, "Autoradiographische Untersuchungen zur Dotterbildung," Naturwissenschaften, *49*, 332-333.

Bier, K., 1963a, "Autoradiographische Untersuchungen uber die Leistungen des Follikelepithels un der Nahrzellen bei der Dotterbildung und Eiweissynthese im Fliegenovar," W. Roux' Arch. Entwicklungsmech. Org., *154*, 552-575.

Bier, K., 1963b, "Synthese, interzellularer Transport und Abbau von Ribonucleinsäure im Ovar der Stubenfliege *Musca domestica*," J. Cell Biol., *16*, 436-440.

Bier, K., 1964, "Gerichteter Ribonukleinsäuretransport durch das Cytoplasma," Naturwissenschaften, *51*, 418-438.

Bier, K., 1970, "Oogenesetypen bei Insekten und Vertebrate, ihre Bedeutung für die Embryogenese und Phylogenese," Zool. Anz., Suppl., *33*, 7-29.

Bier, K., Kunz, W., and Ribbert D., 1969, "Insect oogenesis with and without lampbrush chromosomes," Chromosomes Today, *2*, 107-115.

Bluemink J..G., 1969, "Are yolk granules related to lysosomes?" Zeiss Inform., *73*, 95-99.

Boctor, F. N., and Kamel, M. Y., 1976, "Purification and characterization of two lipovitellins from eggs of the tick *Dermacentor andersoni*," Insect. Biochem., *6*, 233-240.

Bottke, W., 1972, "Zur Morphologie des Ovars von *Viviparus contectus* (Millet 1813) (Gastropoda, Prosobranchia). I. Die Follikelzellen," Z. Zellforsch., *133*, 103-118.

Bou-Resli, M., 1974, "Ultrastructural studies on the intercellular bridges between the oocyte and follicle cell in the lizard *Acanthodactylus scutellatus* Hardyi," Z. Anat. Entwick lungsgesch., *143*, 239-254.

Bownes, M., 1980, "The use of yolk protein variations in *Drosophila* species to analyze the control of vitellogenesis," Differentiation, *16*, 109-116.
Boyer, B. C., 1972, "Ultrastructural studies of differentiation in the oocyte of the polyclad turbellarian *Prostheceraeus floridanus*," J. Morphol., *136*, 273-294.
Brookes, V. J., 1969, "The maturation of the oocytes in the isolated abdomen of *Leucophaea maderae*," J. Insect Physiol., *15*, 621-631.
Brown, D. D., and Dawid, I. B., 1969, "Development genetics," Annu. Rev. Genet., *3*, 127-154.
Brown, E. H., and King, R. C., 1964, "Studies on the events resulting in the formation of an egg chamber in *Drosophila melanogaster*," Growth, *28*, 41-48.
Buckley, S. K. L., 1976, "Octopus gonadotropin and its role in the regulation of egg development," Colloq. Int. CNRS, *251*, 161-167.
Bühlmann, G., 1974, "Vitellogenin in adulten Weibchen der Schabe *Nauphoeta cinerea*. Immunologische Untersuchungen uber Herkunft und Einbau," Rev. Suisse Zool., *81*, 642-647.
Büning, J., 1972, "Untersuchungen am Ovar von *Bruchidiusobtectus* Say. (Coleoptera – Polyphaga) zur Klärung des Oocytenwachstums in der Pravitellogenese," Z. Zellforsch., *128*, 241-282.
Carasso, N., Ovtracht, L., and Favard, P., 1971, "Observation en microscopie électronique haute tension, de l'appareil de Golgi sur coupes de 0.5 a 5 μ d'épaisseur," C. R. Acad. Sci. Paris, *D273*, 876-879.
Cave, M. D., 1975, "Absence of ribosomal DNA amplification in the meroistic (telotrophic) ovary of the large milkweed bug *Oncopeltus fasciatus* (Dallas) (Hemiptera: Lygaeidae)," J. Cell Biol., *66*, 461-469.
Charniaux-Cotton, H., 1973, "Introduction a l'ovogenèse chez les invertébres," Ann. Biol. Anim., Biochim., Biophys., *13*, 13-19.
Chmilevsky, D. A., 1970, "Synthesis of DNA in early oogenesis of the ruff *Acerina cernua* L.," Tsitologiya, *12*, 675-678.
Chmilevsky, D. A., 1972, "Peculiarities of protein and polysaccharide synthesis in developing oocytes of the ruff," Tsitologiya, *14*, 277-283.
Clavert, J., 1958, "Contribution à l'étude de la vitellogenèse chez les oiseaux. Phases physiologiques et rôle de la folliculine dans la vitellogenèse," Arch. Anat. Microsc. Morphol. Exp., *47*, 653-675.
Coggins, L. W., 1973, "An ultrastructural and radioautographic study of early oogenesis in the toad *Xenopus laevis*," J. Cell Sci., *12*, 71-93.
Croisille, G., Junera, H., Meusy, J., and Charniaux-Cotton, H., 1974, "The female-specific protein (vitellogenic protein) in crustacea with particular reference to *Orchestia gammarella* (Amphipoda)," Am. Zool., *14*, 1219-1228.

Cummings, M. R., and King, R. C., 1970, "The cytology of the vitellogenic stages of oogenesis in *Drosophila melanogaster*. II. Ultrastructural investigations on the origin of protein yolk spheres," J. Morphol., *130*, 467-478.

Cutting, J. A., and Roth, T. F., 1973, "Changes in specific sequestration of protein during transport into the developing oocyte of the chicken," Biochim. Biophys. Acta, *298*, 951-955.

Davey, K. G., 1981, "Hormonal control of vitellogenin uptake in *Rhodnius prolixus Stäl*," Am. Zool., *21*, 701-705.

Davey, K. G., and Huebner, E., 1974, "The response of the follicle cells of *Rhodnius prolixus* to juvenile hormone and antigonadotropin in vitro," Can. J. Zool., *52*, 1407-1412.

Davidson, E. H., 1969, Gene Activity in Early Development, Academic Press, New York—London.

Davis, G. C., and Wilt, F. H., 1972, "RNA synthesis during oogenesis in the echiuroid worm *Urechis caupo*," Dev. Biol., *27*, 1-12.

Dehn, P. F., and Wallace, R. A., 1973, "Sequestered and injected vitellogenin. Alternative routes of protein processing in *Xenopus* oocyte," J. Cell Biol., *58*, 721-724.

Dejmal, R. K., and Brookes, V. J., 1972, "Insect lipovitellin: chemical and physical characteristics of a yolk protein from the ovaries of *Leucophaea maderae*," J. Biol. Chem., *247*, 869-874.

Dhainaut, A., 1967, "Étude de la vitellogenèse chez *Nereis diversicolor* O. F. Müller (annélide polychète) par autoradiographie à haute résolution," C. R. Acad. Sci. Paris, *D265*, 434-436.

Diehl, P. A., 1970, "Zur Oogenese bei *Ornithodorus moubata* Murray (Ixodoidea: Argasidae) unter besonderer Berucksichtigung des Vitellogenese," Acta Trop., *27*, 301-355.

Dolphin, P. J., Ansari, A. Q., Lazier, C. B., Munday, K. A., and Akhtar, M., 1971, "Studies on the induction and biosynthesis of vitellogen, an estrogen-induced glycolipophosphoprotein," Biochem. J., *124*, 751-758.

Droller, M. J., and Roth, T. F., 1966, "An electron microscope study of yolk formation during oogenesis in *Lebistes reticulatus* guppyi," J. Cell Biol., *28*, 209-232.

Droz, B., 1966, "Elaboration de glycoproteines dans l'appareil de Golgi des cellules hépatiques chez le rat; étude radioautographique en microscopie electronique après injection de galactose-^{3}H," C. R. Acad. Sci. Paris, *D262*, 1766-1768.

Dumont, J. N., 1967, "Oocyte differentiation in the annelid *Enchytraeus albidus*," J. Cell Biol., *35*, Pt. 2, 35A.

Dumont, J. N., 1969, "Oogenesis in the annelid *Enchytraeus albidus* with special reference to the origin and cytochemistry of yolk," J. Morphol., *129*, 317-343.

Dumont, J. N., and Anderson, E., 1967, "Vitellogenesis in the horseshoe crab *Limulus polyphemus*," J. Microsc., *6*, 791-806.

Dumont, J. N., and Wallace, R. A., 1972, "The effects of vinblastine on isolated *Xenopus* oocytes," J. Cell Biol., *53*, 605-610.

Durchon, M., 1975, "Modalités du déterminisme hormonal de la maturation sexuelle chez les Nereidiens (Annélides, Polychètes)," Pubbl. Staz. Zool. Napoli, *39*, Suppl. 1, 510-531.

Eckelbarger, K. J., 1975, "A light and electron microscope investigation of gametogenesis in *Nicolea zostericola* (Polychaeta: Terebellidae)," Mar. Biol., *30*, 353-370.

Engelman, F., 1970, The Physiology of Insect Reproduction, Pergamon Press, Oxford.

Engels, W., 1973, "Das zeitliche und räumliche Muster der Dottereinlagerung in die Oocyte von *Apis mellifica*," Z. Zellforsch., *142*, 409-430.

Favard, P., 1969, "The Golgi apparatus," in: Handbook of Molecular Cytology, Vol. 15, A. Lima-de-Faria (ed.), Elsevier, Amsterdam-London, pp. 1131-1155.

Favard, P., and Carasso, N., 1958, "Origine et ultrastructure des plaguettes vitellines de la Planorbe," Arch. Anat., Microsc. Morphol., *47*, 211-229.

Filosa, S., and Taddei, C., 1976, "Intercellular bridges in lizard oogenesis," Cell Differ., *5*, 199-206.

Filosa, S., Taddei, C., and Andreuccetti, P., 1979, "The differentiation and proliferation of follicle cells during oocyte growth in *Lacerta sicula*," J. Embryol. Exp. Morphol., *54*, 5-51.

Fischer, A., 1975, "The structure of symplasmic early oocytes and their enveloping sheath cells in the polychaete *Platynereis* dumerilli," Cell Tissue Res., *160*, 327-343.

Flickinger, Ch. J., 1971, "Alterations in the Golgi apparatus of amoeba in the presence of an inhibitor of protein synthesis," Exp. Cell Res., *68*, 381-387.

Follett, B. K., and Redshaw, M. R., 1974, "The physiology of vitellogenesis," in: Physiology of Amphibia, Academic Press, New York—London, 2, 219-308.

Follett, B. K., Nicholls, T. J., and Redshaw, M. R., 1968, "The vitellogenic response in the South African clawed toad (*Xenopus laevis*, Daudin)," J. Cell Physiol., *72*, Suppl., 91-102.

Franchi, L. L., and Mandl, A. M., 1962, "The ultrastructure of oogonia and oocytes in the fetal and neonatal rat," Proc. R. Soc. London, *B157*, 99-114.

Frey-Wyssling, A., 1973, Comparative Organellography of the Cytoplasm, Springer Verlag, Wien-New York.

Fujitani, N., 1961, "Cytological and cytochemical investigations on the rabbit ova during oogenesis and fertilization," Bull. Exp. Biol., *10*, 5-22.

Gabaeva, N. S., 1970, "Comparative histology of the follicle epithelium in a number of vertebrates," Arkh. Anat., Gistol. Embriol., *58*, 20-37.

Gabaeva, N. S., 1980, "On regular patterns of histogenesis of the ovary follicular epithelia in a phylogenetical sequence of vertebrates," Arkh. Anat., Gistol. Embriol., *79*, 18-35.
Gabaeva, N. S., Levai, I. V., and Tumanova, N. L., 1971, "On the ingrowth of the follicle epithelium in the oocytes of some rock lizards," Arkh. Anat., Gistol. Embriol., *60*, 45-49.
Gaginskaya, E. R., and Gruzova, M. N., 1969, "Peculiarities of the oogenesis of Fringilla (Aves)," Tsitologiya, *11*, 1241-1251.
Gall, J. G., 1969, "The genes for ribosomal RNA during oogenesis," Genetics, *61*, Suppl., 121-132.
Ganion, L. R., and Kessel, R. G., 1972, "Intracellular synthesis, transport, and packaging of proteinaceous yolk in oocytes of *Orconectes immunis*.," J. Cell Biol., *52*, 420-437.
Gondos, B., 1973, "Germ cell degeneration and intercellular bridges in the human fetal ovary," Z. Zellforsch., *138*, 1, 23-30.
Greenfield, M. L., 1966, "The oocyte of the domestic chicken shortly after hatching, studied by electron microscopy," J. Embryol. Exp. Morphol., *15*, 297-316.
Gruzova, M. N., and Zaichikova, P., 1967, "The karyosphere in oogenesis of the leech *Glossiphonia complanata*," Tsitologiya, *9*, 388-396.
Gruzova, M. N., Zaichikova, Z. P., and Sokolov, I. I., 1972, "The structural and functional organization of the oocyte nucleus of *Chrysopa* (Neuroptera). I. The nucleolar apparatus and extrachromosomal DNA," Tsitologiya, *14*, 269-276.
Guraya, S. S., 1968, "Cytochemistry of yolk elements in the egg of the tunicate *Molgula*," Z. Zellforsch., *86*, 505-510.
Guraya, S. S., 1975, "Histochemical observations of the juxtanuclear complex of organelles in the hamster oocyte," Acta Anat., *93*, 335-343.
Gureeva, M. A., 1972, "'Sorites' and oogenesis in endemic Baikal sponges," Tsitologiya, *14*, 32-45.
Hagedorn, H. H., 1974, "The control of vitellogenesis in the mosquito *Aedes aegypti*," Am. Zool., *14*, 1207-1217.
Hagedorn, H. H., O'Connor, J. D., Fuchs, M. S., Sage, B., Schlaeger, D. A., and Bohm, M. K., 1975, "The ovary as a source of α-ecdysone in an adult mosquito," Proc. Natl. Acad. Sci. USA, *72*, 3255-3259.
Hallberg, R. L., and Smith, D. C., 1975, "Ribosomal protein synthesis in *Xenopus laevis* oocytes," Dev. Biol., *42*, 40-52.
Hanaoka, K., and Hagedorn, H. H., 1980, "Brain hormone control of ecdysone secretion by the ovary in a mosquito," in: Progress in Ecdysone Research, J. A. Hoffmann (ed.), Elsevier/North-Holland Biomedical Press, Amsterdam, pp. 467-480.
Handler, A. M., and Postlethwait, J. H., 1977, "Endocrine control of vitellogenesis in *Drosophila melanogaster*: ef-

fects of the brain and corpus allatum," J. Exp. Zool., *202*, 389-401.

Handler, A. M., and Postlethwait, J. H., 1978, "Regulation of vitellogenin synthesis in *Drosophila* by ecdysterone and juvenile hormone," J. Exp. Zool., *206*, 247-257.

Hausman, S. J., Anderson, L. M., and Telfer, W. A., 1971, "The dependence of cecropia yolk formation in vitro on specific blood proteins," J. Cell Biol., *48*, 303-313.

Hayward, M. A., and Shapiro, D. J., 1981, "A middle-affinity estrogen-specific binding protein in livers of vitellogenic and nonvitellogenic *Xenopus laevis*," Dev. Biol., *88*, 333-340.

Heald, P. J., and McLachlan, P. M., 1965, "The synthesis of phosvitin in vitro by slices of liver from the laying hen," Biochem. J., *94*, 32-39.

Heesen, D., and Engels, W., 1973, "Elektrophoretische Untersuchungen zur Vitellogenese von *Brachydanio rerio* (Cyprinida, Teleostei)," W. Roux' Arch. Entwicklungsmech. Org., *173*, 46-59.

Hertig, A. T., 1968, "The primary human oocyte: some observations on the fine structure of Balbiani's vitelline body and the origin of the annulate lamellae," Am. J. Anat., *122*, 107-137.

Hertig, A. T., and Adams, E. C., 1967, "Studies on the human oocyte and its follicles. I. Ultrastructural and histochemical observations on the primordial follicle stage," J. Cell Biol., *34*, 647-675.

Highnam, K. C., Lusis, O., and Hill, L., 1963, "The role of the corpora allata during oocyte growth in the desert locust *Schistocerca gregarina* Forsk," J. Insect Physiol., *9*, 587-596.

Hinsch, G. W., and Cone, M. V., 1969, "Ultrastructural observations of vitellogenesis in the spider crab *Libinia emarginata* L.," J. Cell Biol., *40*, 336-342.

Hirose, K., 1971, "The ultrastructure of the ovarian follicle of medaka, *Oryzias latipes*," Z. Zellforsch., *123*, 316-329.

Holland, Ch. A., and Dumont, J. N., 1975, "Oogenesis in *Xenopus laevis* (Daudin). IV. Effects of gonadotrophin, estrogen, and starvation on endocytosis in developing oocytes," Cell Tissue Res., *162*, 177-184.

Holland, N. D., 1976, "The fine structure of the yolk nucleus in oocytes of *Antedon bifida* (Pennant) (Echinodermata: Crinoidea)," J. Mar. Biol. Assoc. U.K., *56*, 59-63.

Hope, J., 1965, "The fine structure of the developing follicle of the rhesus ovary," J. Ultrastruct. Res., *12*, 592-610.

Hope, J., Humphries, A. A., and Bourne, G. H., 1963, "Ultrastructural studies on developing oocytes of the salamander *Triturus viridescens*. I. The relationship between follicle cells and developing oocytes," J. Ultrastruct. Res., *9*, 302-324.

Hoser, P., 1975, "The cytology and cytochemistry of extrachromosomal DNA in oogenesis of beetles of the family *Gyrinidae* (Coleoptera)," Folia Histochem. Cytochem., *13*, 102-103.

Hubert, J., 1970, "Données preliminaires sur l'ultrastructure des ovocytes et du follicule ovarien chez le lézard vivipare (*Lacerta vivipara* Jacquin) quelques mois après la naissance," C. R. Acad. Sci. Paris, *D270*, 2674-2677.

Hubert, J., 1971, "Aspects ultrastructuraux des relations entre les couches folliculaires et l'ovocyte depuis la formation du follicule jusqu'au début de la vitellogenèse chez le lézard *Lacerta vivipara Jacquin*," Z. Zellforsch., *116*, 240-249.

Heubner, E., and Anderson, E., 1972, "A cytological study of the ovary of *Rhodnius prolixus*. II. Oocyte differentiation," J. Morphol., *137*, 385-415.

Hughes, M., and Berry, S. J., 1970, "The synthesis and secretion of ribosomes by nurse cells of *Antheraea polyphemus*," Dev. Biol., *23*, 651-664.

Jacob, J., and Sirlin, J. L., 1959, "Cell function in the ovary of *Drosophila*. I. DNA classes in nurse cell nuclei as determined by autoradiography," Chromosoma, *10*, 210-228.

Jalaja, M., 1974, "Complete inhibition of vitellogenesis after extirpation of median neurosecretory cells in *Dysdercus cingulatus*," Curr. Sci., *43*, 286-287.

Jamieson, J. D., and Palade, G. E., 1971, "Synthesis, intracellular transport, and discharge of secretory proteins in stimulated pancreatic exocrine cells," J. Cell Biol., *50*, 135-158.

Jared, D. W., and Wallace, R. A., 1969, "Protein uptake in vitro by amphibian oocytes," Exp. Cell Res., *57*, 454-457.

Jukes, T. H., and Kay, H. D., 1932, quoted from T. P. Roth and K. R. Porter, 1964.

Kalt, M., 1963, "Cytoplasmic transmission of a vanadium compound in a tunicate oocyte, visible with electron microscopy," Acta Embryol. Morphol. Exp., *6*, 289-303.

Kalt, M. R., 1972, "Germ cell development and premeiotic rDNA amplification in *Xenopus laevis*," J. Cell Biol., *55*, Pt. 2, 128A.

Kanaev, I. I., 1952, Hydra [in Russian], Izd. Akad. Nauk SSSR, Moscow—Leningrad.

Karasaki, S., 1963a, "Studies on amphibian yolk. 1. The ultrastructure of the yolk platelet," J. Cell Biol., *18*, 135-151.

Karasaki, S., 1963b, "Studies on amphibian yolk. 5. Electron microscopic observations on the utilization of yolk platelets during embryogenesis," J. Ultrastruct. Res., *18*, 225-247.

Karasaki, S., 1967, "An electron microscope study on the crystalline structure of the yolk platelets of the lamprey egg," J. Ultrastruct. Res., *18*, 377-390.

Karasaki, S., and Komoda, T., 1958, "Electron micrographs of a crystalline lattice structure in yolk platelets of the amphibian embryo," Nature (London), *181*, 407-408.

Kato, K., 1968, "Cytochemistry and fine structure of elimination chromatin in *Dytiscidae*," Exp. Cell Res., *52*, 507-522.

Kerr, M. S., 1969, "The hemolymph proteins of the blue crab *Callinectes spidus*. II. A lipoprotein serologically identical to oocyte lipovitellin," Dev. Biol., *20*, 1-17.

Kessel, R. G., 1966a, "Some observations on the ultrastructure of the oocyte of *Thyone briareus* with special reference to the relationship of the Golgi complex and endoplasmic reticulum in the formation of yolk," J. Ultrastruct. Res., *16*, 305-319.

Kessel, R. G., 1966b, "Electron microscope studies on the origin and maturation of yolk in oocytes of the tunicate *Ciona intestinalis*," Z. Zellforsch., *71*, 525-544.

Kessel, R. G., 1968a, "Electron microscope studies on developing oocytes of a coelenterate medusa with special reference to vitellogenesis," J. Morphol., *126*, 211-248.

Kessel, R. G., 1968b, "Mechanisms of protein yolk synthesis and deposition in crustacean oocytes," Z. Zellforsch., *89*, 17-38.

Kessel, R. G., 1968c, "Annulate lamellae," J. Ultrastruct. Res., Suppl. *10*, 1-82.

Kessel, R. G., 1968d, "An electron microscope study of differentiation and growth in oocytes of *Ophioderma panamensis*," J. Ultrastruct. Res., *22*, 63-89.

Kessel, R. G., 1971, "Cytodifferentiation in the *Rana pipiens* oocyte. II. Intramitochondrial yolk," Z. Zellforsch., *112*, 313-332.

Kessel, R. G., and Beams, H. W., 1969, "Annulate lamellae and 'yolk nuclei' in oocytes of the dragonfly *Libellula pulchella*," J. Cell Biol., *42*, 185-201.

Kessel, R. G., and Kemp, N. E., 1962, "An electron microscope study on the oocyte, test cells, and follicular envelope of the tunicate, *Molgula manhattensis*," J. Ultrastruct. Res., 57-76.

Kessel, R. G., and Panje, W. R., 1968, "Organization and activity in the pre- and postovulatory follicle of *Necturus maculosus*," J. Cell Biol., 39 1-34.

King, R. C., 1974, "lnsect gametogenesis," J. Med. Entomol., *11*, 1-7.

King, R. C., 1975, "The cell cycle and cell differentiation in the *Drosophila* ovary," in: Results and Problems in Cell Differentiation, Vol. 7, Springer-Verlag, Berlin—Heidelberg—New York, pp. 85-109.

King, R. C., and Aggarwal, S. K., 1965, "Oogenesis in *Hyalophora cecropia*," Growth, *29*, 17-83.

Knight, P. F., and Schechtman, A. M., 1954, "The passage of heterologous serum proteins from the circulation into the ovum of the fowl," J. Exp. Zool., *127*, 271-304.

Koch, E. A., and King, R. C., 1966, "The origin and early differentiation of the egg chamber of *Drosophila melanogaster*," J. Morphol., *119*, 283-303.

Koch, E. A., Smith, P. A., and King, R. C., 1967, "The division and differentiation of *Drosophila* cystocytes," J. Morphol., *121*, 55-70.
Koeppe, J. K., Hobson, K., and Wellman, S. E., 1980, "Juvenile hormone regulation of structural changes and DNA synthesis in the follicular epithelium of *Leucophaea maderae*," J. Insect Physiol., *26*, 229-240.
Korfsmeier, K. H., 1966, "Zur Genese des Dottersystems in der Oocyte von *Brachydanio rerio*. Autoradiographische Untersuchungen," Z. Zellforsch., *71*, 283-296.
Korotkova, G. P., and Aizenshtadt, T. B., 1976, "Research on oogenesis in rhe marine sponge *Halisarca dujardini*. Origin of the oogonia and the early stages of development in oocytes," Tsitologiya, *18*, 549-555.
Kress, A., and Spornitz, U. M., "Ultrastructural studies of oogenesis in some European amphibiana. I. *Rana esculenta* and *Rana temporaria*," Z. Zellforsch., *128*, 438-456.
Kunz, W., 1969, "Die Entstehung multipler Oocytennukleolen aus akzessorischen DNS-Körpern bei *Gryllus domesticus*," Chromosoma, *26*, 41-75.
Kurosumi, K., 1961, "Electron microscopic analysis of the secretion mechanism," Int. Rev. Cytol., *11*, 1-124.
Lanzavecchia, G., 1961, "The formation of the yolk in frog oocytes," Proc. Eur. Regional Conf. Electron Microsc., Vol. 2, Delft, p. 746.
Lanzavecchia, G., 1965, "Structure of yolk in *Rana esculenta* L.," J. Ultrastruct. Res., *12*, 147-159.
Lanzrein, B., 1974, "Influence of a juvenile hormone analog on vitellogenin synthesis and oogenesis in larvae of *Nauphoeta cinerea*," J. Insect Physiol., *20*, 1855-1871.
Larsen, J. R., and Bodenstein, D., 1959, "The humoral control of egg maturation in the mosquito," J. Exp. Zool., *140*, 343-381.
Laverdure, A. M., 1972, "L'évolution de l'ovaire chez la femelle adulte de *Tenebria molitor*, la vitellogenèse," J. Insect. Physiol., *18*, 1369-1385.
Lea, A. O., 1963, "Some relationships between environment, corpora allata, and egg maturation in *Aedine* mosquitos," J. Insect Physiol., *9*, 793-809.
Leblond, C. P., and Bennett, G., 1974, "Elaboration and turnover of cell coat glycoproteins," in: Cell Surface in Development, Wiley, New York, pp. 29-49.
Lender, T., 1965, "Rôle du complexe céphalique au cours de la vitellogenèse de *Tenebrio molitor* (Coléoptère)," C. R. Acad. Sci. Paris, *261*, 557-559.
Leonard, D. A., and LaMarca, M. J., 1975, "In vivo synthesis and turnover of cytoplasmic ribosomal RNA by stage 6 oocytes of *Xenopus laevis*," Dev. Biol., *45*, 199-202.
Leonard, R., Deamer, D. W., and Armstrong, P., 1972, "Amphibian yolk platelet ultrastructure visualized by freeze-etching," J. Ultrastruct. Res., *40*, 1-24.

Lewis, W., 1931, "Pinocytosis," Bull. Johns Hopkins Hospital, *49*, 17-23.

Limatola, E., Filosa, S., and Ghiara, G., 1972, "Osservazioni sperimentali mediante impiego di perossidasi sui rapporti tra micropinocitosi e vitellogenesi," Boll. Zool., *39*, 631-632.

Loof, A., 1971, "Synthesis and deposition of oocyte envelopes in the Colorado beetle *Leptinotarsa decemlineata* Say," Z. Zellforsch., *115*, 351-360.

Loof, A., and Lagasse, A., 1970, "Resorption of the terminal oocyte in the allatectomized colorado beetle, *Leptinotarsa decemlineata* Say," Proc. Koninkl. Nederl. Akad. Wet., *C73*, 284-297.

Lui, Ch. W., and O'Connor, J. D., 1976, "Biosynthesis of lipovitellin by the crustacean ovary. II. Characterization of and in vitro incorporation of amino acids into the purified subunits," J. Exp. Zool., *195*, 41-52.

Macgregor, H. C., and Stebbings, H., 1970, "A massive system of microtubules associated with cytoplasmic movement in telotrophic ovarioles," J. Cell Sci., *6*, 431-449.

Mahowald, A. P., 1972, "Ultrastructural observations on oogenesis in *Drosophila*," J. Morphol., *137*, 29-48.

Mairy, M., and Denis, H., 1970, "Synthèse et accumulation du RNA pendant l'oogenèse de *Xenopus laevis*," Arch. Int. Physiol. Biochim., *78*, 599-601.

Mancuso, V., 1964, "Ultrastructural changes in the cytoplasm of *Ciona intestinalis* oocytes," Acta Embryol. Morphol. Exp., *7*, 269-295.

Mansueto, C., 1964, "Sulla riproduzione per divisione mitotica delle cellule testali delle ascidie," Atti Accad. Naz. Lincei, Cl. Sci. Fis. Mat. Nat. Rend., *36*, 683-689.

Masler, E. P., Fuchs, M. S., Sage, B, and O'Connor, J. D., 1980, "Endocrine regulation of ovarian development in the autogenous mosquito *Aedes atropalpus*," Gen. Comp. Endocrinol., *41*, 250-259.

Massover, W. H., 1971a, "Intramitochondrial yolk-crystals of frog oocytes. I. Formation of yolk-crystal inclusions by mitochondria during bullfrog oogenesis," J. Cell Biol., *48*, 266-279.

Massover, W. H., 1971b, "Nascent yolk platelets of anuran amphibian oocytes," J. Ultrastruct. Res., *37*, 574-591.

Massover, W. H., 1971c, "Intramitochondrial yolk-crystals of frog oocytes. II. Expulsion of intramitochondrial yolk-crystals to form single membrane bound hexagonal crystalloids," J. Ultrastruct. Res., *36*, 603-620.

Matsuzaki, M., 1965, "Electron microscopic studies on the early developments of various insect eggs. II. On the changes of the nurse cells and the nutritive substances in telotrophic ovaries," New Entomol., *14*, 49-57.

Matsuzaki, M., 1975, "Ultrastructural changes in developing oocytes, nurse cells, and follicular cells during oogenesis in the teletrophic ovarioles of *Bothrogonia japonica* Ishihara (Homoptera, Tettigellidae), Jpn. J. Entomol., *43*, 75-90.

Matuszewski, B., and Kloc, M., 1976, "Gene amplification in oocytes of the rove beetle *Creophilus maxillosus* (Staphylinidae Coleoptera Polyphaga)," Experientia, *32*, 34-36.

May, F. E. B., and Knowland, J., 1981, "Estrogen receptor levels and vitellogenin synthesis during development of *Xenopus laevis*," Nature (London), 292 853-855.

May, F. E. B., Westley, B. R., and Knowland, J., 1981, "Vitellogenin synthesis and characterization of the liver estrogen receptor in the neotenous salamander *Ambystoma mexicanum*," Dev. Biol., *82*, 350-357.

Melius, M. E., and Telfer, W. H., 1969, "An autoradiographic analysis of yolk deposition in the cortex of the cecropia moth oocyte," J. Morphol., *129*, 1-16.

Melton, C. G., and Smorul, R. P., 1974, "Functional volume of frog eggs: equivalence of metabolite diffusion space in chemically demembranated embryos and aqueous phase (non-yolk) volume," J. Exp. Zool., *187*, 239-247.

Meusy, J. J., 1980, "Vitellogenin, the extraovarian precursor of the protein yolk in *Crustacea*: a review," Reprod., Nutr., Dev., *20*, 1-21.

Mills, R. R., Greenslade, F. C., and Couch, E. F., 1966, "Studies on vitellogenesis in the American cockroach," J. Insect Physiol., *12*, 767-779.

Morré, D. J., Mollenhauer, H. H., and Bracker, C. E., 1971, "Origin and continuity of Golgi apparatus," in: Origin and Continuity of Cell Organelles, Vol. 2, Springer-Verlag, Berlin—Heidelberg—New York, pp. 82-126.

Mukai, H., and Watanabe, H., 1976, "Studies on the formation of germ cells in a compound ascidian *Botryllus primigenus* Oka," J. Morphol., *148*, 337-362.

Neaves, W. B., 1971, "Intercellular bridges between follicle cells and oocyte in the lizard *Anolis carolinensis*," Anat. Rec., *170*, 285-302.

Neutra, M., and Leblond, C. P., 1966, "Synthesis of the carbohydrate of mucus in the Golgi complex, as shown by electron microscope radioautography of goblet cells from rats injected with glucose-^{3}H," J. Cell Biol., *30*, 119-136.

Neutra, M., and Leblond, C. P., 1969, "The Golgi apparatus," Sci. Am., *220*, 100-107.

Nieuwkoop, P. D., and Faber, J., 1956, Normal Table of *Xenopus laevis* (Daudin), North-Holland, Amsterdam.

Nørrevang, A., 1968, "Electron microscopic morphology of oogenesis," Int. Rev. Cytol., *23*, 113-186.

O'Dor, R. K., and Wells, M. J., 1973, "Yolk protein synthesis in the ovary of *Octopus vulgaris* and its control by the optic gland gonadotropin," J. Exp. Biol., *59*, 663-674.

O'Dor, R. K., and Wells, M. J., 1975, "Control of yolk protein synthesis by *Octopus* gonadotropin in vivo and in vitro (ef-

fects of *Octopus* gonadotropin)," Gen. Comp. Endocrinol., *27*, 129-135.
O'Dor, R. K., and Wells, M. J., 1978, "Reproduction versus somatic growth: hormonal control in *Octopus vulgaris*," J. Exp. Biol., *77*, 15-31.
Oie, M., Takahashi, S. Y., and Ishizaki, H., 1975, "Vitellogenin in the eggs of the cockroach *Blattela germanica*: purification and characterization," Dev., Growth, Differ., *17*, 237-246.
Olmo, E., and Taddei, C., 1974, "Histophotometric measurements of the DNA content in the ovarian follicle cells of *Lacerta sicula* Raf.," Experientia, *30*, 1331-1332.
Opresko, L., Wiley, H. S., and Wallace, R. A., 1980, "Differential postendocytotic compartmentation in *Xenopus* oocytes as mediated by a specifically bound ligand," Cell, *22*, 47-57.
Osaki, H., 1972, "Electron microscope studies on developing oocytes of the spider *Plexippus paykulli*," Annot. Zool. Jpn., *45*, 187-200.
Painter, T. S., and Reindorp, E. C., 1939, "Endomitosis in the nurse cells of the ovary of *Drosophila melanogaster*." Chromosoma, *1*, 276-283.
Paschma, M., and Ogorzalek, A., 1972, "O interrelacjach jadrowo-cytoplasma-tycznych w oogenezie," Prz. Zool., *16*, 11-20.
Pasteels, J. J., 1973, "Yolk and lysosomes," in: Lysosomes in Biology and Pathology, Vol. 3, Elsevier, Amsterdam—London—New York, pp. 216-234.
Paulson, J. L., and Rosenberg, M. D., 1972, "The function and transposition of lining bodies in developing avian oocytes," J. Ultrastruct. Res., *40*, 25-43.
Paulson, J. L., and Rosenberg, M. D., 1974, "Formation of lining bodies and oocyte bodies during avian oogenesis," Dev. Biol., *40*, 366-371.
Postlethwait, J. H., and Handler, A. M., 1978, "Nonvitellogenic female sterile mutants and the regulation of vitellogenesis in *Drosophila melanogaster*," Dev. Biol., *79*, 202-213.
Postlethwait, J. H., Bownes, M., and Jowett, T., 1980a, "Sexual phenotype and vitellogenin synthesis in *Drosophila melanogaster*," Dev. Biol., *79*, 379-387.
Postlethwait, J. H., Laugé, G., and Handler, A. M., 1980b, "Yolk protein synthesis in ovariectomized and genetically agametic [X^{87}] *Drosophila melanogaster*," Gen. Comp. Endocrinol., *40*, 385-390.
Press, N., 1964, "An unusual organelle in avian ovaries," J. Ultrastruct. Res., *10*, 528-546.
Rahil, K. S., and Narbaitz, R., 1973, "Ultrastructural studies on the relationship between follicular cells and growing oocytes in the turtle *Pseudemys scripta elegans*," J. Anat., *115*, 175-186.
Ramamurty, P. S., 1963, "Über die Herkunft der Ribonukleinsäure

in den wachsenden Eizellen der Skorpionsfliege *Panorpa communis* (lnsecta, Mecoptera)," Naturwissenschaften, *50*, 383-384.

Ramamurty, P. S., 1964a, "On the contribution of the follicle epithelium to the deposition of yolk in the oocyte of *Panorpa communis* (Mecoptera)," Exp. Cell Res., *33*, 601-605.

Ramamurty, P. S., 1964b, "Intercellular bridges in the ovarian follicles of *Panorpa communis* L. (Mecoptera-Insecta)," Curr. Sci., *33*, 493-495.

Raven, Ch., 1961, "Oogenesis: the storage of developmental information," Pergamon Press, Oxford.

Rebhun, L. I., 1961, "Some electron microscope observations on membranous basophilic elements of invertebrate eggs," J. Ultrastruct. Res., *5*, 208-225.

Recourt, A., 1961, "Een elektronenmicroscopisch onderzoek naar de oogenese bij *Limnaea stagnalis* L.," Dissertation, Utrecht.

Reger, J. F., 1970, "A study on the origin and fine structure of yolk granules in oocytes of the arachnid *Leiobunum* sp. (Phalangid; Harvestman)," J. Submicrosc. Cytol., *2*, 1-12.

Ries, E., and van Well, P. B., 1934, "Die Eibildung der Kleiderlaus, untersucht an lebenden, vital gefärbten und fixierten Präparaten," Z. Zellforsch., *20*, 565-618.

Roth, T. F., and Jackson, R., 1972, "Protein transport driven uniquely by glycolysis in in vitro cultured oocytes," J. Cell Biol., *55*, Pt. 2, 221.

Roth, T. F., and Porter, K. R., 1964, "Yolk protein uptake in the oocyte of the mosquito *Aedes aegypti* L.," Cell Biol., *20*, 313-332.

Roth, T. F., Woodward, M. P., and Woods, J., 1977, "Comparison and selective dissociation of coated vesicles isolated from brain and oocytes," J. Cell Biol., *75*, Pt. 2, 372a.

Rudack, D., and Wallace, R. A., 1968, "On the site of phosvitin synthesis in *Xenopus laevis*," Biochim. Biophys. Acta, *155*, 299-301.

Rué, G., and Bierne, J., 1980, "Contrôle endocrinien de l'oogenèse chez l'hoplonémerte *Amphiporous lactifloreus*," Bull. Soc. Zool. Fr., *105*, 155-163.

Ruthmann, A., 1964, "Zellwachstum und RNA-Synthese in Ei-Nährzellverband von *Ophryotrocha puerilis*," Z. Zellforsch., *63*, 816-829.

Ryffel, G. U., Wyler, T., Muellender, D. B., and Weber, R., 1980, "ldentification, organization, and processing intermediates of the putative precursors of *Xenopus* vitellogenin messenger RNA," Cell, *19*, 53-61.

Satch, N., 1974, "Intercellular bridges between oocytes in the developing ovary of the teleost *Oryzias latipes*," Annot. Zool. Jpn., *47*, 215-220.

Sawada, N., Noda, Y., and Ochi, O., 1968, "An electron microscope study on the oogenesis of *Golfingia ikedai*," Mem. Ehime Univ. Sci. Ser. B, *6*, 25-39.

Schjeide, O. A., Galey, F., Grellert, E. A., Lin, R., I-San, de Vellis, J., and Mead, J. F., 1970, "Macromolecules in oocyte maturation," Biol. Reprod., Suppl., *2*, 14-43.
Schjeide, O. A., Hanzely, L., Holshouser, S. J., and Briles, W. E., 1974, "Production fates of unique organelles (transosomes) in ovarian follicles of *Gallus domesticus* under various conditions," Cell Tissue Res., *156*, 47-59.
Schmick, E., 1971, "Über die Kongruenz von Plasmaphysiologie und Kolloidchemie. Ein Beitrag zur Morphogenese des endoplasmatischen Retikulums," Prax. Naturwiss., *20*, 207-213.
Schuel, H., Wilson, W. L., Wilson, J. R., and Bressler, R. S., 1975, "Heterogeneous distribution of 'lysosomal' hydrolases in yolk platelets isolated from unfertilized sea urchin eggs by zonal centrifugation," Dev. Biol., *46*, 404-412.
Schuetz, A. W., Wallace, R. A., and Dumont, J. N., 1974, "Inhibition of protein incorporation by isolated amphibian oocytes: steroid dose and specificity relationships," J. Cell Biol., *63*, 305.
Selman, K., and Wallace, R. A., 1978, "An autoradiographic study of vitellogenesis in the squid *Loligo pealei*," Tissue Cell, *10*, 599-608.
Selwood, L., 1968, "Interrelationships between developing oocytes and ovarian tissues in the chiton *Sypharochiton septentriones* (Ashby) (Mollusca, Polyplacophora)," J. Morphol., *125*, 71-104.
Selwood, L., 1970, "The role of the follicle cells during oogenesis in the chiton *Sypharochiton septentriones* (Ashby) (Polyplacophora, Mollusca)," Z. Zellforsch., *104*, 178-192.
Skalko, R. G., Kerrigan, J. M., Ruby, J. R., and Dyer, R. F., 1972, "Intercellular bridges between oocytes in the chicken ovary," Z. Zellforsch., *128*, 31-41.
Sportnitz, U. M., and Kress, A., 1973, Ultrastructural studies of oogenesis in some European amphibians. II. *Triturus vulgaris*," Z. Zellforsch., *143*, 387-407.
Srdić, Z., Reinhardt, C., Beck, H., and Gloor, H., 1979, "Autonomous yolk protein synthesis in ovaries of *Drosophila* cultured in vivo," W. Roux' Arch. Dev. Biol., *187*, 255-266.
Srivastava, M. D. L., 1965, "Cytoplasmic inclusions in oogenesis," Int. Rev. Cytol., *18*, 73-98.
Stahl, A., and Luciani, J. M., 1975, "Heterochromatin, micronucleoli, and nucleolus organizers at the first meiotic prophase in the human oocyte," Proc. 10th Int. Congr. Anat. and 8th Annual Meet. Jpn. Assoc. Anatomists, Tokyo, p. 456.
Stang-Voss, Ch., 1970, "Zur Entstehung des Golgi-Apparates. Elektronenmikroskopische Untersuchungen an Spermatiden von *Eisenia foetida* (Annelidae)," Z. Zellforsch., *109*, 287-296.
Stay, B., 1965, "Protein uptake in the oocytes of the cecropia moth," J. Cell Biol., *26*, 49-62.
Stay, B., Friedel, T., Tobe, S. S., and Mundall, E. C., 1980, "Feedback control of juvenile hormone synthesis in cock-

roaches: possible role for ecdysterone," Science, *207*, 898-900.

Stay, B., and Tobe, S. S., 1981, "Control of the corpora allata during a reproductive cycle in a viviparous cockroach," Am. Zool., *21*, 663-674.

Steopoe, I., Ionescu, V. M., and Vlad, M., 1970, "Présence d'une calotte nucleohistonique extrachromosomique dans les noyaux des oocytes des téleosteens et des acipenserides," Acta Histochim., *38*, 311-317.

Stockem, W., and Wohlfarth-Bottermann, K. E., 1969, "Pinocytosis (Endocytosis)," in: Handbook of Molecular Cytology, Vol. 15A, A. Lima-de-Faria (ed.), Elsevier, Amsterdam—London, pp. 1373-1400.

Sundararaj, G. I., and Nath, P., 1981, "Steroid-induced synthesis of vitellogenin in the catfish *Heteropneustes fossilis* (Bloch)," Gen. Comp. Endocrinol., *43*, 201-210.

Taddei, C., 1972, "Significance of pyriform cells in ovarian follicle of *Lacerta sicula*," Exp. Cell Res., *72*, 562-566.

Tanaka, A., and Ishizaki, H., 1974, "Immunohistochemical detection of vitellogenin in the ovary and fat body during a reproductive cycle of the cockroach *Blattella germanica*," Dev., Growth, Differ., *16*, 247-255.

Tata, J. R., 1976, "The expression of the vitellogenin gene," Cell, *9*, 1-14.

Tata, J. R., 1978, "Regulation of vitellogenesis by estrogen," in: Hormone and Cell Regulations, Vol. 2, J. Dumond and J. Nunez (eds.), Elsevier/North-Holland Biomedical Press, Amsterdam, pp. 37-54.

Tedesco, J. L., Courtright, J. B., and Kumaran, A. K., 1981, "Ultrastructural changes induced by juvenile hormone analog in oocyte membranes of apterous *Drosphila melanogaster*," J. Insect Physiol., *27*, 895-902.

Telfer, W. H., 1954, "Immunological studies of insect metamorphosis. II. The role of a sex-limited blood protein in egg formation by the cecropia silkworm," J. Gen. Physiol., *37*, 539-558.

Telfer, W. H., 1960, "The selective accumulation of blood proteins by the oocytes of saturniid moths," Biol. Bull., *118*, 338-351.

Telfer, W. H., 1965, "The mechanism and control of yolk formation," Annu. Rev. Entomol., *10*, 161-184.

Telfer, W. H., 1975, "Development and physiology of the oocyte-nurse cell syncytium," Adv. Insect Physiol., *11*, 223-319.

Telfer, W. H., and Anderson, L. M., 1968, "Functional transformations accompanying the initiation of a terminal growth phase in the cecropia moth oocyte," Dev. Biol., *17*, 512-535.

Tuzet, O., 1962, "L'origine de cellule germinale et gametogenèse chez de l'Eponge," in: L'origine de la lignée germinale chez les Vertébrés et chez quelques groups d'Invertébrés, Paris.

Ulrich, E., 1969, "Étude des ultrastructures au cours de l'ovogenèse d'un poisson téléosteen, le Danio *Brachydanio rerio* (Hamilton-Buchanan), J. Microsc., *8*, 447-478.

Urbani, E., 1969, "Cytochemical and ultrastructural studies of oogenesis in the *Dytiscidae*," Monit. Zool. Ital., *3*, 55-87.

Verhey, C. A., and Moyer, F. H., 1967, "Fine structural changes during sea urchin oogenesis," J. Exp. Zool., *164*, 195-225.

Vincent, W. S., Halvorson, H. O., Chen, H.-R., and Shin, D., 1969, "A comparison of ribosomal gene amplification and uni- and multinucleolate oocytes, Exp. Cell Res., *57*, 240-250.

Vitaioli, L., and Materazzi, G., 1972, "Studi citochimici comparativi sulle cellule testacee di alcuni ascidiacei," Riv. Biol., *65*, 137-145.

Wallace, R. A., 1963, "Studies on amphibian yolk. IV. An analysis of the main-body component of yolk platelets," Biochim. Biophys. Acta, *74*, 505-518.

Wallace, R. A., 1965, "Resolution and isolation of avian and amphibian yolk-granule proteins using TEAE-cellulose," Anal. Biochem., *11*, 292-311.

Wallace, R. A., and Dumont, J. N., 1968, "The induced synthesis and transport of yolk proteins and their accumulation by the oocyte in *Xenopus laevis*," J. Cell Physiol., *72*, 73-89.

Wallace, R. A., and Jared, D. W., 1968, "Estrogen-induced lipophosphoprotein in serum of male *Xenopus laevis*," Science, *160*, 91-92.

Wallace, R. A., and Jared, D. W., 1969, "Studies on amphibian yolk. VIII. The estrogen-induced hepatic synthesis of a serum lipophosphoprotein and its selective uptake by the ovary and transformation into yolk platelet proteins in *Xenopus laevis*," Dev. Biol., *19*, 498-526.

Wallace, R. A., and Jared, D. W., 1976, "Protein incorporation by isolated amphibian oocytes. V. Specificity for vitellogenin incorporation," J. Cell Biol., *69*, 345-351.

Wallace, R. A., Jared, D. W., Dumont, J. N., and Sega, M. W., 1973, "Protein incorporation by isolated amphibian oocytes. III. Optimum incubation conditions," J. Exp. Zool., *184*, 321-333.

Wallace, R. A., Jared, D. W., and Nelson, B. L., 1970, "Protein incorporation by isolated amphibian oocytes. I. Preliminary studies," J. Exp. Zool., *175*, 259-269.

Wallace, R. A., Nickol, J. M., Ho, T., and Jared, D. W., 1972, "Studies on amphibian yolk. X. The relative roles of autosynthetic and heterosynthetic processes during yolk protein assembly by isolated oocytes," Dev. Biol., *29*, 255-272.

Wallace, R. A., Salter, D. W., Ho, T., and Jared, D. W., 1973, "Protein incorporation by isolated amphibian oocytes. IV. The role of follicle cells and calcium during protein uptake," Exp. Cell Res., *82*., 287-295.

Wangh, L. J., and Knowland, J., 1975, "Synthesis of vitellogenin in cultures of male and female frog liver regulated by estradiol treatment in vitro," Proc. Natl. Acad. Sci. USA, *72*, 3172-3175.

Ward, R. T., 1962, "The origin of protein and yolk in *Rana pipiens*. II. Electron microscopical and cytochemical observations of young and mature oocytes," J. Cell Biol., *14*, 309-341.

Ward, R. T., 1965, "Formation of Golgi bodies during maturation of oocytes in *Rana pipiens*," Anat. Rec., *151*, 430.

Wartenberg, H., 1964, "Experimentelle Untersuchungen über die Stoffaufnahme durch Pinocotyse während der Vitellogenese des Amphibien oocyten," Z. Zellforsch., *63*, 1004-1019.

Wegmann, I., and Götting, K-J., 1971, "Untersuchungen zur Dotterbildung in den Oocyten von *Xiphophorus helleri* (Heckel, 1948) (Teleostei Poeciliidae.," Z. Zellforsch., *199*, 405-433.

Wells, M. J., and Wells, J., 1959, "Hormonal control of sexual maturity in octopus," J. Exp. Biol., *36*, 1-33.

Wilhelm, R., and Lüscher, M., 1974, "On the relative importance of juvenile hormone and vitellogenin for oocyte growth in the cockroach *Nauphoeta cinerea*," J. Insect Physiol., *20*, 1887-1894.

Williams, D. L., 1977, "Processing and secretion of avian vitellogenin," J. Cell Biol., *75*, 366a.

Winter, H., 1974, "Ribonucleoprotein-Partikel aus dem telotroph-meroistischen Ovar von *Dysdercus intermedius* Dist. (Heterop tera, Pyrrhoc.) und ihr Verhalten im zellfreien Proteinsyn thesesystem," W. Roux' Arch. Entwicklungsmech. Org., *175*, 103-127.

Wolin, E. M., Laufer, H., and Albertini, D. F., 1973, "Uptake of the yolk protein, lipovitellin, by developing crustacean oocytes," Dev. Biol., *35*, 160-170.

Woodruff, R. I., and Telfer, W. H., 1973, "Polarized intercellular bridges in ovarian follicles of the cecropia moth," J. Cell Biol., *58*, 172-188.

Yamamoto, K., and Oota, J., 1967, "Fine structure of yolk globules in the oocyte of the zebrafish *Brachydanio rerio*," Annot. Zool. Jpn., *40*, 20-27.

Yamamoto, M., Ohkawa, M., and Ishida, J., 1973, "An electron microscopic study of oogenesis in the echiuroid *Urechis unicinctus*," J. Fac. Sci. Univ. Tokyo, Sec. 4, *13*, 39-54.

Yu, J.Y.-L., Dickhoff, W. W., Swanson, P., and Gorbman, A., 1981, "Vitellogenesis and its hormonal regulation in the Pacific hagfish *Eptatretus stouti* L.," Gen. Comp. Endocrinol., *43*, 492-502.

Yung Ko Ching, M., 1930, "Contribution a l'étude cytologique de l'ovogenèse, due développement et de quelques organes chez les céphalopodes," Ann. Inst. Océanogr. Monaco, *7*, 301-364.

Zamboni, L., Smith, D. M., and Thompson, R. S., 1972, "Migration of follicle cells through the zona pellucida and their sequestration by human oocytes in vitro," J. Exp. Zool., *181*, 319-338.

Zelenin, A. V., 1962, "Pinocytosis," Usp. Sovrem. Biol., *53*, 364-374.

Zihler, J., 1972, "Zur Gametogenese und Befruchtungsbiologie von *Hydra*," W. Roux' Arch. Entwicklungsmech. Org., *169*, 239-267.

Zinsmeister, P. P., and Davenport, R., 1971, "An autoradiographic and cytochemical study of cellular interactions during oogenesis in the milkweed bud *Oncopeltus fasciatus*," Exp. Cell Res., *67*, 273-278.

Chapter 2

THE NUCLEUS DURING OOGENESIS WITH SPECIAL REFERENCE TO EXTRACHROMOSOMAL STRUCTURES

M. N. Gruzova

In studies of oocytes the nuclear structure has not only been used for studying oogenesis proper, but also for solving a number of general cytological problems. In recent times, attention has been concentrated on lampbrush chromosomes, the nucleoli, the karyosphere, and the nuclear envelope. Electron microscopy studies led to the discovery of a structure specific to germ cells, namely, the synaptonemal complex (SC), and this has made it possible to explain the mechanism of chromosome conjugation during meiosis at the molecular level. Detailed investigations into amplification, a phenomenon first discovered on ribosomal genes, have revealed the functional significance of the chromatin bodies described much earlier in the nuclei of many oocytes (see Gall, 1969).

At present, the comprehensive use of morphological and biochemical methods has led to the accumulation of a great many facts, which has made it possible to interpret many structural and functional aspects of the main nuclear components of oocytes. However, insufficient attention has been given to the interaction of these components and, therefore, our general conceptions of the organization of nuclei are by no means conclusive. Further, the link between the ultrastructural organization of the nucleus and the type of oogenesis and the specific nature of the nutritional apparatus has not been established.

A new trend has emerged over the last few years. This is indicated by data which show that morphogenetically active factors necessary for early embryogenesis accumulate in the nucleoplasm during oogenesis (see Chapter 3).

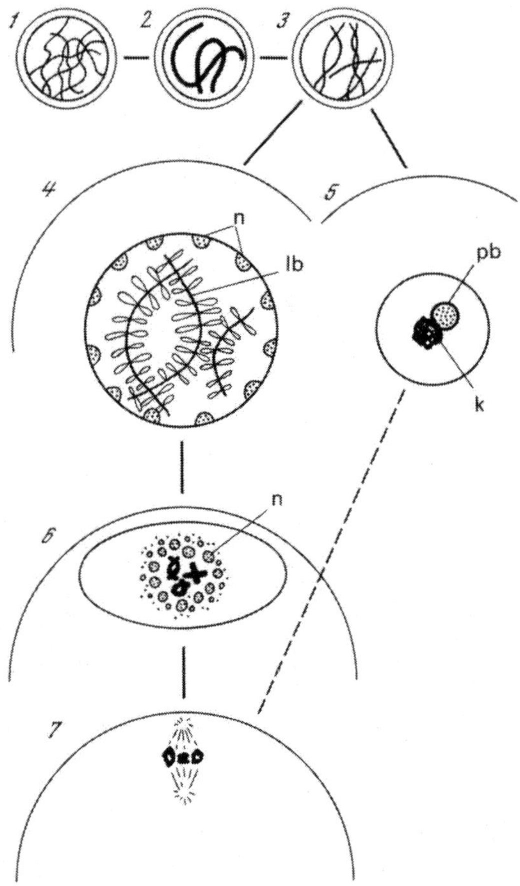

Fig. 14. Correlation of the stages of meiosis and oogenesis (Gruzova, 1975a): 1-3 is the period of slow growth (generative phase of oogenesis); 4-7 is the period of rapid growth (vegetative phase of oogenesis). 1) Leptotene; 2) zygotene and pachytene; 3) early diplotene; 4, 5) middle diplotene [4, solitary type; 5, nutrient type]; 6) diakinesis; 7) metaphase I. pb) protein body; k) karyosphere; lb) lampbrush chromosomes; n) nucleoli.

In this review, we dwell briefly on the advances made in the study of extrachromosomal nuclear structures, such as nucleoli, nucleoli-like formations, extrachromosomal DNA bodies, the synaptonemal complex, and the karyosphere capsule; we also touch upon nucleocytoplasmic interactions and some newly discovered but little-studied nuclear structures. Some problems connected with nuclear matrix investigation in oocytes will also be dealt with.

At the beginning of the review, the events taking place in the oogonia are briefly described; the differentiation of the oogonial cells into oocytes and trophocytes is illustrated using insects as an example. A general scheme of the oocyte growth period is considered, and the characteristic changes in the chromosomal and nucleolar apparatus during oogenesis are indicated. Variations in the different types of oogenesis are borne in mind (Fig. 14).

2.1. OOGONIA AND THEIR DIFFERENTIATION INTO OOCYTES AND TROPHOCYTES (NURSE CELLS)

The origin of the primordial germ cells is not considered here. Modern approaches to this problem have been dealt with in a number of reviews (Aizenshtadt, 1975, 1984; Eddy, 1975). Oogenesis itself begins with the oogonial divisions, which occur in the gonads of many higher invertebrates and vertebrates. However, in a number of invertebrates (Wolf, 1962; Loseva, 1970; Gruzova and Dzyuba, 1973) oogonial divisions have not been found in the gonads. The pattern of the gonial divisions right up to the emergence of the secondary gonia is the same for both oogenesis and spermatogenesis (Wilson, 1925). The primary oogonia are subject to a series of consecutive divisions (the number of these divisions varies in different animal species) and turn into secondary oogonia (Fig. 15).

Cytophotometry and autoradiography show that the amount of DNA doubles in secondary gonia in the premeiotic interphase and that the gonocytes enter meiosis with a tetraploid amount of DNA associated with a diploid number of chromosomes. However, compared to the mitotic S phase, the premeiotic phase of DNA synthesis

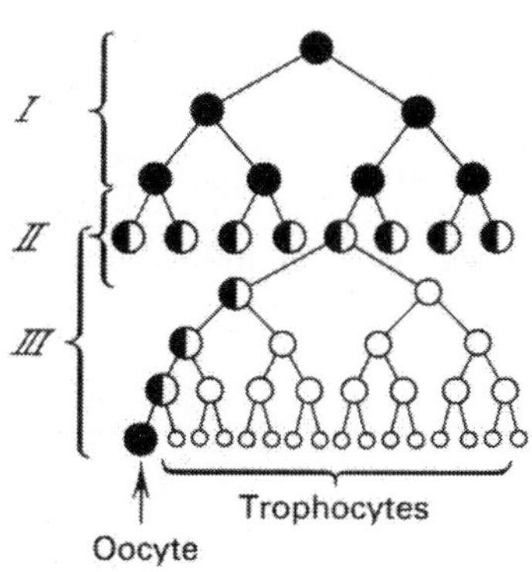

Fig. 15. The oogonial divisions of *Dysticus marginalis* (from Wilson, 1925). During division the extrachromosomal DNA body is transferred only to the oocyte: I) primary oogonia; II) secondary oogonia; III) postgonial divisions (differential mitoses).

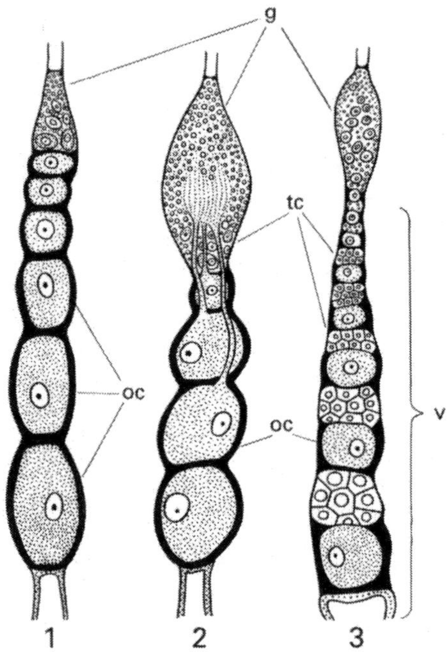

Fig. 16. Types of ovariole structure in insects (Ries and Gersch, 1953): 1) panoistic; 2) telotrophic; 3) polytrophic. v) vitellaria; g) germaria; oc) oocyte; tc) trophocytes.

is characterized by a number of specific features, namely, longer duration (in the microsporocytes of the waterlily it is six times that of the mitotic S phase), greater amount of DNA packed within the chromocenters, and, lastly, in the premeiotic S phase the replication of euchromatin and heterochromatin which is separated by a period during which DNA synthesis is practically absent in the nucleus (Rasmussen and Holm, 1980). rDNA amplification begins in the oogonia of some organisms, and is morphologically expressed by the appearance of extra-DNA bodies (Gall, 1969; see Section 2.4).

The young oocytes differ mainly from the interphase gonial cells in that axial structures providing for the beginning of synaptonemal complex formation appear in their nuclei (Al-Mukhtar-Kawakib and Webb, 1971; Baker and Franchi, 1972).

In insects with meroistic ovarioles (Fig. 16, 2, 3), in the upper part of the ovariole (germarium) the secondary oogonia give rise to the sibling cluster cells. In the course of these differential mitoses the fate of the future oocytes and trophocytes is determined. In a number of cases nuclear markers allow these two cell types to be distinguished. Thus, in 1901, Giardina found that a chromatin body appears in each secondary oogonium in Dytiscus which is then transferred only to one of sixteen cells during subsequent mitoses; it is this cell which becomes the oocyte, while the remaining fifteen become trophocytes (see Fig. 15). It has been shown by hybridization of nucleic acids that this body is a result of the amplification of ribosomal genes (Gall et al., 1969). A similar situation is observed in the insects *Tipula* (Bauer, 1933; Lima-de-Faria, 1969) and *Chrysopa* (Gruzova et al., 1972a, b; Rousset, 1977), the only difference being that the DNA body enters two cells, the two pro-oocytes. Only one of these becomes a true oocyte, while the other becomes a nurse cell possessing bodies containing rDNA, unlike the other trophocytes (Gaginskaya and Gruzova, 1975). However, many insects do not have marker nuclear structures similar to those described above. Serial reconstruction, based on ultrathin sections, enabled King et al. to make a thorough study of the morphology of oocytes and trophocytes in the germaria of *Drosophila* (Koch and King, 1966; Koch et al., 1977) and *Habrobracon* (Cassidy and King, 1972). It appeared that, as a result of differential mitoses, the two cells having the greatest number of intercellular bridges (four bridges in the case of *Drosophila* and five in that of *Habrobracon*) become pro-oocytes. These cells take part in meiosis and form synaptonemal complexes. In the remaining cells, endoreduplication occurs and they become polyploid trophocytes (see Chapter 1).

With rare exceptions, the number of trophocytes in the egg chambers of a polytrophic ovariole may be described by the $2^n - 1$ rule, where n is the species specific number of mitotic divisions. It is suggested that the synchronism of mitotic divisions of the oogonia is maintained by annular channels and forms the basis for the $2^n - 1$ rule (see Telfer, 1975). An example to this rule was found in the development of the chambers in the germarium of *Chrysopa perla* (Rousset, 1978a, b). In this case, a reconstruction using semithin sections established that *Chrysopa* egg chambers consist of an unstable number of cells (12-14), among which three categories of cystocytes were observed, designated A, B, and C. It was found that mitotic synchronism is lost already during the second cleavage, when the lateral cells (type C), without initiating meiosis, undergo endopolyploidy and become early trophocytes. The division of the medial cells leads to the formation of A- and B-type cells. In the A cells, the two pro-oocytes, a DNA body appears. All the A- and B-type cells initiate meiosis; however, the B cells undergo meiotic arrest without reaching pachytene. Pachytene is completed only in the A-type cells (Fig. 17). The precise number of annular channels connecting all these cells was not established; however, in Rousset's opinion, this is not a determining factor for the initiation of meiosis, as was supposed for *Drosophila* (King, 1970).

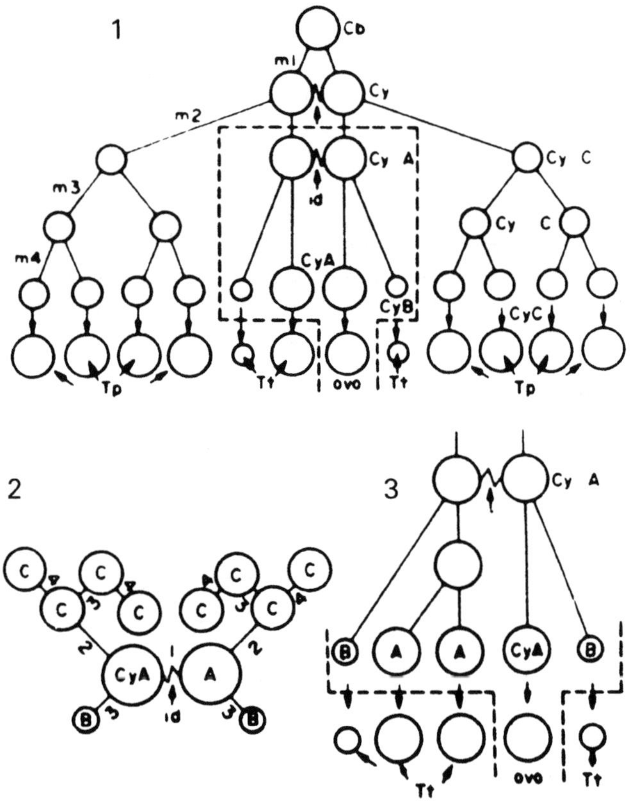

Fig. 17. 1) The distribution of *Chrysopa perla*. Cells enclosed by the dotted line enter the meiotic prophase. In this case, the number of cystocytes (n) = 12; 2) arrangement of intercellular bridges in a group of cells depicted in one plane (n = 12); 3) Variation in number of cystocytes. When n = 13, three A-type cystocytes are formed (middle row). For details, see text. Cb) cystoblast; Cy) cystocyte; id) interdigitation; m) mitosis; ovo) oocyte; Tp) protrophocytes; Tt) definitive trophocytes (Roussett, 1978a).

There are insects in which all the cells of the sibling cluster enter meiosis synchronously, as, for example, in *Laspeyresia pomonella* and *Eurygaster integriceps* (Kozhanova and Gruzova, 1975). In the germaria of *Laspeyresia* all eight cells reach the pachytene stage of meiosis; in the pro-oocyte this stage lasts for some time, while in the trophocytes the chromosomes are rapidly spiralized and form short bivalents. After this, the nurse cells undergo a series of DNA endoreduplications (Gruzova, 1974). In *Eurygaster* the early differentiation of oocytes and trophocytes occurs somewhat earlier: the chromosomes in the trophocytes have no zygotene stage, which is only characteristic of oocytes.

Until recently, the process of differentiation of the oocytes and trophocytes in telotrophic ovarioles remained unknown. The development of the syncytium and the trophic core in the germarium hindered the observation of cellular interrelations within this system (Telfer, 1975). However, it became possible to observe this process in the beetle *Creophilus maxillosus* (Kloc and Matuszewski, 1977), the presence of DNA bodies in the oocytes facilitating the observation of the fate of the cells. In the pupal germaria the terminal oogonial cells (chordoblasts = cytoblasts) undergo a series of synchronous mitoses. The division of these cells is perpendicular to the ovariolar long axis. As a result, each chordoblast produces a linear chain of chordocytes (cells interlinked by intercellular bridges; see Fig. 18). In each line of cells the synthesis of extra-DNA is limited to the most terminal cell, the future oocyte.

Even this brief outline of oocyte and trophocyte differentiation in insect ovaries indicates that morphological criteria alone are not sufficient. A comparison of data obtained in different insect species demonstrates that neither DNA bodies nor the number of intercellular bridges (in *Creophilus* the oocyte has only one of two possible bridges) determine the pathway of cellular development. For the time being, it may be assumed that a decisive factor in the determination of the cells is the position of the cell within the system and its contact with the surrounding tissues. The chemical and physiological factors of oocyte and trophocyte differentiation are still awaiting clarification, and research continues (see Chapter 1).

2.2. OOCYTES AND TROPHOCYTES (NURSE CELLS)

The main stages of oocyte development are shown in Fig. 11. The diagram also demonstrates the correlation between the stages of meiosis and oogenesis. Clarification is necessary because there is some confusion in the literature over the terminology used to designate these stages. In particular, Raven (1961) calls the early stages of the meiotic prophase (leptotene, pachytene, diplotene) *premeiotic* transformations of the nuclear structures, despite the fact that it is precisely in this period that such extremely important meiotic events occur as the conjugation of the chromosomes, the formation of the synaptonemal complexes, and crossing-over, without which further meiotic processes would be unthinkable.

In the initial period of meiosis, the replicative phase, transformations of the chromosomes occur in the oocyte and, consecutive-

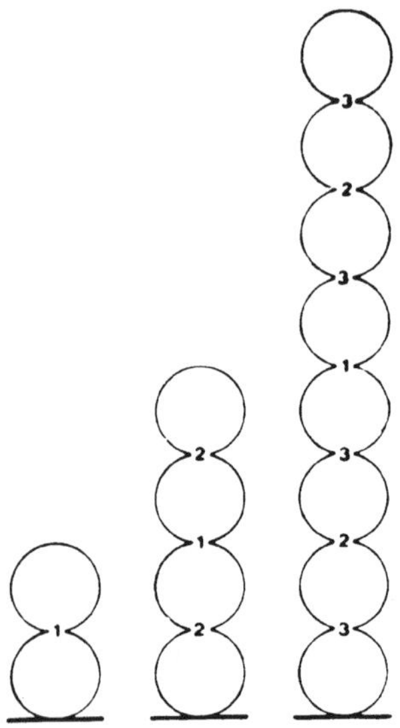

Fig. 18. Diagram showing the positions of successive generations of putative intercellular bridges within the sibling strings formed at the 1st, 2nd, and 3rd chordocyte division. Although the number of chordocyte divisions that the chordoblast undergoes in the development of the *Creophilus* ovary is still unknown, it seems likely that at least five or six chordocyte divisions take place in this species (Kloc and Matuszewski, 1977).

ly, leptotene, zygotene, pachytene, and diplotene take place (see Fig. 14, 1-3). Synaptonemal complex formation and pairing of homologous chromosomes begins at the zygotene stage of meiosis. These processes are completed in the pachytene, and under a light microscope the bivalent looks like a thick thread. Thus, a supposed reduction of the number of chromosomes is achieved. In some animals extrachromosomal nuclear DNA functions in oogenesis and appears at leptotene in the form of a dense body. As a rule, intense DNA synthesis is observed at the pachytene stage. The DNA body increases in size considerably and is intensely labeled by ^{3}H-thymidine (see Section 2.6). The formation of lampbrush chromosomes and the dispersion of extra-DNA throughout the nucleus begins at late pachytene and early diplotene.

The growth phase of the oocyte begins at diplotene. During the first half of this period, the nucleus and cytoplasm grow intensively and the lampbrush chromosomes and nucleoli reach maximum development and actively synthesize RNA (see Fig. 14, 4). Vitellogenesis takes place in the second half of this period and a decrease in RNA synthesis is observed in the nucleus. Frequently the chromosomes and the nucleoli unite into a single complex, forming a karyosphere. The lampbrush chromosomes lose their loops and become much shorter; the stage of diakinesis (see Fig. 14, 6) begins, after which the nucleus disintegrates, and the chromosomes form a metaphase plate of the maturation division (see Fig. 14, 7).

The intensive functioning of lampbrush chromosomes and nucleoli is typical of oocytes, which do not have specialized nurse cells (solitary and follicle-type oogenesis). Where there are trophocytes (nutrimentary-type oogenesis), the period of lampbrush chromosome functioning is shorter or they are absent. In this case, there is a tendency toward nucleoli reduction, and the chromosomes form the karyosphere early (see Fig. 14, 5).

It is quite obvious that the scheme proposed above for the correspondence of the nuclear organization with the type of oogenesis is somewhat limited in application. In nature we find both cases and a number of examples of intermediate forms and deviations from the scheme.

The ovarioles of many panoistic and meroistic insects (see Fig. 16) well illustrate the above scheme. In panoistic ovarioles each oocyte is only surrounded by a follicle (see Fig. 16, 1), there are no trophocytes, and the oocyte nucleus synthesizes all types of RNA (Bier et al., 1967). The cricket *Acheta domesticus* is a typical example of the organization of the nuclei in growing oocytes in panoistic ovarioles; lampbrush chromosomes function actively, and complex nucleoli (for details see Section 2.6) develop on the basis of the amplification of ribosomal cistrons.

In addition to the follicular cells surrounding the oocyte, the meroistic ovarioles of insects contain specialized trophocytes (nurse cells), which synthesize the main bulk of RNA and transfer it to the oocyte (see Fig. 16, 2, 3). In the oocytes of polytrophic ovarioles, for example in many *Diptera* and *Lepidoptera*, the chromosomes become inactive very early and form the karyosphere, and the nucleolus rapidly disintegrates (Gruzova, 1960, 1971, 1974; Bier et al., 1967; King, 1970). Endopolyploidization of the entire set of chromosomes intensifies the function of the trophocytes (Nagl, 1978). The amount of DNA in the trophocytes varies among species. The trophocytes of the polytrophic ovarioles of the *Drosophila* are highly polyploid; 512n-1024n (Painter and Reindorp, cit. in Nagl, 1978). Much lower degrees of ploidy are observed in the trophocytes of telotrophic ovarioles. Thus, the DNA in the stinkbug *Gerris* reaches 32 C (Choi and Nagl, 1977) and in the trophocytes of the *Polyphaga* beetles 3-5 cycles of endoreduplication occur (Büning, 1979; King and Büning, 1984). The amount of DNA in cells is directly correlated to their ploidy when the entire genome is equally replicated. However, underreplication and amplification in the trophocytes may be combined with endopolyploidization, and then the amount of DNA in the nuclei of trophocytes is not an accurate index of their ploidy; such a situation was observed in *Chrysopa perla* (Zaichikova, 1976). There is little information on the morphology of trophocyte polyploidization. It has been demonstrated that during the first cycles of endoreplication the chromosomes in the trophocytes tend to polytenization, but later the chromatids become diffuse and form a network of chromonemas in the nucleus (Painter and Reindorp, 1939; Bier, 1959, cit. in Nagl,

1978). Cryptopolytenic chromosomes (a combination of polytenic morphology with lampbrush loops) were described in the trophocytes of *Rhynchosciara* (Basile, 1969).

Nucleolar functioning in the trophocytes is intensified both by polyploidization and by the amplification of nucleolar genes (Renkawitz and Kunz, 1975). These processes relate to characteristic nuclear structures. For example, in every nurse cell of *Laspeyresia pomonella* a single large spherical DNA body develops (Plate IX, I), which is the result of the endoreduplication of XY-bivalent and possibly of the amplification of the nucleolar organizer of this bivalent. The chromatin bodies incorporate ^{3}H-thymidine and ^{3}H-uridine and contain rDNA. The nucleolar material is often in contact with the surface of these bodies. It may be assumed that this DNA body functions as a single large organizer of numerous nucleoli; its fragmentation gives rise to multiple complex nucleoli in the nucleus (Gruzova, 1974). Typical nucleoli are not found in the trophocyte nuclei of several beetle species from the Tenebrionidae, but numerous nucleolar-like formations (NLF) are located in the perinuclear zone of the cytoplasm (Gruzova, 1979a). A detailed study of the trophocytes of *Blaps lethifera*, using in situ nucleic acid hybridization and electron microscopic autoradiography, showed that, just inside the nuclear membrane, complexes emerge consisting of blocks of nucleolar heterochromatin and accumulations of large perichromatin-like granules (Plate IX, 2). The complexes contain rDNA. The NLF are extruded from the nucleus and contain freshly synthesized RNA derived from the area of nuclear complexes (Plate IX, 3). The NLF contain RNA and DNA and consist of fibrillar and, less often, granular (ribosomal-type granules) components. They may be regarded as "cytoplasmic nucleoli" (Gruzova, 1979a, 1985). Later on it was shown in young trophocytes that only a single ring-shaped DNA body was involved in production of numerous nucleolar chromatin blocks presumably a result of amplification. The results obtained by the Ag-staining technique (Howell and Black, 1980) and data provided by karyotype research (Gruzova and Mamaeva, 1984) enabled us to regard the above DNA body as sex chromatin composed of $X_1X_2X_3X_4$ chromosomes (Gruzova, 1985). Thus, the situation appears to be similar to that in *Laspeyresia*.

Biochemical investigations show that the trophocytes of the telotrophic ovarioles of the stinkbug *Dysdercus* supply the oocyte with stable populations of RNP-particles containing RNA of every class. The RNP-particles remain in the oocyte right up to the beginning of embryogenesis. Throughout the entire growth of the oocyte the germinal vesicle is not active in RNA synthesis. However, shortly before oviposition, RNA is synthesized briefly by the oocyte itself; this RNA is an unstable heterogeneous mRNA (Duspiva, 1981).

Among the insects with polytrophic ovarioles we find cases which are, as it were, intermediate ones between the solitary and the alimentary type; this is observed in certain beetles (Bier et al., 1967; Matuszewski and Hoser, 1975; Matuszewski and Kloc, 1976; Ray and Ramamurty, 1979) and in the neuropteran *Chrysopa perla*, where the ultrastructural organization of the nuclei of oocytes and trophocytes has been studied fully (Gruzova et al., 1972b; Zaichikova and Gruzova, 1975). In this species it appears that, in spite of the early cessation of RNA synthesis on chromosomes and the formation of a karyosphere, a large, actively functioning nucleolar apparatus develops in the nuclei of the oocytes (Gruzova et al., 1972a, b) as a result of rDNA amplification (Gaginskaya and Gruzova, 1975). The high degree of polyploidization characteristic of the trophocytes of other insects, such as Dipterans, is not found in the nuclei of *Chrysopa* trophocytes. The amount of DNA in the imago is due only to the amplification of the nucleolar organizers and reaches 65-135 C (Zaichikova, 1976). It is obvious that in the oocyte-trophocyte system of *Chrysopa perla* the insufficient activity of the trophocyte nuclei during rRNA synthesis is supplemented by the oocyte's own nucleolar apparatus.

2.3. THE SYNAPTONEMAL COMPLEX

In 1956, using electron microscopy (cit. by Moses, 1968), Moses was the first to observe a tripartite, ribbon-like structure between the conjugating homologs in the spermatocytes of crayfish. This structure became known as the "synaptonemal complex" (SC). It consists of two lateral ribbons (lateral elements) stretched along the homologs, and a central ribbon (medial element) united with the lateral elements by fine crosswise fibrils. A typical SC is shown in Plate VII, 1. The SC is a universal structure. It is found in almost all the gametocytes of those plants and animals that have been studied. It belongs exclusively to the meiotic chromosomes and has never been found in the somatic conjugation of chromosomes. Extensive literature has been devoted to the SC, including several reviews (Moses, 1968, 1969; Westergaard and Wettstein, 1972; Bogdanov, 1975; Rasmussen and Holm, 1980). Here we shall deal with general features of the SC. The reader is referred to the above-mentioned literature for more details.

2.3.1. General Characteristics of the Synaptonemal Complex

As mentioned above, the SC is a tripartite structure occupying the space between the conjugating homologs. A three-dimensional schematic diagram of the SC structure is shown in Fig. 19. The overall thickness of the SC in various forms is 160-240 nm, i.e., about 0.2 μm. The lateral elements are 30-60 nm in longitudinal section, and the central elements are 10-40 nm. The thickness of the cross-sectional fibrils uniting the lateral elements with the central one is 2-5 nm. These fibrils may either be

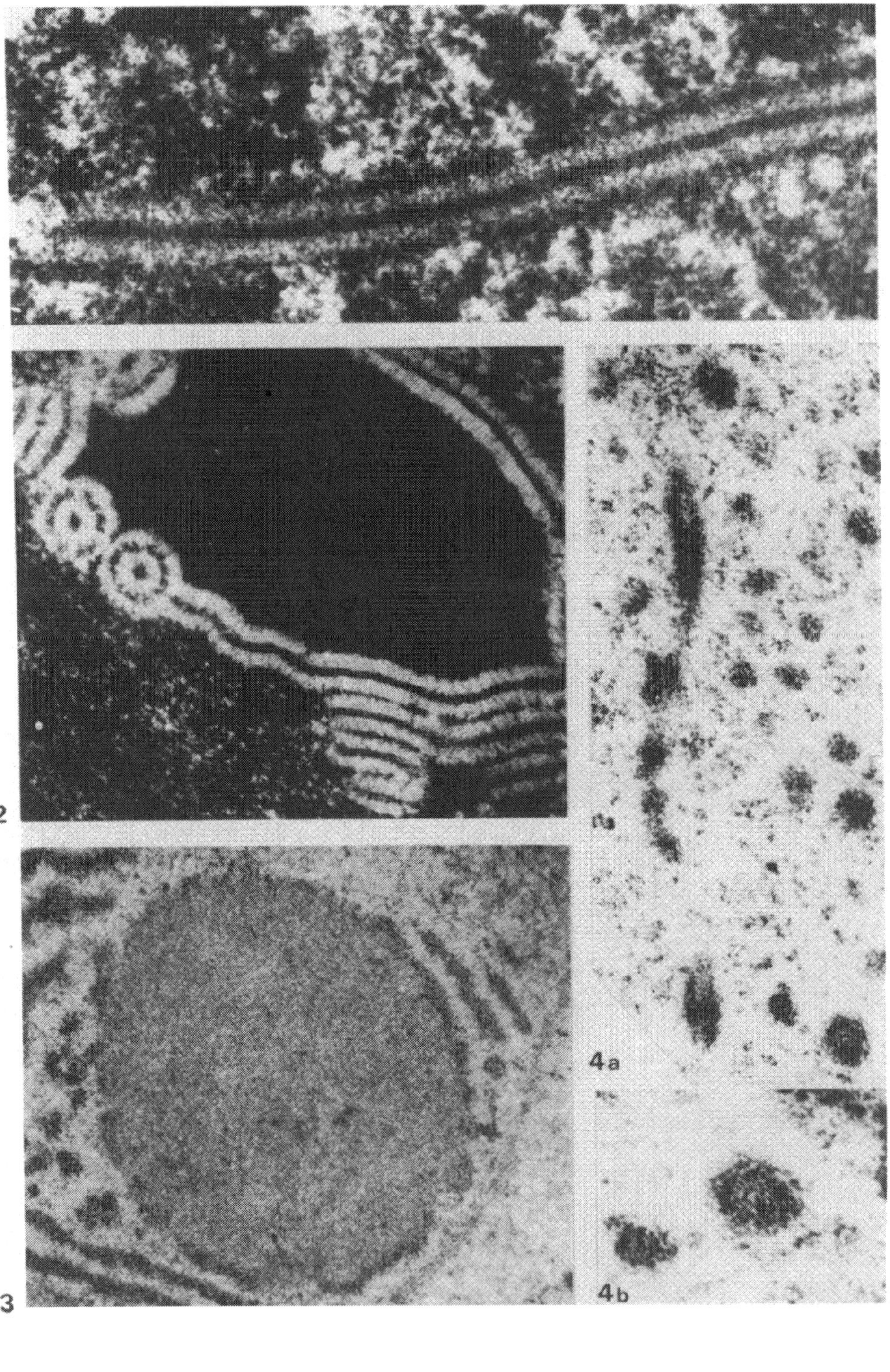

1

2

3

4a

4b

located strictly alternatively (see Fig. 19) or unevenly. The parameters and the fine structure of the elements composing the SC are highly species specific in form. Thus, for example, in some insects (*Acheta*, *Panorpa*), in the lateral elements the two parts differ from one another in the packaging of the fibrillar material composing them (see Fig. 19).

The following events take place during SC formation. In the leptotene chromosome an axial cord appears between the sister chromatids to become a lateral element of the SC. The telomeres of the chromosomes and the ends of the axial cords attach to the nuclear membrane. In zygotene the telomeres of the homologs draw together, sliding along the nuclear membrane. When the distance between the homologs becomes no more than 300 nm, cross fibrils emerge between the axial cords, and a central element is formed. Whereas in the majority of animals examined the formation of the SC begins, as a rule, on the telomere of the chromosomes, in plants (*Lilium*) the SC appears in several loci along the chromosomes (Rasmussen and Holm, 1980). The development of the SC reaches a maximum during pachytene: the length of the SC is equal to the length of the chromosomes and the conjugation of the homologs becomes complete. As the homologs begin to disjoin, the SC disappears. Separate pieces of the SC, frequently abnormal in form, are observed for considerably longer periods only at the sites of chiasma (Westergaard and Wettstein, 1972).

Thus, observations of SC development during zygotene and in diplotene have shown that the SC is not an organic component of the chromosomes. Roth (1966) was the first to show that the SC detaches from the homologs when they disjoin. As a result, individual pieces, or polymerized SC fragments in the form of so-called polycomplexes, may exist in the nucleoplasm independently of the chromosomes (for details see Section 2.3.2).

Cytochemical studies have shown that the SC consists mainly of proteins. Observations indicate that these proteins are of a

---

Plate VII. The synaptonemal complex (SC) and its derivatives. 1. SC in the pachytene oocytes of *Acheta domesticus* (Jaworska and Lima-de-Faria, 1969). 2. Concentrically located polycomplexes in the DNA body in the oocytes of *Acheta domesticus* (Jaworska and Lima-de-Faria, 1969). 3. Fibrillar cords (modified SC) separating from the surface of the nucleoli in the oocytes of *Chrysopa perla* (Zaichikova and Gruzova, 1973); material of the karyosphere capsules appears in contact with the strands (arrows). 4. Ultrathin sections of the fibrillar strand in the nucleus of the oocyte of *Chrysopa perla* (Gruzova and Zaichikova, 1972). a) Cross section showing the tubular structure; b) greatly magnified cross section.

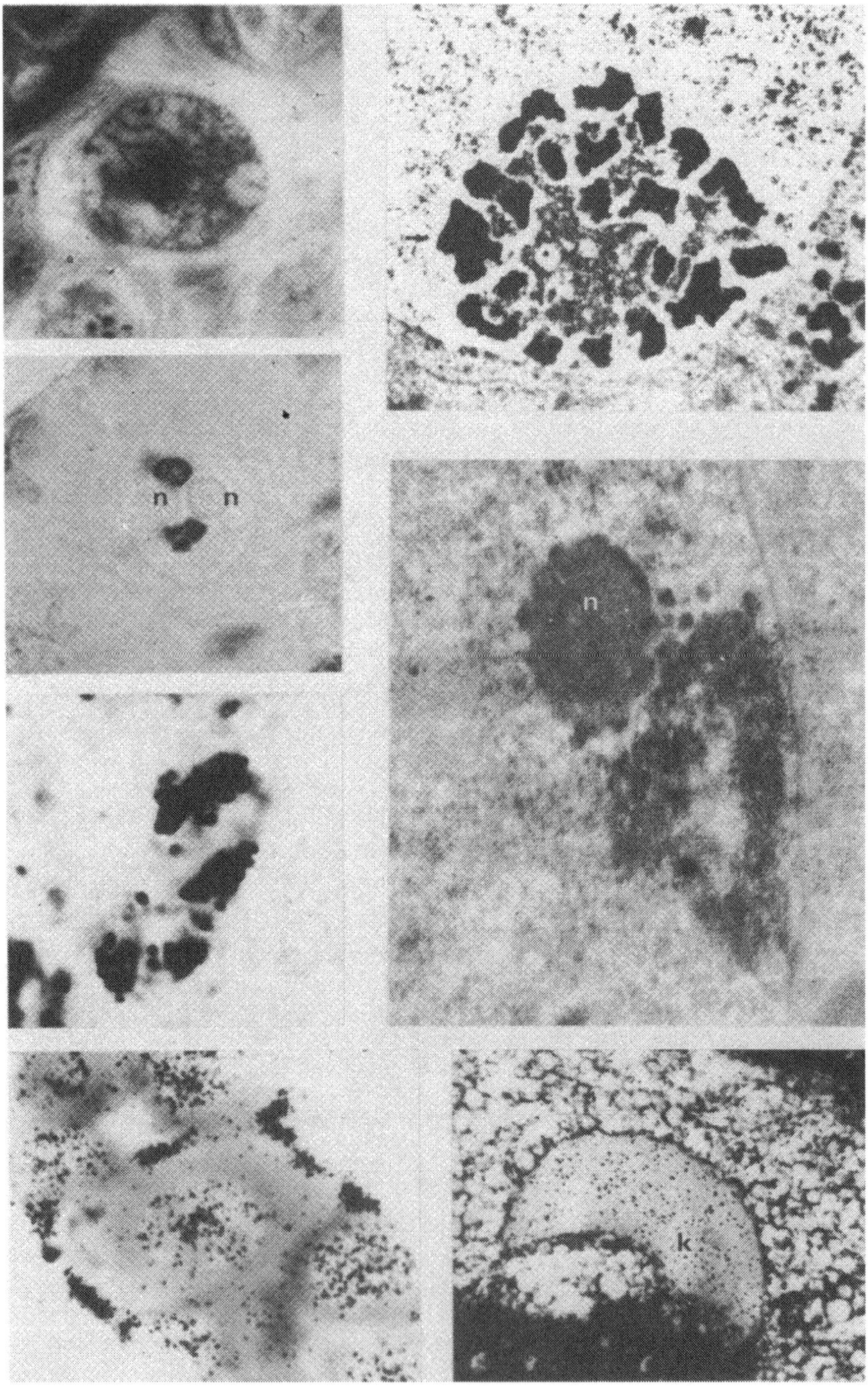

1 2 3 4 5 6 7

basic nature (Sheridan and Barnett, 1969). DNase and RNase do not destroy the SC structure, but SCs isolated from the cell are completely destroyed by protease. In a number of cases (Monneron and Bernhard, 1969), RNA has been detected in the lateral elements of the SC. DNA has been found in the SC in negligible quantities. Traces of DNA in the form of thin, poorly visible threads can be followed in the central space and in the central element of the SC (Moses, 1969). There is also, apparently, a certain amount of DNA in the lateral elements of the SC. When pachytene chromosomes isolated from the spermatocytes of birds and mammals were treated with DNase, part of the lateral elements of the SC were removed along with the chromatin (Solary, 1972). On the basis of his observations, Solary suggests that each lateral element contains one longitudinally situated DNA molecule. Taking a number of examples, it was calculated that the SC contains 0.3-0.6% of all the DNA (Westergaard and Wettstein, 1972). Biochemical experiments with cultures of lily microsporocytes showed that the pairing of the homologs and SC formation requires the synthesis of a minor fraction of DNA during zygotene (zDNA); this fraction makes up about 0.3% of all the DNA synthesized in the premeiotic interphase. The zDNA fraction represents a unique sequence. The inhibition of the synthesis of this DNA during zygotene disrupts the pairing and integrity of the chromosomes. These observations and other data pertaining to the biochemistry of meiosis in plants and animals are reviewed by Stern and Hotta (1977). The synthesis of another DNA fraction also occurs during pachytene. This is pDNA, distinct from zDNA, and its synthesis is of a reparative rather than replicative character. This synthesis is apparently associated with DNA-nicking and reparation in definite loci during crossing-over. The loci consist of mean repeated sequences in 1500-2000 base pairs (Bouchard and Stern, 1980).

Plate VIII. Extrachromosomal nuclear DNA. 1. Pachytene oocyte of a sturgeon (Raikova, 1968); DNA "cap" is similar to the extra-DNA cap seen in amphibian oocytes. 2. DNA body in the oocytes of *Acheta domesticus* (Lima-de-Faria, 1969); electron-dense islets are nucleolar material. 3. Two DNA bodies with two nucleoli (n) in the oocytes of *Chrysopa perla* (Feulgen staining). 4. DNA body near the nuclear envelope in the oocytes of *Chrysopa perla* (Gruzova et al., 1972); in contact with the body is a fibrillar nucleolus (n). 5. DNA bodies in the nucleus of a young oocyte of *Chrysopa perla*, labeled with ^{3}H-rRNA as a result of in situ hybridization (Gaginskaya and Gruzova, 1975). 6. Oocyte of *Chrysopa perla* showing nucleus; karyosphere is "stained" with ^{3}H-actinomycin D; the remaining label is connected with extra-DNA diffused throughout the nucleus. 7. ^{3}H-Thymidine incorporation in the nucleus of the *Chrysopa perla* oocyte at the stage of vitellogenesis 5 h after the precursor injection; karyosphere (k) is not labeled. Scale: 0.5 μm.

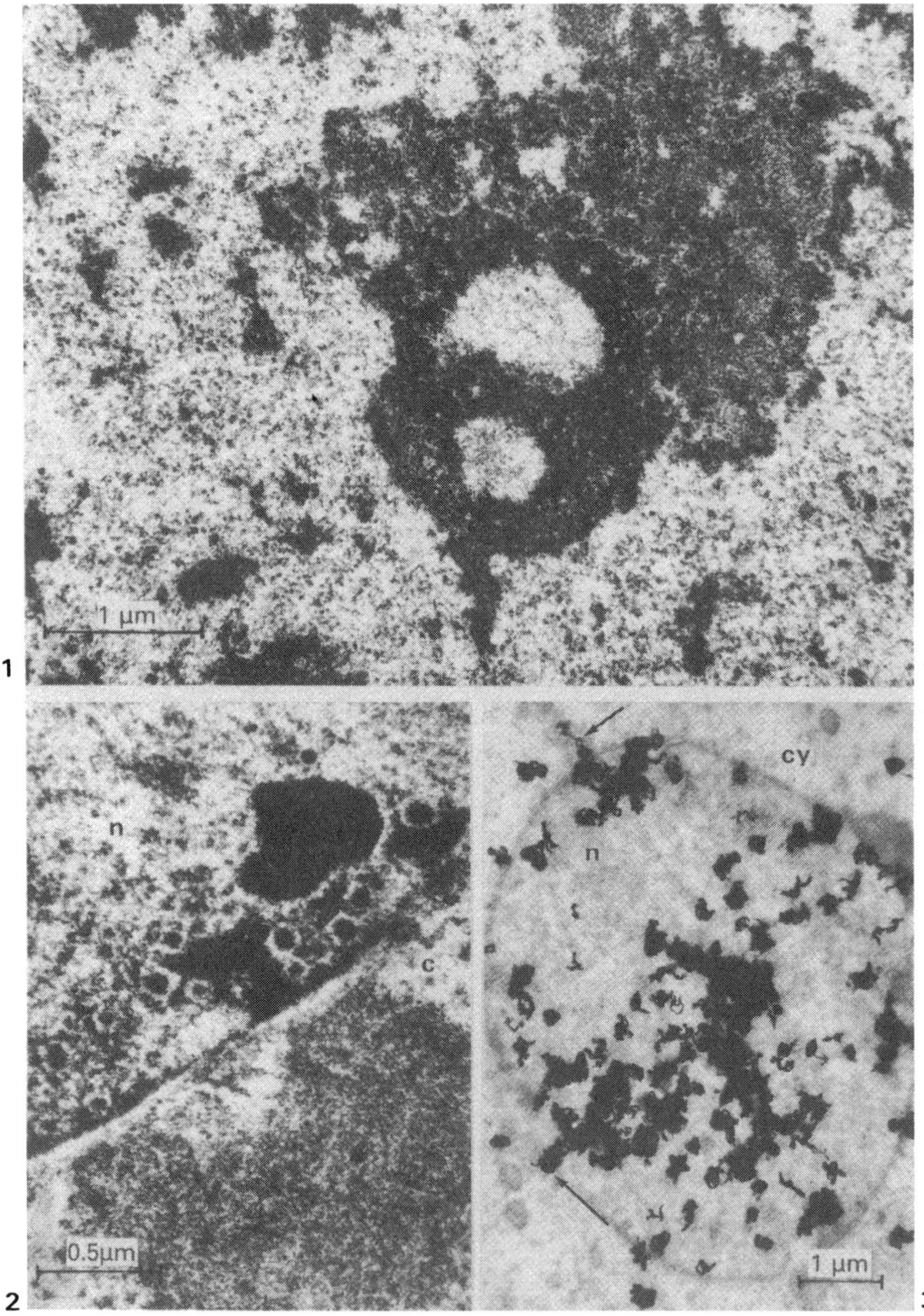
1 µm
1
n
c
0.5µm
2
cy
n
1 µm
3

Associated with the pairing of chromosomes and zDNA synthesis is the synthesis and activation of a complex of proteins. This group of proteins contains an endonuclease responsible for DNA-nicking, an R-protein which facilitates the reassociation of single-stranded DNA into a duplex, and an unwinding "U-protein," which unwinds the limited duplex areas and attaches to its terminals or at the loci of DNA-nicking (Hotta and Stern, 1978). The mechanism regulating the activity of these proteins is unclear so far, but what is clear is that it operates at the level of pairing homologs and its activity depends on the effectiveness of their pairing. It has been demonstrated that, in achiasmatic deployed hybrids of the lily, the activity of endonuclease and R-protein is very low, and it increases considerably in colchicine-induced tetraploid chiasmatic microsporocytes (Hotta et al., 1979).

Questions regarding the genetic control of SC formation are not examined in this section. We shall only note that SC formation is a genetically determined event.

We must also emphasize that, for crossing-over to proceed normally, the formation of an SC is necessary but not sufficient. Thus, *Bombyx mori* females show achiasmatic meiosis and the absence of crossing-over, even though a normal SC forms in their oocytes (Rasmussen and Holm, 1980).

The basic function of the SC is apparently to ensure homologous matching, while the stabilization and effectiveness of pairing, as well as the fulfillment of crossing-over, depend in all probability on the involvement in these processes of so-called recombination nodules (RN). The RN represent a new class of structure discovered between conjugating homologs in the meiosis of *Drosophila* females (Carpenter, 1975), and in other organisms as well (Holm and Rasmussen, 1981). The RN are ovoid or rounded structures 30-50 nm in diameter. They appear in the zygotene and are associated with the central element of the SC. During pachytene

---

Plate IX. 1. A region of the nucleus of a young trophocyte of *Laspeyresia pomonella*. A vacuolized DNA body (center) is in contact with the nucleolus. 2. A fragment of the nucleocytoplasmic zone of a trophocyte of *Blaps lethifera*. In the nucleus (n) blocks of nucleolar chromatin and large perichromatin-like granules adjoin the nuclear envelope. In the cytoplasm (c) a nucleolus-like RNP body adjoins the nuclear envelope. 3. ^{3}H-Uridine incorporation (2 h after precursor injection) in the nucleus of the trophocytes of *Blaps lethifera*. A cytoplasmic nucleolus-like globule (cy) is also labeled.

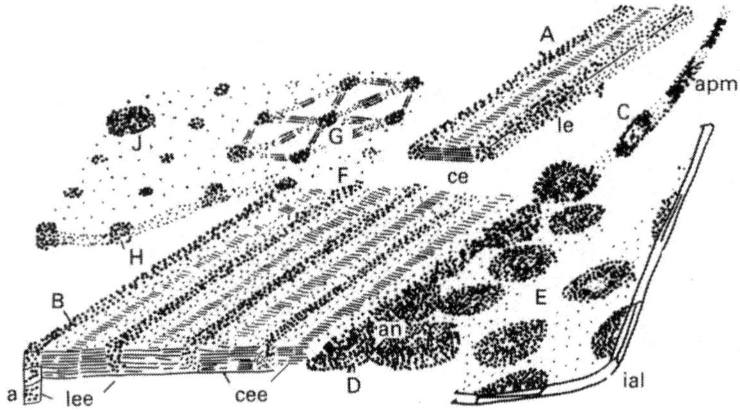

Fig. 19. Diagrams of modified synaptonemal complexes in *Aedes aegypti* (Jiil and Moens, 1973). There is a normal complex at A with lateral (le) and central (ce) elements. The lateral elements are connected by transverse sheets of filaments. The polycomplex B has lateral element equivalents (lee) and central element equivalents (cee). The lateral element equivalents break up into short bars at F. The bars are arranged in hexagonal bars (G) maintaining the normal spacing between lateral elements. The transverse elements between the bars form sheets (H). Some bars give rise to annuli (J). Annuli (an) also form on the lateral element equivalents of the polycomplexes (D). The annuli dissociate from the polycomplex and then reassociate with wider spacing into annulated pseudomembranes (apm) at C and E. Some of these become intranuclear annulate lamellae (ial).

the number of RNs stabilizes and corresponds to the number of chiasmata. No RNs are found in organisms with achiasmatic meiosis or in meiotic mutants with a reduced frequency of crossing-over (Carpenter, 1979; Rasmussen and Holm, 1980). It is assumed that these structures are involved in crossing-over and provide the enzymatic and structural foundation for recombination, as substantially confirmed by the inclusion in the RN of ^{3}H-thymidine (Carpenter, 1981). It is also possible that, along with the remnants of the SC, the RNs constitute the structural basis of the chiasmata (Carpenter, 1975; Holm and Rasmussen, 1981).

2.3.2. *Abnormal Synaptonemal Complexes*

Distinct additional structures similar to the SC emerge during gametogenesis in many animals (particularly in insects). Abnormal

SCs may vary considerably in their spatial organization, but always have regular alternation of their lateral and central elements, or structures similar to these elements (see reviews by Moses, 1968, 1969; Bogdanov, 1975; Gruzova, 1975a, b).

Abnormal complexes may appear at different stages of gametogenesis, but must be present in the postsynaptic period when the conjugating homologous chromosomes begin to disjoin.

It is not clear whether the abnormal SC complexes come into being by simple polymerization of the SC segments which have separated from the chromosomes, or whether these structures are formed *de novo* on the pre-existing matrices.

The appearance of polycomplexes in the oocytes of the mosquito *Aedes aegypti* was studied by Roth (1966). In this species a few polycomplexes emerge during early diplotene in contact with the SC bivalents, which are connected with the nuclear envelope. As the homologs contract and disjoin, the SC disappears from the bivalent and polycomplexes appear in the nucleoplasm in an ever-increasing number and lose their bond with the chromosomes. These observations led Roth to suggest that the polycomplexes appear as a result of the slipping of the SC from the chromosomes and the adhesion of the SC fragments to the lateral elements. Further observations by Fiil and Moens (1973) indicated a more complicated pattern, showing that the abnormal complexes appeared around the karyosphere in such large numbers that their *de novo* formation seems likely. Besides the ordinary polycomplexes, there were also polycomplexes in contact with structures similar to the pore complexes of the nuclear envelope. The latter sometimes formed intranuclear annulate lamellae (Fig. 20).

Giant polycomplexes consisting of 300-400 units have been discovered in the pachytene oocytes of the cricket *Acheta domesticus* (see Plate VII, 2) (Jaworska and Lima-de-Faria, 1969). The authors note that the polycomplexes emerge on the surface of the DNA-containing bodies during rDNA amplification.

Abnormal SC are frequently observed in connection with the nucleolus, or appear between the nucleolus and the extra-DNA bodies (see Moses, 1968). A tubular form of the abnormal SC has been discovered in the oocytes of the insect *Chrysopa perla* (Gruzova, 1975a). These tubular strands exfoliate concentrically from the surface of the pronucleoli (see Plate VII, 3, 4a, b).

Abnormal complexes have been described in spermatogenesis more often than in oogenesis. One possible reason is the fact that the material of the abnormal SC can be so heavily modified by the features associated with oocyte growth that it loses its obvious similarity with the SC and is rather difficult to analyze. Moses

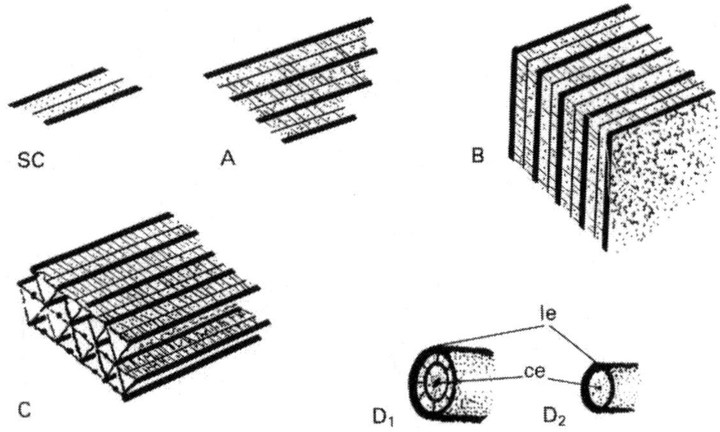

Fig. 20. Diagram showing various configurations and arrays in which anomalous complexes related to the synaptonemal complex have been observed (Moses, 1968). A. Side-by-side array of fused complex (polycomplex). B. Laminated plates of multiple complex. C. Multiple plates of fused complexes in offset register. D_1, D_2. Tubular assemblies characteristic of nucleolar complex. SC) tripartite synaptonemal complex; ce) central element; le) lateral element.

(1968) examined variations in the abnormal complexes in various species and isolated approximately five types of these structures (Fig. 21). He postulated several conditions necessary for the formation of SC-type structures. 1) There must be a definite DNA fraction; the SC material aggregates with the DNA. In autosomes this DNA is limited to the axial region of the chromosome; in extrachromosomal DNA bodies, or in the nucleolus, this condition applies to the structure as a whole. 2) Two threads of homologous DNA are needed. While in pairing homologs this DNA is located sequentially and linearly along the chromosome, in the nucleolus and extra-DNA bodies it is dispersed. 3) A critical distance must be attained between the DNA threads to form the regularly alternating SC-type complexes. 4) The material of the lateral elements in the abnormal structures is, to a considerable extent, determined by the structures on which it comes into being, and may thus be substantially changed in form. 5) Contact with the nuclear envelope by the abnormal SC themselves, or by the structures on which they emerge, is necessary.

The appearance of pore complexes on the lateral elements of SC-type structures indicates that the material of the nuclear envelope also takes part in the formation of the abnormal complexes. This is confirmed by what we know about the involvement of the

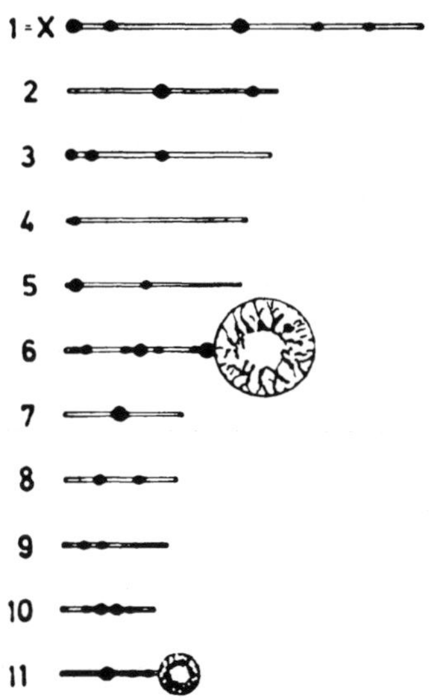

Fig. 21. Diagram showing pachytene chromosomes in *Acheta* oocytes. Only the chromomeres used as markers are shown. Chromosomes 6 and 11 have major chromomeres, DNA bodies.

nuclear membrane elements in the formation of a normal SC (Beçak and Beçak, 1972; Engelhardt and Pusa, 1972). A number of authors note the hexagonal packing of both the pore-like structures (Fiil and Moens, 1973; Chentsov and Polyakov, 1974; Gruzova and Parfenov, 1976) and the elements of the abnormal complexes (Sotelo and Wettstein, 1964, 1969; Wettstein and Sotelo, 1965).

Thus, the study of the abnormal SC has made it possible to realize, first, that the SC is not an organic structure of the chromosomes, but assembles on the chromosomes when specific conditions are attained (Moses, 1969); second, that the material of the SC (its lateral and central elements) is synthesized or becomes aggregated in different parts of the nucleus (Westergaard and Wettstein, 1972; Fiil and Moens, 1973). One of these sites is the nucleolus (Guenin, 1965; Stockert et al., 1970; Sotelo et al., 1973; Rasmussen, 1973, 1974; Gillies et al., 1974; Gruzova, 1975a). There are reasons to believe that the nucleolus favors the formation of SC material during experimental (Snow, 1972) or physiological (Gruzova, 1975a, b) inhibition of rRNA synthesis.

An abnormal SC may exist in the nucleus for a long time, sometimes directly up to metaphase I (Fiil, 1974). In certain cases, the complexes are found in the cytoplasm, where they evidently enter during the first (I) and second (II) maturation divisions (Moses, 1968; Moens, 1969). What happens to them after that is not known. The extreme case of the emergence of polycomplexes independent of the chromosomes was described in the cytoplasm of gametocytes (Bogdanov, 1977; Fiil et al., 1977). It is surprising

Table 2.1. Extrachromosomal DNA in the Nuclei of the Oocytes of Various Animals

Species	Method of determining extrachromosomal DNA	Reference
AMPHIBIA Anura		
Bufo americanus	Morphology (staining)	King, 1908
B. valliceps	Feulgen stain	Painter and Taylor, 1942
B. americanus	Feulgen photometry	Gall, 1969
Rana temporaria (fusca)	Feulgen stain	Brachet, 1940; Guyenot and Danon, 1953
Xenopus laevis	Feulgen photometry, TdR-^{3}H autoradiography	Perkowska et al., 1968; Macgregor, 1968
	Satellite DNA	Brown and Dauid, 1968
	RNA–DNA hybridization	Evans and Birnstiel, 1968; Gall, 1969
	Electron microscopy	Van Gansen and Schram, 1972
	Electron microscopy of isolated molecules of rDNA	Hourcade et al., 1973
	As above, combined with TdR-^{3}H autoradiography	Bird et al., 1974
Urodela		
Ambystoma mexicanum	Satellite DNA, RNA–DNA hybridization	Brown and Dawid, 1968
Necturus maculosus	As above	
Plethodon vehiculum	DNase digestion of nucleoli	Kezer, 1965
P. cinereus	As above	
Triturus pyrrhogaster	As above	Miller, 1964, 1966
T. viridescens	Electron microscopy of nucleolar DNA, isolated in deionized media	Miller and Beatty, 1969
T. viridescens	^{3}H-actinomycin D autoradiography	Ebstein, 1967
	RNA–DNA hybridization	Brown and Dawid, 1968
OSTEICHTHYES		
Roccus saxatilis	Determination of satellite DNA, RNA–DNA hybridization	Vincent et al., 1968
Acerina cernua	Feulgen stain, TdR-^{3}H autoradiography	Chmilevskii, 1973
Acipenser ruthenus A. guldenstadti	Feulgen stain, Feulgen photometry, TdR-^{3}H autogradiography	Raikova, 1976

Table 2.1 (continued)

Species	Method of determining extrachromosomal DNA	Reference
MOLLUSCA		
Spisula solidissima	RNA–DNA hybridization	Brown and Dawid, 1968
ECHIURIDA		
Urechis caupo	As above	Brown and Dawid, 1968
INSECTA		
Orthoptera		
Gryllus campestris	Morphology	Buchner, 1909
Nemobius fasciatus	Feulgen stain	Johnson, 1938
Acheta (Gryllus)	As above	Nilsson, 1966
domesticus	As above	Heinonen and Halkka, 1967
	TdR-^{3}H autoradiography	Kunz, 1967b; Kunz, 1969; Bier, Kunz and Ribbert, 1967; Lima-de-Faria et al., 1968
	Cytochemistry and electron microscopy	Allen and Dave, 1969
	RNA–DNA hybridization	Cave, 1973
	A combination of morphological and biochemical methods	Lima-de-Faria, 1974a
	Electron microscopy of nucleolar DNA isolated in deionized media	Trendelenburg et al., 1973, 1977
	Determination of satellite DNA, RNA–DNA hybridization	Lima-de-Faria et al., 1969
Locusta migratoria	Digestion of the nucleoli by DNase	Kunz, 1967a
Diptera		
Tipula oleracea	Feulgen stain, TdR-^{3}H-autoradiography	Lima-de-Faria, 1962
	Feulgen stain, electron microscopy	Lima-de-Faria and Moses, 1966
Neuroptera		
Chrysopa perla	Rivanol and auramine SO_2	Gruzova et al., 1972
	Feulgen stain	Zaichkova and Gruzova, 1973
	Electron microscopy and ^{3}H-actinomycin D autoradiography	Gruzova, 1975a, b
	RNA–DNA hybridization	Gaginskaya and Gruzova, 1975
Coleoptera		
Dytiscus marginalis	Morphology	Giardina, 1901

Table 2.1 (continued)

Species	Method of determining extrachromosomal DNA	Reference
Dytiscus marginalis	Feulgen staining, ^{3}H-thymidine autoradiography	Urbani and Russo-Caia, 1964
	Feulgen staining	Bier et al., 1967
	Feulgen staining, electron microscopy	Kato, 1968
	RNA–DNA hybridization, electron microscopy of denatured molecules of rDNA	Gall and Rochaix, 1974
	Electron microscopy of nuclear DNA, isolated in deionized media	Trendelenburg, 1974 Trendelenburg et al., 1977
Colymbetes fuscus	RNA–DNA hybridization	Gall et al., 1969
	Electron microscopy of denatured molecules of rDNA	Gall and Rochaix, 1974
Cyrinus natator	Cytochemistry	Matuszewski and Hoser, 1975
Creophilus maxillosus	As above	Matuszewski and Kloc, 1976
	^{3}H-thymidine autoradiography	Kloc, 1980
Necrophorus hyumator	Cytochemistry	Matuszewski et al., 1977
Silpha sinuata	Cytochemistry	"
S. atrata	Cytochemistry	"
S. thoracica	Cytochemistry	"
Necrodes littorales	Cytochemistry	"

Note. For literature up to 1969 inclusive, see Gall, 1969.

that the polycomplexes appeared in the cytoplasm of gametocytes before the onset of chromosome conjugation and disappeared at the beginning of this process. These observations show that the SC material is obviously not only synthesized in the cytoplasm, but also self-assembles under certain conditions. Hardly anything is known about the functional role of abnormal SC. In a number of cases it has been shown that they are used as material for the construction of the complex karyosphere capsules (see Section 2.7).

2.4. EXTRACHROMOSOMAL NUCLEAR DNA OF OOCYTES

The synthesis of large amounts of ribosomal RNA in oocytes is ensured by a special mechanism of amplification of ribosomal cistrons. This phenomenon is widespread (see Table 2.1). For the

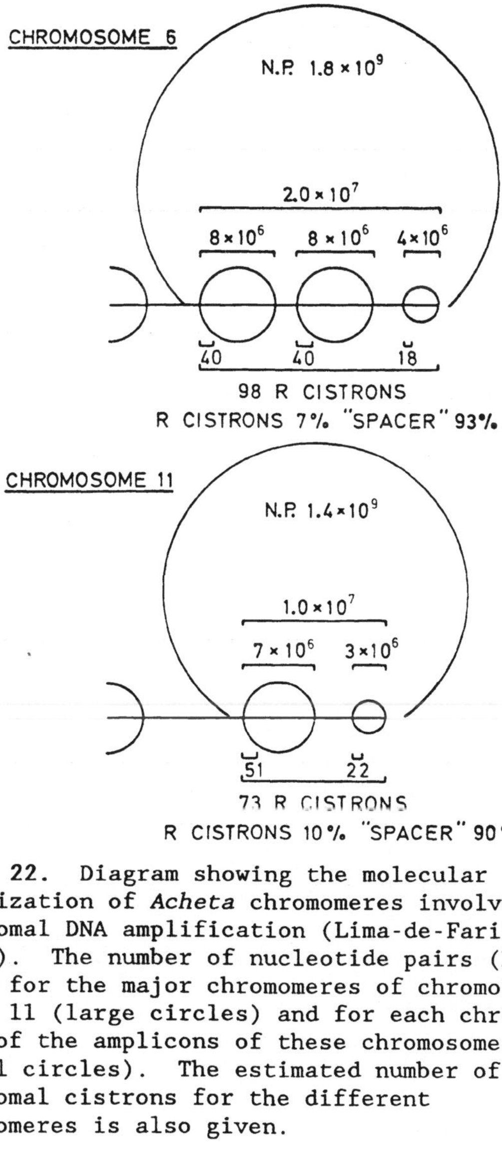

Fig. 22. Diagram showing the molecular organization of *Acheta* chromomeres involved in ribosomal DNA amplification (Lima-de-Faria, 1974a). The number of nucleotide pairs (NP) is given for the major chromomeres of chromosomes 6 and 11 (large circles) and for each chromomere of the amplicons of these chromosomes (small circles). The estimated number of ribosomal cistrons for the different chromomeres is also given.

oogenetic strategy and gene amplification in different animals, see also the review by Macgregor (1982). Amplification has been studied most thoroughly in the oocytes of insects and amphibians. The cricket *Acheta domesticus* is a convenient subject for the study of the morphological aspect of this phenomenon, since its oocytes have only eleven large bivalents (2n = 22). This made it possible to carry out a detailed analysis of the chromosomes on squash preparations (see Lima-de-Faria, 1974a). Five DNA bodies, which appear as a result of amplification, are observed on the chromosomes in the young oocytes. The excess rDNA copies in these bodies have been convincingly demonstrated by ^{3}H-rRNA hybridization with the DNA of pachytene chromosomes in situ (Cave, 1973; Ullman et al., 1973).

Detailed morphological and cytochemical analysis has also been made of the largest DNA bodies of chromosomes 6 and 11 (Fig. 22). The chromosomes' locus, in which amplification occurs, is called the "amplicon" (Lima-de-Faria, 1974a). It has been shown by chromomere analysis that the amplicon of chromosome 6 consists of three small chromomeres, and that of chromosome 11, of two chromomeres. The large size of the chromosomes made it possible to measure, by ultraspectrophotometry, the quantity of DNA not only in DNA bodies, but also in individual chromomeres; the quantity of DNA in the large body was 100 times that in the amplicon. It has also become possible, by combining biochemical analysis and ultraspectrophotometry, to calculate the number of ribosomal cistrons not only in the amplicons of chromosomes 6 and 11, but also in each chromomere of the amplicon of these chromosomes.

The number of ribosomal cistrons in somatic cells in which there is no amplification and in oocytes has been determined by preparative centrifugation in a CsCl gradient and molecular hybridization of nucleic acids (Lima-de-Faria, 1974a). The nuclear organizers of the chromosomes in the somatic cells contain 342 ribosomal cistrons, while there were four or five times more in the oocyte (1600 repeats). When rRNA was isolated, it was distributed in the zone of the satellite DNA (which differs from the main mass of nuclear DNA by its higher density) but forms only an insignificant part of it. According to data of various authors, the quantity of ribosomal DNA in the satellite fraction of DNA in *Acheta* varies considerably, apparently depending on the method of determination (1% according to Allen and Cave, 1973; 8% according to Lima-de-Faria, 1974a; 21% according to Trendelenburg et al., 1974). The nature of the remaining part of the satellite DNA is yet unknown. The cistrons coding for 18S- and 28S-rRNA are represented in rDNA in equal quantities (see review by Lima-de-Faria, 1974a).

All these data are in good agreement with the data on the quality and structure of ribosomal cistrons obtained on other species of insects and amphibians (see below).

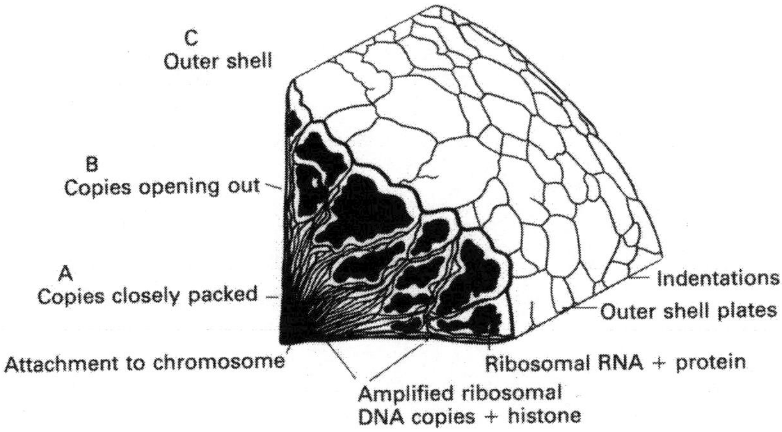

Fig. 23. Diagram to illustrate the three-dimensional organization of the amplified ribosomal DNA of *Acheta* complexed with ribosomal RNA and protein in the major chromomere of chromosome 6 at pachytene (Lima-de-Faria, 1974b).

As the oocyte grows and enters the diplotene, the DNA bodies are subject to "puffing" from the dense structure and they become transformed into looser formations. This process is accompanied by RNA synthesis in the peripheral region of the body. Thus, the DNA body (a giant chromomere) becomes a typical nucleolus (Fig. 23; Plate VIII, 2). At the same time, the previously synthesized DNA copies, together with the nucleolar RNP, separate from the chromosomes and free nucleoli appear in the nucleoplasm (Fig. 24). Each of the five loci where the amplification and formation of DNA bodies occurs behaves in a different way during oogenesis. The quantity of DNA in the giant chromomere of chromosome 6 remains constant throughout the pachytene stage, while in the large chromomere of chromosome 11 it has already become noticeably less by the late pachytene, and the three smaller bodies have already disappeared by the early pachytene. Electron microscopy combined with autoradiography showed that the extrachromosomal DNA, together with fragments of the nucleoli, enter the cytoplasm by intracytosis (Fig. 25) and are preserved there for a certain time. A question arises as to the significance and further fate of these DNP—RNP complexes and the possibility of amplification and transcription of this DNA in the cytoplasm.

The observations of Lima-de-Faria et al. led to several important conclusions: 1) The DNA bodies are the product of a certain locus of the chromosome, the amplicon, which consists of several chromomeres; in other words, there is a superchromomere level of

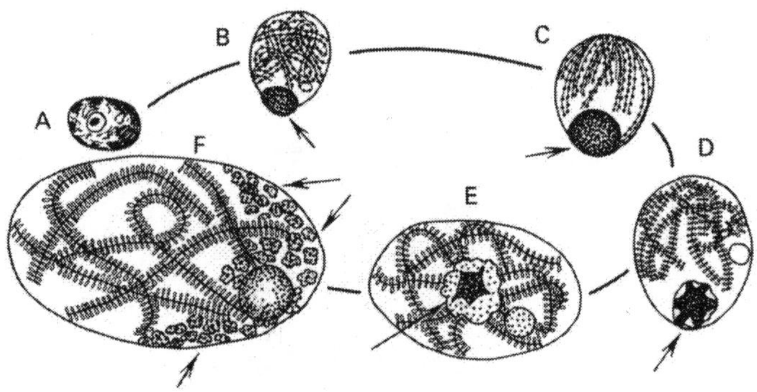

Fig. 24. Development during oogenesis of the primary nucleoli (arrows); these emerge from small lumps of extrachromosomal DNA (Allen and Cave, 1969). A. Oogonia. B, C. Pachytene. D-F. Diplotene. The secondary nucleolus is indicated by the light-colored circle.

organization of the chromosome. 2) When several DNA bodies emerge, each of them has its own individual fate during oogenesis; i.e., the amplified copies of each chromosomal locus are used at a certain time of the oocyte growth. 3) The separating copies in the form of small lumps of DNA with material from the nucleoli enter the cytoplasm and possibly continue to function there (Lima-de-Faria, 1974a, b).

Another very useful species for the study of rDNA amplification in oocytes is *Xenopus laevis* (for literature, see Table 2.1). In *Xenopus laevis*, as in all species of amphibians studied in this respect, pachytene oocytes predominate in the ovary at the stage immediately following metamorphosis. It has been shown that Feulgen-positive "caps" are formed in the nucleus of these oocytes (see Plate VIII, 1). This material rapidly increases, and it soon occupies more than half the volume of the nucleus. Several small nucleoli appear in the "cap." When the oocyte enters into the diplotene, the material of the "cap" becomes dispersed throughout the nucleus, and Feulgen-positive granules can also be seen in the early diplotene which are distributed more or less uniformly under the nuclear envelope; for some time the nucleolar granules do not have the heterochromatin granules connected with them. Microdensitometry has shown that, at this time, the nucleus of *Xenopus laevis* contains up to 42 pg of DNA, which is 14 times greater than the haploid quantity of DNA in this species. As the diplotene develops and the oocyte grows, a marginal nucleolus is formed in connection with each Feulgen-positive granule. At this time there is already a 20- to 30-fold excess of ribosomal genes in the large nuclei of the *Xenopus laevis* oocytes (Macgregor, 1972). At the end of oogenesis in amphibians, when the germinal vesicle breaks

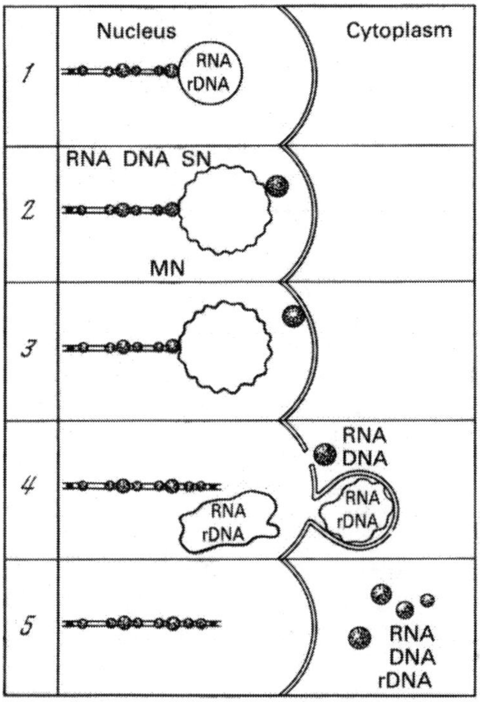

Fig. 25. Diagram illustrating the release of DNA, ribosomal DNA (rDNA), and RNA into the cytoplasm of the oocyte in *Acheta* (Lima-de-Faria, 1974a). Chromosome 6 is shown at every stage of diplotene (1-5). The major chromomere of chromosome 6 differentiates (at stage 2) into a main nucleolus (MN) and a secondary nucleolus (SN). Both move into the cytoplasm by disrupting the nuclear envelope or by carrying a piece of it into the cytoplasm (stage 4). From in situ hybridization studies it is known that the main nucleolus contains rDNA. No hybridization has so far been carried out with the secondary nucleolus, but it is thought to contain DNA.

down, the excess nucleolar DNA is discovered in the form of Feulgen-positive granules, which are preserved for a long time at the animal pole in the cytoplasm (Brown and Blackler, 1972).

The prime advances in studying amplification in amphibian oocytes relate to the molecular structure of ribosomal cistrons. It has been shown by preparative centrifugation and nucleic acid hybridization that rDNA isolated from the nucleoli consists of repeated sequences, each with a molecular weight of approximately $8 \cdot 10^6$ daltons (13,000 base pairs). Each rDNA repeat consists of two regions; one of them contains the nucleotide sequences coding for 18S- and 28S-rRNA (this region is transcribed into 40S-particles, the precursors of rRNA); the remaining, greater part of the ribosomal cistron consists of a spacer which has not been accounted for, and whose function is unclear. Having treated the oocyte nucleoli of newt in low ionic strength solutions, Miller and Beatty (1969) isolated nucleolar DNA and prepared electron microscope specimens, thus demonstrating that the DNA from the nucleoli is a ring-shaped molecule to which the synthesizing molecules of rRNA are attached.

In each ribosomal cistron there are about 100 rRNA molecules. The ring-like DNA molecules contain a different number of segments with a matrix, ranging from one to several dozen and even several hundred (Miller and Beatty, 1969). The areas covered by the matrix correspond to the transcribing genes of rRNA, while the bare areas correspond to the nontranscribing spacer. The 28S + 18S + spacer complex is shown below:

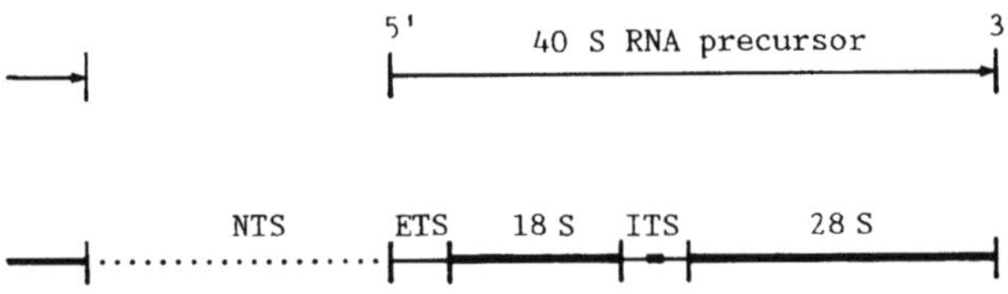

This diagram shows the arrangement of a transcribed and nontranscribed region of the ribosomal gene complex in *Xenopus laevis*. Here NTS is the nontranscribed spacer, ETS is the external transcribed spacer, and ITS is the internal transcribed spacer. One whole repeat extends from the left end of the NTS to the right end of the 28S gene (Macgregor, 1982). Miller's method was used to obtain ring-like rDNA molecules from the oocytes of *Acheta domesticus* and *Dytiscus marginalis* (Trendelenburg, 1974; Trendelenburg et al., 1977). Small ring-like molecules contained 3-5 ribosomal cistrons. They differed from one another in their very long, nontranscribed spacers (no less than half the circle of the molecule); they are twice the length of the rRNA molecules (18S + 28S).

2.4.1. The Amplification Mechanism

Despite the data suggesting that the copying of the chromosomal nucleolar organizer in the oocytes of amphibians involves the cascade-like reproduction of copies (Brown and Blackler, 1972), for a long time the real mechanism of amplification remained unclear. The main difficulty arose from the uneven number of units of amplification in the amplified rings of the nucleolar DNA of the newt *Triturus viridescens*, while the nucleolar organizer on the chromosome had 450 such units (Miller and Beatty, 1969). The largest rings were longer than the chromosomal nucleolar organizer, and the smallest were shorter. Several speculative hypotheses were proposed to explain the amplification mechanism (Brown and Tocchini-Valentini, 1972; Mahdavi and Crippa, 1972). The first direct observations to throw light on the mechanism of amplification were recorded in 1973-1974. The isolated, purified rDNA from pachytene oocytes of *Xenopus laevis* was spread on an aqueous surface and studied by electron microscopy. By measuring the overall length of the rings and by partial denaturation, it was shown that the ring-like molecules of rDNA contain one, two, three, four, or more copies of the main rDNA sequences. The latter observation, and also autoradiography, demonstrated that amplification occurs by a rolling circle mechanism (Hourcade et al., 1973) (Fig. 26). These observations have been confirmed by Bird et al. (1974) who showed, by electron autoradiography, that the "tails" of the rings do indeed represent the replicating molecules. Similar results were obtained using rDNA from the oocytes of two species of *Dytiscus* (Gall and Rochaix, 1974). A considerable percentage of the rDNA molecules were ring-like (in *Colymbetes*, 14%; in *Dytiscus*, 19%). But here "tailed" rings were rarely found. This may be explained by the fact that the rDNA at the moment of isolation in the imago beetle was not undergoing amplification (see below). If the amplification mechanism of the rolling-circle type does indeed take place, then some basic molecular problems of the first replication events are unresolved, such as formation of the first ring of rDNA and selection of certain ribosomal sequences for amplification (see review by Macgregor, 1982).

2.4.2. Time of Amplification

In frogs, the main synthesis of nucleolar DNA takes place in oocytes at the pachytene state, immediately after metamorphosis (Gall, 1968; Macgregor, 1972). However, by using nucleic acid hybridization, Gall and Pardue (1971) demonstrated that the nuclei of oogonia already bind 5-20 times more rRNA than the nuclei of somatic cells. Thus, amplification of rDNA commences in oogonia, continues at a low level until pachytene, and then increases sharply. rDNA synthesis ceases completely in the early diplotene. By injecting ^{3}H-thymidine into the tadpoles of *Xenopus laevis* just

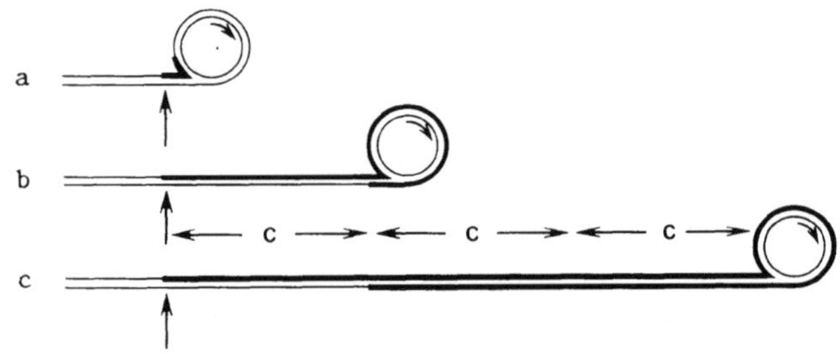

Fig. 26. Diagram illustrating the progressive stages in the labeling of a rolling circle (Bird et al., 1974). Vertical arrows mark the beginning of a pulse labeling of DNA with ^{3}H-thymidine. "c" is the circumference of the circle. a) At this point the fork has replicated only a small part of the circle since the beginning of the pulse. The two labeled strands lie close together, each labeled in one strand of the duplex. b) The fork has now copied slightly more than one circumference of the circle. Most DNA is labeled in one strand. However, since the start of the second circle, a small region of doubly labeled duplex has appeared. c) The fork has now traveled around the circle three times. Since the first cycle was completed, all the DNA appearing in the tail has been doubly labeled. The circle remains labeled in one strand, as does a section at the distal end of the tail equal in length to "c."

before metamorphosis, and by counting labeled nuclei in the gonads, it has been established that the intensive synthesis of amplified rDNA continues for 15 days, and the overall time of DNA synthesis during meiosis is 20 days (Watson-Coggins and Gall, 1971; Bird and Birnstiel, 1971). The primordial germ cells of *Xenopus laevis* do not show amplification of rDNA (Kalt and Gall, 1974). These authors showed that amplification of ribosomal cistrons begins simultaneously with the beginning of gonial mitoses and sexual differentiation. The amplification continues with gonial mitotic activity. On the average, the gonial cells undergo 4-9 rounds of divisions. From the beginning of meiotic prophase the germ cells lose the amplified copies of rDNA; there is complete loss of the copies in the spermatogonia, but only partial loss in the oogonia.

In other animals and, in particular, insects, the stages of oogenesis at which rDNA amplification takes place have not been studied in detail. Here, too, however, some observations (Giar-

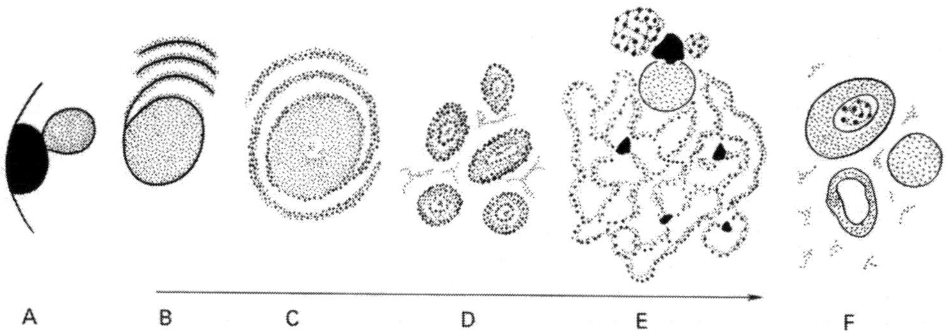

Fig. 27. Transformation of nucleoli during oogenesis in *Chrysopa perla*. A-D. Pre-vitellogenesis. E, F. Vitellogenesis. For full explanation, see text.

dina, 1901, cit. by Telfer, 1975) show that the chromatin body in *Dytiscus*, which contains amplified copies of ribosomal genes (Gall et al., 1969), is already formed in the mitotically dividing oogonia (see Fig. 15).

The process of rDNA amplification is, at least in some insects, not restricted to the early stages of oogenesis, and may occur during the later growth phase of the oocyte. As already mentioned, there are extrachromosomal DNA bodies in the young oocytes of *Chrysopa perla* (see Plate VIII, 3, 4). In situ hybridization of nucleic acids has shown that these bodies contain numerous copies of rRNA genes (Gaginskaya and Gruzova, 1975) (see Plate VIII, 5). But amplification is evidently already completed in the early meiotic oocytes of imago; here the DNA bodies are not labeled by ^{3}H-thymidine. Apparently there is no DNA synthesis in the pachytene oocytes of beetle imago (*Dytiscus*, *Colymbetes*), which contain large DNA bodies with amplified nucleolar genes (Gall and Rochaix, 1974). Obviously, in insects with complete metamorphosis the amplification of rRNA genes is terminated in ovarioles of the pupae. However, in *Chrysopa perla* imago a noticeable DNA synthesis is observed in the nucleus of the growing diplotene oocytes, right up to the stage of late vitellogenesis (see Plate VIII, 7) (Gruzova et al., 1972b). In this period, extrachromosomal DNA is dispersed throughout the nucleus. The appearance of the ^{3}H-thymidine label in the nucleus coincides with the onset of the transformation of the ring-like nucleoli into giant nucleolonemas (Fig. 27). The reserves of nucleolar DNA amplified at the early stages of oogenesis are apparently only sufficient for the synthesis of several dozen ring-like nucleoli. Obviously, a second round of ribosomal DNA is required for the further increase in the nucleolar material in the oocytes of *Chrysopa perla*. Similar data on extra-DNA synthesis in the diplotene oocytes have also been obtained in other insects (Magakian and Matuszewski, personal communications).

Obviously, the amplification of rRNA genes is extremely common in oogenesis. The question arises: Is this phenomenon specific to the ribosomal cistrons of oocytes, or does it occur generally? Some information is available on the synthesis of extra-DNA in somatic cells (Da Cunha et al., 1969; Roberts, 1975; Scali and Montanelli, 1975), but in these cases the genetic nature of the amplified chromosome loci still remains unclear. It may be suggested that the mechanism of amplification which intensifies the work of individual chromosomal loci is preserved during oogenesis not only due to its ability to meet the oocyte's greater need for ribosome synthesis, but also in connection with the fact that the amplified genes, which are breaking away from the chromosomes, do not prevent the meiotic chromosomal transformations (Lima-de-Faria, 1974b).

2.5. OOCYTE NUCLEOLI

The formation of numerous nucleoli not connected with chromosomes is a feature of the oocytes of many animals as distinct from somatic cells. Brown and Ris (1959) showed in amphibians that the nucleoli of the oocytes do not differ in principle from those of the somatic cells, whose main function is rRNA synthesis. The nucleoli of the oocytes and, in particular, those of amphibians, are extremely suitable for the study of their chemical and structural organization due to their large size and number (500 to more than 1000). Relevant data are given in a number of detailed reviews (Macgregor, 1972, 1982; Chentsov and Polyakov, 1974; Smetana and Busch, 1974). The study of oocyte nucleoli led to the isolation and visualization of nucleolar genes (Miller and Beatty, 1969) and the discovery of the phenomenon of rRNA amplification (see review by Gall, 1969).

During oogenesis, in addition to the true nucleoli, various nucleoli-like formations (see Section 2.6) usually appear. It is therefore appropriate here to note the characteristics of true nucleoli: these are structures rich in, and actively synthesizing, rRNA. They thus must contain fibrillar and granular components. Granules (150-200 Å) contain 45S RNA and reflect the formation of proribosomal particles in the nucleolus. The nucleoli of germ and somatic cells also contain a fibrillar center (see review of Goessens and Lepoint, 1979; Fakan and Hernandez-Verdun, 1986). This component of the nucleolus constitutes part of the nucleolus-organizer region (NOR) formed when the nucleolus is connected with the chromosome, or the rDNA body in case of an extrachromosomal nucleolus. The fibrillar center consists of protein fibrils (20-30 Å thick) and DNA strands. High resolution autoradiography (using

^{3}H-uridine) demonstrates that rDNA transcription takes place at the border between the fibrillar center and the dense fibrillar area; subsequently the label transfers to the granular area (Mirré and Stahl, 1978). It is well established that rDNA transcription ("Christmas tree" formation) takes place in the dense fibrillar region, whereas the fibrillar center contains nontranscribed rDNA. It should be emphasized that the granules are characteristic of the nucleoli when they are actively functioning; in periods of initiation or abatement of their activity, there are usually fibrillar structures either not synthesizing or slowly synthesizing RNA.

Different aspects of structural–functional relationships in the nucleolus of somatic and germ cells have been actively discussed recently. Attention was paid mainly to the following topics:

1) A skeletal meshwork (nucleolar matrix) specific to the nucleolus (see Scheer et al., 1982).
2) Proteins of NORs.
3) Organization, distribution, and transcriptional activity of rDNA and its relation to proteins of NOR.
4) NOR and fibrillar centers during gametogenesis.

One of the useful tools of the investigation of all these questions is the silver-staining procedure (Howell and Black, 1980) revealing proteins bound to rDNA (Ag–NOR-proteins). Ag-proteins have been repeatedly observed during mitosis and meiosis, as well as in both the fibrillar centers and their peripheral dense fibrillar components. Although their exact biochemical nature is still controversial, it appears that they are probably regulatory proteins of rDNA transcription. The argentophilia of the NOR proteins appears to be a preferential cytochemical marker of both the rDNA and the level of actual or potential transcription of the rDNA (see a review by Fakan and Hernandez-Verdun, 1986; Ploton et al., 1986).

Nucleoli also contain larger granules (350-500 Å) of the perichromatin type (PG) (see Section 2.6). The presence of a small number of such granules has been described in the nucleoli of the oocytes of *Rana ridibunda* (Parfenov, 1979). Great aggregates of very large granules (500-700 Å) appear in connection with nucleolar chromatin in the marginal nucleoli of the trophocyte of the beetle *Blaps lethifera* (Gruzova, 1979a, 1985) (for details, see Section 2.2). Perichromatin granules have also been described in the nucleoli of somatic cells both under experimental conditions (Puvion et al., 1977; Moyne et al., 1977) and in normal organisms (Deltour et al., 1979). Experiments with inhibitors to retard the processing of RNA in the nucleolus confirm the assumption that PG formation is associated not with mRNA synthesis, but with the slowing down of the process and abnormal rRNA storage (Puvion et al., 1981).

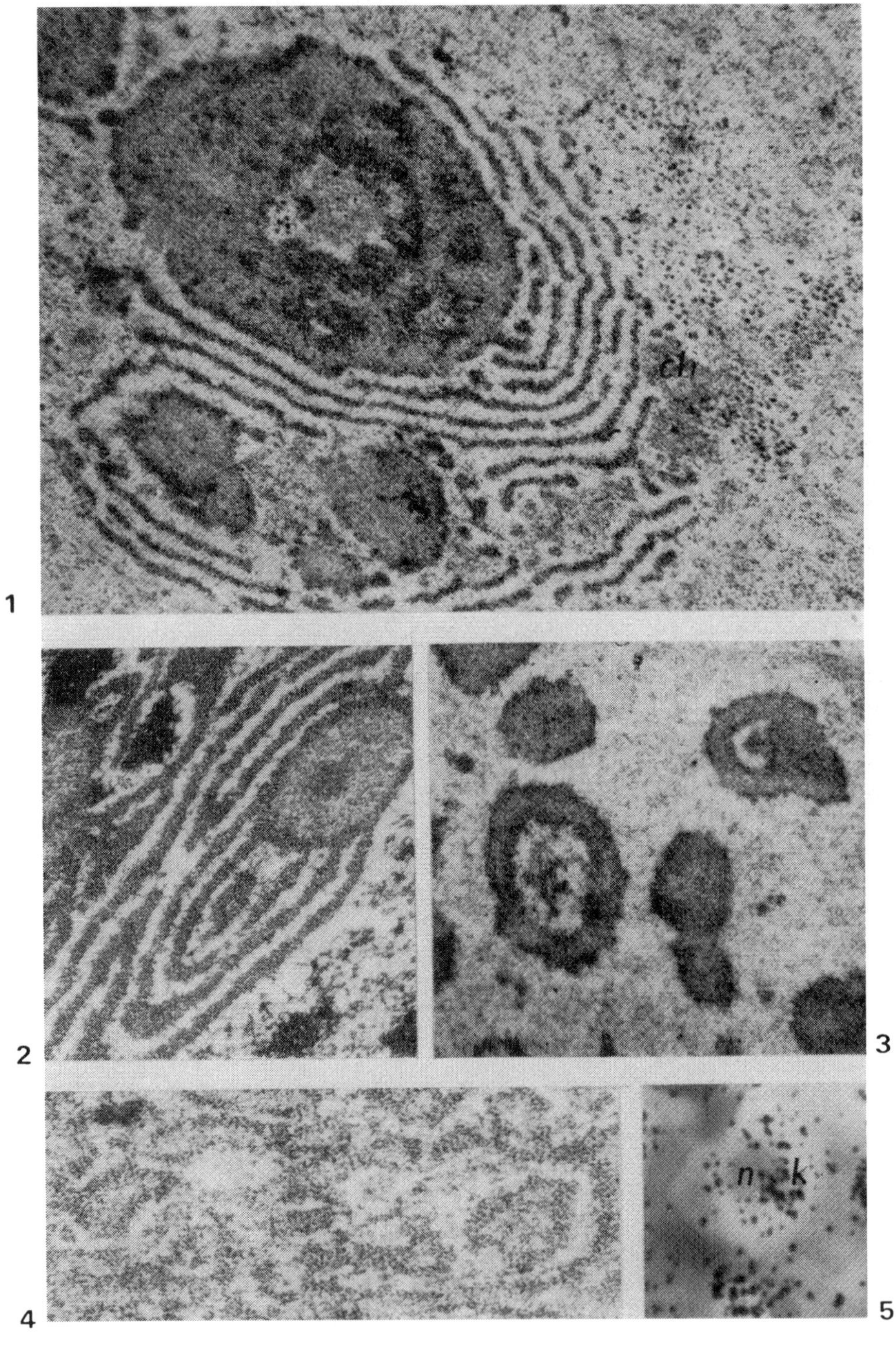
1
ch
2
3
n
k
4
5

During oogenesis, along with large nucleoli, numerous micronucleoli are formed in the germinal vesicle. However, these have not been studied in detail (see Section 2.6).

The morphology of oocyte nucleoli is extremely diverse. Single large nucleoli are found in the oocytes of many invertebrates as well as mammals. The nucleolus of the oocytes of sea urchins and some worms (Bal et al., 1969) are good examples. These nucleoli never lose their link with the chromosomes. In the early stages of oogenesis, heterochromatin, which incorporates ^{3}H-thymidine, appears around them (Bal et al., 1968).

In the oocytes of *Urechis* and *Arbacia* rDNA amplification leads to the emergence of numerous micronucleoli in the nucleus. It has been shown by nucleic acid hybridization that the nucleus of the *Urechis caupo* oocytes contains amplified rDNA, in an amount 24 times that in the spermatozoa (Dawid and Brown, 1970). In the oocytes of the starfish *Asterias forbesii* there is one large nucleolus and amplification of the rDNA has not been observed (Vincent et al., 1969); the single nucleolus suggests that there is no excess rDNA in the germinal vesicle.

Distinct, often single, nucleoli, the so-called amphinucleoli (Jorgensen, 1913), have been found in the oogenesis of molluscs and some crustaceans. Appearing at the onset of oogenesis as dense homogeneous structures intensely synthesizing DNA, by the middle or end of oogenesis these nucleoli become double structures; one part contains primarily protein and the other RNA (Romanova and Gazaryan, 1966; Bottke, 1973a, b). It is not clear whether the amphinucleoli are a striking example of the natural segregation of the nucleolar material as synthetic activity ceases (Romanova, 1974) or whether they are the result of the combination of a protein

---

Plate X. The nucleoli of the oocytes and trophocytes of *Chrysopa perla* (*Orthoptera*). 1. Complex nucleolus of the trophocyte (Zaichikova and Gruzova, 1975); the fibrillar core surrounded by concentric circles of granular strands and the nucleolar chromatin (ch) with perichromatin granules can be seen. 2. Part of the complex nucleolus from the trophocyte at higher magnification (Zaichikova and Gruzova, 1975); granules can be seen in the strands. 3. Ring-shaped nucleoli in the nucleus of a previtellogenic oocyte (Gruzova and Zaichikova, 1972). 4. Part of the oocyte nucleus filled with granular nucleolonema (Gruzova et al., 1972). 5. Nucleus of the oocyte after "staining" with ^{3}H-actinomycin D; the karyosphere (k) is labeled: tracks can be seen on the periphery of the nucleolus (n), and the rest of the label is connected with the extra-DNA dispersed throughout the nucleus.

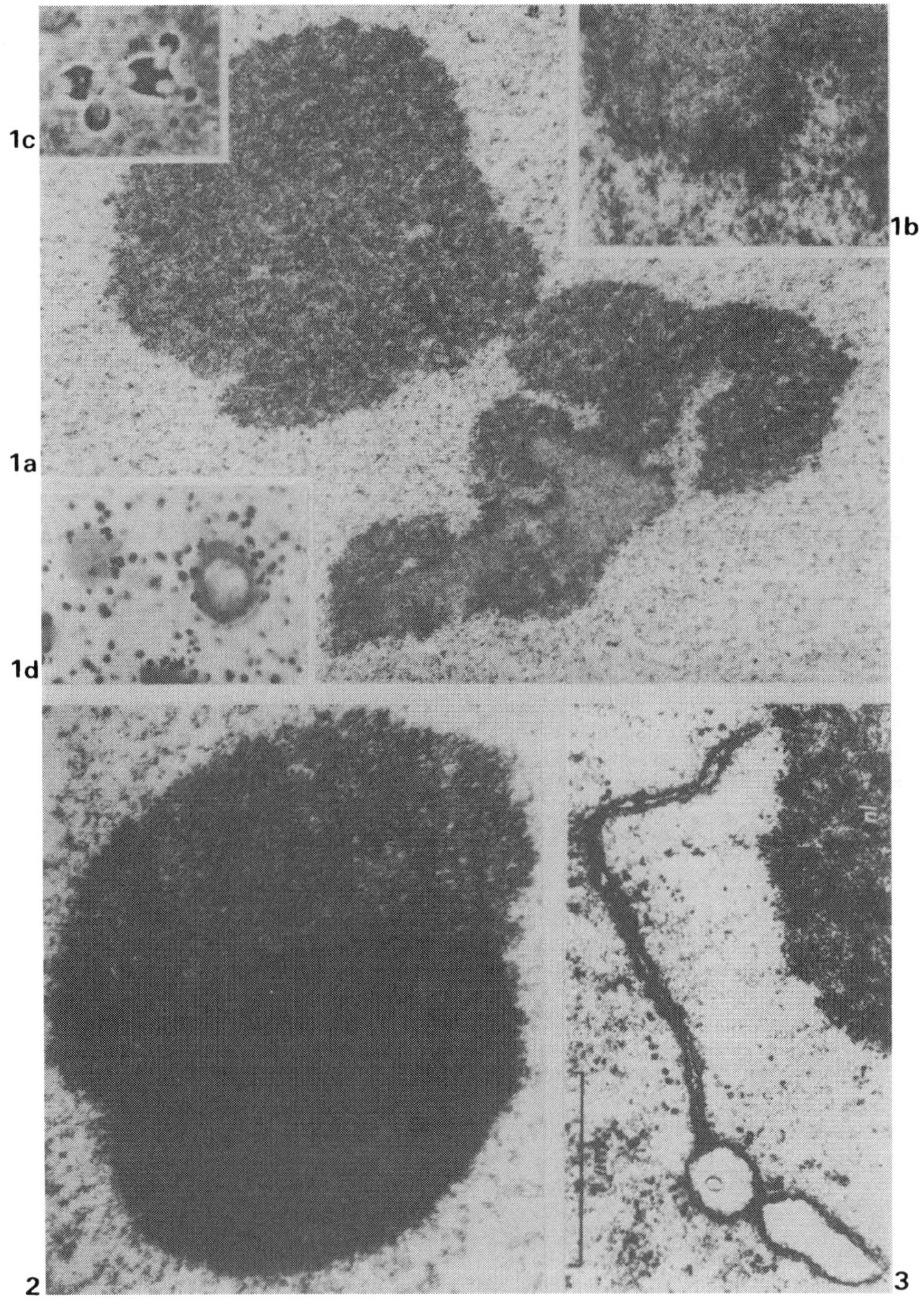
1c
1b
1a
1d
2
3

body (Binnenkörper type) and the true nucleolus (Bottke, 1973a). Numerous marginal nucleoli develop in the oocytes of fish and amphibians (Raikova, 1976; for review, see Macgregor, 1972) and also in a number of invertebrates: in some molluscs (Kubinskii, 1974), holothurians (Nizovskaya and Arronet, 1975), crabs (Pakhomova, 1976), and ascidians (Kessel, 1966a, b). The marginal nucleoli in the oocytes of invertebrates usually closely adjoin the nuclear membrane and whole marginal nucleoli or their fragments may be transported to the cytoplasm (Kessel, 1966a, b). It has been shown that the marginal nucleoli in oogenesis in fish and amphibians, and also numerous nucleoli in the oocytes of a number of insects, are the result of rDNA amplification (for literature, see Table 2.1).

In many insects complex nucleoli with an extensive network of nucleolonemas (Bier et al., 1967; Kato, 1968; Dapples and King, 1970; Gruzova et al., 1972a, b; Gruzova, 1974) develop both in the oocytes (where the ovarioles have a panoistic structure) and in trophocytes (where the ovarioles are of the polytrophic type). The nucleolonema, frequently in the form of concentric layers, branches out from the fibrillar center of the nucleolus (Plate X, 1, 2).

The form of the nucleoli is rather specific. The configuration of the nucleolus as a whole, and the distribution of its fibrillar and granular components, depends on the location and degree of amplification of the nucleolar DNA (Lane, 1967; Allen and Cave, 1972). The oocytes in insects serve as a vivid example of the morphological differences in the nucleoli in closely related species. Thus, nucleoli of different types (Cave and Allen, 1974) are described for the representatives of three subfamilies of the *Gryllidae* family. An extrachromosomal DNA body was discovered in the pachytene oocytes of all representatives of this family. However, numerous small nucleoli derived from this DNA are found in

---

Plate XI. Nucleoli in the oocytes of amphibia. 1. Nucleoli in late oocytes of *Rana temporaria* (Gruzova and Parfenov, 1976): a) a fibrillar nucleolus; the segregation of the material into the fibrillar center and the electron dense cortex can be seen; b) a magnified part of the nucleolus with the fibrillar center and cortex; c) a part of the oocyte nucleus "stained" with ^{3}H-actinomycin D: the label can be seen not only over the nucleoli, but between them as well; d) fragmentation of the nucleolus, the total preparation of the isolated nucleus (phase contrast). 2. A double fibrillar micronucleolus of *Rana temporaria* (Gruzova and Parfenov, 1976). 3. Part of the nucleolus of *Plethodon cinereus* with a "nucleolar tail" in the form of a membranous tubular structure with two vesicles (Kezer et al., 1971).

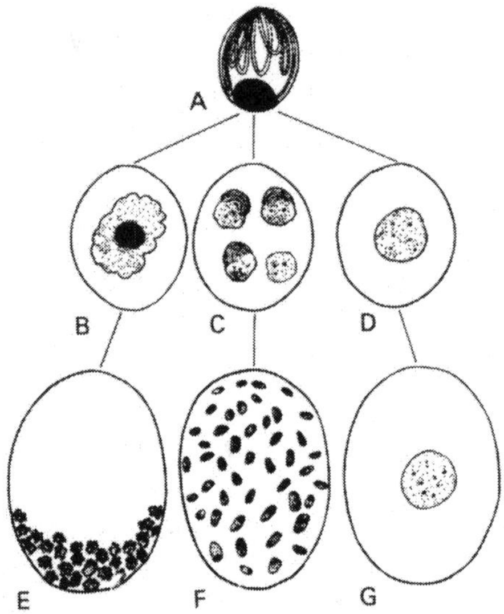

Fig. 28. Diagram showing the relationship of extrachromosomal DNA to nucleolar structure in oocytes of representatives from three different subfamilies of crickets (Cave and Allen, 1974). A, B, E. *Acheta domesticus*, subfamily Gryllinae. A, C, F. *Oecanthus nigricornis*, *O. celerincticus*, *O. quadripunctatus*, subfamily Oecanthinae. A, D, G. *Scapteriscus acletus*, subfamily Gryllotalpinae. The DNA body is a regularly occurring feature within the nucleus of early prophase oocytes, when the chromosomes form a "bouquet"-type arrangement. On entering diplotene, the chromosomes (not shown) take on a lampbrush appearance and are not visualized by ordinary cytochemical procedures. Nucleolar material is represented by the stipple, extrachromosomal DNA in solid black. A. Pachytene stage oocyte. B-D. Early diplotene stage oocytes. E-G. Late diplotene stage oocytes.

the nuclei of the late oocytes in the representatives of two subfamilies (*Gryllinae* and *Oecanthinae*), whereas in the *Gryllotalpinae* one large nucleolus containing amplified DNA (Fig. 28) was preserved throughout oogenesis.

Membrane components associated with nucleoli (Plate XI, 3) have been discovered in the oocytes of some animals (Kessel and Beams, 1968; Kezer et al., 1971) (for more detail, see below). However, this structural peculiarity is apparently not specific to the germ cells, since membrane formations are sometimes observed in the nucleoli of somatic cancerous cells (Babai et al., 1969).

Throughout oogenesis the nucleoli of the developing oocytes undergo extremely complex morphological and functional transformations. Two examples selected from invertebrates and vertebrates are given.

As already mentioned, complex nucleoli develop in the oocytes of many insects with panoistic ovarioles (Bier et al., 1967). In insects with meroistic ovarioles similar nucleoli are formed more often than not in the trophocytes (Dapples and King, 1970; Gruzova, 1974). As an exception to the rule, in the polytrophic ovarioles of *Chrysopa perla* complex nucleoli develop simultaneously, both in the trophocytes and in the oocytes (see Plate X, 1-4). The emergence and destiny of the complex nucleoli in the oocytes of this insect has been traced in detail, and is depicted in Fig. 27. At the early diplotene stage several fibrillar pronucleolar bodies appear in the nuclei of this insect, in contact with the small lumps of extra-DNA (see Fig. 27a; Plate VIII, 4). The pronucleoli do not incorporate ^{3}H-uridine. Fibrillar tubular strands (see Fig. 27b; Plate VII, 3, 4) soon begin to detach from their surface, which contains nucleolar DNA (see Plate X, 5). The strands are abnormal synaptonemal complexes (see Fig. 23, 5; Plate VII, 3, 4) and take part in the formation of the karyosphere capsule (see Plate VII, 3). After the beginning of RNA synthesis 1-2 large nucleoli become replaced with a multitude of small ring-shaped nucleoli. This transformation of nucleoli coincides with the "scattering" of extra-DNA throughout the nucleus. The ring-shaped nucleoli also take part in producing granular—fibrillar strands and gradually disappear, while the nucleus becomes filled with a gigantic network of granular strands – the nucleolonema (see Fig. 27d, Plate X, 4).*

It should be noted that in *Chrysopa* the ring-shaped nucleoli are intermediate, forming during the replacement of weak nucleolar activity by intensive RNA synthesis. A similar phenomenon is observed in somatic cells, during the experimental inhibition of rRNA synthesis. However, the ring-shaped nucleoli appear here when the highly active state gives way to the weakly active state (Smetana and Likovsky, 1978, 1980; Raska and Smetana, 1978; Raska et al., 1980). By the end of vitellogenesis the nucleolonema disappear from the nucleus, and large nucleolus-like protein bodies appear around the karyosphere; they consist of fibrillar material

*Details of this transformation of the nucleoli remain unclear. Studying a similar process of transformation in the oocytes of *Acheta* and *Dytiscus*, Trendelenburg et al. (1977) suggest that the emergence of the numerous circular nucleoli is due to the disaggregation of the small circular rDNA molecules, which compose the extra-DNA body.

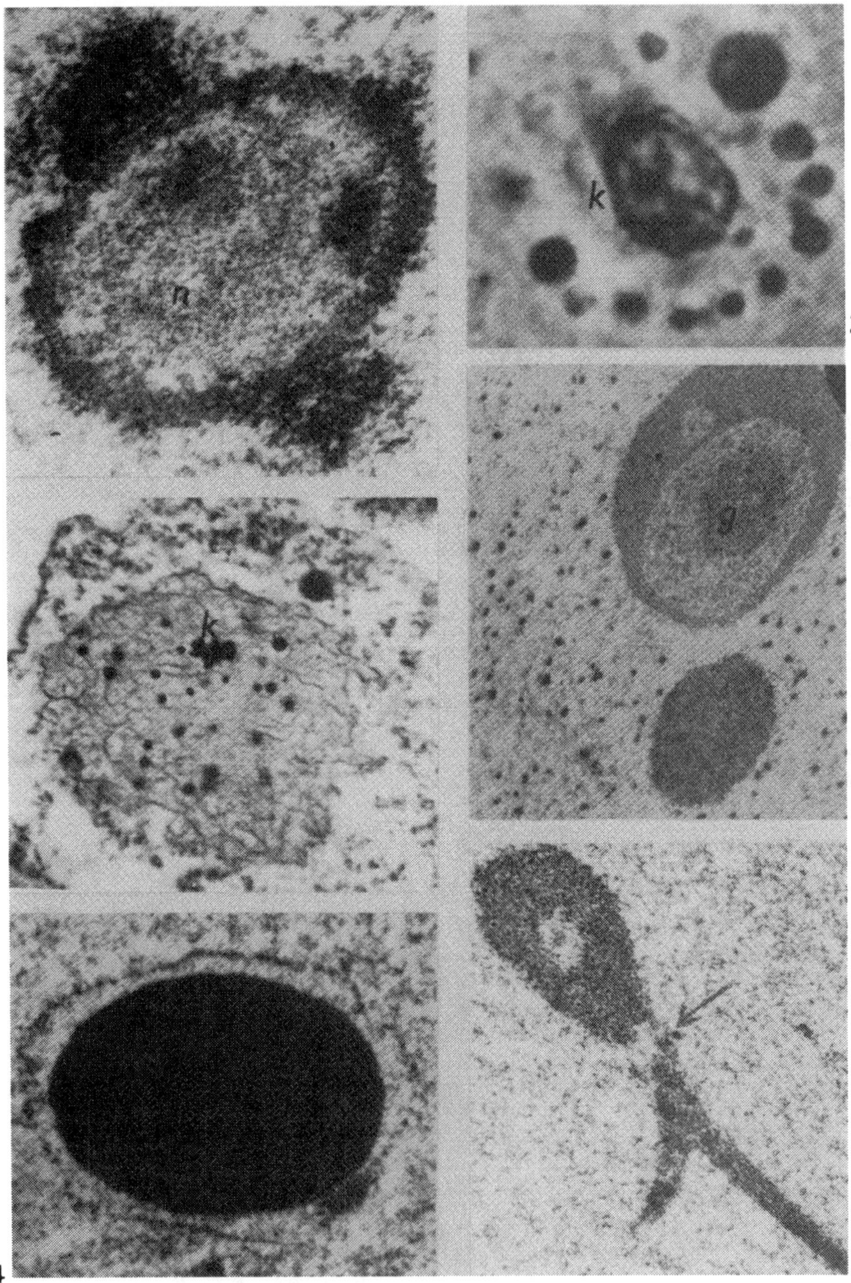

1 2a 2b 3 4 5

similar to that of karyosphere capsules and do not have proribosomal particles. Sometimes there is an accumulation of granules 500 Å in diameter (see Fig. 27e) (Plate XII, 2) in the center of the circular structure. These formations are no longer reminiscent of nucleoli, but rather of nuclear bodies (see Section 2.6).

During oogenesis in amphibians (Callan, 1966; Lane, 1967) and in fish (Raikova, 1976; Raikova et al., 1979) the nucleoli undergo a complex transformation; these authors demonstrate that the numerous marginal nucleoli, which emerge on the granules of extra-DNA, are dense fibrillar formations. From the onset of RNA synthesis in the nucleoli and the appearance of proribosomal granules, the nucleoli begin to change their shape and are transformed into a circular structure resembling a necklace of beads (Fig. 29). Changes in the distribution of the granular component, and the incorporation of ^{3}H-uridine, indicate the redistribution of the nucleolar DNA which is thus responsible for the morphology of the nucleolus. In maturing newt and salamander oocytes the nucleoli again become compact fibrillar granular structures. At this time, the nucleoli are already detached from the nuclear envelope and gather in the center of the nucleus around the knot of the chromosomes, forming the karyosphere capsule. In some amphibians, in particular *Rana temporaria* and *R. ridibunda*, the change does not stop there. In late vitellogenic oocytes the nucleoli which have already achieved a compact round shape and are part of the karyosphere capsule (Plate XIII, 1) unfold and assume a complex coiled shape (see Plate XI, 1a, d). They no longer have ribosome-like granules, and consist of fibrillar material, which is intensely fragmented (see Plate XI, 1d); segregation of the fibrillar center and the dense fibrillar component occurs (see Plate XI, 1a, b). During this period, at the end of oogenesis, the nucleoli do not synthesize RNA. Electron microscopy and the "staining" of the

---

Plate XII. Nuclear bodies. 1. Nuclear body in the nucleus of a late oocyte of *Rana temporaria* (Gruzova and Parfenov, 1976); the nucleolar fibrillar material (n) is enclosed in a capsule consisting of fibrils 150 Å thick. 2. Nuclear bodies in the nucleus of the late oocyte of *Chrysopa perla* (Gruzova et al., 1972): a) nuclear bodies gathered around the karyosphere (k) (araldite, semithin section); b) nuclear body at higher magnification; accumulation of granules (g) 500 Å in diameter is surrounded by a fibrillar capsule. 3-5. Nuclear structures in the oocytes of *Blaps lethifera*. 3) Thread-like zone around the karyosphere (k); nucleolus-like bodies (nlb) can be seen among the threads; 4) nlb surrounded by the fibrillar capsule; 5) part of the strand of the karyosphere capsule (below) and the granular nlb; the strand consists of a bundle of longitudinally oriented fibrils linked (arrow) with the granular body.

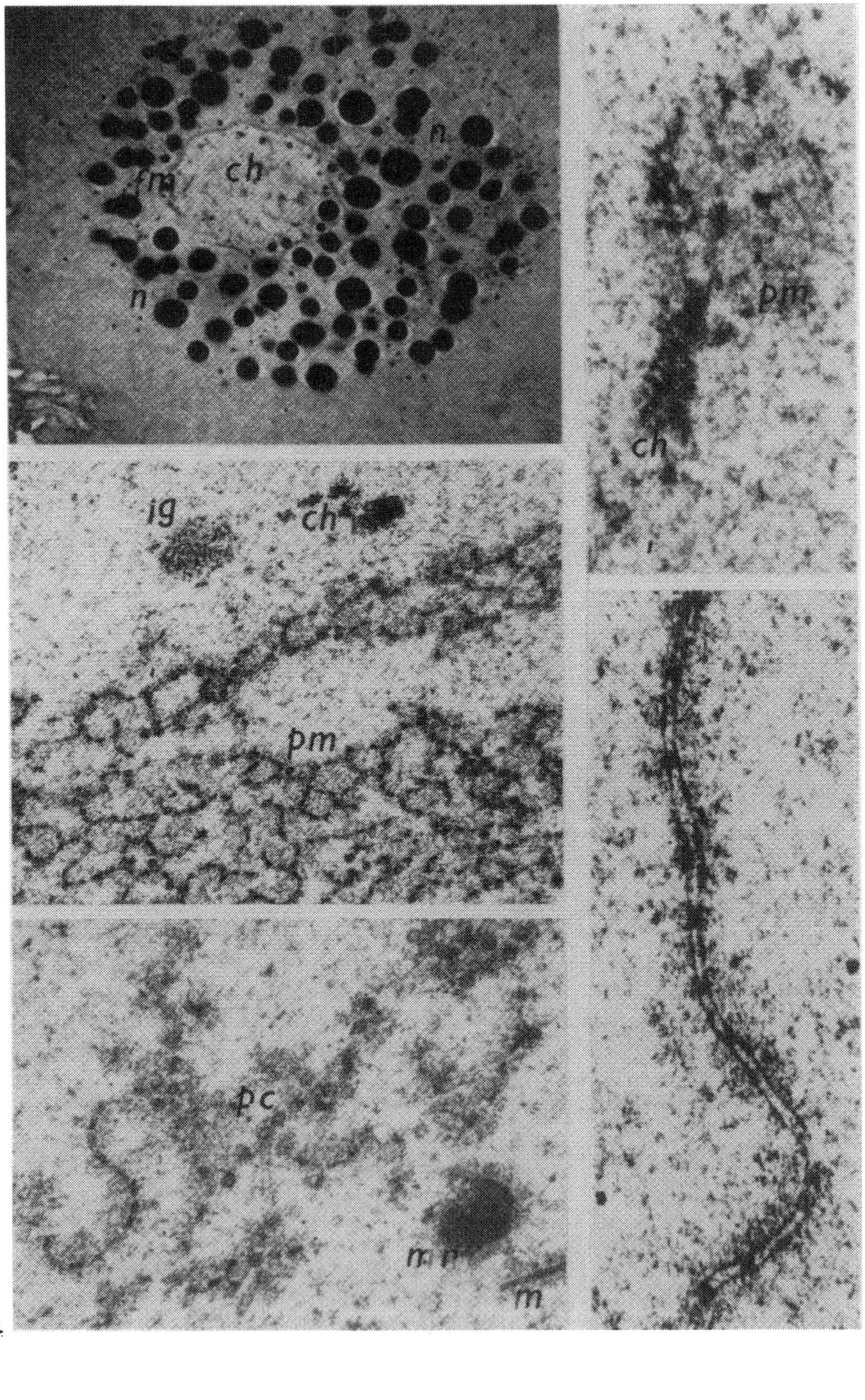

1 2 3 4 5

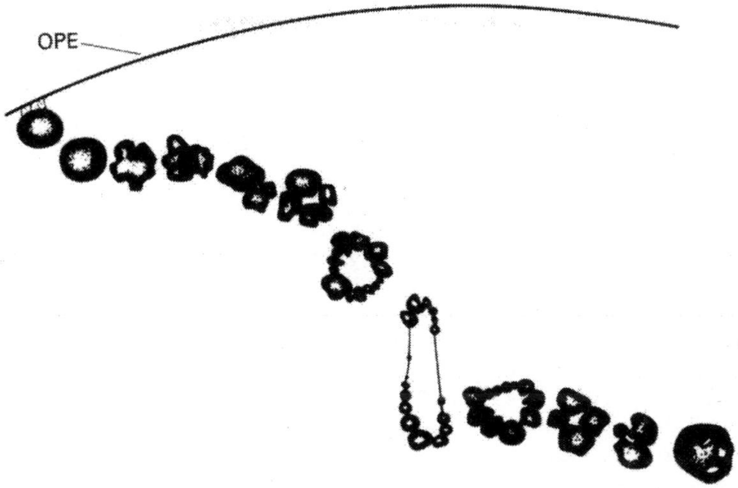

Fig. 29. Changes in the form of nucleoli during oogenesis in *Triturus viridescens* (Lane, 1967).

preparations with ^{3}H-actinomycin D (see Plate XI, 1c) suggest that during this period the nucleoli lose a considerable quantity of DNA, which is found in the nucleoplasm within the numerous micronucleoli (see Plate XI, 1c, 2) and nuclear bodies. Before the destruction of the nucleus, nucleoli fuse to form a massive conglomerate around the chromosomes (on the participation of the nucleolar apparatus in the morphogenesis of the karyosphere capsule; see Section 2.7).

In general, similar transformations of the nucleoli have been observed in the oogenesis of *Xenopus laevis* (van Gansen and Schram, 1972). Parallel to the change in the morphology, two phases are

---

Plate XIII. The karyosphere capsule in oocytes of *Rana temporaria* (Gruzova and Parfenov, 1976). 1. General view of the nucleus of a late oocyte: the centrally located chromosomes (ch) are surrounded by fibrous material (fm) and nucleoli (n). 2. Accumulation of pseudomembranes (pm) from the central zone of the karyosphere capsule: chromatin (ch) and interchromatin granules (ig) can be seen. 3. Chromatin from the central region of the karyosphere in close contact with the pseudomembranes (pm). 4. Accumulation of pseudomembranes in the middle zone of the karyosphere capsule; the pore complexes (pc), micronucleoli (mn), and a section of the membrane (m) can be seen; note the fibrillar matrix around the pseudomembranes, the micronucleoli, and the membrane section. 5. Part of the intranuclear annulate lamellae from the peripheral region of the karyosphere capsule.

noted in the active RNA synthesis in the nucleoli; these coincide with the enrichment of the cytoplasm in ribosomes and the intensive growth of the oocyte.

2.6. NUCLEOLUS-LIKE AND NUCLEAR BODIES

In the nuclei of growing oocytes both in invertebrates and vertebrates, apart from the nucleolus there is, as a rule, one or several nucleolus-like bodies. Cytochemical analysis shows that these bodies contain acid proteins and some RNA (Bier et al., 1967; Gruzova, 1971; Arronet, 1975). They differ from the typical nucleoli by having a low content of RNA, the absence or a low level of incorporation of labeled RNA precursors, and the absence of pro-ribosomal granules. Very different terms are used to designate these nuclear structures (prenucleoli, secondary nucleoli, pronucleolar bodies, protein bodies, protein spheres, karyosphere nucleoli, nucleolus-like formations, and others). They usually appear on the chromosomes, rarely in contact with the nucleolus. Hardly anything is known about the role of these structures in the germ cells.

Purely on the basis of their external similarity to nucleoli, we call all these nuclear structures "nucleolus-like formations" or "bodies" (NLB). However, a more detailed examination of the NLB in the oogenesis of various animals shows them to be neither structurally nor functionally identical to nucleoli. We have singled out several classes of NLB on the basis of their structure and origin.

1. NLB appearing on the chromosomes in nucleolus organizer regions (NOR). They contain rRNA, yet differ from real nucleoli by their low RNA metabolism. Thus, the oocytes of *Urechis* (Das and Alfert, 1973) show numerous NLBs in contact with the chromosomes. They differ from true nucleoli by a low RNA content, their fibrillar structure, and poor uptake of RNA precursors. Biochemical analysis of RNA isolated from the fraction of these structures indicates that the NLBs contain small quantities of rRNA precursors which are characteristic of true nucleoli. The authors suggest that the accumulation of NLBs on the chromosomes may be the result of insufficiently active functioning of the nucleolar organizer and compare them to the pronucleolar bodies, which appear in early normal oogenesis before the onset of the nucleolar organizer function in rRNA synthesis (Karasaki, 1965; Kiknadze and Belyaeva, 1965; Jacob, 1969) or in mutant embryos of *Xenopus* (Hay and Gurdon, 1967). This assumption was confirmed by Miller and Gonzales (1976), who studied the embryos of the partial mutant in regard to NOR. They demonstrated that the structure of the nucleolus (segregation and distribution of granular and fibrillar components) and the number of fibrillar NLBs depend on the intensity of NOR function and rRNA synthesis.

Evidently, a similar situation occurs in the alimentary type of oogenesis, when the nucleolus in the oocytes either does not develop or rapidly degenerates. Some authors suggest that this is a result of the competitive relation between the nucleolar organizers of the oocytes and the trophocytes (Dapples and King, 1970). In fact, in the oocytes of many insects and other invertebrates, where there are actively functioning nucleoli in the trophocytes, numerous NLBs appear on the chromosomes instead of the true nucleoli (Bier et al., 1967; Gruzova, 1971). These structures appear to be accumulating nucleolar proteins in the absence of intensive rRNA synthesis. It was found that the dense fibrillar NLBs appearing on the chromosomes in the karyosphere in the oocytes of the beetle *Blaps lethifera* and the butterfly *Laspeyresia pomonella* contain argentophilic material, indicating that they contain proteins associated with rDNA (Gruzova and Mamaeva, 1986).

The same class of structures apparently includes the numerous micronucleoli which appear in the oocytes during the diplotene. rDNA-Containing micronucleoli have been found and described in the oogenesis of mammals (Wolgemuth-Jarashow, 1977). They contain ribosome-like granules and large perichromatin-like granules (Chouinard, 1973, 1975; Zybina, 1975b; Zybina and Grishchenko, 1977). In amphibian oogenesis the micronucleoli appear at the end of vitellogenesis as a result of segregation and fragmentation of large nucleoli (Van Gansen and Shram, 1972; Gruzova and Parfenov, 1977; Parfenov, 1979). In *Rana ridibunda* the convoluted and ring-shaped micronucleoli consist of fibrils similar to the dense fibrillar component of the nucleoli and contain DNA (Parfenov and Gruzova, 1982). As already mentioned, the function of the NLBs and the micronucleoli is unclear. In some cases it has been demonstrated that the fibrillar NLBs and micronucleoli are involved in intranuclear morphogenetic processes, namely in the formation of the material of the karyosphere capsule (Gruzova et al., 1972b; Gruzova and Parfenov, 1977; Gruzova, 1979a, b; Parfenov and Gruzova, 1982) (see Section 2.7).

According to Halkka and Halkka (1968), the secondary fibrillar nucleoli of the dragonfly take part in the metabolism of the low-molecular-weight RNA.

2. Another NLB class is characterized by the presence of large perichromatin-like granules. For example, the secondary nucleolus of the cricket (Plate XIV, 4) contains, in addition to the fibrillar protein component, a chromatin lump and granules 500 Å in diameter (Jaworska and Lima-de-Faria, 1973a, b). The authors suggest that the secondary nucleolus of the cricket participates in the processing of the ribosomal particles. Similar structures have been described in the oocytes of other insects (Büning, 1972; Cassidy and King, 1972; Gruzova et al., 1972b). The ratio and packing of the granular and fibrillar components of these structures appear to vary greatly. In the tsetse fly the accumulation of

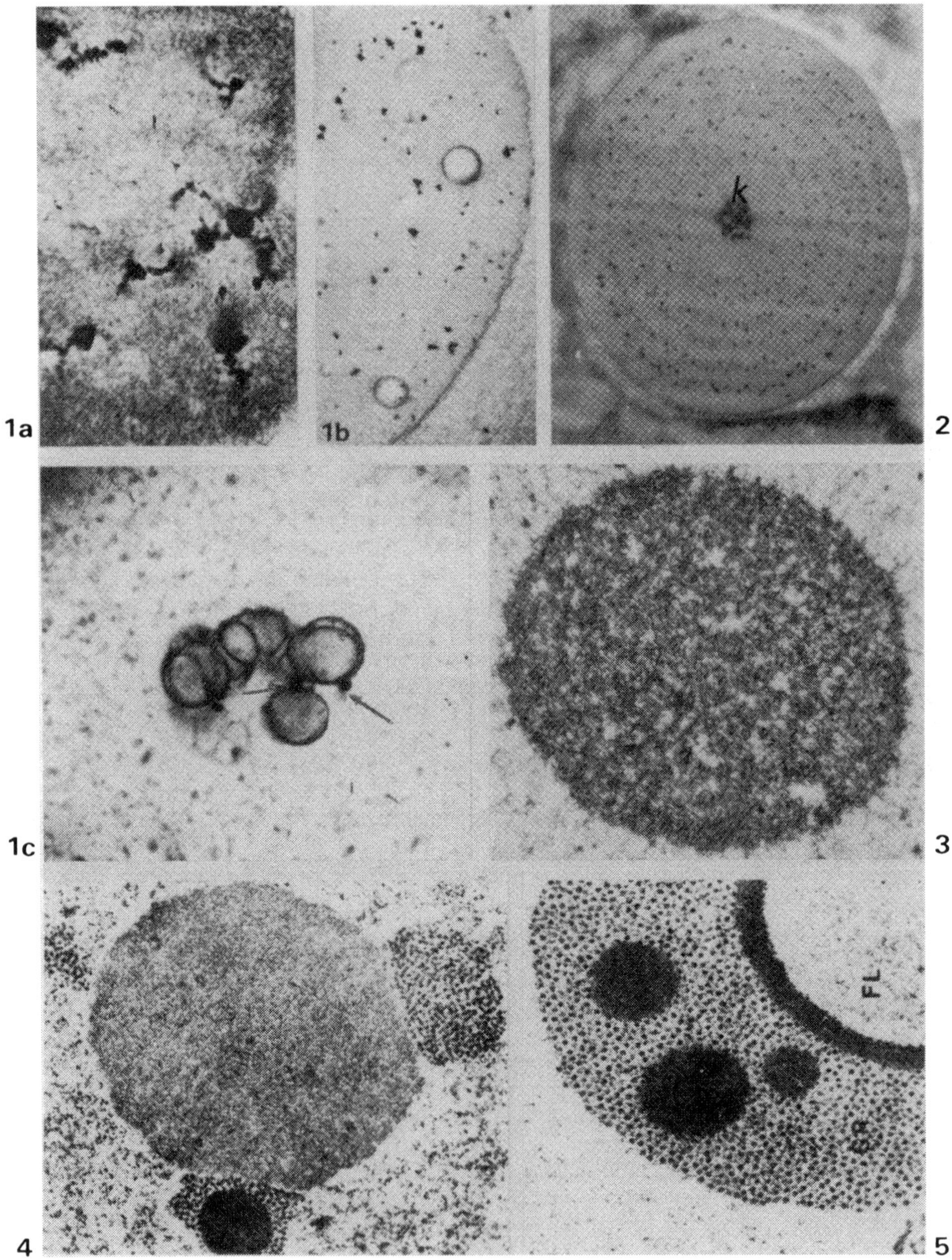
1a
1b
2
k
1c
3
4
FL
5

400-Å granules forms a large circular structure (see Plate XIV, 5) (Heubner et al., 1975). Recent data on the possible appearance of perichromatin-like granules in the nucleoli in connection with the slowdown of the ribosome processing and rRNA storage (see references in Section 2.5) have been discussed; these NLBs may in fact be nucleoli. However, this assertion requires experimental confirmation.

3. Protein spheres have been described in the oocytes of amphibians and fishes (Callan, 1965; Mancino et al., 1969; Raikova, 1968; Nardi et al., 1972) where numerous nucleoli develop. They appear on specific loci of the chromosomes and consist of fibrillar material which does not differ from the fibrillar component of the nucleolus (see Plate XIV, 3) (Brown and Ris, 1959; Raikova, 1974). In some amphibians repeated sequences are associated with the point of attachment of the protein sphere to the chromosomes (Lacroix and Loones, 1971). It is difficult to say to what extent the protein spheres in the oocytes of birds are similar to those in fishes and amphibians (Gaginskaya and Gruzova, 1969; Gaginskaya, 1972). For a long time these structures were thought to be nucleoli, but they neither synthesize nor contain noticeable quantities of RNA (see Plate XIV, 1b). The spheres remain attached to the chromosomes on which they form (see Plate XIV, 1, a, c).

At a submicroscopic level, the organization of the oocyte nuclei is complex. Granules of various types [for example, interchromatin (IG) and perichromatin (PG) granules], and so-called nuclear bodies, are observed in the nuclei of the oocytes, just as in the somatic cells. Several reviews have been devoted to the functional organization of the nuclei of the somatic cells at the ultrastructural level (Monneron and Bernhard, 1969; Smetana, 1974;

Plate XIV. Nucleolus-like bodies. 1. Protein spheres in the oocytes of the chaffinch *Fringilla coelebs* (Gaginskaya and Gruzova, 1969); a) part of the nucleus of the oocyte with spheres attached to the lampbrush chromosomes; b) ^{3}H-uridine is not included in the protein spheres; c) protein spheres in the process of merging into the karyosphere; shortened chromosomes (arrows) can be seen between the spheres. 2. Nucleus of the oocyte of *Eremias velox* (Kulikova, 1963); numerous protein bodies leave the karyosphere (k). 3. Protein sphere in the nucleus of the sterlet *Acipenser ruthenus* oocyte (Raikova, 1974). 4. Secondary nucleolus in *Acheta domesticus* oocyte (Jaworska and Lima-de-Faria, 1973a); the electron-dense DNA-containing block is surrounded by granules 500 Å in diameter, and in the center is the fibrillar part of the nucleolus. 5. Part of the granular body in the nucleus of the tsetse fly *Glossina austeni* (Heubner et al., 1975); dense fibrillar inclusions can be seen.

Bouteille et al., 1974), in which particular attention is paid to the nuclear bodies. Nuclear bodies were first described by Bernhardt in 1966 (according to Bouteille et al., 1974). These are formations 0.5-1.5 μm in diameter, hardly visible under a light microscope. They greatly vary in form and are observed in metabolically active cells. The number and size of these bodies increase markedly in the pathological states of the cells (viral infection, carcinogenesis). Simple nuclear bodies consist of amorphous protein material of nucleoli (Bouteille et al., 1974). Complex nuclear bodies usually have a protein capsule and include fibrils and granules of RNP and DNP, glycogen, lipids, and other material. Rod-like nuclear bodies consisting of bundles of protein fibrils have been described. In most cases, the origin of these bodies is uncertain. The study of the chemical composition is complicated by the small size of the nuclear bodies and the difficulties of cytochemical analysis at the electron-microscopic level. The authors of the review (Bouteille et al., 1974) suggest that nuclear bodies are mediators in the exchange between the main nuclear components and play an important part in nucleocytoplasmic interactions. In particular, the complex nuclear bodies may act as a temporary store of genetic information transported from the nucleus to the cytoplasm.

As yet, oocyte nuclear bodies have not received sufficient attention. However, it is evident that in oocytes (very large cells with an extremely active metabolism) the nuclear bodies may be even more complex and much larger than in somatic cells. These features of the nuclear bodies of oocytes suggest that they are more suitable for studying the genesis, chemical composition, and possible function of nuclear bodies in general.

Complex nuclear bodies, which have a protein capsule and include granular or fibrillar materials, have been found in the late oocytes of *Chrysopa perla* (see Plate XII, 2a, b) and in the common frog (see Plate XII, 1). The appear around the karyosphere of the oocytes (for more detail, see Section 2.7) and may be regarded as giant complex nuclear bodies (Plate XV).

Additional structures, which take the form of strands, platelets, or paracrystalline formations, have repeatedly been found in oocytes. They may be compared to the rod-like nuclear bodies of the somatic cells (Bouteille et al., 1974; Bertel and Gouranton, 1981), but in oocytes these structures are hypertrophied. Thus, distinctive structures, the "paracrystalline bodies," are described in the gametocytes of *Panorpa communis*. These structures are a bundle of 80-Å threads united by fine fibrils. They form in contact with the nuclear envelope and the nucleolus, and serve as centers for the accumulation and packing of the RNP material (Gassner, 1967; Welsch, 1974).

Similar bundles of fibrils arranged in long strands have been found in the oocytes of darkling beetles (see Plate XII, 4, 5). By the middle of oogenesis the strands anastomose and occupy a large zone around the chromosomes (see Plate XII, 3). Contact between the strands and the chromatin-like and nucleolus-like bodies is observed (see Plate XII, 5). Around the nucleolus-like bodies the fibrillar material frequently forms capsules (see Plate XII, 4).

Paracrystalline structures in the form of spirally coiled strands consisting of granules have been described in the oocytes of the marine worm *Priapulus caudatus* (Nørrevang, 1963), in the insect *Wachtiella persicaria* (Kunz et al., 1970), in the dragonfly (Halkka et al., 1969), and in the crustacean *Cypridopsis vidua* (Reger et al., 1965). In the latter case, it has been shown that the bands consist of 100- to 150-Å granules. Similar structures have also been found in the cytoplasm near the nuclear membrane. It is suggested that these structures relate to the conservation and storage of ribosomes or polysomes. Perhaps these paracrystalline bodies may be compared with the coiled bodies described in the nuclei of both somatic (Monneron and Bernhard, 1969; Smetana, 1974) and germ cells of vertebrates (Chouinard, 1973; Zybina and Grishchenko, 1977) and invertebrates (Romanova, 1982). The function of these coiled bodies in the oocytes remains unknown.

It cannot be excluded that some protein bodies appear as a result of a change in the normal or experimentally induced transport of protein molecules. Thus, the rod-like paracrystalline structures reappear near the nucleoli in the oocytes of amphibians, in which rRNA synthesis has been inhibited by actinomycin D (Lane, 1969; Snow, 1972). The authors suggest that the transport of ribosomal proteins is also disrupted at the same time, resulting in proteins which become arranged in the structures described above.

It has already been noted that accumulations of IG, PG, and aggregates of larger granules have been described in the oocyte nuclei of both vertebrates and invertebrates, and also in somatic cells (Gruzova et al., 1972b; Parfenov, 1979; Palombi and Viron, 1977). Some progress has been made in the study of their cytochemical composition and role in the metabolic processes of the cell. Kloetzei et al. (1981) isolated the nuclear material of newt oocytes by differential centrifugation and analysis in a sucrose gradient, and revealed three basic components: 1) a population of RNP granules with a sedimentation constant of 30-120S, containing a rapidly labeling heterogeneous RNA (hnRNA); 2) aggregates of protein particles shaped as beads; and, 3) a fibrillar mesh containing 10-nm and smaller microfilaments. The authors believe that the fibrillar mesh constitutes a nuclear matrix (although the proteins of the mesh have not yet been identified), closely interacting with which are the RNP particles of chromosomal

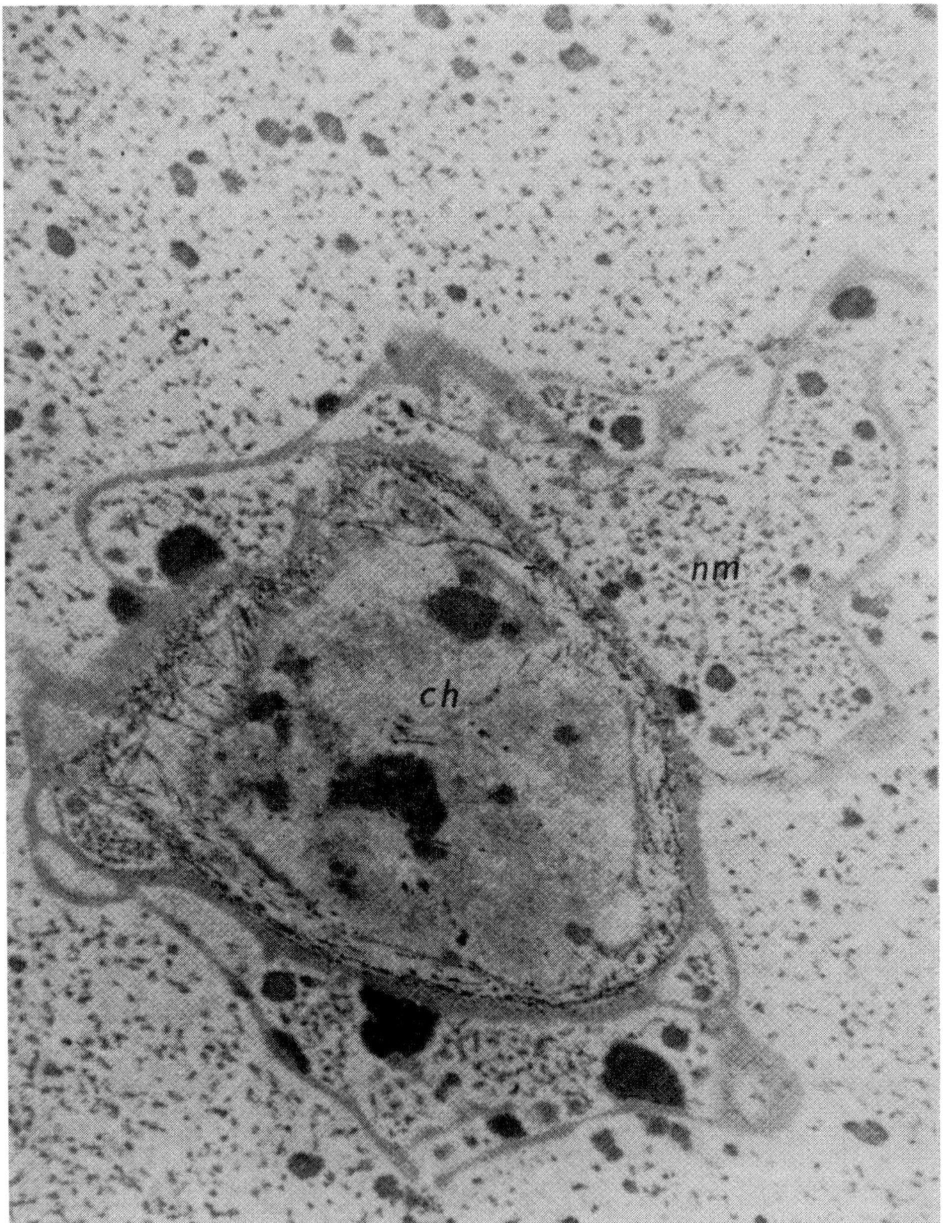
nm
ch

origin; this structure facilitates metabolic processes between the RNP particles, the fibrillar–granular protein component of the mesh, and low-soluble components.

Thus, oogenesis in most animals is accompanied by the appearance of certain NLBs and accumulations of granules and fibrils in the oocyte nuclei. Further research into the ultrastructure, chemical composition, and peculiarities of their metabolism will eventually allow a more precise identification of these nuclear formations. At the moment, it can only be said in general that the NLBs are reflections of an unusually active protein metabolism and complex nucleocytoplasmic interactions in oocytes. Current biochemical and autoradiographic studies have shown that by no means can all protein molecules penetrate into the nucleus of the oocyte. There are separate classes of nuclear and cytoplasmic proteins. Having penetrated into the nucleus from the cytoplasm, some proteins cannot migrate back, obviously binding with some kind of nuclear component (Bonner, 1975a, b). The protein bodies appear to be the morphological expression of these relations between the nucleus and the cytoplasm.

2.7. THE KARYOSPHERE

Characteristic of oogenesis in many animals is the withdrawal of the chromosomes to the center of the nucleus to form a condensed mass known as the karyosphere. The formation of a karyosphere is common to gametogenesis in many vertebrates and invertebrates; it has been described in more than 120 species from 12 classes of animals (Gruzova, 1975b). Apparently this nuclear body is characteristic only of gametocytes. For detailed information on the formation and characteristics of the karyosphere, see reviews by Gruzova (1975a, b; 1979b). Here we give brief additional information not given in the previous reviews (referring particularly to the karyosphere capsule).

Typical karyospheres are formed in the oocytes of insects with the nutrimentary type of oogenesis; when there are actively functioning trophocytes, the karyosphere appears at the onset of diplotene and remains right up to the end of oogenesis.

Plate XV. The karyosphere of the oocyte of *Chrysopa perla* (Gruzova and Zaichikova, 1972). The chromosomes (ch) are surrounded by a complex capsule, which is lined inside with electron-dense strands; the lobes of the capsule are filled with nucleolonema (nm), and lumpy extra-chromosomal DNA closely adjoins the material of the capsule.

In different organisms the karyosphere appears at different periods during diplotene. A number of common properties characteristic of the karyospheres of different animals allows this nuclear body to be defined as a complex structure containing the entire set of chromosomes of the germ cell and reflecting a state of relative inactivation of RNA synthesis in the genome combined with active protein metabolism (incorporation of labeled amino acids). Morphologically this is expressed in the formation on the chromosomes of protein nucleolar-like bodies and granules which are detached from the karyosphere.

Typically, the karyospheres of some animals possess a multilayered capsule. There are numerous electron microscope studies of the karyosphere and its capsule (Gruzova and Zaichikova, 1968; Gruzova et al., 1972b; Fiil and Moens, 1973; Gruzova and Parfenov, 1976, 1977; Fiil, 1974, 1976, Fiil and Moens, 1973; Zbarsky and Filatova, 1979; Gruzova, 1979, 1982; Parfenov, 1979; Filatova et al., 1982; Filatova, 1986). Initial studies showed that the capsule material of the karyosphere in oocytes originates as a result of complex interaction between the main nuclear structures: the chromosomes, the nucleoli, and the nuclear envelope. Complex morphogenetic processes take place in the nucleus, leading to the appearance of a structural complex around the compacted chromosomes. Interactions of the synaptonemal complexes, the nuclear envelope, and their derivatives have been described during the formation of the karyosphere capsule in mosquitos (see Fig. 20) (Fiil and Moens, 1973; Fiil, 1976). An ultrastructural analysis of the karyosphere of the oocytes of *Chrysopa perla* and the beetles *Blaps lethifera*, *B. mortisaga*, and *Gnaptor spinimanus* reveals the special part played by micro- and pronucleoli in the formation of the capsule (see Plate VIII, 3; Plate XII, 3-5) and by bodies of extra-DNA (Plate XV, 1) (Gruzova et al., 1972a, b; Zaichikova and Gruzova, 1973; Gruzova, 1979b, 1982). Despite the specific features of karyosphere formation in these insects, common to them all is the fact that the main material of the karyosphere capsule, the strands, originate in contact with the chromosomes, the pronucleoli, and the extra-DNA-containing bodies (Plate VII, 3; Plate XI, 1, 4, 5). We regard these strands as abnormal elements of the synaptonemal complex.

Among vertebrates the karyosphere capsule is formed in the oocytes of some amphibians, reptiles, and birds (see Fig. 30) (Gruzova, 1975b). Let us dwell in greater detail on the peculiarities in the formation of the karyosphere and its capsule during late oogenesis in frogs. Early studies by Wagner (1923) showed that in the vitellogenic oocytes of *Rana temporaria* an extensive capsule is formed around the chromosomes. This capsule consists of numerous nucleoli and a fibrous substance of unknown nature (see Plate XIII, 1). When the karyosphere of *R. temporaria* was isolated in vitro by Callan's method (1965), it appeared that its structures were tightly interconnected by an elastic fibrous material (Parfenov,

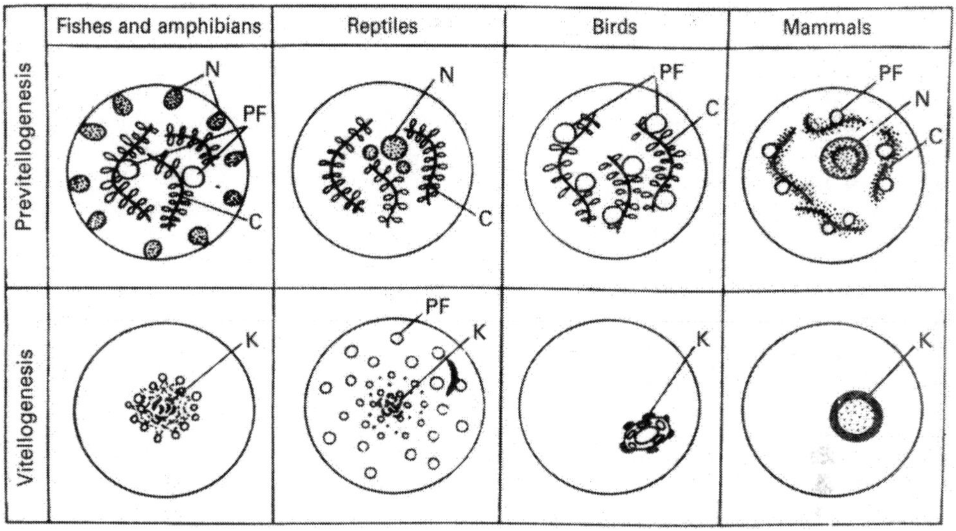

Fig. 30. Changes in functional states of nuclear structures in the oocytes of vertebrates. N) Nucleus; PF) protein formations; C) chromosomes; K) karyosphere.

1974). Electron microscopy showed that this fibrous material consists of a network of pseudomembranes consisting of annuli similar to the pore complexes of the nuclear membrane, and also consists of variations of the membrane structures (see Plate XIII, 2, 4, 5). At the same time, there is close contact between the chromatin of the karyosphere and the pseudomembranes (see Plate XIII, 3) and a connection between the pseudomembranes and the micronucleoli. Intranuclear annulate lamellae (see Plate XIII, 5) and membrane tubular formations have been discovered in the peripheral zone of the capsule and beyond its boundaries (Gruzova and Parfenov, 1977). In *Rana temporaria*, besides pseudomembrane in the capsule, strands consisting of longitudinal bundles of fibrils have been discovered (Chentsov and Polyakov, 1974; Zbarsky and Filatova, 1979). By analogy with beetles these latter may be regarded as abnormal material of the synaptonemal complex (SC).

In the spring oocytes of *Rana ridibunda*, as in *R. temporaria*, the collapsing lampbrush chromosomes are found in the center of the nucleus surrounded by nucleoli moving from the periphery; some of these nucleoli have long strands of membranous origin, so-called "nucleolar tails," similar to those described in two species of *Urodela* (Kezer et al., 1971). The karyosphere in *R. ridibunda* differs from that of *R. temporaria* by the absence of fibrillar material in the capsule. Instead of a capsule there is a large central protein body, and all chromosomes are anchored to its surface. Ultrastructural analysis reveals that this central body in *R. ridibunda*, similarly to the fibrillar material in the capsule of *R. temporaria*, consists of an agglomeration of annuli similar to autonomous pore complexes (Parfenov, 1979). Thus, it appears that the karyosphere capsule in *R. ridibunda* seems to be inverted. The role of fixing and retaining the chromosomes together is played here by the central body, rather than the walls of the capsule surrounding the chromosomes, as in *R. temporaria*. However, the material making up these components is identical in both cases, consisting of autonomous pore-like complexes. Stages of the central body formation and participation of different nuclear structures in this process have been studied. In autumn-winter oocytes the central body is absent. At the same time, among the chromosomes in the center of the karyosphere there are polymorphous and ring-shaped structures containing rDNA which appear in connection with fragmentation of nucleoli (Gruzova and Parfenov, 1973; Parfenov et al., 1983). We regard these structures as micronucleoli. In the karyosphere of early spring oocytes several small protein formations first appear in contact with the micronucleoli. Then in spring oocytes one large central body (up to 20 μm in diameter) is formed as a result of the fusion of small ones; bivalents are seen to be attached to the central body surface. The ultrastructural analysis has revealed a direct contribution of the micronucleolar material to the central body formation. As has been demonstrated by means of DNA, RNA, and histone extraction from the nucleus, the basic component of the micronucleoli and central body consists mainly of remnant proteins of the nuclear matrix (Parfenov and Gruzova, 1984).

A comparison of electron microscopic findings concerning the organization of the karyosphere capsule in insects and amphibians reveals that the material of the capsule consists of different combinations of derivatives of the SC (different variants of polycomplexes) and of the nuclear envelope (mainly autonomous pore-like complexes). There occurs both the reutilization of the material of "spent" SC, and apparently the *de novo* formation of modified SC out of an excess of SC precursors existing in the nucleus or cytoplasm. How the autonomous pore complexes find themselves in the central zone of the nucleus remains unclear. Individual pore complexes may get into the karyosphere capsule, together with the chromosomes or nucleoli that have detached from the nuclear envelope, during the formation of the karyosphere; but the large accumulation of pore complexes in the capsule testify to their *de novo* formation off the nuclear envelope. Thus, investigating the karyosphere capsule not only helps to elucidate certain problems of intranuclear morphogenesis, but leads to the discovery of new properties of such widely investigated structures as the pore complexes and SC elements; it also provides a further opportunity for studying their development. The close contact between these struc-

Table 2.2. Dependence of Karyosphere Capsule Formation on Dimension of Oocyte Nucleus and Presence of Extra-DNA

	Species	Diameter of nucleus (in μm)	Extra-DNA	Karyosphere capsule	Reference
1.	Chrysopa perla	220 × 140	+	+	Gruzova, 1960; Gruzova et al., 1972
2.	Gnaptor spinimanus	450	+	+	Gruzova and Batalova, 1979
3.	Blaps lethifera	300	+	+	Gruzova, 1962, 1979
4.	Blaps mortisaga	350	+	+	Gruzova, 1979
5.	Culex pipiens	400 × 140	?	+	Fiil and Moens, 1973
6.	Aedes aegypti				
7.	Rana temporaria	400-500	+	+	Wagner, 1923
8.	Rana ridibunda	500-700	+	+	Gruzova and Parfenov, I973; Parfenov, 1979
9.	Gyrinus natator	300 × 16	+	+	Matuszewski and Hoser, 1975
10.	Necrophorus humator	950 × 150	+	+	Matuszewski et al., 1977
11.	Silpha sinuata	500	+	+	Matuszewski et al., 1977
12.	Silpha atrata	500	+	+	Matuszewski et al., 1977
13.	Silpha thoracica	500	+	+	Matuszewski et al., 1977
14.	Necrodes littorales	580	+	+	Matuszewski et al., 1977
15.	Creophilus maxillosus	700 × 160	+	+	Matuszewski and Kloc, 1976
16.	Mikiola fagi	20 × 30	+	+	Matuszewski, 1962
17.	Oligotrophus schmidti	20 × 30	+	+	Matuszewski, 1966
18.	Aphidoletes aphidimysa	20 × 30	+	+	Gruzova et al., 1987
19.	Calliphora erythocephala	40	-	-	Gruzova, 1967
20.	Musca domestica	40	-	-	Bier et al., 1967
21.	Rhogogaster picta	35	-	-	Bobrova and Gruzova, 1967
22.	Eurygaster integriceps	40-50	-	-	Kozhanova, 1974
23.	Laspeyresia pomonella	50-70	-	-	Kozhanova et al., 1973

Table 2.2 (continued)

Species	Diameter of nucleus (in μm)	Extra-DNA	Karyosphere capsule	Reference
24. Panorpa communis	50-80	-	-	Gruzova, 1962
25. Glossosiphonia complanata	45-50	-	-	Gruzova and Zaichikova, 1967
26. Drosophila melanogaster	10-15	-	-	Koch et al., 1967

tures and their mutual interpenetrations point to an organic relationship between the SC material and the pore complexes of nuclear membranes (Fiil and Moens, 1973).

Since the karyosphere capsule is not formed in oocytes of all the animals studied, it was of interest to elucidate which distinctive features of oogenesis, and of the nuclear organization in particular, are responsible for its appearance. Table 2.2 presents invertebrate and vertebrate species in the oocytes of which a karyosphere is formed. It can be seen that the capsule formation is commonly concomitant with the large germinal vesicle size (above 100 μm in diameter) and the constant presence of extrachromosomal DNA in the nucleus. The only exception to this rule are gallflies (Nos. 16-18) with small oocyte nuclei (20-30 μm). Nonetheless, here the lamellar capsule separates somatic (S) chromosomes and eliminated (E) chromosomes. At the same time, Ag-staining procedure (Howell and Black, 1980) shows that in *A. aphidimysa* E chromosomes are enriched in rDNA (Gruzova et al., 1987).

Further attempts to determine which structures in the nucleus take part in the formation of the capsule material led to the conclusion that the capsule elements appear in immediate contact with DNA-containing nuclear structures: chromosomes, micro- and macronucleoli, pronucleoli, and lumps of extra-rDNA. The movement of nucleoli and pronucleoli during karyosphere capsule formation is shown in Fig. 31. These observations suggest that a certain DNA fraction ("tropho-DNA") is taking part in the formation both of the capsule as a whole and of its individual elements which is subsequently eliminated from the nucleus. It may be thought that this DNA either has a ribosomal nature or accompanies ribosomal genes.

According to the topography of the parts, "external" and "inverted" capsules may be distinguished. In the latter case the material by which the chromosomes are held together in a small volume in the nucleus is concentrated in the center of the karyosphere. The central body of the "inverted" capsules may have a different structural base: pore complexes (lake frog), protein

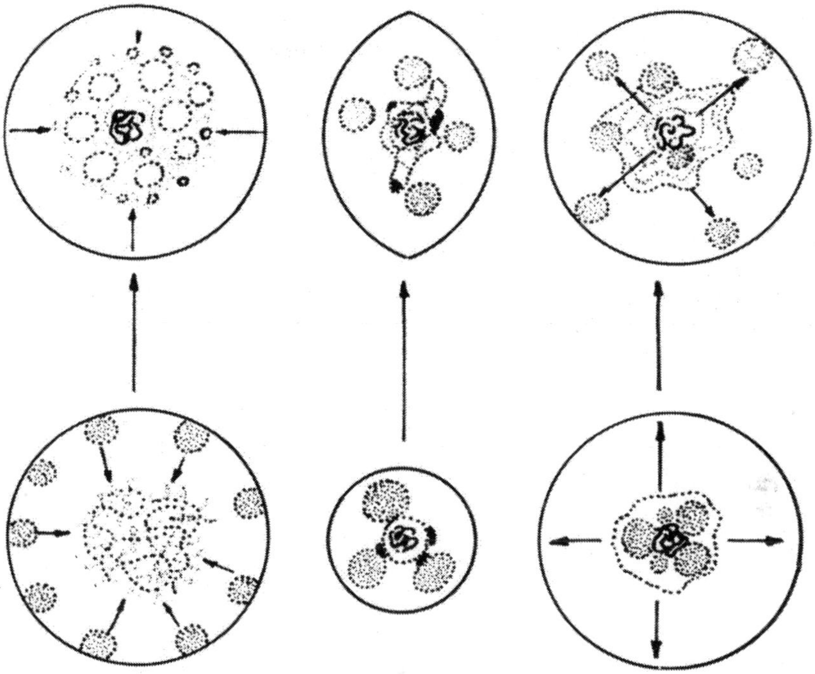

III II I

Fig. 31. The movement of nucleoli, or pronucleoli, and karyosphere capsule formation. Type I (beetles): along with the formation of the capsule, the pronucleolar bodies move from the karyosphere to the periphery of the nucleus. Type II (lacewing flies, mosquitoes): in the period of capsule formation the nucleoli, extrachromosomal DNA bodies, and the karyosphere are in close proximity. Type III (frogs): in the period of karyosphere formation the nucleoli move from the periphery to the center. Arrows indicate direction of nucleoli movement.

NLB (some birds) (Gaginskaya and Gruzova, 1969; Gaginskaya, 1972; Tsvetkov and Gaginskaya, 1983), or a "spent" nucleolus in mammals (Kiknadze et al., 1980; Zybina et al., 1980) and human oocytes (Kurilo, 1982, 1985; Parfenov et al., 1984). This also makes clear the purpose of the capsule in the nucleus. The chromosomes detaching from the nuclear membrane during the formation of the karyosphere must not only be held together during the intensive growth of the nucleus, but in the period of preparation for the reductional division must have a fixed position in the nucleus relating to the topography of the egg, the proximity to its surface, and the position of the micropyle (in insects), etc. Thus, the construction of the capsule provides a mechanical framework in the nucleus by which the metaphase plate is kept in a definite position during maturation division. In each individual case relevant (nearby) structures are used for shaping such a framework, whether they be pore complexes or the remnants of the nucleolus. The role of the capsule in the oocyte is not, however, limited to this. The "external" capsule probably achieves the compartmentalization of the nucleus essential for the storage of synthesized products and control of the metabolic processes between the nucleus and the cytoplasm. The capsule, usually surviving immediately up to metaphase I, also has a protective function ("screens the chromosomes") in the period immediately after the germinal vesicle breakdown. Thus, in the mosquito, in the absence of suitable conditions for oviposition, maturation divisions do not occur and the disintegrating capsule is replaced by a new one (Fiil, 1976).

In light of the latest data on the nuclear matrix in somatic cells, the karyosphere capsule may be regarded as a special derivative of the nuclear matrix of the germ cell (Zbarsky and Filatova, 1979; Parfenov, 1979; Filatova et al., 1982; Engelhardt et al., 1982). A number of works have demonstrated that the matrix includes a dense lamina firmly bound to pore complexes, the protein framework of the nucleoli, and SC material (in the case of spermatocytes), i.e., the same structures from which the karyosphere capsule is built; in all the matrix components of different cells common proteins with different molecular weights have been discovered (Comings and Okada, 1976; Berezney and Coffey, 1977; Zbarsky and Filatova, 1979; Zbarsky et al., 1980; Kuzmina et al., 1981). These data shed further light on the complex interaction of the ultrastructures observed in the karyosphere capsules of different organisms. Particles of 25-30 nm in size, making up the basic component of the nuclear matrix of somatic and germ cells, are supposed to be involved in interaction and interpenetration of all these structures (Engelhardt et al., 1982).

The karyosphere capsule is found to contain, along with proteins with molecular weights of 220-180, 145-135, and 60-70 kD, glycoproteins (Filatova, 1986) and lipoproteins (Nikolaenko et al., 1985). It is not inconceivable that actin filaments associated with micronucleoli and nucleoli are the essential component of the capsule responsible for its contractile and rigid properties (Drosdov and Parfenov, 1983; Parfenov and Galaktionov, 1987). Actin in globular and filamentous form occurs also in oocyte nuclei of other amphibian species (Clark and Rosenbaum, 1979; Scheer et al., 1984).

2.8. THE NUCLEAR ENVELOPE

The nuclear envelope in oocytes, as in all other cells, is a barrier controlling nucleocytoplasmic interactions. There have been many investigations of the fine organization of the nuclear

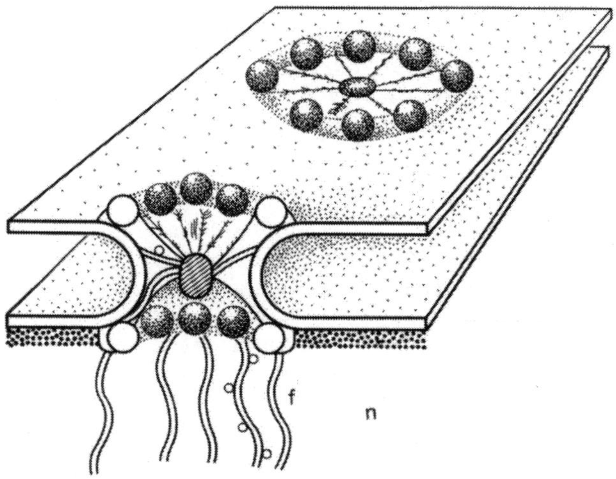

Fig. 32. A model of the pore complex of the nuclear envelope (Franke and Scheer, 1970a). The central granule (shaded) is connected with lateral granules by fibrils. Fibrils (f) are attached to lateral granules on the side of the nucleus and the granules on them form a cylindrical structure (annulus).

envelope, its components, chemical composition, and functions (see reviews: Zbarsky, 1972, 1978; Kay and Johnston, 1973; Chentsov and Polyakov, 1974; Franke, 1974; Franke and Scheer, 1974; Maul, 1977; Zbarsky et al., 1980). In this section, emphasis is placed on the pore complexes of the nuclear envelopes of oocytes. No special features of the ultrastructural and chemical organization typical only of the nuclear envelope of oocytes have yet been found. Detailed research into the nuclear envelope began with electron microscopy of the nuclear envelope of the germinal vesicle in amphibians (Callan and Tomlin, 1950, cit. according to Franke, 1974). These authors showed that the nuclear envelope consists of an outer and inner layer, and that both layers are perforated by nuclear pores at specific intervals. The permanent existence of pores is a characteristic feature of the nuclear envelope of oocytes at all stages of oogenesis (Anderson, 1974). Already the first ultrastructural studies of the nuclear envelope revealed that the nuclear pore is not just an opening. Special fibrillar and granular material is connected with the pores; it fills the pore itself, and part of this material is arranged in the form of short cylinders protruding from both sides of the nuclear envelope into the nucleus and the cytoplasm (Gall, 1959, 1967). These cylinders have come to be called "annuli." All the material connected with the pore is called a "pore complex" (Fig. 32) (Franke, 1974). It is noteworthy that well-defined nuclear "annuli" are only typical of oocytes (Anderson, 1974). The discovery of nuclear pores and the pore complexes connected with them has stimulated studies of this component of the nuclear envelope and its role in the transport of substances from the nucleus to the cytoplasm.

The study of oocytes, primarily the oocytes of amphibians, has contributed substantially to our conceptions of the organization and function of the nuclear envelope (Gall, 1967; Scheer and Franke, 1969; Franke and Scheer, 1970a, b; Scheer, 1972, 1973). The large size of the nuclear envelope of oocytes in amphibians makes it a convenient object for the application of preparational methods of isolation and subsequent structural and chemical analysis of the envelope components. The area of the germinal vesicle envelope in *Xenopus laevis* is equal to the total surface area of 4500 nuclei of its liver cells. One of the characteristic features of the nuclear envelope is its plication; this increases its surface, possibly creates special conditions for the penetration of substances, and obviously allows a very rapid increase in the volume of the nucleus during periods in the oogenesis. Thus, for example, in the newt the area of the nuclear envelope during the growth of the nucleus from 200 to 300 μm in diameter may increase by 500,000 μm^2 (Kezer et al., 1971). The nuclear envelope is an extremely dynamic formation; during oogenesis in the Alpine newt the number of pores per nucleus increases from 1.7 to 57 million (Franke and Scheer, 1970b). Every minute 500 new pore complexes may come into being (Franke, 1974). Franke and Scheer (1970a) studied the ultrastructure of the pore complexes in six species of urodelean and anuran amphibians and created a pore complex model (for explanation, see caption to Fig. 32), as a universal component of the nuclear envelope. Without going into the details of the fine organization of the pore complex, we merely note that throughout oogenesis in the newt the pore complex with all its subunits remains unchanged, but as the oocyte grows the frequency of the pores containing central granules drops (from 61% in the young oocyte of larvae to 36% in the fully grown oocyte) (Franke and Scheer, 1970b). The authors develop their view on the pore complex as a structure in which a large reserve of RNP material is stored in bound, inactive form; RNP concentration in the pore complexes is 60 times higher than in the nucleus (Scheer, 1972). At the same time, in the active phase of oogenesis, at the lampbrush chromosome stage, each pore complex transports into the cytoplasm an average of eight molecules of ribosomal RNA per minute (Scheer, 1973).

The existence of DNA in the pore complex, and the part it plays in the formation of the inner nuclear envelope, is under discussion. This question relates to the fact that in many somatic cells the nuclear membrane is lined with heterochromatin (Chentsov and Polyakov, 1974; Franke, 1974). A number of authors suggest that the fibrils connected with the ring of pores on the side of the nucleus represent highly dispersed chromatin (Du Praw, 1970). An alternative hypothesis regards these fibrils as RNP material (possibly of polysomal nature) (Franke and Scheer, 1970b). Using ^{3}H-actinomycin D staining of nuclear envelope preparations and Bernhard's method of staining ultrathin sections (Bernhard, 1969), Scheer and Franke could not detect DNA in the pore complex of the nuclear envelope of amphibian oocytes. They calculated (tentatively assuming the total DNA of the chromosomes to be in contact with the nuclear membrane) that even if there is some DNA in the pore complex, it is present in quantities too low (0.3-$1.6 \cdot 10^{-18}$) for detection with available methods. The close link between the chromatin and the components of the nuclear membrane in some pathological states of the cell (Babai et al., 1969; Franke and Scheer, 1971; Franke et al., 1972) and in the particular cases of oogenesis (Gruzova and Parfenov, 1977) prompts the suggestion that self-assembly of the nuclear membrane components sometimes takes place, possibly with the participation of a particular fraction of the chromatin (see Section 2.7). These particular cases suggest the possibility of the autonomous existence of the pore complexes outside the nuclear envelope. This is indicated by observations of the behavior of the pore complexes under experimental conditions, which reveal their particular resistance to detergents (Chentsov and Polyakov, 1974; Maul, 1977). Some authors believe the pore complexes have the properties and strength to allow them to exist independently (Monroe et al., 1967; Franke and Scheer, 1970a).

2.9. NUCLEOCYTOPLASMIC INTERACTIONS

At present, the nuclear envelope is regarded as part of a single membrane system in the cell and, as already mentioned, plays an extremely important part in the transport of substances between the nucleus and the cytoplasm. In oocytes, cells which grow to tremendous size and in which many structural and chemical components are stored up for the future embryo (Davidson, 1969), the nucleocytoplasmic interrelations are particularly marked and are frequently in a highly distinctive form. The extensive observations which have been accumulated on nuclear and cytoplasmic interactions in oocytes make it possible to discern some forms and mechanisms of cell component transport.

2.9.1. Transport of Nuclear Material through Pores

The transport of nuclear material through pores has been repeatedly observed in the oogonia and young oocytes of invertebrates and vertebrates (Kessel, 1966a, b, 1969; Scharrer and Wurzelman, 1969a, b; Eddy and Ito, 1971; Franke and Scheer, 1974; Raikova, 1974; Halkka and Halkka, 1975; Zaichikova and Gruzova, 1975). The

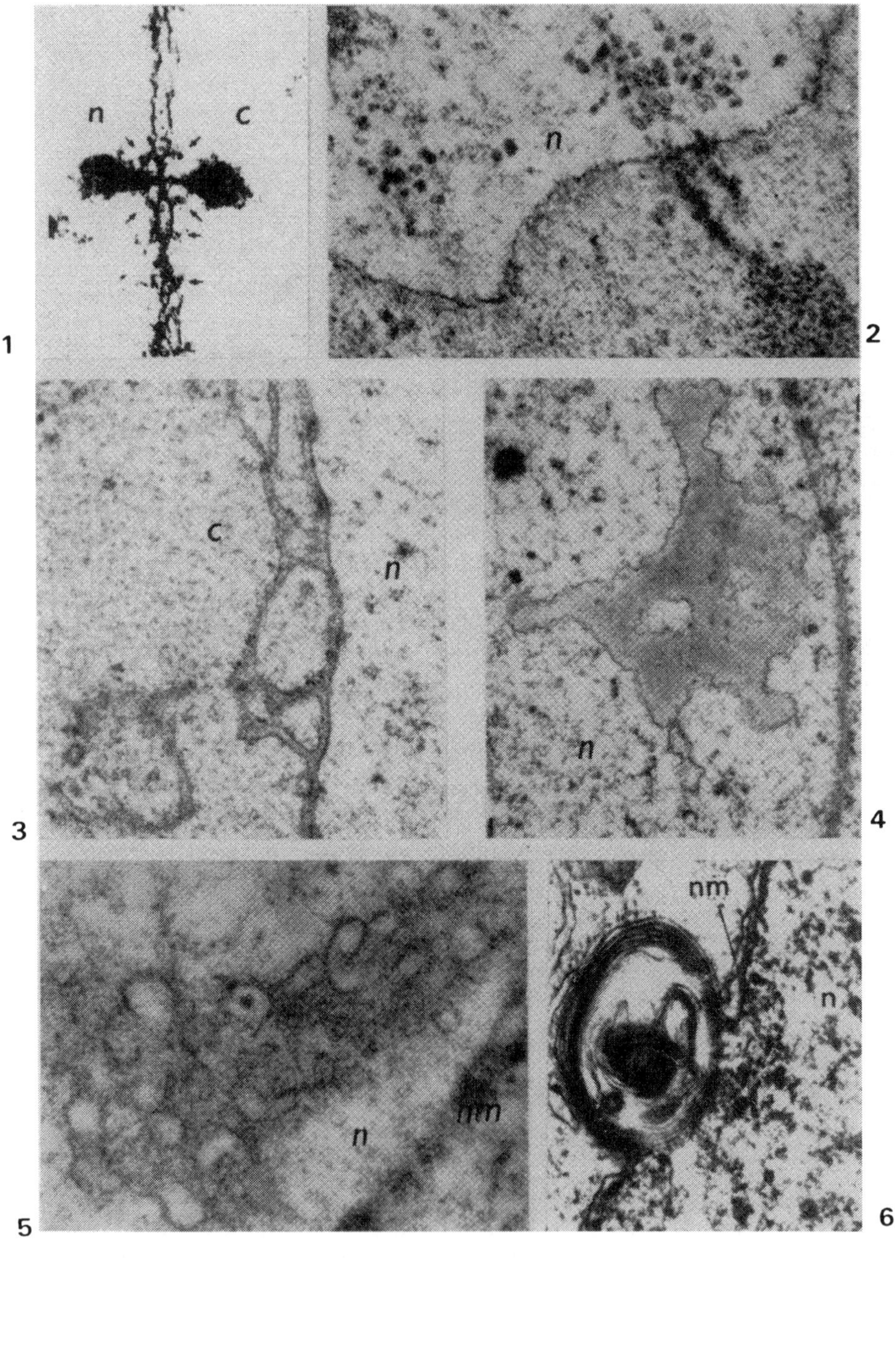
n
c
1
n
2
c
n
3
n
4
n
nm
5
nm
n
6

material which penetrates through the pores may be of both chromosomal and nucleolar origin. Kessel (1966a) observed the transport of ribosome-like granules in the oocytes of ascidians. The transport of fibrillar masses connected with the nucleolus, which occurs through the pores in newts (Franke and Scheer, 1970b) and lungfishes (Scharrer and Wurzelman, 1969a), has been traced in detail. These authors observed successive stages of transport and note that, on leaving the nucleolus, the fibrils come into contact with the pore complex, mainly with the internal granules of the annulus. As a result, dumbbell-like structures are observed on both sides of the nuclear membrane, connected by a fine bridge through the pore canal (Plate XVI, 1).

Some authors note the transport of large (350-450 Å) granules connected with chromatin (possibly perichromatin granules) (Eddy and Ito, 1971; Halkka and Halkka, 1975; Zaichikova and Gruzova, 1975) through the pores into the cytoplasm. Sometimes it may be observed that these granules, like those in somatic cells (Monneron and Bernhard, 1969; Smetana, 1974), first uncoil themselves into fibril-like formations, and penetrate the pore in this form (see Plate XVI, 2). The fate of the fibrogranular dense accumulations of nuclear origin, and their contact with other cytoplasmic structures, is described in detail by Eddy (1975) and Aizenshtadt (1975). The granules connected with the chromatin not only penetrate the pores but, obviously, the chromatin itself. The exit of the chromatin-like mass into the cytoplasm is observed in the oocytes of *Trematoda* (Sasha, 1965) and trophocytes of *Blaps* (Gruzova, 1979a).

2.9.2. Transport by "Blebbing"

Blebbing is a common phenomenon in the oocytes of invertebrates and vertebrates (Merriam, 1959; Kessel, 1964, 1965; Szöllösi, 1965;

Plate XVI. Nucleocytoplasmic interrelations. 1. Fragment of the isolated nuclear envelope of *Triturus alpestris* (Franke, 1974): a dumbbell-like figure of fibrillar material penetrates into the cytoplasm (c) from the nucleus (n) via a pore. 2. Exit of granular material from the nucleus (n) of trophocytes into the cytoplasm in *Chrysopa perla* (Zaichikova and Gruzova, 1976). 3. Cytoplasmic blebs separating from the external nuclear membrane in rabbit oocytes (Zybina, 1975a). 4. Intranuclear blebbing in the oocytes of the beetle *Gnaptor spinimanus*. 5. Intranuclear tubular structures in the late oocytes of *Blaps lethifera*. 6. Platelet-like myeloid body going from the nucleus into the cytoplasm in the bat oocyte (Ruby and Webster, 1972). c) cytoplasm; n) nucleus; nm) nuclear membranes.

Hsu, 1967; Scharrer and Wurzelmann, 1969a; Zybina, 1975a). Scharrer and Wurzelmann (1969a) observed swellings of the perinuclear space and the subsequent detachment of the vesicles from the outer nuclear membrane in the oocytes of lungfish, in places where the nucleolus comes into contact with the inner membrane; these vesicles contained fibrillar and granular material. In rabbit oocytes, a vast network of lacunae and little canals is formed, which spreads out from the outer nuclear membrane (see Plate XVI, 3) (Zybina, 1975a). Intranuclear blebbing involving the inner nuclear membrane has been described in oocytes of a number of vertebrates as a stage in the formation of intranuclear annulate lamellae (Merriam, 1959; Kessel, 1964, 1965; Hsu, 1967). Sometimes vast lacunae of irregular shape are formed as, for instance, in the young oocytes of the darkling beetle (see Plate XVI, 4). At times, a system of coiled tubules (see Plate XVI, 5) is formed in the nucleus from the inner nuclear membrane. The intranuclear and cytoplasmic vesicles are, as a rule, filled with some kind of material whose nature is in most cases unclear; characteristically, the nuclear or cytoplasmic material penetrates through the pores into the perinuclear space, and only after this does it enter the nucleus or cytoplasm by the invagination of one of the nuclear membranes and the subsequent formation of a closed vesicle. Different kinds of blebbing in which both the nuclear membranes take part have been observed in mammalian oocytes. In this case, invagination of the nuclear membrane occurs either inside the nucleus (Dyer et al., 1973) or in the cytoplasm (Baker and Franchi, 1969). A special case of blebbing has been discovered in the embryonic cells of bats; spirally coiled membrane bodies are transported from the nucleus to the cytoplasm by invagination of the nuclear membrane and further separation of membrane vesicles form the nucleus (see Plate XVI, 6) (Ruby and Webster, 1972). The authors suggest that the membrane layers produced by the nuclear membrane possess informational properties and may control the subsequent morphogenesis by cytomembranes.

The tremendous potential for nuclear membrane proliferation in oocytes is observed at times also in destructive processes. The formation of a vast intranuclear labyrinth of membranes and accumulation of vesicles filled with regularly packed protein fibrillar material coming apparently from the cytoplasm through ruptures in the nuclear membrane in *Rana ridibunda* oocytes may be regarded as a variant of blebbing (Parfenov and Gruzova, 1982).

Actually it is by blebbing that accessory nuclei appear in the oocytes of some insects (Meng, 1968; Cassidy and King, 1972). It has been demonstrated that in the dipteran *Phryne* the accessory nuclei are formed by the protrusion and proliferation of the two-layered nuclear membrane with the subsequent detachment of the vesicle from the nucleus. In other cases, parallel to the nuclear membrane there appear membranous discs which subsequently close up into vesicles. The accessory nuclei store up a great amount of proteins and RNA, while DNA has not been detected; the great number of pores points to intensive metabolic processes between the accessory nuclei and the oocyte's cytoplasm (Meyer et al., 1979).

Although blebbing is chiefly a normal phenomenon characteristic of oocytes, in somatic cells it frequently takes place in pathological conditions (see Franke, 1974). Thus, the transport of structural and chemical components from the nucleus to the cytoplasm and back is effected by blebbing. Moreover, these components are also preserved immediately up to certain stages in the oogenesis, since the material surrounded by the membrane can either be slowly transformed or remain inactive and inaccessible to any kind of cellular reaction. Another function of blebbing is the creation of a phospholipid pool of cellular membranes and, finally, it is not impossible that blebbing plays a role in the genetic control of nuclear or cytoplasmic morphogenesis (Hsu, 1967; Ruby and Webster, 1972).

2.9.3. Extrusion of Nucleoli

This may be regarded as a variation of blebbing. Facts regarding the extrusion of whole nucleoli from the nucleus into the cytoplasm of the oocytes have long been known (Kroning, 1923; Sareen, 1965, 1967). But for a long time most researchers regarded the description of the extrusion of whole nucleoli or their fragments from the nucleus into the cytoplasm on fixed preparations as an artifact connected with the histological treatment of the preparations. However, electron microscopy confirms that extrusion of nucleoli does in fact take place. When it is a case of nucleolar fragments, it becomes possible, in many instances, to trace the successive stages of the process: the protrusion of the nuclear membrane with nuclear material in it, followed by the separation of this membranous vesicle from the nucleus (variation of blebbing) (Kessel and Beams, 1963; Kessel, 1966a, b; Jaworska and Lima-de-Faria, 1973b; Szöllösi, 1965). Lima-de-Faria suggests that this process should be called "intracytosis." Frequently, large structures identical to the nucleoli (up to 6-10 μm) are found in the cytoplasm (Sasha, 1965; Jaworska and Lima-de-Faria, 1973b; Kurilo, 1981). Jaworska and Lima-de-Faria were able to demonstrate that the nucleoli emerge in the ooplasm because of the extensive destruction of the nuclear membrane (see Fig. 25). The breakdown of the nuclear membrane and the extrusion of the nucleolus from the nucleus into the cytoplasm obviously occurs very rapidly. By isolating nuclei from living sea urchin oocytes, Lev (1966) observed an instantaneous "discharge" of the nucleolus, after which neither the nucleus nor the nuclear membrane lost their integrity.

2.9.4. Annulate Lamellae

One aspect of nucleocytoplasmic interactions is the formation of annulate lamellae. Without touching on cytoplasmic annulate lamellae, we shall dwell on the formation of intranuclear annulate lamellae (IAL). These emerge from the inner nuclear membrane and contain material of the perinuclear space. Unlike the cytoplasmic annulate lamellae, the IAL never form parallel stacks, but are arranged randomly at various angles to the nuclear envelope (Kessel, 1968).

Stages of IAL formation have been traced in ascidians (Kessel, 1964, 1968). To begin with, numerous vesicles separate from the inner nuclear membrane; some of them stick to the walls, and during protraction a bridge is formed in which a pore complex appears. However, in a number of cases the IALs emerge by invagination or narrow swellings of the nuclear membrane, in which pore complexes subsequently appear (Hsu, 1963, 1967). Sometimes the IALs come into contact with the nucleolus (Dhainaut, 1970). Intranucleolar membranous structures are also described in crab oocytes (Kessel and Beams, 1968). A special case of tubular membranous formations (possibly abnormal IAL) has been described in the nuclei of two species of amphibians (Kezer et al., 1971). These membranous formations ("nucleolar tails") appear at the moment when the nucleoli move from the periphery to the center of the nucleus. The authors point to contact between the membranous structures and the DNA-containing components of the nucleoli (see Plate XI, 3). Similar intranuclear membrane structures have also been found in the oocytes of *Rana ridibunda* and *R. temporaria* (Plate XIII, 5). In *R. temporaria* a number of intermediate structures between the IAL and the tubular structures were discovered which did not have pore complexes (Gruzova and Parfenov, 1977). The specific role of the IAL and also the "nucleolar tails" is not clear. It may be that these structures promote the transport of substances from the cytoplasm (across the perinuclear space) to the nucleus at certain stages of oogenesis. In the case of amphibians, the "nucleolar tails" are canals directly connecting the perinuclear space with the central area of the enormous (500 μm in diameter) germinal vesicle, where the chromosomes are concentrated. In the frog they are structural components of the karyosphere capsule and possibly play a mechanical part in keeping the chromosomes in the center of the germinal vesicle. Finally, the intranuclear membrane structures form a reserve of membrane which may be used to increase the nuclear envelope in the period of very rapid nuclear growth (Kezer et al., 1971).

2.10. CONCLUSIONS

Of all the examined nuclear structures, only the synaptonemal complex and the karyosphere are specific to gametogenesis. The remaining structures described in the nuclei of oocytes and their combinations may also be observed in the nuclei of somatic cells with active metabolism, especially in pathological states such as carcinogenesis and infections.

Research into the structural and functional organization of the nuclei of somatic cells (Monneron and Bernhard, 1969; Smetana, 1974; Bouteille et al., 1974) shows that additional nuclear formations in the shape of various types of granules and nuclear bodies appear in the nuclei of actively metabolizing cells; they come into being as a result of the complex interactions between the main nuclear structures themselves, and also between the nucleus and the cytoplasm. However, the functions of many of these structures have not yet been interpreted because of the difficulty of using modern biochemical methods to analyze the submicroscopic components. The methods of electron microscopic cytochemistry are not sufficiently developed, and their application is limited. Electron autoradiographic methods are the most promising from this point of view (Bouteille et al., 1974). These difficulties also arise in studying the nuclei of the germ cells, in particular of oocytes. Much factual material is still being accumulated in this field.

One of the tasks of this brief review was to outline research into the structures and functions of nuclear components and the organization of oocyte nuclei as a whole. It is obvious that the complex nature of the oocyte nuclei depends, first, on the active growth and the enormous size of egg cells in which a considerable part of the synthesized products are earmarked not for direct utilization but for ensuring the subsequent stages of oogenesis and embryogenesis. Second, the germ cell needs to "resolve" the problem of the inactivation and the packing of certain nuclear products, not only from a biochemical, but also from a structural point of view. At the same time, the oocytes are meiotic cells in which the active synthetic processes are coupled with meiotic transformations of the chromosomes (conjugation, formation of the synaptonemal complex, crossing-over, etc.). Amplification is one of the genome's convenient forms of "work" in the oocytes which do not obstruct these processes. Obviously, for the purpose of preserving the chromosome set of the egg cell, in certain situations accessory capsules emerge around the karyosphere, forming a "nucleus within a nucleus," as it were. Such original material as the derivatives of the synaptonemal complex (in mosquitos and *Chrysopa perla*), and pore complexes in pseudomembranes (in mosquitos and frogs) is used to form the capsule. It is obvious that not only are spent nuclear elements utilized, but they are also reproduced *de novo*. The giant size of the oocytes and their nuclei makes it possible to approach the solution of problems concerning intranuclear morphogenesis, and reveal a number of unusual interrelations of the nuclear components which are concealed or difficult to assess in somatic cells.

REFERENCES

Aizenshtadt, T. B., 1975, "Modern ideas about determinants of cells of the germ path," Ontogenez, *6*, 379-386.

Aizenshtadt, T. B., 1984, Cytology of Oogenesis [in Russian], Nauka, Moscow.

Allen, E. R., and Cave, M. D., 1969, "Cytochemical and ultrastructural studies of ribonucleoprotein-containing structures in oocytes of *Acheta domesticus*," Z. Zellforsch., *101*, 63-71.

Allen, E. R., and Cave, M. D., 1972, "Nucleolar organization in oocytes of Gryllid crickets: subfamilies *Gryllinae* and *Nemobiinae*," J. Morphol., *137*, 433-448.

Al-Mukhtar-Kawakib, A. K., and Webb, A. C., 1971, "An ultrastructural study of primordial germ cells, oogonia, and early oocytes in *Xenopus laevis*," J. Embryol. Exp. Morphol., *26*, 195-217.

Anderson, E., 1974, "Comparative aspects of the ultrastructure of the female gamete," Int. Rev. Cytol., Suppl., *4*, 1-72.

Arronet, V. N., 1975, "Protein granules and spheres in oocyte nuclei of reptiles," Tsitologiya, *17*, 137-143.

Babai, F., Tremblay, G., and Dumont, A., 1969, "Intranuclear and intranucleolar tubular structures in Novikoff hepatoma cells," J. Ultrastruct. Res., *28*, 125-130.

Baker, T. G., and Franchi, L. L., 1969, "The origin of cytoplasmic inclusions from the nuclear envelope of mammalian oocytes," Z. Zellforsch., *93*, 45-55.

Baker, T. G., and Franchi, L. L., 1972, "The structure of oogonia and oocytes in the rhesus monkey (*Macaca mulatta*)." Z. Zellforsch., *126*, 53-74.

Bal, A. K., Jubinville, F., and Cousineau, G. H., 1968, "Nuclear activity during oogenesis of sea urchins. I. Incorporation of ^{3}H-thymidine in oocytes of *Arbacia punctulata*," in: The Nucleus, Academic Press, New York—London, pp. 244-250.

Bal, A. K., Jubinville, F., and Cousineau, G. H., 1969, "Nuclear activity during oogenesis in sea urchins. II. Fine structural changes and patterns of RNA synthesis during meiotic prophases of *Arbacia punctulata* oocytes," Z. Zellforsch., *100*, 180-188.

Basile, R., 1969, "Nucleic acid synthesis in nurse cells of *Rhynchosciara angelae* Nonato and Pavan," Genetics, *61*, 261-273.

Bauer, H., 1933, "Die wachsenden Oocytenkerne einiger Insecten in ihrem Verhalten zur Nuclearfärbung," Z. Zellforsch., *18*, 252-296.

Beçak, M. L., and Beçak, W., 1972, "Origin of the synaptonemal complex," Experientia, *28*, 1367-1368.

Berezney, R., and Coffey, S., 1977, "Nuclear matrix. Isolation and characterization of a framework structure from rat liver nuclei," J. Cell Biol., *73*, 616-637.

Bernhard, W., 1969, "A new staining procedure for electron microscopical cytology," J. Ultrastruct. Res., *27*, 250-256.

Bertel, C., and Gouranton, J., 1981, "The intranuclear filamentous inclusions of a human glioma. Their relation with nuclear bodies," Eur. J. Cell Biol., *25*, 36-45.

Bier, K., Kunz, W., and Ribbert, D., 1967, "Struktur und Funktion der Oocytenchromosomen und Nukleolen sowie der Extra-DNS während der Oogenese panoistischer und meroistischer Insekten," Chromosoma, *23*, 214-254.

Bird, A. P., and Birnstiel, M. L., 1971, "A timing study of DNA amplification in *Xenopus laevis* oocytes," Chromosoma, *35*, 300-309.

Bird, A. P., Rochaix, J. D., and Bakken, A. H., 1974, "The mechanism of gene amplification in *Xenopus laevis* oocytes," in: Molecular Cytogenetics, Plenum Press, New York, pp. 49-58.

Bobrova, I. G., and Gruzova, M. N., 1967, "Morphological and cytochemical studies of the oocyte nuclei in the sawfly in relation to the problem of the karyosphere," Tsitologiya, *9*, 27-33.

Bogdanov, Yu. F., 1975, "The ultrastructure of chromosomes in meiosis and the synaptonemal complex," in: Cytology and Genetics of Meiosis [in Russian], Nauka, Moscow, pp. 58-136.

Bogdanov, Yu. F., 1977. "Formation of cytoplasmic synaptonemal-like polycomplexes at leptotene and normal synaptonemal complexes at zygotene in *Ascaris suum* male meiosis," Chromosoma, *61*, 1-21.

Bonner, W. M., 1975a, "Protein migration into nuclei. I. Frog oocyte nuclei in vivo accumulate microinjected histones, allow entry to small proteins, and exclude large proteins," J. Cell Biol., *64*, 421-430.

Bonner, W. M., 1975b, "Protein migration into nuclei. II. Frog oocyte nuclei accumulate a class of microinjected oocyte cytoplasmic proteins," J. Cell Biol., *64*, 431-437.

Bottke, W., 1973a, "Lampenbürstenchromosomen und Amphinukleolen in Oocytenkernen der Schnecke *Bithynia tenteculata* L.," Chromsoma, *42*, 175-190.

Bottke, W., 1973b, "Zur Ultrastruktur des Ovars von *Viviparus contectus* (Millet, 1813) (*Gastropoda, Prosobranchia*). II. Die Oocyten," Z. Zellforsch., *138*, 239-259.

Bouchard, R. A., and Stern, H., 1980, "DNA synthesized at pachytene in *Lillium*: a nondivergent subclass of moderately repetitive sequences," Chromosoma, *81*, 349-364.

Bouteille, M., Laval, M., and Dupuy-Coin, A. M., 1974, "Localization of nuclear functions as revealed by ultrastructural autoradiography and cytochemistry," in: The Cell Nucleus, Academic Press, New York—London, pp. 5-71.

Brown, C. A., and Ris, H., 1959, "Amphibian oocyte nucleoli," J. Morphol., *104*, 377-402.

Brown, D. D., and Blackler, A. W., 1972, "Gene amplification proceeds by a chromosome copy mechanism," J. Mol. Biol., *63*, 75-83.

Brown, D. D., and Dawid, J. B., 1968, "Specific gene amplification in oocytes," Science, *160*, 272-280.

Brown, R. D., and Tocchini-Valentini, G. P., 1972, "On the role of

RNA in gene amplification," Proc. Natl. Acad. Sci. USA, *69*, 1746-1748.

Büning, J., 1972, "Untersuchungen am Ovar von *Bruchidius obtectus* Say. (*Coleoptera*, *Polyphaga*) zur Klärung des Oocytenwachstums in der Prävitellogenese," Z. Zellforsch., *128*, 241-282.

Büning, J., 1979, "The trophic tissue of telotrophic ovarioles in polyphaga Coleoptera," Zoomorphologie, *93*, 33-50.

Callan, H. G., 1963, "The nature of lampbrush chromosomes," Int. Rev. Cytol., *15*, 1-34.

Callan, H. G., 1965, "The structure of lampbrush chromosomes," Genet. Today, *2*, 323-324.

Callan, H. G., 1966, "Chromosomes and nucleoli of the axolotl *Ambystoma mexicanum*," J. Cell Sci., *2*, 85-108.

Callan, H. G., 1967. "The organization of genetic units in chromosomes," J. Cell Sci., *2*, 1-7.

Carpenter, A. T. C., 1975, "Electron microscopy of meiosis in *Drosophila melanogaster* females. II. The recombination nodule — a recombination-associated structure at pachytene? Proc. Natl. Acad. Sci. USA, *72*, 8, 31-86.

Carpenter, A. T. C., 1979, "Synaptonemal complex and recombination nodules in recombination-deficient mutants of *Drosophila melanogaster*," Chromosoma, *75*, 259-292.

Carpenter, A. T. C., 1981, "EM autoradiographic evidence that DNA synthesis occurs at recombination nodules during meiosis in *Drosophila melanogaster* females," Chromosoma, *83*, 59-80.

Cassidy, J. D., and King, R. C., 1972, "Ovarian development in *Habrobracon juglandis* (Ashmead) (*Hymenoptera*, *Braconidae*). I. The origin differentiation of the oocyte-nurse cell complex," Biol. Bull., *143*, 483-505.

Cave, M. D., 1973. "Synthesis and characterization of amplified DNA in oocytes of the house cricket *Acheta domesticus* (*Orthoptera*, *Gryllidae*)," Chromosoma, *42*, 1-22.

Cave, M. D., and Allen, E. R., 1974, "Nucleolar DNA in oocytes of crickets: representatives of the subfamilies *Oecanthinae* and *Gryllotalpinae* (*Orthoptera*, *Gryllidae*)," J. Morphol., *142*, 379-394.

Chentsov, Yu. S., and Polyakov, V. Yu., 1974, Cell Nucleus [in Russian], Nauka, Moscow.

Chmilevskii, D. A. 1973, "Biosynthetic activity of the nucleus and cytoplasm in developing oocytes of *Acerina cernua* L. (Teleostei)," Author's abstract of Candidate's dissertation, Leningrad.

Choi, W. C., and Nagl, W., 1977, "Patterns of DNA and RNA synthesis during the development of ovarian nurse cells in *Gerris najas* (*Heteroptera*)," Dev. Biol., *61*, 262-272.

Clark, G., and Rosenbaum, J. L., 1979, "An actin-filament matrix in hand-isolated nuclei of *X. laevis* oocytes," Cell, *18*, 1101-1108.

Chouinard, L. A., 1973, "An electron-microscope study of the extranucleolar bodies during growth of the oocyte in the prepubertal mouse," J. Cell Sci., *12*, 55-70.

Chouinard, L. A., 1975, "A light- and electron-microscope study of the oocyte nucleus during development of the antral follicle of the prepubertal mouse," J. Cell Sci., *17*, 589-616.

Comings, D. E., and Okada, T. A., 1976, "Nuclear proteins. III. The fibrillar nature of the nuclear matrix," Exp. Cell Res., *103*, 341-360.

Crone, M., Levy, E., and Peters, H., 1965, "The duration of the premeiotic DNA synthesis in mouse oocytes," Exp. Cell Res., *39*, 678-688.

Da Cunha, A. B., Pavan, C., Morgante, J. S., and Garrido, M. C., 1969, "Studies on cytology and differentiation in *Sciaridae*. II. DNA redundancy in salivary gland cells of *Hybosciara fragilis*," Genetics, *61*, Suppl., 335-344.

Dapples, C. C., and King, R. C., 1970, "The development of the nucleolus of the ovarian nurse cell of *Drosophila melanogaster*," Z. Zellforsch., 103, 34-47.

Das, N. K., and Alfert, M., 1973, "Prenucleolus bodies of *Urechis* oocytes," J. Cell Sci., *12*, 781-787.

Davidson, E., 1969, Gene activity in Early Development, Academic Press, New York—London.

Dawid, I. B., and Brown, D. D., 1970, "The mitochondrial and ribosomal DNA components of oocytes of *Urechis caupo*," Dev. Biol., *22*, 1-14.

Deltour, R., Gautier, A., and Fakan, J., 1979, "Ultrastructural cytochemistry of the nucleus in *Zea mays* embryos during germination," J. Cell Sci., *40*, 43-62.

Dhainaut, A., 1970, "Étude cytochimique et ultrastructurale de l'évolution ovocytaire de *Nereis pelagica* L. (Annelide, Polychete). I. Ovogenèse naturelle," Z. Zellforsch., *104*, 375-389.

Drozdov, A., and Parfenov, V., 1983, "Actin-filaments in oocyte nuclei of *Rana temporaria*," Dokl. Akad. Nauk SSSR, *273*, 1237-1240.

Du Praw, E. J., 1970, DNA and Chromosomes, Holt, Reinhart and Winston, New York.

Duspiva, F., 1981, "RNP-particles in insect eggs, candidates for the establishment of a morphogenetic gradient," Fortschr. Zool., *26*, 87-99.

Dyer, R. F., Rucby, J. N., and Skalko, R. G., 1973, "Ultrastructural observations on membranous structures in developing mouse oocytes," Z. Zellforsch., *137*, 159-166.

Eddy, E. M., and Ito, S., 1971, "Fine structural and radioautographic observations on dense perinuclear cytoplasmic material in tadpole oocytes," J. Cell Biol., *49*, 90-108.

Eddy, E. M., 1975, "Germ plasm and the differentiation of the germ cell line," Int. Rev. Cytol., *43*, 229-280.

Englehardt, P., and Pusa, K. 1972, "Nuclear pore complex: 'presstud' elements of chromosomes in pairing and control," Nat., New Biol., *240*, 163-166.

Engelhardt, P., Plagens, U., Zbarsky, I. B., and Filatova, L., 1982, "Granules 25-30 nm in diameter — a basic constituent of the nuclear matrix, chromosome scaffold, and nuclear envelope," Proc. Natl. Acad. Sci. USA, *78*, 6937-6940.

Fakan, S., and Hernandez-Verdun, D., 1986, "The nucleolus and the nucleolar organizer regions," Biol. Cell, *56*, 189-206.

Fiil, A., 1974, "Structural and functional modifications of the nucleus during oogenesis in the mosquito *Aedes aegypti*," J. Cell Sci., *14*, 51-68.

Fiil, A., 1976, "Oogenesis in the malaria mosquito – *Anopheles gambiae*," Cell Tissue Res., 167, *1*, 23-35.

Fiil, A., and Moens, P., 1973, "The development structure and function of modified synaptonemal complexes in mosquito oocytes," Chromosoma, *41*, 37-62.

Fiil, A., Goldstein, P., and Moens, P. B., 1977, "Precocious formation of synaptonemal-like polycomplexes and their subsequent fate in female *Ascaris lumbricoides* var. *suum*," Chromosoma, *65*, 21-36.

Filatova, L. S., 1986, "Ultrastructure and protein composition of karyosphere of *Rana temporaria* oocytes," Author's Abstract of Candidate's Dissertation, Moscow.

Filatova, L. S., Stolyarov, S. D., and Zbarsky, I., 1982, "The ultrastructure of the oocyte karyosphere and nuclear envelope of grass frog *Rana temporaria* after nuclease treatment and extraction of chromatin and alkali-soluble proteins," Tsitologiya, *24*, 131-136.

Franke, W. W., 1974, "Structure, biochemistry, and functions of the nuclear envelope," Int. Rev. Cytol., 4, Suppl., 72-211.

Franke, W. W., and Scheer, U., 1970a, "The ultrastructure of the nuclear envelope of amphibian oocytes: a reinvestigation. I. The mature oocyte," J. Ultrastruct. Res., *30*, 288-316.

Franke, W. W., and Scheer, U., 1970b, "The ultrastructure of the nuclear envelope of amphibian oocytes: a reinvestigation. II. The immature oocyte and dynamic aspects," J. Ultrastruct. Res., *30*, 317-327.

Franke, W. W., and Scheer, U., 1971, "Some structural differentiation in the HeLa cell: heavy bodies, annulate lamellae, and cotte de maillet endoplasmic reticulum," Cytobiologie, 4, 317-329.

Franke, W. W., and Scheer, U., 1974, "Structure and function of the nuclear envelope," in: The Cell Nucleus, Vol. 1, Academic Press, New York–London, pp. 220-328.

Franke, W. W., Scheer, U., and Fritsch, H., 1972, "Intranuclear and cytoplasmic annulate lamella in plant cells," J. Cell Biol., *53*, 823-827.

Gaginskaya, E. R., 1972, "The nuclear structures of oocytes in adult birds. Protein bodies and the karyosphere," Tsitologiya, *14*, 569-578.

Gaginskaya, E. R., and Gruzova, M.N., 1969, "Peculiarities of the oogenesis of *Fringilla* (*Aves*)," Tsitologiya, *11*, 1241-1251.

Gaginskaya, E. R., and Gruzova, M. V., 1975, "Detection of the amplified rDNA in ovarial cells of some insects and birds by hybridization of nucleic acids in situ," Tsitologiya, *17*, 1132-1137.

Gall, J. G., 1959, "The nuclear envelope after $KMnO_4$ fixation," J. Biophys. Biochem. Cytol., *6*, 115-117.
Gall, J. G., 1967, "Octagonal nuclear pores," J. Cell Biol., *32*, 391-399.
Gall, J. G., 1968, "Differential synthesis of the genes for ribosomal DNA during amphibian oogenesis," Proc. Natl. Acad. Sci. USA, *60*, 553-560.
Gall, J. G., 1969, "The genes for ribosomal RNA during oogenesis," Genetics, *61*, Suppl., 121-132.
Gall, J. G., Macgregor, H. C., and Kidston, M. E., 1969, "Gene amplification in the oocytes of Dytiscid water beetles," Chromosoma, *26*, 169-187.
Gall, J. G., and Pardue, M. L., 1971, "Nucleic acid hybridization in cytological preparation," in: Methods in Enzymology, Vol. 21, Academic Press, New York–London, pp. 470-483.
Gall, J. G., and Rochaix, J.-D., 1974, "The amplified ribosomal DNA of Dytiscid beetles," Proc. Natl. Acad. Sci. USA, *71*, 1819-1823.
Gassner, G., 1967, "A paracrystalline body in primary spermatocytes and primary oocyte nurse cells of the scorpion fly *Panorpa nuptialis* (Mecoptera)," J. Cell Biol., *35*, 166A.
Gillies, C. B., Rasmussen, S. W., and Wettstein, D. von, 1974, "The synaptonemal complex in homologous and nonhomologous pairing of chromosomes," Cold Spring Harbor Sympos. Quant. Biol., *38*, 9-18.
Goessens, G., and Lepoint, A., 1979, "The nucleolus-organizing regions (NORs): recent data and hypotheses," Biol. Cell, *35*, 211-220.
Grishchenko, T. A., 1972, "Morphological and autoradiographical studies of DNA synthesis in rabbit oocytes in early oogenesis," Tsitologiya, *14*, 558-566.
Gruzova, M. N., 1960, "Formation of karyosphere during oogenesis of *Chrysopa* (*Neuroptera*)," Tsitologiya, *2*, 519-527.
Gruzova, M. N., 1962a, "The formation of the karyosphere during oogenesis of *Panorpa* (*Mecoptera*)," Tsitologiya, *4*, 150-159.
Gruzova, M. N., 1962b, "The karyosphere in oogenesis of *Blaps lethifera* (*Insecta*: *Coleoptera*)," Tsitologiya, *4*, 335-338.
Gruzova, M. N., 1971, "The functional morphology of nuclear structures in relation to various types of oogenesis," Usp. Sovrem. Genet., *3*, 206-212.
Gruzova, M. N., 1974, "Nuclear structures in the ovarioles of the moth *Laspeyresia pomonella*. Sex chromatin in trophocytes," Ontogenez, *5*, 548-558.
Gruzova, M. N., 1975a, "Some aspects of meiosis in oogenesis," in: Cytology and Genetics of Meiosis [in Russian], Nauka, Moscow, pp. 113-136.
Gruzova, M. N., 1975b, "The karyosphere during oogenesis," Tsitologiya, *17*, 219-237.
Gruzova, M. N., 1979a, "Nuclear structures in telotrophic ovarioles of darkling beetles (*Tenebrionidae*, *Polyphaga*). I. Trophocytes of *Blaps lethifera*," Ontogenez, *10*, 13-27.

Gruzova, M. N., 1979b, "Functional morphology of the nuclear structures in connection with karyosphere formation," Author's Abstract of Doctoral Dissertation, Leningrad.

Gruzova, M. N., 1982, "Ultrastructure of the karyosphere in darkling beetles (*Tenebrionidae Coleoptera. Polyphaga*)," Monit. Zool. Ital. (N.S.), *16*, 231-246.

Gruzova, M. N., 1985, "Sex chromatin and evolution of nucleoli in trophocytes of the beetle *Blaps lethifera* (*Tenebrionidae*)," Tsitologiya, *27*, 368-375.

Gruzova, M. N., and Batalova, F. M., 1979, "Nuclear structures in telotrophic ovarioles of darkling beetles (*Tenebrionidae, Polyphaga*). II. The nucleus of oocytes of *Blaps lethifera* and *Gnaptor spinimanus*. Light microscopic data," Ontogenez, *10*, 323-331.

Gruzova, M. N., and Dzyuba, S. M., 1973, "A solitary type of oogenesis, examined in *Mizuhopecten yessoensis* (*Mollusca, Lemellibranchia*)," Tsitologiya, *15*, 283-290.

Gruzova, M. N., and Mamaeva, S. E., 1984, "Revealing the sex chromosome complex during male meiosis in the beetle *Blaps lethifera* Movsh. (*Tenebrionidae*) using silver staining," Tsitologiya, *26*, 802-806.

Gruzova, M. N., and Mamaeva, S. E., 1986, "The nucleoli- and nucleolus-like bodies in oogenesis as revealed with $AgNO_3$ staining," Tsitologiya, *28*, 1190-1195.

Gruzova, M. N., and Parfenov, V. N., 1973. "The karyosphere in late oogenesis of frogs," Monit. Zool. Ital., *7*, 225-242.

Gruzova, M. N., and Parfenov, V. N., 1976, "The nuclear structures in the late oogenesis of *Rana temporaria*. III. Electron microscope data," Tsitologiya, *18*, 261-274.

Gruzova, M. N., and Parfenov, V. N., 1977, "Ultrastructure of late oocyte nuclei in *Rana temporaria*," J. Cell Sci., *28*, 1-13.

Gruzova, M. N., and Zaichikova, Z. P., 1967, "The karyosphere in oogenesis of the leech *Glossiphonia complanata*," Tsitologiya, *9*, 387-396.

Gruzova, M. N., and Zaichikova, Z. P., 1968, "The fine structure of karyosphere in oocytes of *Chrysopa perla* (*Insecta*)," Tsitologiya, *10*, 1180-1182.

Gruzova, M. N., Zaichikova, Z. P., and Sokolov, I. I., 1972a, "The structural and functional organization of the oocyte nuclei of *Chrysopa* (Neuroptera). I. The nucleolar apparatus and extrachromosomal DNA," Tsitologiya, *15*, 269-276.

Gruzova, M. N., Zaichikova, Z. P., and Sokolov, I. I., 1972b, "Functional organization of the nucleus of *Chrysopa perla* L. (*Insecta, Neuroptera*)," Chromosoma, *37*, 353-386.

Gruzova, M. N., Batalova, F. M., and Ermolaeva, L. V., 1986, "Oogenesis of the raptorial gallfly *Aphidoletes aphidimyza*. I. Ultrastructure of young oocytes of pupae and imago," Tsitologiya (in press).

Guenin, H. A., 1965, "Observations sur la structure submicroscopique du complex axial dans les chromosomes méiotiques chez *Gryllus*

campestris L. et *Gr. bimaculatus* de Geer (Orthoptera, Gryllidae)," J. Microsc., 4, 748-758.
Halkka, L., and Halkka, O., (1968), "RNA and protein in nucleolar structures of dragonfly oocytes," Science, *162*, 803-805.
Halkka, L., and Halkka, O., 1975, "Accumulation of gene products in the oocytes of the dragonfly *Cordulia aenea*, L. I. The nematosomes," J. Cell Sci., *19*, 103-116.
Halkka, O., Halkka, L., and Nyholm, M., 1969, "Lammellate bodies in the nuclei of dragonfly (*Odonata*) oocytes," Z. Zellforsch., *94*, 524-541.
Hay, E. D., and Gurdon, J. B., 1967, "Fine structure of the nucleolus in normal and mutant *Xenopus* embryos," J. Cell Sci., *2*, 151-162.
Henderson, S. A., 1971, "Grades of chromatid organization in mitotic and meiotic chromosomes. II. Their interpretation in terms of the master—slave model," Chromosoma, *35*, 41-57.
Heubner, E., Tobe, S. S., and Davey, K. G., 1975, "Structural and functional dynamics of oogenesis in *Clossina austeni*: general features, previtellogenesis, and nurse cells," Tissue Cell, *7*, 297-317.
Holm, P. B., and Rasmussen, S. W., 1981, "Chromosome pairing, crossing over, chiasma formation, and disjunction as revealed by three-dimensional reconstructions," Int. Cell Biol., 1980-1981, Springer—Verlag, Berlin—Heidelberg—New York, pp. 195-204.
Hotta, Y., and Stern, H., 1978, "DNA unwinding protein from meiotic cells of *Lilium*," Biochemistry, *17*, 1872-1880.
Hotta, Y., Bennet, M. D., and Toledo, L. A., 1979, "Regulation of R protein and endonuclease activities in meiocytes by homologous chromosome pairing," Chromosoma, *72*, 191-201.
Hourcade, D., Dressler, D., and Wolfson, J., 1973, "The amplification of ribosomal RNA genes involves a rolling circle intermediate," Proc. Natl. Acad. Sci. USA, *70*, 2916-2920.
Howell, W. M., and Black, D. A., 1980, "Controlled silver staining of nucleolus organizer regions with protective colloidal developer: a 1-step method," Experientia, *36*, 104.
Hsu, W. S., 1963, "The nuclear envelope in the developing oocytes of the tunicate *Boltenia villosa*," Z. Zellforsch., *58*, 660-678.
Hsu, W. S., 1967, "The origin of annulate lamellae in the oocyte of the ascidian *Boltenia villosa* Stimpson," Z. Zellforsch., *82*, 376-390.
Jacob, J., 1969, "Changes in nucleolar ultrastructure during amphibian development," Exp. Cell Res., *54*, 281-284.
Jaworska, H., and Lima-de-Faria, A., 1969, "Multiple synaptonemal complexes at the region of gene amplification in *Acheta*," Chromosoma, *28*, 309-327.
Jaworska, H., and Lima-de-Faria, A., 1973a, "Amplification of ribosomal DNA in *Acheta*. VI. Ultrastructure of two types of nucleolar components associated with ribosomal DNA," Hereditas, *74*, 169-186.
Jaworska, H., and Lima-de-Faria, A., 1973b, "Amplification of ri-

bosomal DNA in *Acheta*. VII. Transfer of DNA–RNA assemblies from the nucleus to the cytoplasm," Hereditas, *74*, 187-204.
Jörgensen, M., 1913, "Zellenstudien. I. Morphologische Beiträge zur Problem des Eiwachstum," Arch. Zellforsch., *10*, 1-126.
Kalt, M., and Gall, J. G., 1974, "Observations on early germ cell development and premeiotic ribosomal DNA amplification in *Xenopus laevis*," J. Cell Biol., *62*, 460-472.
Karasaki, S., 1965. "Electron microscopic examination of the sites of nuclear RNA synthesis during amphibian embryogenesis," J. Cell Biol., *26*, 937-958.
Kato, K., 1968, "Cytochemistry and fine structure of elimination chromatin in Dytiscidae," Exp. Cell Res., *52*, 507-522.
Kay, R. R., and Johnston, I. R., 1973, "The nuclear envelope: current problems of structure and of function," Subcell. Biochem., *2*, 127-166.
Kessel, R. G., 1964, "Intranuclear annulate lamellae in oocytes of the tunicate *Styela partita*," Z. Zellforsch., *63*, 37-41.
Kessel, R. G., 1965, "Intranuclear and cytoplasmic annulate lamellae in tunicate oocytes," J. Cell Biol., *24*, 471-487.
Kessel, R. G., 1966a, "An electron microscope study of nuclear-cytoplasmic exchange in oocytes of *Ciona intestinalis*," J. Ultrastruct. Res., *15*, 181-196.
Kessel, R. G., 1966b, "Some observations on the ultrastructure of the oocyte of *Thyone briareus* with special reference to the relationship of the Golgi complex and endoplasmic reticulum in the formation of yolk," J. Ultrastruct. Res., *16*, 305-309.
Kessel, R. G., 1968, "Annulate lamellae," J. Ultrastruct. Res., *10*, Suppl., 1-81.
Kessel, R. G., 1969, "Cytodifferentiation in the *Rana pipiens* oocyte. I. Association between mitochondria and nucleolus-like bodies in young oocytes," J. Ultrastruct. Res., *28*, 61-77.
Kessel, R. G., and Beams, H. W., 1963, "Nucleolar extrusion in oocytes of *Thyone briareus*," Exp. Cell Res., *32*, 612-615.
Kessel, R. G., and Beams, H. W., 1968, "Intranucleolar membranes and nuclear-cytoplasmic exchange in young crayfish oocytes," J. Cell Biol., *39*, 735-741.
Keyl, H. G., 1975, "Lampbrush chromosomes in spermatocytes of *Chironomus*," Chromosoma, *51*, 75-91.
Kezer, J., Macgregor, H. C., and Schabrach, E., 1971, "Observations on the membranous components of amphibian oocyte nucleoli," J. Cell Sci., *8*, 1-18.
Kiknadze, I. I., and Belyaeva, E. S., 1965, "The structure of the nucleolus in early embryogenesis," Genetika, *3*, 11-14.
Kiknadze, I. I., Zybina, T. G., Zybina, E. V., and Zhelezova, A. L., 1980, "Peculiarities of the nuclear structure and karyosphere formation in the oogenesis of the mink," Tsitologiya, *22*, 127-133.
King, R. C., 1970, "The meiotic behavior of *Drosophila* oocyte," Int. Rev. Cytol., *28*, 125-165.

King, R. C., and Büning, J., 1984, "The origin and functioning of insect oocytes and nurse cells," in: Comprehensive Insect Physiology, Biochemistry, and Pharmacology, Pergamon Press, Oxford–New York, Vol. 1, pp. 91-145.

Kloetzei, P.-M., Whitfield, W., and Sommerville, J., 1981, "Composition and structural organization of nuclear ribonucleoprotein in amphibian oocytes," in: International Cell Biology, 1980-1981. Springer–Verlag, Berlin–Heidelberg–New York, pp. 66-71.

Kloc, M., 1980, "Extrachromosomal DNA and its activity in RNA synthesis in oogonia and oocytes in the pupal ovary of *Creophilus maxillosus* (*Staphylinidae*, *Coleoptera-Polyphaga*)," Eur. J. Cell Biol., *21*, 328-334.

Kloc, M., and Matuszewski, B., 1977, "Extrachromosomal DNA and the origin of oocytes in the telotrophic-meroistic ovary of *Creophilus maxillosus* (L) (*Staphylinidae*, *Coleoptera-Polyphaga*)," W. Roux's Arch. Dev. Biol., *183*, 351-368.

Koch, E. A., and King, R. C., 1966, "The origin and early differentiation of the egg chamber of *Drosophila melanogaster*," J. Morphol., *119*, 283-303.

Koch, E. A., Smith, P. A., and King, R. C., 1967, "The division and differentiation of *Drosophila* cystocytes," J. Morphol., *121*, 55-70.

Kozhanova, N. I., 1974, "Effects of thiotef treatment on oogenesis of *Eurygaster integriceps* Put., Tr. Vses. Inst. Zashch. Rast., *40*, 94-104.

Kozhanova, N. I., and Gruzova, M. N., 1975, "Differentiation of oocytes and nurse cells in insect ovaries with different ovariole structures," Tsitologiya, *17*, 607-613.

Kozhanova, N. I., Gruzova, M. N., and Bulyginskaya, M. A., 1973, "Effects of alkylating agents on the ovarioles of the codling moth *Laspeyresia pomonella*," Tsitologiya, *15*, 985-989.

Kroning, F., 1923, "Studien zur Chromatinreifung der Keimzellen. Die Tetradenbildung und die Reifeteilungen bei einigen Nematoden," Arch. Zellforsch., *17*, 63-85.

Kubinskii, O. Yu., 1974, "The sex cycle, oogenesis, and regulation of fertility in *Testudinalia tesselata* (Mull.) (Gastropoda) from a littoral zone of the Bering Sea," Author's Abstract of Candidate's Dissertation.

Kulikova, V. M., 1963, "Comparative studies of morphology of nuclear structures in oogenesis of some reptile species," Tsitologiya, *5*,, 648-680.

Kunz, W., Trepte, H. H., and Bier, K., 1970, "On the function of the germline chromosomes in the oogenesis of *Wachtiella persicaria* (*Cecidomyiidae*)," Chromosoma, *30*, 180-192.

Kurilo, L. F., 1981, "Extrusion of nucleolus-like bodies from human oocyte nuclei," Tsitologiya, *23*, 894-899.

Kurilo, L. F., 1982, "About some aspects of human oogenesis," in: Genetics, Biochemistry, and Cytology of Meiosis [in Russian], Nauka, Moscow, pp. 88-92.

Kurilo, L. F., 1985, "Morpho-functional character of mammals and human oogenesis," Author's Abstract of Doctoral Dissertation, Moscow.

Kuzmina, S. T., Buldyaeva, L., Troitskaya, L., and Zbarsky, I., 1981, "Characterization and fractionation of rat liver nuclear matrix," Eur. J. Cell Biol., *25*, 225-232.

Lacroix, J. C., and Loones, M. T., 1971, "Fragmentation par les rayons X de l'organisateur d'une différentiation de chromosome en écouvillon (lampbrush), chez *Pleurodeles waltlii*," Chromosoma, *36*, 112-118.

Lane, N. J., 1967, "Spheroidal and ring nucleoli in amphibian oocytes. Patterns of uridine incorporation and fine structural features," J. Cell Biol., 35, *2*, 421-434.

Lane, N. J., 1969, "Internuclear fibrillar bodies in actinomycin D-treated oocytes," J. Cell Biol., *40*, 286-291.

Lev, A. A., 1966, "The exit of the nucleolus from isolated nuclei in oocytes of sea urchin," Tsitologiya, *8*, 748-749.

Lima-de-Faria, A., 1969, "DNA replication and gene amplification in heterochromatin," in: Handbook of Molecular Cytology, Elsevier, Amsterdam—London, pp. 278-324.

Lima-de-Faria, A., 1974a, "The molecular organization of the chromosomes of *Acheta* involved in ribosomal DNA amplification," Cold Spring Harbor Sympos. Quant. Biol., *38*, 559-571.

Lima-de-Faria, A., 1974b, "The three-dimensional organization of the amplified ribosomal DNA complex in *Acheta*," Proc. Natl. Acad. Sci. USA, *71*, 4778-4782.

Loseva, L. M., 1970, "Observations of oogenesis in Actinia. I. Oogenesis in *Bunodactis stella* (Verr.)," Vestn. Leningr. Gos. Univ., *21*, 48-61.

Macgregor, H. C., 1972, "The nucleolus and its genes in amphibian oogenesis," Biol. Rev. Cambridge Philos. Soc., *47*, 177-210.

Macgregor, H. C., 1982, "Ways of amplifying ribosomal genes," in: Nucleolus, E. G. Jordan and C. A. Cullis (eds.), Cambridge Univ. Press, pp. 129-152.

Mahdavi, V., and Crippa, M., 1972, "An RNA—DNA complex intermediate in ribosomal gene amplification," Proc. Natl. Acad. Sci. USA, *69*, 1749-1752.

Mancino, G., Barsacchi, G., and Nardi, B., 1969, "The lampbrush chromosomes of *Salmonella salamandra* (L.) (Amphibia, Urodela)," Chromosoma, *26*, 365-387.

Matuszewski, B., 1966, "Nuclear origin of the mitotic spindle in the oogenesis of *Oligotrophus* Schmidti (*Cecidomyiidae*; *Diptera*)," Chromosoma, *19*, 194-207.

Matuszewski, B., 1982, Animal Cytogenetics. Diptera I: Cecidomyiidae, Gebrüder Borntraeger, Berlin—Stuttgart.

Matuszewski, B., and Hoser, P., 1975, "Gene amplification and its effect on the structure and function of the oocyte nucleus in the whirligig beetle *Gyrinus natator* (*Gyrinidae*, *Coleoptera-Adiphaga*)," Experientia, *31*, 431-433.

Matuszewski, B., and Kloc, M., 1976, "Gene amplification in oocytes of the rove beetle *Creophilus maxillosus* (*Staphylinidae*, *Coleoptera-Polyphaga*)," Experientia, *32*, 34-36.

Matuszewski, B., Hoser, P., Hoser, G., and Michalak, M., 1977, "Organization of the oocyte nucleus in Silphid beetles (*Silphidae, Coleoptera-Polyphaga*)," Experientia, *33*, 883-885.
Maul, G. G., 1977, "The nuclear and the cytoplasmic pore complex: structure, dynamics, distribution, and evolution," Int. Rev. Cytol., Suppl., *6*, 76-186.
Maundrell, K., 1975, "Proteins of the newt oocyte nucleus: analysis of the nonhistone proteins from lampbrush chromosomes, nucleoli, and nuclear sap," J. Cell Sci., *17*, 579-588.
Meng, C., 1968, "Strukturwandel und histochemische Befunde insbesondere am Oosom während der Oogenese und nach der Ablage des Eies von *Pimplo furionellae* L. (*Hymenoptera, Ichneumonidae*)," W. Roux's Arch. Entwicklungsmech. Org., *161*, 162-208.
Merriam, R. W., 1959, "The origin and fate of annulate lamellae in maturing sand-dollar eggs," J. Biophys. Biochem. Cytol., *5*, 117.
Meyer, G. F., Sokoloff, S., Wolf, B. E., and Brand, B., 1979, "Accessory nuclei (nuclear membrane balloons) in the oocytes of the dipteran *Phryne*," Chromosoma, *75*, 89-99.
Miller, O. L., and Beatty, B. R., 1969, "Extrachromosomal nucleolar genes in amphibian oocytes," Genetics, *61*, Suppl., 133-144.
Miller, O. L., and Gonzales, F., 1976, "The relationship of ribosomal RNA synthesis to the formation of segregated nucleoli and nucleolus-like bodies," J. Cell Biol., *71*, 939-950.
Mirré, C., and Stahl, A., 1978, "Peripheral RNA synthesis of fibrillar center in nucleoli of Japanese quail oocytes and somatic cells," J. Ultrastruct. Res., *64*, 377-387.
Moens, P. B., 1969. "Multiple pore complexes in grasshopper spermatocytes and spermatids," J. Cell Biol., *40*, 542-551.
Monneron, A., and Bernhard, E., 1969, "Fine structural organization of the interphase nucleus in some mammalian cells," J. Ultrastruct. Res., *27*, 266-288.
Monroe, J. H., Schidlovsky, G., and Chandra, S., 1967, "Membrane pores and herpes virus-type particles in negatively stained whole cells," J. Ultrastruct. Res., *21*, 134-144.
Moses, M. J., 1968, "Synaptonemal complex," Annu. Rev. Genet., *2*, 363-412.
Moses, M. J., 1969, "Structure and function of the synaptonemal complex," Genetics, *61*, 41-52.
Mott, M. R., and Callan, H. G., 1975, "An electron-microscope study of the lampbrush chromosomes of the newt *Triturus cristatus*," J. Cell Sci., *17*, 241-261.
Moyne, G., Nash, R. E., and Puvion, E., 1977, "Perichromatin granules in isolated hepatocytes treated with cortisol and cyclohexamide," Biol. Cell, *30*, 5-16.
Müller, W. P., 1974, "The lampbrush chromosomes of *Xenopus laevis* (Daudin)," Chromosoma, *47*, 283-296.
Nagl, W., 1978, Endopolyploidy and Polyteny in Differentiation and Evolution, North-Holland Publishing Co., Amsterdam—New York—Oxford.
Nardi, I., Ragghianti, M., and Mancino, G., 1972, "Characterization

of the lampbrush chromosomes of the marbled newt *Triturus marmaratus*," Chromosoma, *37*, 1-22.

Nikolaenko, N., Parfenov, V. N., and Gruzova, M. N., 1985, "Intranuclear lipids in late oocytes of *Rana temporaria*," Tsitologiya, *27*, 1129-1136.

Nizovskaya, L. V., and Arronet, V. N., 1975, "Specific features of morphological changes in the nuclear structures in *Holothurioidea* oogenesis," Tsitologiya, *17*, 238-244.

Nørrevang, A., 1973, "Helical structures in the nucleus of early oocytes of *Priapulus*," Exp. Cell Res., *31*, 603-606.

Pakhomova, N. A., 1976, "The peculiarities of oogenesis and embryogenesis in the crab *Hays araneus* (Linn.)," Author's Abstract of Candidate's Dissertation, Leningrad.

Palombi, F., and Viron, A., 1977, "Nuclear cytochemistry of mouse oogenesis. I. Changes of extranucleolar RNP components through meiotic prophase," J. Ultrastruct. Res., *61*, 10-20.

Parfenov, V. N., 1974, "Some observations of the karyosphere in the late oogenesis of *Rana temporaria*," Tsitologiya, *16*, 1023-1025.

Parfenov, V. N., 1979, "The karyosphere during late oogenesis in *Rana ridibunda*," Eur. J. Cell Biol., *19*, 102-108.

Parfenov, V. N., and Galaktionov, K., 1987, "Intranuclear actin filaments in oocytes of *Rana temporaria*," Tsitologiya, *29*, 2.

Parfenov, V. N., and Gruzova, M. N., 1982, "Organization of oocyte nuclei from atretic follicles of *Rana ridibunda*," Tsitologiya, *24*, 528-535.

Parfenov, V. N., and Gruzova, M. N., 1984, "Peculiarities of the karyosphere formation in *Rana ridibunda*. Electron microscope data," Tsitologya, *26*, 165-168.

Parfenov, V. N., Pochukalina, G. N., and Gruzova, M. N., 1983, "Peculiarities of the karyosphere formation in *Rana ridibunda*. Light microscope data," Tsitologiya, *25*, 1243-1251.

Parfenov, V. N., Dudina, L. M., Kostynchek, D. F., and Gruzova, M. N., 1984, "The ultrastructure of oocyte nuclei of human antral follicles. The karyosphere formation," Tsitologiya, *26*, 1343-1350.

Ploton, D., Menager, M., Jeannesson, P., Himber, G., Pigeon, F., and Adnet, J., 1986, "Improvement in the standing and in the visualization of the argyrophilic proteins of the nucleolar organizer region at the optical level," Histochem. J., *18*, 5-14.

Puvion, E., Viron, A., and Bernhard, W., 1977, "Unusual accumulation of ribonucleoprotein constituents in the nucleus of cultured rat liver cells after hypothermal shock," Biol. Cell, *29*, 81-88.

Puvion, E., Puvion-Dutilleul, F., and Leduc, E. H., 1981, "The formation of nucleolar perichromatin granules," J. Ultrastruct. Res., *76*, 181-190.

Raikova, E. V., 1968, "The morphology of the nucleoli during the oocyte growth in *Acipenseridae*," Zh. Obshch. Biol., *29*, 316-333.

Raikova, E. V., 1974, "Ultrastructure of the sterlet oocytes during early vitellogenesis. I. Nuclear ultrastructure," Tsitologiya, *16*, 679-683.
Raikova, E. V., 1976, "Evolution of the nucleolar apparatus during oogenesis in *Acipenseridae*," J. Embryol. Exp. Morphol., *35*, 667-687.
Raikova, E. V., Steinert, G., and Thomas, C., 1979, "Amplified ribosomal DNA in meiotic prophase oocyte nuclei of Acipenserid fishes," W. Roux' Arch. Dev. Biol., *186*, 81-85.
Raska, J., and Smetana, K., 1978, "A further contribution on nucleoii of human lymphocytes," Folia Haematol. (Leipzig), *105*, 200-215.
Raska, J., Novak, J., and Smetana, K., 1980, "Further studies on ring-shaped nucleoli and micronucleoli of human peripheral lymphocytes," Folia Haematol. (Leipzig), *107*, 185-192.
Rasmussen, S. W., 1973. "Ultrastructural studies of spermatogenesis in *Drosophila melanogaster* Meigen," Z. Zellforsch., *140*, 125-144.
Rasmussen, S. W., 1974, "Studies on the development of the synaptonemal complex in *Drosophila melanogaster*," C. R. Trav. Lab. Carlsberg, *39*, 443-468.
Rasmussen, S. W., and Holm, P. B., 1980, "Mechanics of meiosis," Hereditas, *93*, 187-216.
Raven, P., 1961, Oogenesis: The Storage of Developmental Information, Pergamon Press, Oxford—London—Paris.
Ray, A., and Ramamurty, P. S., 1979, "Sources of RNA supply to the oocytes in *Crynodes peregrinus* Fuessly (*Coleoptera, Chrysomelidae*)," Int. J. Insect Morphol. Embryol., *8*, 113-122.
Reger, J. E., Schultz, P. W., and Rouiller, G. C., 1965, "Observations on intranuclear and cytoplasmic paracrystalline structures on differentiating oocytes of the ostracod, *Cypridopsis vidua*," Exp. Cell Res., *39*, 607-612.
Renkawitz, R., and Kunz, W., 1975, "Independent replication of the ribosomal RNA genes in the polytrophic-meroistic ovaries of *Calliphora erythrocephala*, *Drosophila hydei* and *Sarcophaga barbata*," Chromosoma, *53*, 131-140.
Roberts, B., 1975, "The volume and DNA content of extrachromosomal inclusions in the dorsal footpad nuclei of *Tricholioproctia impatiens* (*Sarcophagidae, Diptera*)," Chromosoma, 52, *1*, 49-58.
Romanova, L. G., 1982, "Ultrastructure of the oocyte nucleus and cytoplasm of the snail *Littorina* saxatilis. II. Fine structure of different nuclear bodies," Tsitologiya, *24*, 522-527.
Romanova, L. G., and Gazaryan, K. G., 1966, "Studies of RNA and protein metabolism in the nucleoli of the mollusk's oocytes," Tsitologiya, *8*, 648-652.
Roth, T. F., 1966, "Changes in the synaptonemal complex during meiotic prophase in mosquito oocytes," Protoplasma, *61*, 346-386.
Rousset, A., 1977, "Formation and development of extrachromosomal heterochromatin in the nucleus of the female germal cells in *Chrysopa perla*," C. R. Acad. Sci. Paris, Ser. D., *285*, 65-68.

Rousset, A., 1978a, "La formation des cystes dan l'ovariole de *Chrysopa perla* (L.) (*Neuroptera*: *Chrysopidae*)," Int. J. Insect Morphol. Embryol., *7*, 45-57.
Rousset, A., 1978b, "La différentiation des cellules germinales dans l'ovariole de *Chrysopa perla* (L.) (*Neuroptera*: *Chrysopidae*)," Int. J. Insect Morphol. Embryol., *7*, 59-71.
Ruby, J. R., and Webster, R. M., 1972, "Origin of the Golgi complex in germ cells in the developing ovary of the bat," Z. Zellforsch., *133*, 1-12.
Sareen, M. L., 1965, "Morphological and cytochemical analysis of the female germ cells of the ticks *Argas persicus* (Oken.) and *Hyalomma aegyptium* (Linn.)," Res. Bull. Panjab. Univ., Sci., *16*, 31-44.
Sareen, M. L., 1967, "Nucleoli in the oocytes of some Arachnids," Cytologiya, *32*, 64-71.
Sasha, K., 1965, "Ultrastructure of differentiating oocytes in the trematode *Grogoderina attenuata*. I. The 'nucleolus-like' body and some lamellar membrane systems," Dev. Biol., *12*, 248-268.
Scali, V., and Montanelli, E., 1975, "DNA—RNA bodies in midgut cell of the stick insect *Bacillus rossius*," J. Exp. Zool., *193*, 361-368.
Scharrer, B., and Wurzelmann, S., 1969a, "Ultrastructural study on nuclear—cytoplasmic relationships in oocytes of the African lungfish *Protopterus aethiopicus*. I. Nucleolocytoplasmic pathways," Z. Zellforsch., *96*, 325-343.
Scharrer, B., and Wurzelmann, S., 1969b, "Ultrastructural study on nuclear—cytoplasmic relationship in oocytes of the African lungish *Protopterus aethiopicus*. II. The microtubular apparatus of the nuclear envelope," Z. Zellforsch., *101*, 1-12.
Scheer, U., 1972, "The ultrastructure of the nuclear envelope of amphibian oocytes. IV. On the chemical nature of the nuclear pore complex material," Z. Zellforsch., *127*, 127-148.
Scheer, U., 1973, "Nuclear pore flow rate of ribosomal RNA and chain growth rate of its precursor during oogenesis of *Xenopus laevis*," Dev. Biol., *30*, 13-28.
Scheer, U., and Franke, W. W., 1969, "Negative staining and adenosine triphosphatase activity of annulate lamellae of newt oocytes," J. Cell Biol., *42*, 519-533.
Scheer, U., Kleinschmidt, J. A., and Franke, W. W., 1982, "Transcriptional and skeletal elements in nucleoli of amphibians," in: The Nucleolus, E. G. Jordan and C. A. Cullis (eds.), Cambridge Univ. Press, pp. 25-42.
Scheer, U., Hinssen, H., Franke, W., and Jockusch, B., 1984, "Microinjection of actin-binding proteins and actin antibodies demonstrates involvement of nuclear actin in transcription of lampbrush chromosomes," Cell, *39*, 111-122.
Sheridan, W. F., and Barnett, R. J., 1969, "Cytochemical studies on chromosome ultrastructure," J. Ultrastruct. Res., *27*, 216-229.

Smetana, K., 1974, "The ultrastructural morphology and formation of nuclear ribonucleoprotein structures," Acta Fac. Med. Univ. Brun., *49*, 155-179.

Smetana, K., and Busch, H., 1974, "The nucleolus and nucleolar DNA," in: The Cell Nucleus, Vol. 1, Academic Press, New York–London, pp. 73-148.

Smetana, K., and Likovsky, Z., 1978, "Studies on nucleoli of pigeon erythroid cells," Cytobiologie, *17*, 146-158.

Smetana, K., and Likovsky, Z., 1980, "Nucleolar and nucleolar effect of experimental endotoxemia on rabbit hepatocytes," Virchows Arch. B, *34*, 143-151.

Snow, M. H. L., 1972, "The effect of actinomycin D in vivo upon peripheral nucleoli and other nuclear organelles in oocytes of *Triturus cristatus*," J. Cell Sci., *10*, 833-856.

Sokolov, I. I., 1966, "The cytological bases of sexual reproduction in multicellular animals," in: Handbook of Cytology, Vol. 2 [in Russian], Nauka, Moscow–Leningrad, pp. 300-460.

Solary, A. J., 1972, "Ultrastructure and composition of the synaptonemal complex in spread and negatively stained spermatocytes of the golden hamster and the albino rat," Chromosoma, *39*, 237-263.

Sotelo, J. R., and Wettstein, R., 1964, "Electron microscope study on meiosis. The sex chromosome in spermatocytes, spermatids, and oocytes of *Gryllus argentinus*," Chromosoma, *15*, 389-415.

Sotelo, J. R., and Wettstein, R., 1969, "Organization of normal chromosomes and postirradiation changes in meiotic cells," Genetics, Suppl., *61*, 67-75.

Sotelo, J. R., Garcia, P. B., and Wettstein, R., 1973, "Serial sectioning study of some meiotic stages in *Scaptericus borelli* (*Grylloidea*)," Chromosoma, *42*, 307-333.

Stern, H., and Hotta, J., 1973, "Biochemical controls of meiosis," Annu. Rev. Genet., *7*, 37-66.

Stern, H., and Hotta, J., 1977, "Biochemistry of meiosis," Philos. Trans. R. Soc. London, Ser. B, *277*, 277-394.

Stockert, J. V., Gimenez-Martin, G., and Isogo, J. M., 1970, "Nucleolus and synaptonemal complexes in pachytene meiocytes of *Allium cepa*," Cytobiologie, *2*, 235-250.

Szöllösi, D., 1965, "Extrusion of nucleoli from pronuclei of the rat," J. Cell Biol., *25*, 545-562.

Telfer, W. H., 1975, "Development and physiology of the oocyte–nurse cell syncytium," Adv. Insect. Physiol., *11*, 223-320.

Trendelenburg, M. F., 1974, "Morphology of ribosomal RNA cistrons in oocytes of the water beetle *Dystiscus marginalis* L," Chromosoma, *48*, 119-135.

Trendelenburg, M. F., Scheer, U., and Franke, W. W., 1974, "Structural organization of the transcription of ribosomal DNA in oocytes of the house cricket *Acheta domesticus*," Nat., New Biol., *245*, 167-170.

Trendelenburg, M. F., Franke, W. W., and Scheer, U., 1977, "Frequencies of circular units of nucleolar DNA in oocytes of two insects, *Acheta domesticus* and *Dytiscus marginalis*, and change of nucleolar morphology during oogenesis," Differentiation, *7*, 133-158.

Tsvetkov, A. G., and Gaginskaya, E. R., 1983, "The nuclear matrix of oocytes of the chaffinch (*Fringilla coelebs* L.)," Tsitologiya, *25*, 649-654.

Ullman, J., Lima-de-Faria, A., Jaworska, H., and Bryngelsson, T., 1973, "Amplification of rDNA in *Acheta*. V. Hybridization of RNA complementary to ribosomal DNA with pachytene chromosomes," Hereditas, *74*, 13-24.

van Gansen, P., and Schram, A., 1972, "Evolution of the nucleoli during oogenesis in *Xenopus laevis* studied by electron microscopy," J. Cell Sci., *10*, 339-367.

Vincent, W. S., Halvorson, H. O., Chen, H. R., and Shin, D., 1969, "A comparison of ribosomal gene amplificaton in uni- and multinucleolate oocytes," Exp. Cell Res., *57*, 240-250.

Vlad, M., and Macgregor, H. G., 1975, "Chromomere member and its genetic significance in lampbrush chromosomes," Chromosoma, *50*, 327-347.

Wagner, K., 1923, "eber die Entwicklung des Froscheies," Arch. Zellforsch., *17*, 1-44.

Watson-Coggins, L., and Gall, J. G., 1971, "The timing of meiosis and DNA synthesis during early oogenesis in the toad *Xenopus laevis*," J. Cell Biol., *52*, 569.

Welsch, B., 1974, "An intranuclear paracrystalloid body in the gametocytes of *Panorpa communis* (Mecoptera)," Chromosoma, *48*, 1-9.

Westergaard, M., and Wettstein, R., 1972, The synaptonemal complex," Annu. Rev. Genet., *6*, 71-110.

Wettstein, R., and Sotelo, R., 1965, "Electron microscope study on the meiotic cycle of *Acanthopachylus aculeatus* (*Arachnida*, *Opiliones*). The composite bodies of the primary spermatocytes," Chromosoma, *17*, 246-257.

Wilson, E. B., 1925, The Cell in Development and Heredity, Macmillan, New York.

Wolf, E. E., 1962, L'Origine de la Lignée Germinale chez les Vertébrés et chez Quelques Groupes d'Invertébrés, Paris.

Wolgemuth-Jarashow, J. D., 1977, "The localization of rDNA in small nucleolus of human diplotene oocytes," Hum. Genet., *36*, 63-68.

Zaichikova, Z. P., 1976, "Cytophotometric and autoradiographic studies of DNA synthesis in the nurse cells of *Chrysopa*," Tsitologiya, *18*, 438-444.

Zaichikova, Z. P., and Gruzova, M. N., 1973, "Structural and functional organization of the oocyte nucleus in *Chrysopa* (*Neuroptera*). III. Extrachromosomal DNA in early oogenesis," Ontogenez, *4*, 595-602.

Zaichikova, Z. P., and Gruzova, M. N., 1975, "The ultrastructure of trophocyte nuclei in the polytrophic ovarioles of *Chrysopa* (*Neuroptera*)," Tsitologiya, *17*, 1253-1262.

Zbarsky, I. B., 1972, "The connection between the nuclear envelope and the structures of the cytoplasm and the mechanisms of nuclear—cytoplasmic interactions," Usp. Sovrem. Biol., *73*, 3-35.

Zbarsky, I. B., 1978, "An enzyme profile of the nuclear envelope," Int. Rev. Cytol., *54*, 295-361.

Zbarsky, I. B., and Filatova, L. S., 1979, Investigation of the karyosphere in *Rana temporaria* oocytes," Ontogenez, *10*, 446-449.

Zbarsky, I., Kuzmina, S. N., Buldyaeva, T. V., Perevoshchikova, K. A., Troitskaya, L. P., and Filatova, L. S., 1980, "Nuclear skeleton structures in some normal and tumor cells" (Abstract), in: Formation of Messenger RNA in Eukaryotic Cells, EMBO-Arolla, Switzerland, p. 124.

Zybina, E. V., 1975a, "The ultrastructure of the rabbit oocyte at the stage of a two-layer follicle," Tsitologiya, *17*, 126-130.

Zybina, E. V., 1975b, "An electron-microscopic study of the lampbrush chromosomes and the products of their activity in the oogenesis of the rabbit," Tsitologiya, *17*, 875-880.

Zybina, E. V., and Grishchenko, T. A., 1977, "The ultrastructure of nucleolus-like bodies in the golden hamster's oocytes at the diplonema stage," Tsitologiya, *19*, 1231-1237.

Zybina, E. V., Zybina, T. G., and Dalmane, A. R., 1980, "A study of the nuclear structure at stages of diploneme and diakinesis in oocytes of *Dyromys nitedula*," Tsitologiya, *22*, 381-386.

Chapter 3

DEVELOPMENT OF THE MATURE EGG ORGANIZATION IN AMPHIBIANS, FISH, AND STARFISH DURING THE CONCLUDING STAGES OF OOGENESIS, IN THE PERIOD OF MATURATION*

T. A. Dettlaff

During the concluding stages of oogenesis, in the period of maturation, the fully grown oocyte turns into an egg capable of fertilization and further development. In this period, meiotic divisions of the nucleus and extensive changes in the structure and properties of the oocyte's cytoplasm take place in vertebrates and the majority of invertebrates. For a long time researchers concentrated their attention on the study of meiosis (see Wilson, 1925). Investigations into the changing properties of the cytoplasm in the process of oocyte maturation in vertebrates are relatively recent, although in some vertebrates they were studied at the beginning of this century. The classical investigations by Delage (1899, 1901), Wilson (1903), and Yatsu (1904, 1905, 1907) produced for the first time data pointing to the involvement of the nucleoplasm of the germinal vesicle (GV-oocyte nucleus), in the emergence of certain properties of the mature egg's cytoplasm (see Wilson, 1925). The subjects of the first investigations were the oocytes of the starfish *Asterias glacialis* and the nemertean *Gerebratulus*, and later the oocytes of the annelid worms *Nereis* (Just, 1915; Costello, 1940) and *Chaetopterus*, excreted into the water by the female at the stage of the full-grown oocyte with the as yet not disrupted GV, which makes them suitable for experimentation at these stages.

Following Delage's attempts (1901) to analyze the nature of the factors introduced by the GV contents, the problem was shelved for many years. Later, Dalcq (1924) initiated investigations in

*Some data relevant to the hormonal regulation of oocyte maturation and ovulation are also examined in this chapter.

the same objects of the influence of different ions on the stimulation of GV breakdown in ovulated oocytes of the same material. Studies of the influence of ions on the GV and the changing properties of the oocyte cytoplasm were continued by Heilbrunn and his colleagues (Heilbrunn and Wilbur, 1937; Goldstein, 1953; Heilbrunn and Wilson, 1955; see Heilbrunn, 1952; Sawada, 1952; and others).

In recent years, the main subjects of research have been the oocytes of amphibians, fish (predominantly sturgeon), starfish, and mammals. The results obtained on mammals are reviewed in Chapter 5.

Active investigations into oocyte maturation in amphibians and fish become possible after it was demonstrated that, in using gonadotropic hormones, it could be induced in vitro, and that oocytes maturing under such conditions are capable of normal development (Heilbrunn et al., 1939; Ryan and Grant, 1940; Good and Daniel, 1943; Tchou-Su, 1948; Tchou-Su and Wang Yu-lan, 1958, 1963b; Dettlaff, 1961; Skoblina, 1968, 1973; Jalabert et al., 1972, 1973; Yamauchi and Yamamoto, 1973; Iwamatsu, 1974; Sakun and Gureyeva-Preobrazhenskaya, 1975; Saat, 1980).

Oocyte maturation outside the body of the female opened up great opportunities for investigating not only the hormonal regulation of maturation and the effect of external conditions on ovulation (Wright, 1945), but also the changes of the oocytes themselves in the process of maturation (Tchou-Su, 1948; Tchou-Su and Wang Yu-lan, 1958, 1959, 1963a). Moreover, it allowed microsurgical methods to be used for investigating the role of the GV and the cytoplasm in the development of the mature egg organization in vertebrates, changes in the properties of the cytoplasm in the maturation process, and how the introduction of different substances, cytoplasm, and nuclei into the oocyte affects its maturation (Dettlaff, 1961; Dettlaff et al., 1964). In starfish, Kanatani (1964) has also initiated this direction of research.

The "period of maturation" may be defined as follows: It begins at the stage when the oocytes, having completed their growth, are enveloped in the follicular epithelium, and acquire, under optimal conditions, the capacity to respond by maturation to a stimulating agent (in vertebrates the gonadotropic hormones of the hypophysis); it ends at the stage when the oocyte nucleus becomes blocked at one or another meiotic phase (in vertebrates at the metaphase of the second maturation division, i.e., metaphase II), and the cytoplasm attains such a state of maturity that the oocytes are capable not only of fertilization but of normal development. From this stage on the oocyte is called an egg.

The completion of metaphase II in the egg of vertebrates is effected by an additional agent (spermatozoon, needle prick, or some other factor), causing the appearance of an egg activation impulse and removing nuclear blockage. This division is completed already at the beginning of the fertilization period, and coincides with the transformation of the sperm nucleus into a pronucleus.

In the following sections we examine some data on the effect of gonadotropic hormones and progesterone on oocyte maturation, in amphibians and fishes, as well as on the effect of 1-methyladenine on oocyte maturation in starfishes, change of oocytes in the process of maturation, the participation of cytoplasm, and the GV in the development of the mature egg organization, and also the molecular nature of the processes underlying the changes of the oocyte in the period of maturation.

3.1. EFFECT OF GONADOTROPIC HORMONES ON OOCYTE MATURATION

3.1.1. Effect of Gonadotropic Hormones of the Hypophysis and Progesterone on Oocyte Maturation

Data on the changes of oocytes in the period of maturation were obtained in vitro by cultivating follicles in Ringer solution with a suspension of hypophysis or with progesterone, and also in vivo by injecting a hypophysial suspension into a female. During the recent decade it was shown that in vivo fish oocyte maturation and ovulation could also be induced by the synthetic mammalian LH-RH and its analogs (see Donaldson and Hunter, 1983). For the most-studied fish species the effective doses proved, however, to be very high. One of the superactive LH-RH analogs could be successfully applied to a few fish species, which were introduced into aquaculture in China, by combining ecological and hormonal stimulation (see Lin, 1982). Sturgeons proved to be very sensitive to superactive LH-RH analogs (Goncharov, 1984a, b, 1985). Many species of amphibians were also shown to be sensitive to LH-RH analogs. A comparative study of the efficiency of hypophysial suspension and of synthetic LH-RH analogs has shown that they are hormonally equal. This presents wide prospects of using the synthetic LH-RH analogs in sturgeon hatcheries and in laboratory studies on amphibians (Goncharov, 1984a, b). LH-RH acts via the gonadotropic hormones of the hypophysis which influence the gonads directly.

It was established in experiments in vitro that the stimulation of oocyte maturation by gonadotropic hormones proceeds in two stages: the gonadotropic hormones induce the follicular cells to synthesize and secrete a progesterone-like substance which stimulates oocyte maturation. The effect of gonadotropic hormones on the oocyte is thus mediated through the cells of the follicular epithelium and, in this case, the reacting system is the follicle, i.e., the oocyte covered by follicular epithelium.

Oocyte maturation in vitro can also be stimulated by progesterone. Unlike gonadotropic hormones, progesterone affects the oocytes of amphibians and sturgeon not through the cells of the follicular epithelium, but directly (Masui, 1967; Schuetz, 1967a-c; Shuetz, 1978; Smith and Ecker, 1968; Dettlaff et al., 1968a; Dettlaff, 1970b). Progesterone affects the oocyte from outside; when injected into the oocyte it is ineffective (Masui and Markert, 1971; Smith and Ecker, 1969a, 1971; Reynhout and Smith, 1973). Progesterone also causes oocyte maturation when bound to a polymer which prevents penetration into the oocyte (Ishikawa et al., 1977; Godeau et al., 1978b). However, the question about the initial site of progesterone action has not been finally solved as yet. It has been shown (Tso et al., 1982) that microinjection of progesterone dissolved, not in water but in paraffin oil, induces the reinitiation of maturation in defolliculated full-grown *Xenopus* oocytes. The authors concluded that the progesterone action is initially intracellular. The same result was then obtained (Tso et al., 1984) on denuded, plasma-membrane-free *Bufo bufo* oocytes embedded in agar and immersed in Ringer solution with progesterone. The cytoplasm of such oocytes acquired the capacity to induce maturation of full-grown oocytes.

It has been established that under the effect of progesterone oocyte maturation is possible over a wider temperature range (Davydova, 1972; Dettlaff and Davydova, 1974; Saat, 1985) and depends less on the conditions in which the female was kept before the oocytes were drawn from her (Schuetz, 1967c; Dettlaff et al., 1968b; Dettlaff and Davydova, 1974) than when maturation is induced by gonadotropic hormones.

These differences reflect the different sensitivity of the oocytes and the cells of the follicular epithelium. At the beginning of maturation the oocytes are considerably less sensitive to changes in external conditions than the cells of the follicular epithelium, and the latter's sensitivity, in all probability, is involved in determining the limits within which both natural breeding and the female's response to hypophysial injection are possible (Dettlaff and Davydova, 1974, 1979).

A lesser dependence on external conditions, higher oocyte maturation percentage (Smith et al., 1968; Subtelny et al., 1968), and the convenience of handling a chemically pure hormone have led to the preferred use of progesterone rather than gonadotropic hormones for maturation studies.

A hormone-dependent period and a period of maturation inertia are distinguished in the maturation of amphibian oocytes under the effect of gonadotropic hormones (Tchou-Su and Wang Yu-lan, 1958). Oocyte maturation during the first period requires the presence of gonadotropic hormones for maturation in vitro or in the ovary when the oocytes mature in the female's body. During the second period they may mature and ovulate outside the female's body, in a Ringer solution not containing hypophysial hormones. Follicles transferred to a hormone-free Ringer solution before the onset of maturation inertia remain at this stage and gradually degenerate. The same considerations apply to follicle maturation in the sturgeon and teleosts (Dettlaff, 1961, 1970a; Skoblina, 1968, 1973).

The duration of the hormone-dependent period may vary from minutes to several hours; it is dependent on when the follicles are taken from the females, and relates to the season of the year; the closer the time to the period of natural breeding, the shorter the hormone-dependent period (Tchou-Su and Wang Yu-lan, 1958; Rudneva, 1968; Smith et al., 1968). The duration of the hormone-dependent period during follicle maturation in vitro may also be affected by the concentration of the hormone in the solution. Therefore, in order to obtain comparable conditions for the oocytes of different females, it is necessary to determine the period's shortest duration by using several concentrations of suspensions of acetonized hypophyses (Goncharov, 1971a, b).

The duration of the hormone-dependent period is the only reliable feature by which follicles can be characterized throughout this period. It is essential to know this value when comparing the results of experiments in which follicles from different females are used. As this will differ with the temperature, it is better to indicate a relative rather than absolute duration (see Section 2.2.5).

One cannot speak of a hormone-dependent period when oocyte maturation is stimulated by progesterone in vitro, since labeled progesterone becomes bound with the oocyte plasma membrane within seconds of placing the oocyte in solution and the further presence of progesterone in the solution is unnecessary. Even a 5-sec immersion of the oocyte in a solution containing 10 μg/ml progesterone is sufficient for its maturation (Smith and Ecker, 1969a).

It is unclear so far whether the release of progesterone (or a progesterone-like substance) by follicle cells and its binding by the oocyte is equally rapid when oocyte maturation is stimulated by gonadotropic hormones. If one assumes that this process is just as rapid, then the oocyte maturation period under the effect of progesterone in vitro will correspond to the maturation inertia period under the effect of gonadotropic hormones. However, if progesterone release and binding during the hormone-dependent period is not an instantaneous process, then their correlation will be somewhat mismatched. Data on the time of GV breakdown and the onset of identical maturation division phases in

Acipenser stellatus oocytes which have matured in a progesterone solution and in a hypophysial suspension favor the second assumption (oocytes mature faster in a progesterone solution than in a hypohysial suspension, but the time lag between the onset of similar maturation phases in the oocytes of both kinds is shorter than the duration of the hormone-dependent period). On the other hand, the effect of gonadotropin on follicles within a period of time shorter than the duration of the hormone-dependent period results in shortening of the oocyte maturation time under the effect of progesterone (Goncharov, unpublished data). Since a similar effect was obtained also by means of oocyte pretreatment with subthreshold doses of progesterone (Marot et al., 1977), one may presume that the effect of gonadotropic hormones during the entire hormone-dependent period is accompanied by the release of small progesterone doses, cumulatively affecting the oocyte.

One should note another important distinction, exhibited when stimulating maturation by hypophysial gonadotropic hormones and exogenous progesterone. In the former case, the progesterone is secreted by the cells of the follicular epithelium and, correspondingly, it affects the oocytes only at those stages when the follicular epithelium enveloping them is capable of responding to the gonadotropic hormones by its excretion, whereas, in the latter case (progesterone administered in a solution), oocytes may go over to maturation at earlier stages, when the follicular epithelium enveloping them does not yet respond to the gonadotropic hormones (Goncharov, 1969a, b, 1971b; Reynhout et al., 1975; Felgenhauer and Chulitskaya, 1977).

It should also be noted that, unlike *Anura* oocytes, the oocytes of some urodeles acquired the capacity to respond to the effect of exogenous progesterone in vitro only after the females had been preliminarily injected with gonadotropic hormones (Barsacchi, 1971; Batistoni and Barsacchi, 1974; Brachet, 1974a, b). The maturation of oocytes from unstimulated females was induced in vitro by progesterone, testosterone, and a synthetic steroid (CDMT) only in *Pleurodeles waltlii* (Skoblina et al., 1984b).

Recent years have seen a great expansion of methods to stimulate oocyte maturation in amphibians, fish, and starfish. It has been found that the effect of progesterone may be imitated by many different substances. At the same time, a large number of antagonists have been used to block oocyte maturation induced by progesterone and its substitutes. (See reviews by Shuetz, 1974; Baulieu et al., 1978; Masui and Clark, 1979; Maller and Krebs, 1980; Morrill et al., 1981; Meijer et al., 1984.) The effect of these substances on the oocytes of amphibians and starfish make it possible to assess events taking place in the oocyte at the initial period of maturation, i.e., during the period of meiotic reinitiation. These investigations are considered below.

3.1.2. Hormone-Induced Molecular Changes in the Cytoplasm of Amphibian Starfish Oocytes during the Initial Stages of Meiotic Reinitiation

Within a few seconds of the binding of progesterone with its receptors on the plasma membrane of amphibian oocytes (Kalimi et al., 1979), a chain of molecular events is initiated which leads to the alteration of the oocyte's properties and structure. These changes lead to the acquisition by the oocyte cytoplasm of the capacity, upon its injection into other full-grown oocytes, to induce the latter's maturation without exposure to hormones. Cytoplasm possessing this property should be called "active cytoplasm" (Dettlaff, 1977b; Dettlaff et al., 1977). This property of "active cytoplasm" is usually explained by the appearance in it of a factor which Masui and Markert (1971) called the maturation-promoting factor (MPF).

On the basis of previous data, Masui and Clarke (1979) suggested a hypothetical working scheme for the initiation of oocyte maturation in *X. laevis*. According to this scheme, the liberation of Ca^{2+} takes place in the plasma membrane under the influence of progesterone. This is followed by the liberation of the Ca-dependent regulatory protein calmodulin which activates phosphodiesterase, the latter causing partial degradation of cyclic AMP (cAMP) and a drop in its level. Thus, cAMP-dependent phosphorylation of the proteins is arrested, with a simultaneous increase in cAMP-independent phosphorylation of the protein precursor of the MPF. In *Xenopus*, MPF is a phosphorylated protein, and its nonphosphorylated form appears as a nonactivated MPF precursor. Moreover, since the MPF is capable of amplification, Masui and Clarke assume that this protein is the phosphorylase kinase which catalyzes the phosphorylation of its own precursor. A protein of this type has, indeed, been discovered in oocytes (Bellé et al., 1976, 1979). According to Wassermann and Masui's hypothesis (1975a), cAMP-independent phosphorylation of the MPF precursor in *Xenopus* oocytes requires the synthesis of a protein (initiator).

In a later review of initial changes taking place in the oocyte during the stimulation of maturation, Morrill et al. (1981) assumed that another sequel of the earliest events occurred. At the end of the intensive growth period in the meiotic prophase, the oocyte contains a higher level of intracellular cAMP which is maintained by a tonic stimulus, probably epinephrine or noradrenaline circulating in the blood and affecting the surface β-receptors. Progesterone initiates the liberation of Ca^{2+} from specific membrane sites, closely associated with the adenylate cyclase β-receptor complex, and removes the tonic stimulus. According to the results of these authors, the initial liberation of Ca^{2+}, occurring during the first 1-2 min, causes a short-lived activation of adenylate cyclase and a rise in the cAMP level. The latter, the authors assume, activates the protein kinase which phosphorylates calmodulin and liberates it from the membrane into the cytoplasm, where it acts in all probability as a regulator of the Ca^{2+}-dependent phosphodiesterase. After a brief rise in cAMP activity, it drops steeply and, at the same time, within 8-12 min, the delivery of intracellular Ca^{2+} begins and the liberation of Ca^{2+} from the pool of slowly metabolizing Ca on the membrane and from its other reserves in the cytoplasm occurs. The increase of Ca^{2+} in the cytoplasm in turn stimulates the activity of guanyl cyclase and an increase in the content of cGMP.

In their work, Morrill et al. (1981) support the view that the changes causing a drop in the cAMP content are connected with the progesterone-induced inactivation of adenylate cyclase and precede an increase in phosphodiesterase activity. They believe that the regulation of adenylate cyclase activity is the key element in the initial changes of oocyte maturation.

A detailed analysis of the biochemical processes going on during this period in the regulation of amphibian oocyte maturation and, particularly, of the role of cAMP phosphoproteins and processes of phosphorylation and dephosphorylation will be found in reviews by Maller and Krebs (1980), by Nielson et al. (1982), and by Maller (1985a, b).

The intensification of protein synthesis and phosphorylation, as well as the appearance of the MPF, take place much later, not long before the GV breakdown. According to Wassermann and Smith (1978b), they represent the oocyte's belated response to the effect of the hormone. They will be discussed in Section 3.5.1.

Currently, numerous papers describe investigations into the process taking place in amphibian oocytes under the effect of progesterone and substances imitating it, and also their inhibitors (Schorderet-Slatkine et al., 1980, 1982; Huchon et al., 1981a, b; Wassermann and Smith, 1981; Cartaud et al., 1981; Hanocq-Quertier and Baltus, 1981, 1982; Sunkara et al., 1981; Bellé et al., 1979; and others). They all agree that the liberation of Ca^{2+}, a drop in the activity of adenylate cyclase, a reduction in cAMP content, a rise in the activity of phosphodiesterase and, later, intensification of protein synthesis and phosphorylation take place both under the effect of progesterone and of substances imitating it, in particular propanol. Substances suppressing the GV breakdown, on the contrary, suppress the liberation of Ca^{2+} and the diminution of adenylate cyclase activity, and maintain a high cAMP level in the oocyte. The inhibitors of Ca-independent phosphorylase also hamper maturation. These data indicate that all the processes described above are, apparently, necessary links in the chain of the initial changes taking place in amphibian oocytes during maturation (see Maller et al., 1977; Masui and Clarke, 1979; Maller and Krebs, 1980; Morrill et al., 1981; Maller, 1985a, b). Events taking place in the *X. laevis* oocyte after its stimulation with progesterone and before GV breakdown also include the increase in pH of the cytoplasm to 7.5-7.6, the 2- to 3-fold increase in protein synthesis, and the 6- to 8-fold increase in phosphorylation of the S6 protein (Houle and Wasserman, 1983). Progesterone also causes a flareup of RNA synthesis in the oocytes during the first 1-3 h; however, this has nothing to do with the GV breakdown (Morrill et al., 1975). Despite the great similarity of the molecular processes taking place in amphibian oocytes, they display substantial specific differences.

It is very interesting to compare the initial changes in the cytoplasm of maturing oocytes of amphibians and starfish. In the latter the gonado-stimulating substance (GSS) from the radial nerves exerts a stimulating effect on the cells of the follicular epithelium and causes the formation in them of a meiosis-inducing substance (MIS) (Kanatani, 1964; Kanatani and Shirai, 1967; Schuetz and Biggers, 1967) which was identified as 1-methyladenine (1-Me-Ade) (Kanatani, 1969; Kanatani et al., 1969). The MIS in turn affects the oocyte and initiates changes (see Meijer and Guerrier, 1984) which lead to the formation of MPF (Kishimoto and Kanatani, 1976).

MIS receptors in starfish oocytes are on the outer surface of the plasma membrane (Kanatani and Hiramoto, 1970), as is the case with the progesterone receptors on amphibian oocytes. The binding of 1-Me-Ade to the receptor begins, apparently, during the first second, since after 2 sec the liberation of Ca^{2+} from the inner surface of the membrane and stimulation of the Na^{+}/K^{+} pump occurs in the oocytes (Moreau et al., 1978). An extract obtained from the isolated cortical layer of the oocyte reacts with 1-Me-Ade with the liberation of Ca^{2+} within 0.1 sec. It responds in the same way to the effect of biological analogs and mimetics of 1-Me-Ade while failing to react to maturation inhibitors (Dorée et al., 1978). Five minutes after placing the oocytes in 1-Me-Ade, a 4- to 7-fold intensification of the inclusion of labeled amino acids into proteins is induced (Johnson and Epel, 1980), and the phosphorylation of cortical and, later, endoplasmic proteins is strongly stimulated (Guerrier et al., 1977a). Theophylline, caffeine, and procaine, substances inhibiting the liberation of Ca^{2+} from the plasma membrane and phosphorylation of proteins induced by 1-Me-Ade, suppress meiotic reinitiation when applied during the hormone-dependent period, but are ineffective if applied later (Dorée et al., 1981a).

When starfish oocytes are incubated with 1-Me-Ade (10^{-7} M) for different periods of time, it has been established (Guerrier and Dorée, 1975) that, at 24°C, the oocytes acquire maturation inertia within 4 min 30 sec, and GV breakdown occurs after 18 min. Moreover, it has been demonstrated that different MIS effects acquire different degrees of receptor engagement; minimum for stimulating the operation of the Na^+/K^+ pump, higher for stimulating protein phosphorylation and meiotic reinitiation, and the highest for the maximum liberation of Ca^{2+} (Guerrier et al., 1977a). If 1-Me-Ade action is arrested within the first 10 sec of the maturation inertia period, there takes place a reassociation of the liberated Ca in the middle of the hormone-dependent period: the change of the plasma membrane potential is fixed at an intermediate level, while the removal of 1-Me-Ade before the end of the hormone-dependent period will arrest phosphorylation. In this case, GV breakdown does not take place (Dorée et al., 1978, 1981a). For in vitro amphibian oocyte maturation there are apparently no data on the intermediate arrest of progesterone action.

In contrast with amphibian oocytes, cAMP content does not change throughout the period of 1-Me-Ade action on starfish oocytes (Dorée et al., 1981b), while a 2- or 3-fold increase of cAMP level during oocyte incubation with cholera toxin or theophylline does not prevent meiotic reinitiation (Mazzei et al., 1981). In addition, the calmodulin content does not change in the process of starfish oocyte maturation (Dorée, 1980). Also, another distinction from amphibians is that neither meiotic initiation nor MPF amplification in starfish appears to require protein synthesis, since these processes take place during the complete suppression of protein synthesis in the oocytes by cycloheximide (Dorée, 1982). Finally, the GV material is required for MPF amplification in starfish oocyte maturation (Kishimoto et al., 1981; Kishimoto, 1982) (see also Section 3.4.2.1).

The natural inducer of starfish maturation, 1-Me-Ade, does not stimulate maturation in holothurians. Meiotic maturation can be induced in holothurians by disulfide-reducing agents, e.g., dithiothreitol, 2,3-mercapto-1-propanol, and Z-cysteine (Kishimoto and Kanatani, 1980; Maruyama, 1980). Kishimoto and Kanatani assume that in the echinoderms, as distinct from amphibians, the oocyte reaction to the hormone excreted by the follicular cells is mediated not by the liberation of Ca^{2+} but by the restoration of disulfide bonds.

3.2. CHANGES IN THE STRUCTURE AND PROPERTIES OF AMPHIBIAN AND STURGEON FISH OOCYTES IN THE PROCESS OF MATURATION

The structure and properties of oocytes at successive stages of maturation in the sturgeon are described by Ginsburg and Dettlaff (1969), Dettlaff and Skoblina (1969), Aizenshtadt and Dettlaff (1972), and Vassetzky (1966, 1967, 1970), and according to the previously unpublished materials of the present author. Corresponding amphibian studies have been made by Tchou-Su and Wang Yu-lan (1958), Dettlaff et al. (1964), Van Gansen (1966), Van Gansen and Schram (1968), Aizenshtadt and Dettlaff (1968), Rudneva (1968), Brachet et al. (1970), Brachet (1974a, b), Masui (1973a, b), Huchon et al. (1981a, b), and Ryabova (1982). Oocyte structure during successive maturation stages is represented in Plates XVII-XIX for *Acipenser güldenstädti* and *A. stellatus*.

3.2.1. Follicle Structure and Properties at the Initial Stage and at the End of the Growth Period

We shall conditionally designate as the initial stage of the maturation period that stage at the end of the growth period when the follicular epithelium enveloping the full-grown oocyte (Plate XVIII, 2) is capable of stimulating oocyte maturation in the presence of hypophysial gonadotropic hormones, and the oocyte is capable of maturation under the effect of a progesterone-like substance released by the cells of the follicular epithelium, and also under the effect of exogenous progesterone. Since oocyte structure practically does not change throughout the hormone-dependent period, oocytes from different females with different durations of the hormone-dependent period exhibit a similar structure. In sturgeons and *Anura* a large round or oval nucleus is situated in the center of the animal region, at a small distance from the oocyte surface (Plate XVIII, 1). The chromosomes are at the diplotene stage of prophase I; they appear as elongated bivalents with several chiasmata (Plates VIII, 3 and XVIII, 3), and form a compact body (see Section 2.8 and also Gruzova, 1975).

In full-grown *Xenopus* oocytes (stage VI, after Dumont, 1972), biochemical methods reveal an intensive synthesis of nonribosomal RNA (Anderson and Smith, 1978), while electron microscopy and phase contrast methods demonstrate transcription loci on the chromosomes, part of which still preserve lateral loops (Martin et al., 1980).

Unlike *Anura* oocytes, the chromosomes in *Urodela* oocytes at the initial stage still have the lampbrush structure where active RNA synthesis occurs (Barsacchi, 1971; Batistoni and Barsacchi, 1974; Brachet, 1974a, b).

The GV in all the investigated species contain numerous nucleoli. In *Anura* they are concentrated around the chromosomes, while in sturgeons they are usually situated close to the nuclear membrane. In sturgeons the GV membrane at this stage forms protrusions on the vegetal side, and cytoplasm adjacent to these

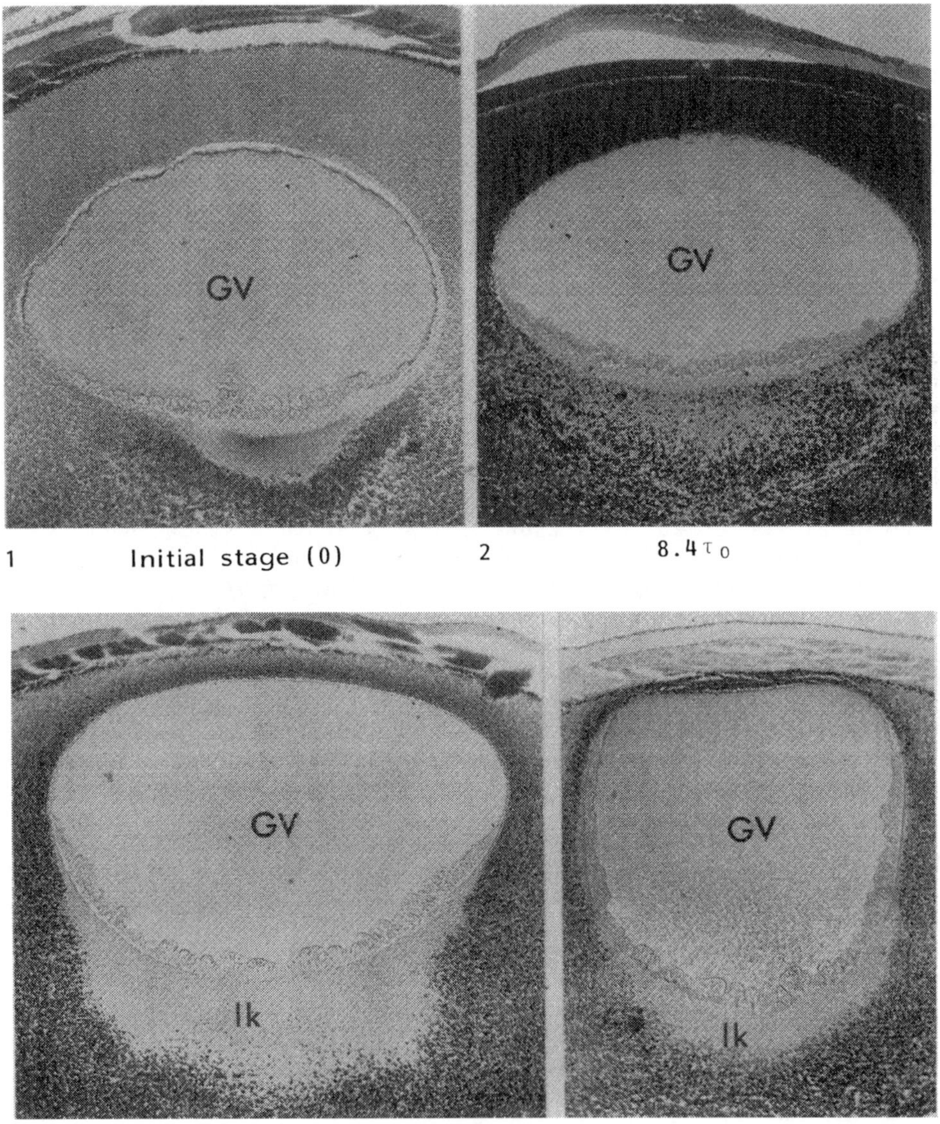

1 Initial stage (0) 2 $8.4\tau_0$

3 $10\tau_0$ 4 $13\tau_0$

protrusions sometimes contains small amounts of nucleoplasm and glycogen granules.

Electron microscopy demonstrates that the perinuclear space adjoins the endoplasmic reticulum (ER) (Aizenshtadt and Dettlaff, 1968). During this period, the number of ER membranes is greater in the perinuclear cytoplasm than in growing oocytes, where the ER is represented by occasional vesicles only. In some cases loose material can be observed in the perinuclear space and in the ER cisternae. The expansion of these cisternae is an indication of the accumulation of substances in them. It is at present difficult to provide a definite answer to the question of whether part of the nuclear material passes into the perinuclear cisternae and into the ER cisternae, or to ascertain if protein synthesis and accumulation take place within these structures at this time.

Situated in the cytoplasm of the animal region of full-grown oocytes are numerous mitochondria, lipid droplets, yolk platelets having a crystalline structure, and a high concentration of ribosomes. In the peripheral layer of the cytoplasm there are accumulations of annulate lamellae (Plate XIX, 1), which are derivatives of nuclear membranes (Kessel, 1968). They are particularly numerous in the oocyte's animal region. The stacks of annulate lamellae in sturgeon oocytes occupy a greater volume than in amphibian oocytes and frequently form concentric figures up to 10 μm in diameter (Aizenshtadt and Dettlaff, 1972).

Lying in the surface layer of the cytoplasm are cortical granules (see Plate XVIII, 1) which contain acid or neutral mucopolysaccharides (Dettlaff, 1961; Kemp and Ystock, 1967). In amphibians they are homogeneous and consist of a low electron-dense material (Van Gansen, 1966), while in sturgeons fibrillar material of relatively low density contains numerous denser inclusions (Ginsburg, 1968). Among the cortical granules (in amphibians) or below them (in sturgeons) there is a layer of pigment granules.

The oocyte plasmalemma in amphibians and sturgeons forms microvilli, situated in the canaliculi of the *zona radiata*. Penetrating the oocyte membranes from outside are cytoplasmic protrusions, macrovilli of the cells of the follicular epithelium closely adjacent to the oocyte.

Plate XVIIa. Sections through the animal region of the Russian sturgeon oocytes at successive stages of in vitro maturation. Sanfelice fixation stained with iron hematoxylin (Dettlaff and Skoblina, 1969). 1) Initial stage at τ_0; 2) at $8.4\tau_0$ (the very beginning of the period of maturation inertia); 3, 4) at 10 and $13\tau_0$ (two stages of the period of maturation inertia).

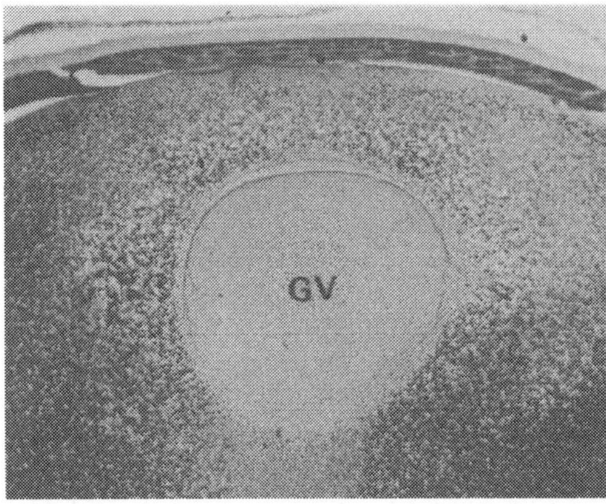

5

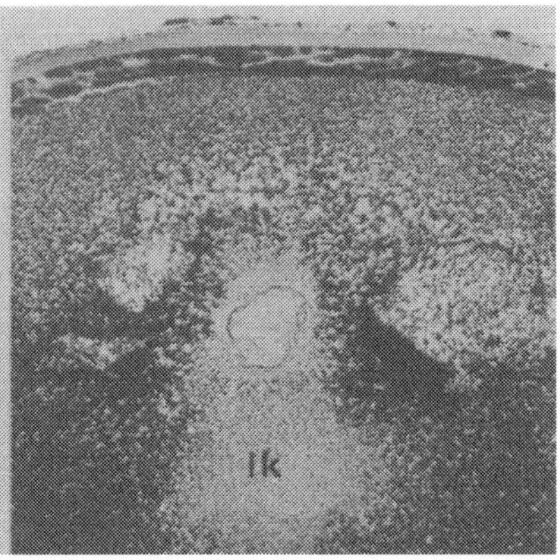

6

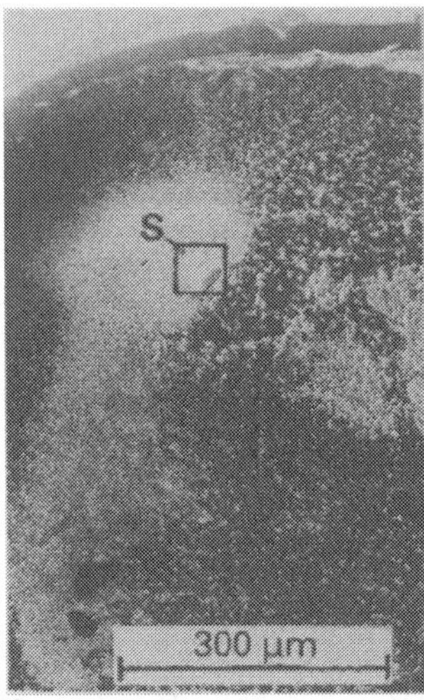

7

7a

8

8a

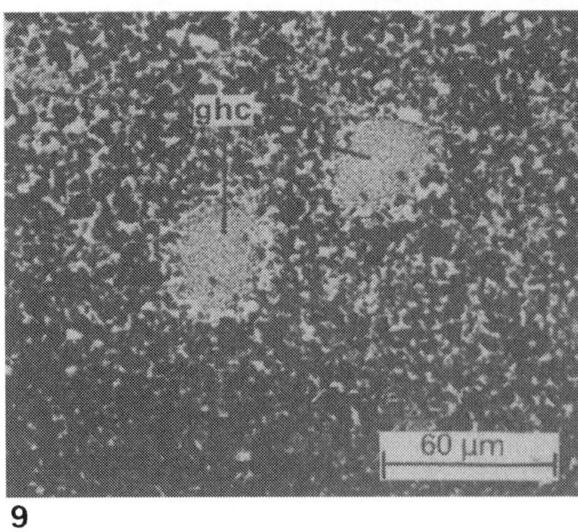

9

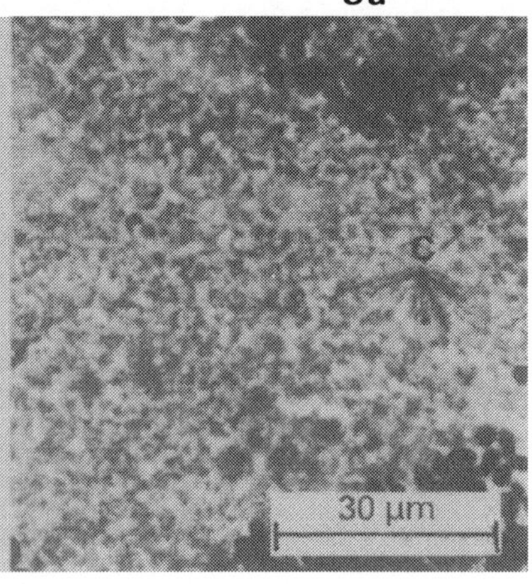

10

The follicular epithelium forms a layer of flat cells. Situated above the follicular epithelium is a connective-tissue layer, the theca, containing collagen fibers; beyond this is the peritoneal epithelium. The structure of the follicular layers of amphibian oocytes has been described in detail by a number of authors (Wischnitzer, 1958, 1966; Van Gansen, 1966; Yatvin and Anderson, 1967; Smith et al., 1968; Anderson and Yatvin, 1970; Schuetz, 1972a, b, 1985).

The surface layer of oocyte cytoplasm during these stages possesses neither elasticity nor contractility. A wound does not close up after a surface incision (see Plate XVIII, 4). Oocytes are permeable by water, and in hypotonic solution or water they undergo plasmolysis or degenerate. The endoplasmic turgor of the oocyte is by no means considerable.

It is easy to remove the GV from the oocyte at the initial stage of maturation by means of a small incision of the surface layer of the cytoplasm in the center of the animal region. After the incision the GV increases in volume slightly and rises to the wound, pushes apart its edges, and emerges (see Plate XVIII, 4). The emergence of the GV can be speeded up by slightly compressing the oocyte surface near the incision. If the GV membrane is ruptured with a needle inside the oocyte, a large lacuna of nucleoplasm remains in place of the GV. In this case, the nucleoplasm does not spread throughout the cytoplasm as a network of tiny lacunae, as it does during normal GV breakdown, and the oocyte undergoes no visible changes. Thus, nucleoplasm–cytoplasm contact following a mechanical disruption of the GV membrane at the initial stage in sturgeon oocytes does not lead to cytoplasm maturation (Dettlaff, 1961; Dettlaff et al., 1964; Davydova, 1968).

3.2.2. Changes in Oocyte Structure and Properties during the Maturation Period

3.2.2.1. Changes in the Oocyte Nucleus

At the beginning of the maturation period the GV shifts toward the animal pole and gradually approaches the surface of the

---

Plate XVIIb. Sections through the animal region of the Russian sturgeon oocytes at successive stages of in vitro maturation. Sanfelice fixation stained with iron hematoxylin (Dettlaff and Skoblina, 1969). 5-7) At $14.4\tau_0$ (oocytes at the stage of GV breakdown and transition of karyoplasm into the cytoplasm); 8) at $15\tau_0$ (stage of maturation division I); 9) globules of hydrophilic colloid (ghc); 10) cytasters (c). GV) Germinal vesicle; lc) lacuna of karyoplasm; S) I meiotic spindle.

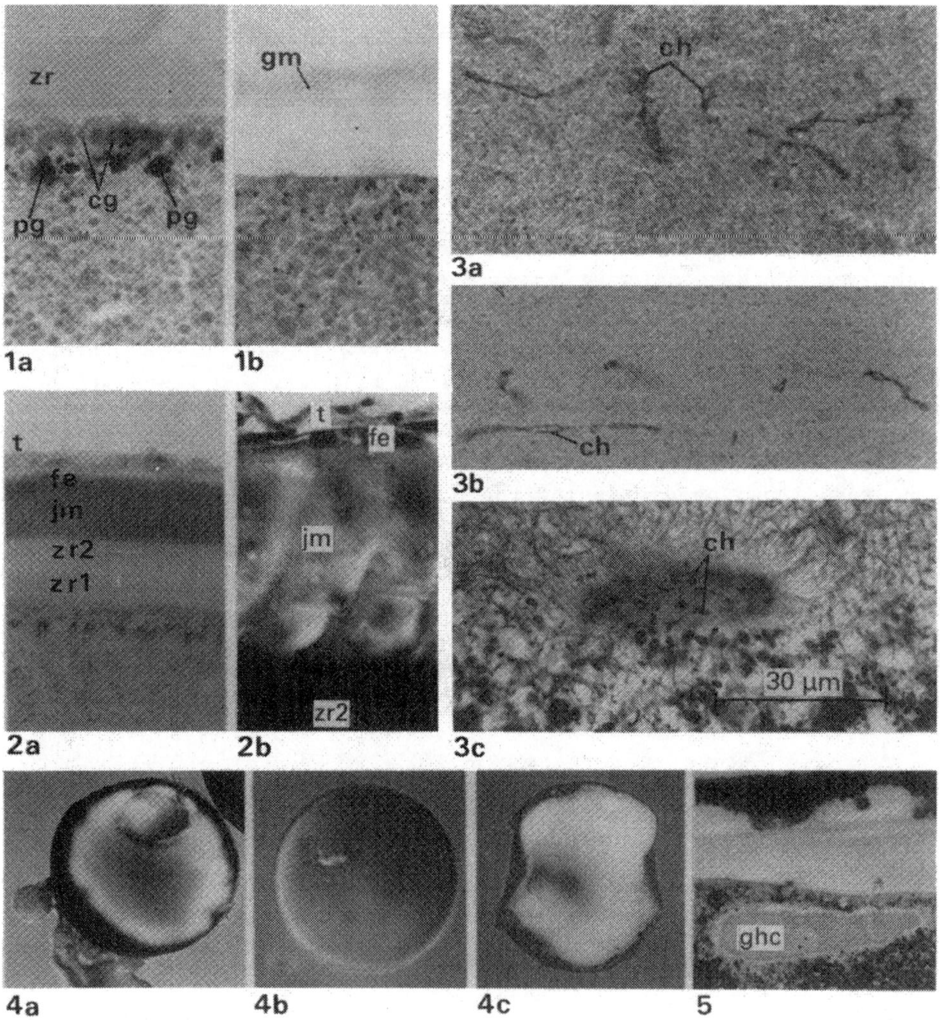
zr
cg
pg
pg
gm
ch
1a
1b
3a
ch
3b
t
fe
jm
zr2
zr1
t
fe
jm
zr2
ch
30 µm
2a
2b
3c
ghc
4a
4b
4c
5

oocyte (see Plate XVII, 2-4). At the same time the number of ramifications of nuclear membrane outgrowths increases on the vegetal side of the GV, and the quantity of nucleoplasm that has passed into the cytoplasm also increases. The nucleoli in the GV gradually become vacuolized and fragmented. In *Rana*, RNA-containing parts of the nucleoli disintegrate and, at the same time, their stainability changes; they lose pyroninophilicity and exhibit a Feulgen-positive reaction (Hanocq-Quertier et al., 1968; Brachet et al., 1970). These nucleolar changes do not occur simultaneously within one and the same nucleus. Apically situated nucleoli do not respond to the Feulgen reagent at the time when fragments of nucleoli situated in the basal part of the GV exhibit a Feulgen-positive response. In *X. laevis* oocytes, shortly before GV breakdown, many, if not all, nucleoli show full transcriptional activity in most of their pre-rRNA genes (Williams et al., 1984).

As the GV moves close to the oocyte surface, it flattens slightly, giving a large area of contact between the GV and the plasma membrane of the oocytes (see Plate XVIII, 4). Subsequently, GV breakdown takes place (see Plate XVII, 5-7). At the lower border of the GV in the area where the nucleoplasm passes into the cytoplasm, and later in the lacunae of the nucleoplasm, there appear a great number of cytasters (see Plate XVII, 10). The nucleoplasm gradually spreads in the cytoplasm of the oocyte animal region, forming a network of small lacunae. In sturgeon oocytes, in addition to the cytasters, globules of hydrophilic colloid may be detected in these lacunae (see Plate XVII, 9). The formation of cytasters in places where nucleoplasm mixes with cytoplasm has been described also in the *X. laevis* oocytes, and cytasters are particularly prominent when oocytes are treated with D_2O (Van Assel and Brachet, 1968; Hanocq-Quertier et al., 1978).

Plate XVIII. 1. Cortical layer of the stellate sturgeon oocyte. Sanfelice fixation, staining by Azan method after Heidenhain (Skoblina, 1968): a) at initial stage; b) after GV removal in the oocyte which was cultivated in body fluid with pituitary suspension and inseminated in water (60 min after insemination); gm) granular material extruded from the egg (inner part of "fertilization membrane"); cg) cortical granules; pg) pigment granules; zr) zona radiata. 2. Membranes of the Russian sturgeon oocyte at initial stage: a, b) at different magnifications; zr1, zr2) two layers of zona radiata; jm) jelly membrane; fe) follicle epithelium; t) theca. 3. Chromosomes (ch) in Russian sturgeon oocytes at different stages of maturation: a) during the hormone-dependent period ($4\tau_0$); b) during the period of maturation inertia ($12\tau_0$); c) at the stage of GV envelope disintegration ($14.4\tau_0$). 4. The toad oocyte in the end of the period of rapid growth: a) section through the intact oocyte fixed in boiling water; b, c) such an oocyte after GV removal: b, external appearance; c, section. 5. A globule of hydrophilic colloid (ghc).

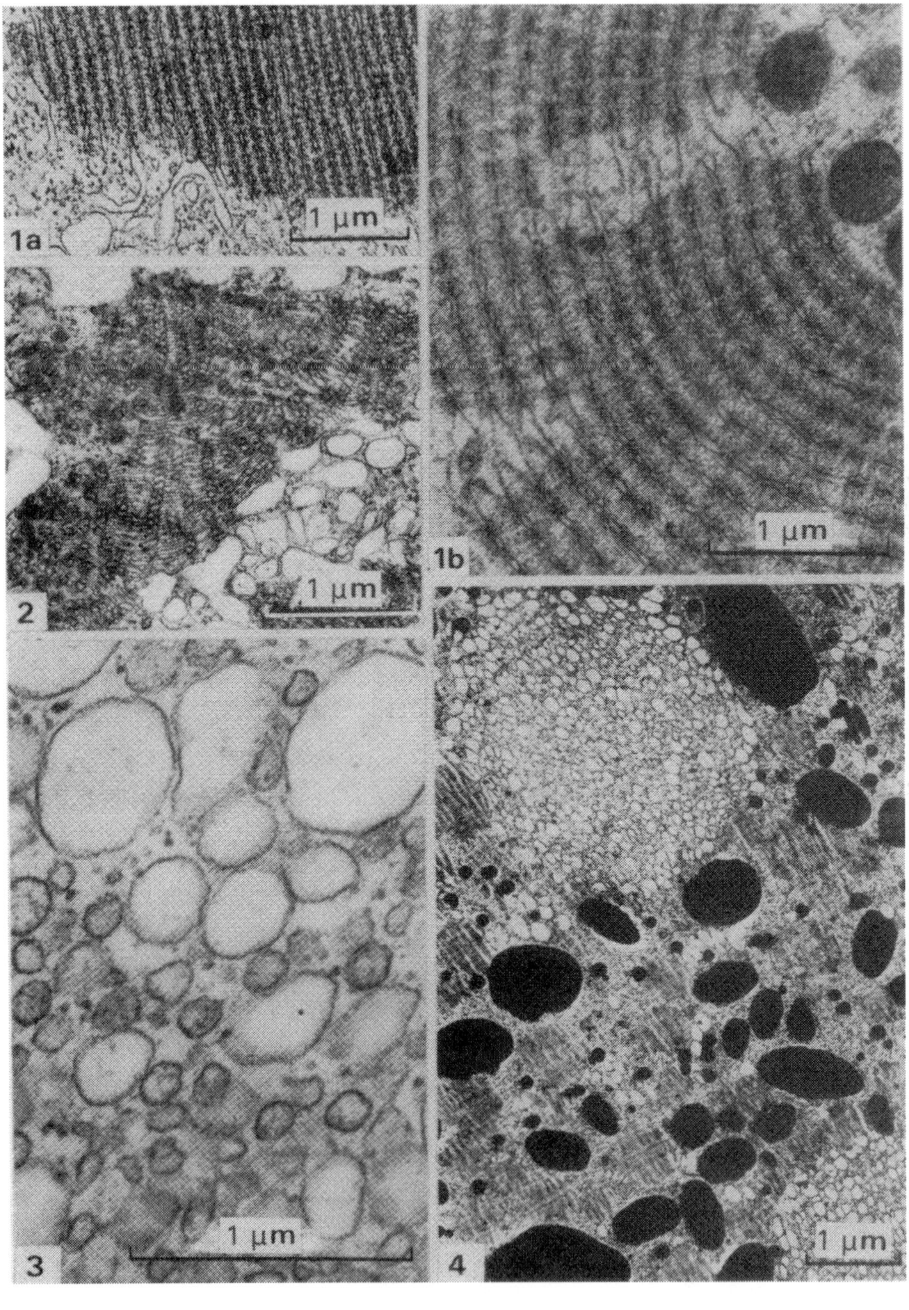
1 μm
1a
1 μm
1b
1 μm
2
1 μm
3
1 μm
4

In amphibian and sturgeon oocytes there is a high glycogen content where the nucleoplasm mixes with the cytoplasm; this is absent in the nucleoplasm of the intact GV (Dettlaff, 1961; Rudneva, 1968; Brachet, 1974b). During this period, along with oocytes which no longer exhibit any traces of the GV, one may encounter oocytes containing partially evacuated GVs (of smaller size) at some distance from the surface of the animal region, or a not-yet-fully disintegrated, collapsed GV membrane with granules resembling nucleolar fragments. The fragments of the nucleoli in the cytoplasm show a Feulgen-positive reaction and may be DNA of the nucleolar organizer (Brachet et al., 1970; Brachet, 1974b). These fragments increase in size, apparently by fusion of individual fragments (Hanocq et al., 1974).

RNA synthesis was demonstrated (Felgenhauer, 1985) in the extrachromosomal nucleoli of the maturing amphibian oocytes after GV breakdown. In *R. temporaria* these nucleoli formed a massive aggregate in the animal region of the oocytes and in *X. laevis* only individual RNA-synthesizing nucleoli were more frequent in the lacunae of nucleoplasm. The changes and migrations of nucleoli, similar to those in *R. temporaria*, were also described as occurring in the maturing oocytes of *Cynops pyrrhogaster* (Imoh, 1981).

In sturgeons, a prometaphase nucleus with strongly but not yet maximally condensed chromosomes is situated in a small karyoplasmic lacuna, alongside the incompletely disintegrated nuclear membrane (see Plate XVIII, 3b). Chromosome condensation takes place in the period when the chromosomes pass into the cytoplasm, since before GV breakdown the chromosomes appear as bivalents with several chiasmata (Ginsburg and Dettlaff, 1969; Dettlaff and Skoblina, 1969).

The condensing effect of amphibian oocyte cytoplasm on chromosomes following GV breakdown has been demonstrated by Gurdon (1967) and Ziegler and Masui (1973) by transplanting somatic cell nuclei into oocytes at the appropriate stage. The same result was

---

Plate XIX. Changes in oocyte cytoplasm ultrastructure at the stage of disintegration of GV envelope in *Rana temporaria* and *Acipenser stellatus*. 1. a) Annulate lamellae in the cytoplasm of *R. temporaria* oocyte in the beginning of maturation period; b) annulate lamellae in the cytoplasm of *A. stellatus*. 2. Appearance of smooth-surfaced vesicles in the region of annulate lamellae at the beginning of their disintegration. 3, 4. Accumulation of smooth vesicles in *A. stellatus* oocytes (3) and *R. temporaria* oocytes after GV breakdown (4).

obtained by Nikitina (1983, 1984) after transplantation of the nuclei from *X. laevis* embryonic cells and vitellogenic oocytes into the cytoplasm of maturing axolotl oocytes.

It should be pointed out that during in vitro maturation of *Urodela* oocytes such chromosome condensation is preceded by the transformation of lampbrush chromosomes into bivalents. This process is accompanied by the retraction of loops and the shortening of the chromosomes, by the loss of their chromomere structure, and the separation of spheres. These changes take place during the first hours of oocyte incubation in a Ringer solution containing progesterone (Barsacchi, 1971; Batistoni and Barsacchi, 1974) and deserve special study (Brachet, 1974a). Chromosomal transformations of this kind occur in *Anura* oocytes during the growth period (Duryee, 1950; Dumont, 1972).

Electron microscopy of the maturing oocytes of amphibians and sturgeons has been concerned mainly with changes in the nuclear membrane and annulate lamellae, whose disintegration takes place simultaneously (Balinsky and Devis, 1963; Van Gansen, 1966; Aizenshtadt and Dettlaff, 1968, 1972; Van Gansen and Shram, 1968; Aizenshtadt and Skoblina, 1970; Brachet et al., 1970).

As the GV moves toward the surface of the animal region, the number of GV membrane areas occupied by complexes of pores and lacking membranous structures connecting them increases. During GV breakdown the membrane itself falls apart into separate blebs (Balinsky, 1966). In the *Xenopus* oocytes during GV breakdown, lamin III, the only lamin present in the nuclear membrane of the oocytes, is solubilized (Stick and Hausen, 1985; Benavente et al., 1985).

In sturgeon, during GV breakdown the annulate lamellae undergo change. These lamellae, consisting of parallel rows of membranes with pores, shift their position in relation to each other, and their end portions widen (Aizenshtadt and Dettlaff, 1972). At the same time, the expanded ER cisternae and end distensions of the annulate lamellae start budding (see Plate XIX, 2, 3, and Fig. 34). This process results in the appearance of large aggregates of smooth-surfaced vesicles (see Plate XIX, 4) which move to the peripheral region of the oocyte cytoplasm.

In Aizenshtadt's opinion, this process of vesicle formation does not differ in any way from the formation of secretory products in other cells, when proteins synthesized in the ER move into the Golgi apparatus where they combine with carbohydrates. This is morphologically expressed as the separation of small smooth-surfaced vesicles forming the dictyosomes of the Golgi complex from the ER cisternae or from the end extensions of the annulate lamellae. The aggregates of small secretory vesicles have been

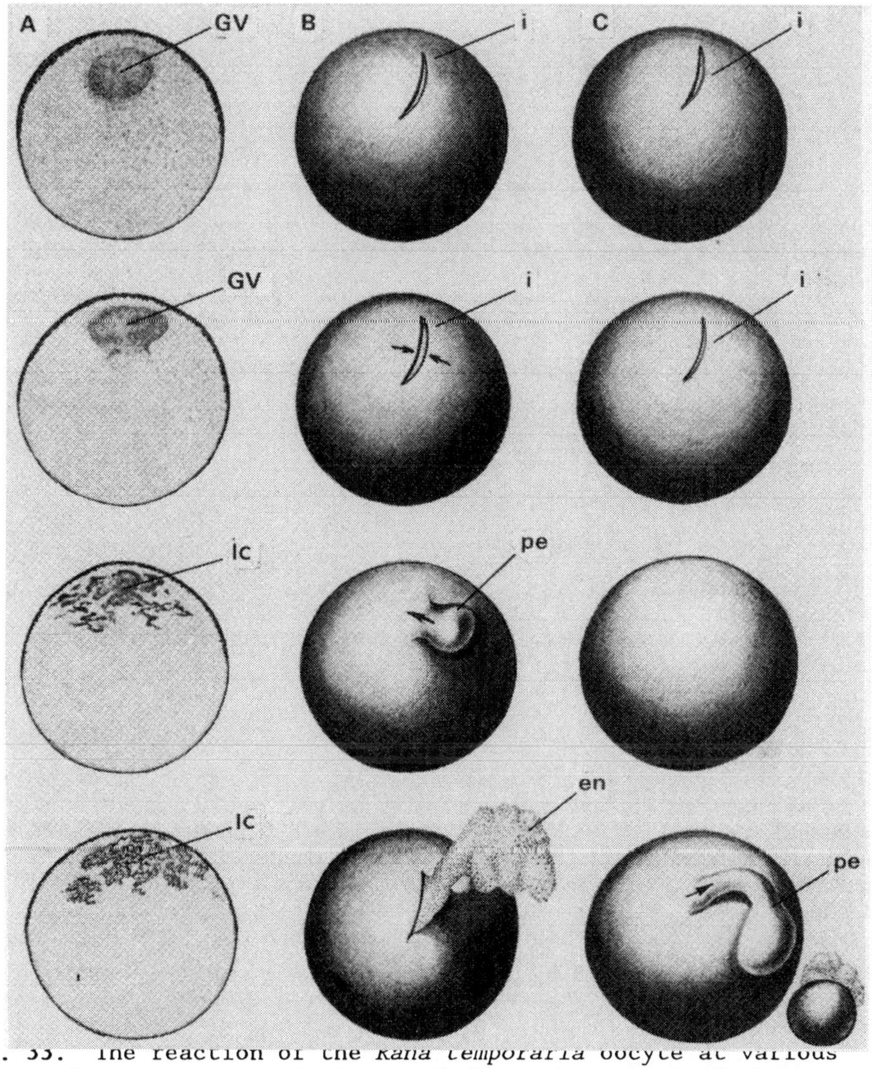

Fig. 33. The reaction of the *Rana temporaria* oocyte at various stages of maturation to incision of the membrane and of the surface layer itself. A. Structure of the oocyte at the moment of incision (diagram). B. External appearance of the oocyte just after incision. C. External appearance of the oocyte 1-2 h later. pe) protuberance of the egg; GV) germinal vesicle; lc) lacuna of karyoplasm; i) incision; en) endoplasm.

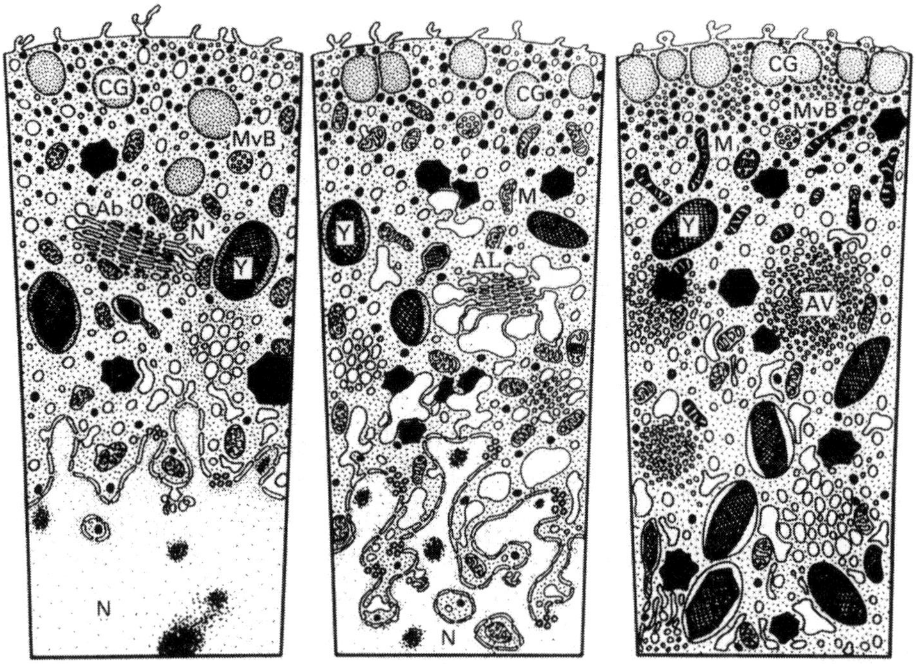

Fig. 34. Ultrastructural changes in *Rana temporaria* oocytes during maturation (after Aizenshtadt and Dettlaff, 1968). AL) annulate lamellae; CG) cortical granules; M) mitochondria; MvB) multivesicular bodies; N) nucleus; Y) yolk; AV) accumulation of vesicles.

described earlier in amphibian eggs as "dense areas of the cytoplasm" (Wartenberg and Schmidt, 1961) or "vacuolar cytoplasm" (Balinsky, 1966; Van Gansen, 1966). However, the question of the origin and role of these structures was not discussed in these earlier works. At the same time, a comparison of light and electron microscopy observations suggests that the spherical aggregates of the small vesicles correspond to the so-called globules of hydrophilic colloid which is excreted into the perivitelline space following egg activation (Dettlaff, 1961, 1962) (see Plate XVIII, 5).

Brachet et al. (1970) have found that individual ribosomes and glycogen granules penetrate the nucleoplasm in the basal part of the GV in the area of nuclear membrane disintegration. Following GV breakdown the nucleoplasm is mixed with a large number of ribosomes, mitochondria, and glycogen granules; small yolk granules and remnants of the nuclear membranes are also found there. There is evidence to suggest that mitochondrial division takes place during this period (Aizenshtadt, 1969).

Following GV breakdown, the forming I meiotic spindle (see Plate XVII, 7, 8) gradually shifts to the animal surface of the oocyte and turns perpendicularly to the surface. The nucleus completes maturation division I and stops at metaphase II (see Plate XXI). In amphibians the movement of the I meiotic spindle toward the oocyte animal pole is accompanied by the movement of Feulgen-positive bodies (nucleolar organizers) in the same direction, but they do not attach themselves to the spindle (Brachet et al., 1970).

According to Huchon et al. (1981a), in *X. laevis* oocytes in areas where the GV membrane breaks down a temporary structure appears. This consists of a dense fibrillary layer and bundles of microtubules situated perpendicularly to it in a fibrillar network. For some time the chromosomes are situated within this fibrillar network and move, together with it, toward the surface of the oocyte. When the GV membrane has almost completely disintegrated, the fibrillar network disappears, while the bivalents remain within a homogeneous formation. The authors name this stage prometaphase I. The subsequent migration of the meiotic spindle to the animal surface of the oocyte in *X. laevis* and its anchoring in the cortex are mediated by microfilaments. In the oocytes treated with cytochalasin B at different stages of maturation, the pigment and yolk granules are displaced inward and a funnel-like zone forms into which the spindle is shifted. Cytochalasin B affects the formation of the spindle and induces the detachment of the already anchored spindle from the surface (Ryabova et al., 1984, 1986).

Nuclear transformations in this period are of tremendous interest not only for cytogenetics, but also for developmental biology, since deviations from normal relations between cytoplasm and nucleus at this time may result in serious hereditary developmental anomalies. In addition, the control of chromosome disjunction during meiotic divisions is a powerful means of genotype regulation. These problems are dealt with in Chapter 4.

3.2.2.2. Changes in the Oocyte Cytoplasm

Changes in the surface layer of the oocyte commence just before GV breakdown, and then rapidly increase. These changes include the development of contractility in the cortical layer of the cytoplasm and an alteration of its permeability; there also appears the capacity for cortical reaction and fertilization. During this period the hydration of the cytoplasm changes, the thermosensitivity of the oocyte increases, and changes take place in the fine structure of the cytoplasm.

3.2.2.2.1. Contractility. The appearance of contractility in the oocyte cortical layer has been studied in detail in *R.*

temporaria oocytes. Soon after the start of maturation inertia, the cortical layer assumes a very weak contractility (Fig. 33). As the GV moves to the animal surface of the oocyte, the contractility of the surface layer increases. After making an incision at this stage, the edges of the wound at first approach each other very slowly, then faster, but leave a scar at the site of their convergence. At the stage of GV breakdown, the contractility increases sharply and a wound closes instantly without any scar (Dettlaff, 1966; Rudneva, 1968; Kasyanov, 1969). If the oocyte surface is slightly pressed during this stage, the portion of the oocyte covered by the cortical layer will protrude through the surrounding membranes, but withdraw at once as soon as the pressure is discontinued.

These changes in surface contractility are correlated with changes in the system of microfilaments (Ryabova, 1982). In a full-grown oocyte the microfilaments form bundles of parallel fibers in the microvilli and a fine haphazard network under the membrane; this is only revealed in the surgically isolated cortex. In an intact oocyte the microfilaments form under the membrane a so-called "felt mat," and a similar "felt mat" is present in the ovulated oocyte. During these stages, if an incision is made on the oocyte surface, the microfilaments form oriented bundles. It has been established that the capacity to form oriented bundles coincides in time with the appearance of contractility in the cortex, and a causal relationship between these phenomena is suggested. Apart from this data, there is scattered information in the literature on the state of microfilaments in the surface layer of the cytoplasm of the amphibian oocyte and egg which, however, have not been compared with the development of contractility. Thus, microfilaments arranged in parallel in microvilli, with only occasional bundles under the membrane, have been described in *X. laevis* at the stage preceding the GV breakdown (Van Gansen, 1966; Brachet et al., 1970). The mature egg of *X. laevis*, just like that of *R. temporaria*, has a "felt mat" under the plasma membrane (Van Gansen, 1966; Selman and Perry, 1970). Actinic formations toward the end of vitellogenesis have been described also in isolated cortices of *Triturus alpestris* oocytes (Franke et al., 1976).

Changes in the contractility of the cortical layer of the cytoplasm also take place during the maturation of sturgeon oocytes. However, these changes are more difficult to reveal by surface-layer incisions because the yolk has a very fluid consistency.

In vivo isolation of the cytoplasmic surface layer fragments of sturgeon oocytes during the GV breakdown stage examined with a scanning microscope demonstrated that the microvilli become shorter and thicker than at earlier stages (see Dettlaff, 1977b). Although it is not ruled out that the appearance of these changes at this stage is somewhat accelerated by the incision, it seems probable that the altered shape and size of the microvilli may be associated with the contractility of the cytoplasm surface layer in normal development.

Still another criterion of cytoplasm surface layer contractility is its local furrowing in response to a microinjection of Ca^{2+}. In experiments carried out with *R. pipiens* oocytes by Hollinger and Schuetz (1976) such furrowing appeared at the site of injection beginning from the GV breakdown stage equally in maturing oocytes with a GV and in oocytes from which the GV had been removed. Cleavage (cytokinesis) cannot serve as a criterion for determining the time when surface layer contractility appears, since, in addition to the cytoplasm surface layer, the endoplasm is also involved, and this assumes a capacity to cleave later than the surface layer (Dettlaff et al., 1964) and then only after mixing with the nucleoplasm.

Thus, the appearance during maturation of contractility in the cytoplasm surface layer of amphibian oocytes is associated with the polymerization and reconstruction of microfilaments. Here, a steep enhancement of contractility in *R. temporaria* coincides with the GV breakdown; however, it is not connected with the movement of nucleoplasm into the cytoplasm, since it appears in enucleated oocytes as well (Skoblina, 1968).

In the starfish cortex contractility waves appear during the formation of the I polar body (Hamaguichi and Hiramoto, 1978). The first wave moves from the vegetal to the animal pole of the oocytes and the second occurs after the separation of the polar body and moves in the reverse direction. Cortex rigidity changes in starfish oocytes in a similar manner. After exposure to 1-Me-Ade it drops steeply, remains at this low level, and then strengthens and forms two peaks in the period of extrusion of the I and II polar bodies (Nakamura and Hiramoto, 1978).

One rapid response of starfish oocytes to 1-Me-Ade is the formation of transient actin-filled spikes on the cell surface. The treatment with 1-Me-Ade results in the conversion and redistribution of cortical G_1- and oligomeric actin into large bundles of F-actin which form the cores of spikes (Schroeder and Stricker, 1983; Otto and Schroeder, 1984a, b). Schroeder and Otto (1984) described the cyclic assembly and disassembly of cortical microtubules during maturation and early development. Specifically, cortical microtubules are present when the nucleus is in the state of GV, in the pronucleus, zygote, and blastomere nucleus. Otto and Schroeder (1984a) indicate that not only actin but fascin and a 220-kilodalton protein undergo major organizational changes in the cortex in response to 1-Me-Ade.

3.2.2.2.2. Permeability of the Oocyte's Plasma Membrane. It has been demonstrated that during maturation the permeability of amphibian and sturgeon fish oocytes drops considerably, and changes in the ionic composition of the environment have less effect on the oocytes. Progesterone-treated amphibian oocytes normally reach the mature egg stage in Ringer solution, whereas in a solution of lower ionic strength they at first become damaged and may normally develop only beginning at metaphase I and ovulation (Wright, 1945; Tchou-Su and Wang Yu-lan, 1956; Rudneva, 1968; Smith et al., 1968). Diminished permeability of the cytoplasmic membrane following GV breakdown and the formation of metaphase I is demonstrated also by experiments with labeled amino acids and inhibitors (Smith et al., 1968; Brachet et al., 1974; Pennequin et al., 1975), by the effect of changes in the ionic composition of the environment (the content of Ca^{++}, Mg^{++}, and K^{+}), and by the effect of ionic strength on the intensity of protein synthesis (Ecker and Smith, 1971b). Permeability change is also indicated by changes in the membrane potential and the ratio of free and bound Na^{+}, K^{+}, Ca^{++}, and Cl^{-} in ovulated oocytes of *R. pipiens* compared to ovarian oocytes (Morrill, 1965; Morrill and Watson, 1966; Morrill et al., 1966; Schlichter, 1983). Whether these characteristics of the membrane change at earlier stages of maturation has not been systematically investigated.

3.2.2.2.3. Cytoplasm Hydration. At some time following the passage of the nucleoplasm into the cytoplasm, cytoplasm hydration takes place (Tchou-Su and Yen Pai-hu, 1950; Morrill, 1965; Dettlaff, 1966; Van Gansen, 1966; Rudneva, 1968). This is expressed as increased oocyte turgor, greater oocyte volume, and wet weight. According to Rudneva (1968), the weight of a *R. temporaria* oocyte at this time increases by 6.9% compared to its initial weight. If the oocyte membrane and cortical layer are incised, the endoplasm is ejected through the incision, and, following a reduction of the internal pressure and the restoration of the cortical layer, exovates covered with the surface layer will emerge through the incision in the membranes and gradually become pinched off (Fig. 34). It has been shown (Rudneva, 1968) that the hydration of the oocyte cytoplasm takes place after the oocyte surface becomes resistant to changes in its osmotic environment. According to Morrill (1965), the water content of ovulated *R. pipiens* oocytes is higher than in oocytes taken from the ovary (ovarian oocytes) by 32%. Lau et al. (1984) measured by cryomicrodissection intraoocytic distribution of water before and during meiotic maturation in *R. pipiens*. Animal ooplasm contained about 10% more water in matured oocytes than in ovarian oocytes. This increase was not dependent on the uptake of extracellular water, occurring even when oocytes matured in a paraffin oil medium. According to the authors, such hydration of animal ooplasm is due to the migration of cytoplasm from the vegetal to animal hemisphere and to the mixing of ooplasm with nucleoplasm during GV breakdown.

In ultrathin sections of frog oocytes at the stage when cortical layer contractility and cytoplasm hydration have appeared, a large aggregate of small smooth-surfaced vesicles is observed in the animal region of the oocyte (Aizenshtadt and Dettlaff, 1972).

It should be added that in *Chaetopterus* and *Spisula* oocytes, unlike those of *Anura* and sturgeon, cytoplasm hydration takes place shortly before GV breakdown (Heilbrunn and Wilson, 1955).

3.2.2.2.4. Capacity for Cortical Reaction and Fertilization. The cortical reaction is a complex process, which includes a number of changes occurring in the mature egg cortex upon fertilization or artificial activation (see Elinson, 1980; Vacquier, 1981; Ginsburg, 1986). We will consider here only the development of the capacity for exocytosis of the cortical granule contents and the propagation of cortical granule breakdown accompanied by the formation of a perivitelline space.

The capacity for cortical reaction in Acipenserid oocytes appears soon after the GV breakdown (Dettlaff, 1961). The appearance of the capacity for cortical reaction has been studied in detail in *A. stellatus* oocytes (Ginsburg and Nikiforova, 1979; Ginsburg, 1986). For this purpose the oocytes were pricked with a glass needle during successive maturation stages. These experiments revealed two specific features of the cortical reaction in the maturing oocytes. The first is that, in some oocytes at the stage when the capacity to cortical reaction is detected for the first time, this reaction proceeds much slower than in mature eggs. A slow cortical reaction is also observed later in a smaller percentage of cases, at prometa-anaphase I. The second feature is an inhibition of the secretion of the contents of the cortical granules in the center of the animal region; as a result, the contact between the oocyte cytoplasm and the *zona radiata*, the region where the micropylar canals are situated, lasts longer. In the case of insemination of not fully mature oocytes, this may result in a considerably higher incidence of polyspermic fertilization.

In sturgeon oocytes covered by follicular envelopes, which are activated by pricking at meta- and anaphase I in a Ringer solution, the contents of the cortical granules excreted under the *zona radiata* do not swell; accordingly, perivitelline space is not formed, and the jelly coat does not become adhesive. If, however, the follicular envelopes are removed from these oocytes, such changes do occur (Dettlaff, 1961). Apparently, for the changes to take place, not only must the contents of the cortical granules be discharged, but water must get under the *zona radiata* (Zotin, 1961); this is prevented at these stages by the follicular envelopes.

Tchou-Su and Wang Yu-lan (1959, 1964), during experimental activation of *Bufo bufo asiaticus* oocytes by pricking with a needle smeared in blood and in tests of inseminating oocytes taken from oviducts at different maturation stages, have found that the capacity for activation arises in oocytes at metaphase I; in the interval between metaphase I and metaphase II, the capacity for normal cleavage appears and, sometime after metaphase II is reached, in the "maturity period," there appears a capacity for far-going development.

In the oocytes of *R. pipiens* (Belanger and Schuetz, 1975), *Bufo bufo asiaticus* (Tchou-Su and Wang Yu-lan, 1959, 1964; Iwao, 1982), and *X. laevis* (Charbonneau and Grey, 1984), the cortical granule breakdown, propagation of the cortical granule breakdown, and appearance of the activation potential upon pricking with a needle take place before complete oocyte maturation at metaphase I or between metaphases I and II. But the wave of cortical granule breakdown spreads much slower at these stages than in the mature egg, at metaphase II [e.g., in *B. bufo*, 8 min instead of 2.5 min (Iwao, 1982)]. Under the effect of ionophore A23187, or injection of Ca^{2+}, these changes appear in the oocytes of all studied species several hours earlier than under the effect of needle pricking. At still earlier stages of maturation, the reaction appears as well, but is less complete: the propagation of cortical granule breakdown is not accompanied by the appearance of the activation potential, or only nonpropagated cortical granule breakdown takes place and small regions of perivitelline space form in some regions. Finally, in the intact full-grown oocytes, high doses of A23187 or Ca^{2+} induce cortical granule breakdown without the formation of perivitelline space. Similar results were obtained on the *A. stellatus* oocytes treated with A23187 or Ca^{2+} at different stages of maturation (Ginsburg, 1986). Thus, the cortical layer reaction to Ca^{2+} undergoes changes during the oocyte maturation. In the *X. laevis* oocytes, the appearance of activation responsiveness coincides with the formation of the cortical ER.

It has been demonstrated (Good and Daniel, 1943; Tchou-Su and Wang Yu-lan, 1964; Katagiri and Moriya, 1976) that amphibian oocytes inseminated before the complete cytoplasmic maturation allow penetration by several spermatozoa, while the transformation of sperm nuclei into pronuclei and karyogamy are retarded. It is interesting that in starfish (as in sturgeons and amphibia), when previously inseminated immature oocytes were treated with 1-MeAde, the formation of the sperm monoaster was observed first during the GV breakdown period, while amphiaster and pronuclei were formed only after formation of the II polar body (Hirai et al., 1981).

Asynchronous changes of the cortical layer and oocyte endoplasm were observed in the oocytes of *B. bufo* and *B. viridis* after nuclear transplantation (Dettlaff et al., 1964). The surface layer of the cytoplasm matures earlier than the endoplasm and upon pricking or spontaneous activation it frequently contracts and forms a cap of very small "blastomeres" on the surface of an uncleaving endoplasm. The capacity for cytokinesis appears later in the endoplasm of the toad oocytes than in the surface layer (see Plate XIX, 5).

3.2.3. The Mature Egg

The mature ovulated egg is characterized by a number of features; it has a smooth, glossy surface; the funnel of the I maturation division can be seen on the surface in the animal region; the egg has turgor and is capable of activation, cleavage, and normal development. The fine structure of the mature egg cytoplasm has been studied in *X. laevis* (Van Gansen, 1966; Van Gansen and Schram, 1968), *R. temporaria* (Ryabova, 1982), and starfish (Longo and Schuetz, 1982; Longo et al., 1982; Schroeder, 1985; Aizenshtadt and Vassetzky, 1986). In amphibians there is a microfilamental "felt" in the surface layer of the cytoplasm under the plasma membrane. There are no longer any microvilli on the surface of the mature egg; instead there remain only small outgrowths or roughness and, consequently, there are no fine bundles of microfilaments formerly present inside the microvilli which protruded into the surface layer of the cytoplasm in the full-grown oocyte. The cytoplasm contains a great number of more or less tiny vesicles and has a looser structure than that of full-grown oocytes. Ribosomes, as in initial oocytes, are scattered throughout the cytoplasm.

With complete maturation the pigmented cytoplasm of the animal region of amphibian oocytes becomes capable of growing over the vegetal region of the egg, so-called pseudo-gastrulation (Holtfreter, 1943; Smith and Ecker, 1969b, 1970a). During prolonged cultivation, the unfertilized egg imitates the epiboly movements of gastrulation at approximately the same time as those in the gastrulation of fertilized control eggs. The capacity for pseudo-gastrulation is inhibited under conditions of anaerobiosis, which inhibits protein synthesis (Smith and Ecker, 1970a), and is also prevented by cycloheximide (Dettlaff et al., 1977; Felgenhauer and Chulitskaya, 1977). The spread of the pigmented cortical cytoplasm of the oocyte's animal region over the nonpigmented cytoplasm of the vegetal region may be related to the phenomenon of "differentiation without cleavage" discovered at the beginning of the century by Lillie (1902). It is of great interest in its own right and deserves fuller investigation in connection with the study of the mechanism of morphogenetic movements during gastrulation.

Eggs retain the capacity for fertilization and normal development only for a limited time, depending upon species and conditions, after which they become overripe and deteriorate (see Tchou-Su and Wang Yu-lan, 1959, 1964; Ginsburg, 1968; Ginsburg and Dettlaff, 1969). Under unfavorable conditions, when ovulation in sturgeons is retarded, such overripe eggs may remain inside the follicular envelopes, but overripeness develops much faster when ovulated eggs are retarded in the female's body cavity (or in the uterus in amphibians), and also in physiological solutions and water (see Ginsburg, 1968). All this should be taken into account when staging tests and assessing their results, since the basic criterion of adequate oocyte maturation is the capacity of the eggs to develop normally after fertilization.

Amphibian oocytes ovulated in vitro have no jelly coat and are, therefore, incapable of fertilization (Kambara, 1953). Insemination requires that they be surrounded in one or another way by a jelly coat (Good and Daniel, 1943; Nadamitsu, 1953; Tchou-Su and Wang Yu-lan, 1956, 1958). Commonly used nowadays for this purpose is a somewhat modified method of Nadamitsu (Smith and Ecker, 1968; Masui, 1973a): ovulated oocytes are stained with neutral red (1 min in a 0.1% solution) or Nile blue (10 min in a 0.2% solution), and are introduced into the body cavity of a recipient female through a small incision in the abdominal wall near the forelimb at a stage when, following the injection of gonadotropic hormones, the ovulation of her own oocytes takes place. The introduced oocytes, together with those of the recipient female, pass through the oviducts, become enveloped in a jelly coat, and accumulate in the uterus. On the next day the female is killed, the eggs are collected from the uterus and inseminated, and the development of the stained eggs is observed (Masui, 1973a). In some cases, in order to reduce the proportion of the recipient female's oocytes before introducing the test oocytes into the female, her body cavity is opened and the ovulated oocytes are washed out; one of the ovaries is removed, after which the stained oocytes are introduced into the body cavity and the edges of the incision are sutured (Smith and Ecker, 1968).

According to Katagiri (1966), eggs stripped of the jelly coat may undergo fertilization in the presence of polyvinylpyrrolidone or substances diffusing out of the jelly coat. Such eggs may be fertilized also by using a thick sperm suspension diluted by egg water or a lyophilized egg water extract (Elinson, 1971, 1973a). Eggs without jelly coats are fertilized after treatment with trypsin or cyanides (Elinson, 1973b).

When working with fish oocytes, the fertilization of oocytes ovulated in vitro presents no difficulty, since their jelly coat is formed by follicular epithelium cells already in the ovary.

Another method of checking the capacity of amphibian eggs for normal development is by transplanting somatic cell nuclei into them (Subtelny and Bradt, 1959). This method is now widely used.

*3.2.4. Ovulation**

Normal oocyte maturation is accompanied by ovulation. It has been established that oocyte maturation and ovulation are two relatively independent processes, whose normal correlation may be upset by artificial stimulation of female maturation under hazardous conditions (Wright, 1945; Tchou-Su and Wang Yu-lan, 1958; Dettlaff, 1970a; Reynhout et al., 1975), and also by using high steroid concentrations (Morrill and Bloch, 1977).

Disruption of the normal correlation of these processes (ovulation of immature oocytes, or maturation and activation and even parthenogenetic cleavage of nonovulated oocytes) not infrequently occurs in the breeding of sturgeon at hatcheries (Kazansky, 1950, 1956, 1957; Dettlaff and Ginsburg, 1950, 1951, 1954; Ginsburg and Dettlaff, 1969; Dettlaff, 1970a). Similar desychronization of oocyte maturation and ovulation has been induced experimentally in amphibians (Kashchenko, 1962; Subtelny et al., 1968; Morrill and Bloch, 1977) and in teleosts (Jalabert et al., 1972; Jalabert, 1976). For example, it has been demonstrated that below-spawning temperatures reversibly suppress oocyte ovulation in amphibians (Wright, 1945; Tchou-Su and Wang Yu-lan, 1958) and sturgeons (Dettlaff, 1970a) and, in the latter, cause the activation of mature oocytes within the follicles.

These and similar facts convincingly show that maturation and ovulation are effected by different mechanisms, although they have gonadotropic hormones and steroids as common inducers.

Although the mechanisms of hormonal regulation of the ovulation have been under investigation for quite a long time, a number of problems remain in this field. In particular, it is not quite clear exactly where gonadotropic hormones act to induce ovulation. According to available data (Yatvin and Anderson, 1967; Anderson and Yatvin, 1970; Wasserman and Masui, 1974), gonadotropic hormones do not affect the follicle cells directly, but act through the cells of the peritoneal epithelium which, in the opinion of the authors, excrete proteolytic enzymes which weaken the mechanical rigidity of the follicular wall. This conclusion is based on indirect data concerning the induction in these cells of a more intensive RNA synthesis under the influence of gonadotropic hormones, and also of the ovulation-inhibiting effect of inhibitors of nucleic and protein syntheses (Dettlaff, 1966; Anderson and Yatvin, 1970). These data were obtained in amphibians where each oocyte is adjacent to the peritoneal epithelium. An analogous

---

*This section has been written with the participation of O. B. Trubnikova. It covers the literature published until 1981. For later literature, see the review by Schuetz (1985).

hypothesis was suggested for explaining the ovulatory mechanism in mammals (Byskov, 1969; Cajander and Bjersing, 1976). However, in teleosts the structure of the ovary is such that the peritoneal epithelium is not in direct contact with each oocyte and, apparently, one should assume in this case the direct action of gonadotropic hormones on the theca or certain theca cells. In teleosts, the cells of the inner theca during the pre-ovulatory period contain many lysosomes which may contain proteolytic enzymes involved in the ovulatory mechanism (Iwamatsu and Ohta, 1981).

The role of each cell layer of the follicular wall in initiating ovulation needs to be investigated.

It has been demonstrated that ovulation, like oocyte maturation, requires the presence of gonadotropic hormones in the medium for a limited time only. It has been found that the hormone-dependent period for ovulation is somewhat longer than for oocyte maturation (for sturgeons, according to unpublished data of Dettlaff, Skoblina, and Goncharov, and for amphibians, from data of Subtelny et al., 1968).

Another question still awaiting final clarification concerns the effect of steroids on ovulation. A study of the steroid action in fishes by means of inhibitors demonstrates that the maturation requires only translation, while ovulation requires also transcription. Moreover, there are two sensitive periods to transcription inhibitors: 1) at the beginning of incubation with steroids and 2) a few hours before ovulation (Georgia, 1980).

The effect of exogenous progesterone on ovulation is not quite clear. According to some authors (Masui, 1967; Schuetz, 1976b), amphibian oocytes do mature under the effect of progesterone, but hardly ever ovulate. According to others (Wright, 1961a, b; Subtelny et al., 1968), oocytes may ovulate under the effect of progesterone, but in a lower percentage than under the effect of gonadotropic hormones. In our laboratory we have obtained the same results on the oocytes of sturgeons and amphibians, and have found a higher percentage of ovulation (just as by Subtelny et al.) during prolonged incubation of the follicles in a Ringer solution containing both gonadotropic hormones and progesterone. In a series of investigations with telosts, Jalabert demonstrated that there was no stimulation of ovulation in the majority of fish species when injected with 17α-20β-dihydroprogesterone only, but there was some degree of stimulation when exposure to progesterone was preceded by exposure to gonadotropic hormones (Jalabert, 1976; Jalabert et al., 1977, 1978a; Montalembert et al., 1978). Studying the molecular mechanisms by introducing polar groups into the progesterone molecule, Morrill and Bloch (1977) demonstrated that such structural modifications differently affect the capacity to cause the specific response of maturation and ovulation: the former is sensitive to changes in the groups at the 19th carbon atom, whereas the latter is sensitive to changes at the 11th carbon atom.

Preparation for ovulation involves changes in the structure of the theca. These include the lysis of its collagen fibers (Yatvin and Anderson, 1967; Anderson and Yatvin, 1970) and the contraction of the macrovilli of the follicular epithelium and the microvilli of the oocyte surface (Van Gansen, 1966; Schuetz, 1972b). In the sturgeon changes also occur in the structure of the follicular epithelium cells; they accumulate a secretory substance which causes their separation from the secondary egg membrane remaining on the oocyte (Kazansky, 1952, 1957). There are grounds to suppose that increased cytoplasmic turgor and oocyte volume also play a role in the ovulatory mechanism (Hirose, 1976; Oshiro and Hibiya, 1981). In frog oocytes these changes occur almost immediately before ovulation, which occurs some time after GV breakdown; at metaphase I, the oocytes usually ovulate. It has been demonstrated (Rudneva, 1968) that oocytes with a normal turgor ovulate in a greater percentage of cases than oocytes with a low turgor. A similar phenomenon has been observed with sturgeon oocytes. When part of the cytoplasm was artificially removed from such oocytes, they matured but did not ovulate.

The liberation of the oocyte from the follicular membrane is accompanied by active contraction of the cells making up this membrane; they assume a cuboid or palisade shape (Larsen et al., 1977; Pendergrass, 1974). Studying ovulation in the frog, Rugh (1935) suggested that smooth muscle cells were present not only in the ovarian wall, but also in the tissue surrounding the oocyte. Subsequent electron microscopy corroborates this assumption (Anderson and Yatvin, 1970). However, data are available which point to the presence in the follicular wall of smooth muscle cells or cells resembling them, and to their involvement in the contraction of follicular membranes during ovulation. These data include such findings as the stimulating effect of prostaglandin F on ovulation and the suppression of ovulation under the effect of indomethacin, a prostaglandin inhibitor (Stacey and Pandey, 1975, Stacey and Peter, 1979; Jalabert and Szöllösi, 1975; Jalabert, 1976; Jalabert et al., 1978a, b; Goetz, 1979; Lemaire et al., 1980). Also, the results of radioimmunological studies demonstrate that the content of prostaglandins increases steeply in the follicle during the pre-ovulatory period (Goetz, 1980).

On the other hand, cytochalasin B, known to affect actomyosin microfilaments, which differ from the microfilaments of smooth muscle cells, also influences ovulation, partially suppressing it in fish (Jalabert, 1976; Schroeder and Pendergrass, 1976) and amphibians (Larsen et al., 1977; Schroeder, 1973). Detailed studies have been carried out on the ultrastructure of the follicle relating to the behavior of microfilaments in ovulation. It has been found that microfilaments are dominant structures in the surface layer of the theca (Kessel and Panje, 1968; Schroeder and Pendergrass, 1976; Larsen et al., 1977; Dumont et al., 1978), and that, as the moment of ovulation approaches, they gather into bundles (Pendergrass and Schroeder, 1976). Similar data have been obtained in our laboratory when studying the microfilamentary system of the follicular wall cells in the sturgeon (Trubnikova and Ryabova, 1988). Thus, ovulation is a complex multifactorial process which involves various cellular layers and structures. In the pre-ovulatory period, the maturation of the oocyte (GV breakdown, contraction of the microvilli, preparation of the follicular wall for rupture and contraction) is ensured, probably by different but related mechanisms which jointly result in the liberation of the mature oocyte from the ovarian tissue.

3.2.5. Chronology of Maturation

The time of onset of the successive stages of oocyte maturation and their duration may be expressed in time units (hours, minutes) or in relative developmental duration units (Dettlaff and Dettlaff, 1961). It has been demonstrated (Dettlaff and Zubova, 1962; Vassetzky, 1966, 1970; Ginsburg and Nikiforova, 1979; Dettlaff et al., 1981) that the duration of the stages of maturation, including the meiotic division phases, depends on temperature in the same way as the duration of different stages of embryonic development, including τ_0, the duration of one mitotic cycle in the period of synchronous cleavage divisions (Dettlaff and Dettlaff, 1961; Dettlaff et al., 1986). It means that the time of onset of successive oocyte maturation stages can be expressed in numbers of τ_0. This allows, by using curves of τ_0 as a function of temperature, which have been plotted for several species of amphibians and fishes (see Dettlaff, 1975; Dettlaff and Vassetzky, in press), calculation of the time of onset of any stage at different (but within optimal) temperatures. This is true of the time characteristics of oocyte maturation both in vitro and in vivo, since it has been demonstrated using sturgeons that the maturation rate of oocytes developing in the body of the female and in vitro is similar at the same temperature.

During in vitro stimulation of oocyte maturation with progesterone, the moment the oocytes are placed in the progesterone solution is taken as the starting point for counting the time.

A study of the relative duration of maturation expressed in τ_0 made it possible to demonstrate that in spring-spawning amphibians, for example *R. temporaria*, maintained under normal conditions and close to spawning, the duration of oocyte maturation varies little, both with regard to the oocytes of a single female and among different females.

On the other hand, research with *A. stellatus*, which enters rivers for spawning, demonstrated that in different females the time from placing the oocytes into a progesterone solution to the GV breakdown may differ by more than a factor of two, depending on their readiness to spawn. This timing of the in vitro response may serve as a criterion for selecting the most promising females (Goncharov, 1976).

For revealing the time regularities of oocyte maturation it would be interesting to obtain a comparable chronology (in τ_0) of oocyte maturation under the effect of progesterone for *X. laevis*, *R. pipiens*, *R. temporaria*, *B. bufo*, and *B. viridis*, and for sturgeons. However, for *R. pipiens*, *B. bufo*, and *B. viridis*, a preliminary determination of τ_0 at different temperatures would have to be made. Such a study is now in progress for *R. pipiens* and *B. viridis*.

The duration of oocyte maturation and individual maturation phases may be expressed not only in τ_0, but also as fractions of the maturation period, which makes it possible to compare data obtained on oocytes from different females of the same species. Wasserman and Masui (1975a) use as a standard unit the period from the placement of the oocytes in a progesterone solution to the moment of complete GV breakdown in 50% of the oocytes in the test. The use of this method reduces the variability of the time characteristics of oocyte maturation on different females. For the special purpose of comparing the duration of identical phases of meiotic and cleavage divisions, it is convenient to express them in fractions of the total duration of active mitotic phases (Vassetzky, 1966, 1970; see Chapter 4).

A relative characteristic based on the duration of the entire oocyte maturation period (from the time of the injection of the female with gonadotropins until ovulation) is of practical importance, since it allows prediction of the time, at different temperatures, when the mature eggs should be received from fish under conditions of sturgeon hatcheries (Dettlaff et al., 1968b, 1981; Igumnova, 1974), and can be used to time egg laying from experimental animals used in the laboratory.

3.3. CHANGES IN OOCYTE STRUCTURE AND PROPERTIES DURING MATURATION IN TELEOSTS

The changes taking place in the oocytes of teleosts during the maturation period have been poorly studied, and the available data are rather fragmentary. Descriptions of the maturing oocytes are available for the goldfish *Carassius auratus* and the black (Amur) bream *Megalobrama terminalis* (Tchou-Su and Wang Yu-lan, 1962), the loach *Misgurnus fossilis* (Belyaeva and Cherfas, 1965; Saat, 1982), the medaka *Oryzias latipes* (Iwamatsu, 1967), and the keta *Oncorhynchus keta* (Osanai et al., 1972).

Although there is great similarity between the changes in teleost oocytes and those of amphibians and sturgeons, there are some characteristic distinctions. It has been established for the Sevan trout (Negonovskaya, 1960) and the loach (Belyaeva and Cherfas, 1965) that the oocytes respond by maturation to the effect of gonadotropic hormone at a stage when their chromosomes are still of the lampbrush type, i.e., at the same stage as in *Urodela* and much earlier than in most of the *Anura* and sturgeons. Moreover, the chromosomes in the loach preserve the same structure also in the period of GV migration to the oocyte's surface (Belyaeva and Cherfas, 1965). According to Sakun (1955), the same is apparently true for the vimba (*Vimba vimba*). Meiotic divisions in teleosts, as in nearly all vertebrates, become blocked at metaphase II (see Ginsburg, 1968; and Chapter 4).

Another feature of nuclear transformations in the oocytes of teleosts is the close association of the nucleoli with the chromosomes. During GV movement to the surface of the oocyte, numerous nucleoli situated in its center begin to undergo vigorous changes: they form processes and become fragmented, then draw together and surround the chromosomes and fuse with one another, forming a dense little body with the chromosomes inside. In this form, surrounded by the nucleoli, the chromosomes, after GV breakdown, pass into the cytoplasm and migrate in the direction of the micropylar channel. During these stages the nucleolar material does not show a Feulgen-positive reaction, a reaction produced only by bivalents (Belyaeva and Cherfas, 1965). The nucleolar material begins to disintegrate into separate parts and gradually disappears prior to metaphase I. These types of nucleolar changes, unlike those in sturgeons, have been described for many species of teleosts (Negonovskaya, 1960; Sakun, 1961; Tchou-Su and Wang Yu-lan, 1962; Belyaeva and Cherfas, 1965; Saat, 1982). However, in the medaka, the nucleoli do not form a dense body and disappear earlier, soon after GV breakdown (Iwamatsu, 1967).

The cytoplasm of the oocytes of teleosts at the stage when they proceed to maturation contains many inclusions such as yolk granules and lipid droplets. In some fish species (for example in many *Cyprinidae*, loaches, and *Macropodus*) the yolk granules do not undergo visible changes during oocyte maturation and are preserved in the form of isolated bodies. In other species fusion of the yolk granules occurs, which is either completed by the end of the maturation period (in many *Salmonidae*, part of the *Percidae*, and others), or it is less complete (for example, in the whitefish, the pike, and in *Clupeidae*) and is completed only after egg activation and fertilization (see Ginsburg, 1968).

The fusion of yolk granules is accompanied by the segregation and movement of the lipid inclusions, and also by the shifting (following GV migration to the surface) of the yolk-free cytoplasm and the excreted nucleoplasm, and the formation of a cortical cytoplasm. Also, during the maturation period, hydration of the yolk occurs (see Ginsburg, 1968).

Following GV breakdown, during metaphase I formation, multiple cytasters appear in the oocytes' cytoplasm (Sakun, 1955; Tchou-Su and Wang Yu-lan, 1964). The capacity for cortical reaction appears in medaka oocytes only after GV breakdown (Iwamatsu, 1967). At prometaphase I there is still only a partial cortical reaction and the breakdown of cortical granules spreads slowly from the site of sperm penetration; at metaphase II, the cortical reaction involves the egg's entire surface and spreads at a normal rate (Iwamatsu et al., 1976). In the loach the capacity for cortical reaction appears later, at anaphase-telophase I (Saat, 1982).

The capacity for fertilization appears in the oocytes of the goldfish and the medaka before complete cytoplasmic maturity, but the spermatozoa in such oocytes do not form asters and do not change (Tchou-Su and Wang Yu-lan, 1964; Iwamatsu, 1965a, b, 1967). According to later data of Iwamatsu et al. (1976), medaka oocytes fertilized at metaphase I can undergo normal cleavage.

A late indication of normal cytoplasmic maturation in teleosts is the capacity for pulsating movements, which manifest themselves after fertilization but are absent after incomplete cytoplasmic maturation (see Iwamatsu, 1967).

It must be added that, as in amphibians and sturgeons, the oocytes of teleosts respond to the effect of progesterone by maturation at earlier stages than under the effect of gonadotropic hormones. In addition, in oocytes responding to both hormones, the fusion of yolk granules takes place earlier under the effect of progesterone than under the effect of gonadotropic hormones (Sakun, 1966; Sakun and Gureyeva-Preobrazhenskaya, 1975).

3.4. THE ROLE OF THE OOCYTE NUCLEUS AND CYTOPLASM IN THE DEVELOPMENT OF MATURE EGG ORGANIZATION DURING THE MATURATION PERIOD

It was earlier assumed (Dettlaff et al., 1964; Dettlaff, 1966) that the initiation of oocyte maturation under the effect of hypophysial gonadotropic hormones was genetically determined by the oocyte nucleus. It was soon discovered, however, that stimulation of the genetic activity of the follicle cells is responsible for this process rather than the oocyte nucleus. It was demonstrated that after removal of the GV the cytoplasm of oocytes exposed to the gonadotropic hormones undergoes changes which, as in intact oocytes, are not blocked by actinomycin D (Skoblina, 1968, 1970). It was further demonstrated that oocytes completely freed from the cells of the follicular epithelium fail to mature under the effect of gonadotropic hormone, yet fully mature under the effect of progesterone (Masui, 1967; Schuetz, 1967a-c; Smith and Ecker, 1968; Dettlaff et al., 1968a; Dettlaff, 1970b). Here, unlike the effect of gonadotropic hormones (Dettlaff, 1966), the effect of progesterone is blocked neither by actinomycin D (Schuetz, 1967c; Skoblina, 1968; Dettlaff and Skoblina, 1969; Dettlaff, 1970b; Smith and Ecker, 1969a, 1970b) nor by exposure to high radiation doses (Masui, 1973a, b; Skoblina et al., 1984b).

The role played in maturation by RNA synthesis induced by gonadotropic hormones in the oocyte nucleus (Brown and Littna, 1964) has not yet been determined (Masui, 1973a; Morrill et al., 1975).

Significant data are available on the participation of the GV nucleoplasm in changing the properties of the oocytes. The first experiments of this kind were carried out on the oocytes of the starfish *Asterias glacialis* (Delage, 1899, 1901) and the nemertean *Cerebratulus lacteus* (Wilson, 1903; Yatsu, 1905). The oocytes of these species ovulate and are released into water before GV breakdown. Delage and Wilson, by means of shaking, divided the oocytes into fragments before, during, and after GV breakdown and then inseminated them. When the oocyte fragments were isolated before GV breakdown, only those parts which contained a GV underwent fertilization and cleavage; however, when the oocytes were separated at stages following GV breakdown, fertilization and cleavage occurred in both nucleated and nonnucleated fragments. In view of this, Delage believed that the difference between cytoplasm capable of fertilization and cleavage and that incapable of it lies in the fact that the former had been penetrated by nuclear sap (Delage, 1901). Chambers (1921) obtained similar results by resorting to microsurgery involving the removal and partition of the GV, and also by injecting nucleoplasm into enucleated starfish oocytes.

Delage believed that cytoplasm maturation coincides with the diffusion of nuclear sap, and that an investigation into the nature of nuclear sap should help to reveal the nature of maturation processes. However, it was only recently, after considerable progress was achieved in studying the mechanisms of hormonal regulation of maturation in echinoderms, and after it was learned that

1-Me-Ade can induce in vitro oocyte maturation in starfish, that investigations commenced on the molecular nature of the maturation processes in starfish oocytes (see Kanatani, 1964, 1972, 1973, 1975, 1979; Katani and Shirai, 1972; Shuetz, 1969). The methods of studying starfish oocytes are being constantly improved (see Hirai et al., 1971; Sekirina et al., 1983; Meijer et al., 1984; Davydov et al., 1986; Vassetzky et al., 1983, 1986; Schroeder, 1986). These experiments are developed in parallel with similar investigations into amphibians and fish oocytes. The first experiments with the removal of the GV and its replacement by somatic cell nuclei in the oocytes of vertebrates were carried out using full-grown oocytes of *B. bufo* and *B. viridis* (Dettlaff et al., 1964). Subsequently, investigations of this kind have been carried out in many laboratories, resulting in ample experimental materials, some of which we shall now try to examine.

3.4.1. Egg Properties Arising in Cytoplasm of Enucleated Oocytes of Bufo viridis, Rana temporaria, Xenopus laevis, and Acipenser stellatus

Before examining these phenomena in brief, it should be pointed out that part of the nucleoplasm may pass from the GV into the perinuclear space and into the ER cisternae before GV removal, and may play some role in the changes taking place during oocyte maturation.

A considerable part of the maturation events in the full-grown oocyte take place at the cytoplasmic level, without the participation of the GV (Skoblina, 1968, 1970; Smith and Ecker, 1968, 1969a, b; Dettlaff and Skoblina, 1969). The surface layer of the cytoplasm matures in enucleated oocytes: contractility and a capacity for cortical reaction are acquired, permeability changes occur, and microvilli disppear from the oocyte surface and it becomes glossy. Oocytes deprived of the GV are capable of activation, which may be judged not only by the discharge of the contents of the cortical granules, but also by the detachment of membranes which enables the eggs within to turn with their animal pole upward. These changes were also observed during insemination of the oocytes of *A. stellatus* (Skoblina, 1970) and *B. bufo* (Katagiri and Moriya, 1976) maturing without the GV participation.

During maturation of enucleated oocytes, the ultrastructure of the cytoplasm undergoes change (Aizenshtadt and Skoblina, 1970): instead of the disappearing annulate lamellae, there appear spheroid aggregates of small secretory vesicles, moving into the surface layer of the cytoplasm. Adjoining these aggregates of vesicles are individual dictyosomes, and the entire structure resembles a Golgi complex with a hypertrophied vesicular element. The size of these aggregates is somewhat smaller than in control oocytes.

The destruction of annulate lamellae in the cytoplasm of enucleated oocytes implies the emergence in the latter of a nuclear membrane disintegration factor. The presence of this factor in the cytoplasm of oocytes which matured without a GV has been demonstrated by experiments in which the cytoplasm from such *R. pipiens* oocytes was injected into nucleated oocytes not treated with hormones. The injected cytoplasm caused GV breakdown in these oocytes in the same percentage of cases as did injected cytoplasm taken from control intact oocytes containing a GV (Masui and Markert, 1971). This result was confirmed by others (Smith and Ecker, 1971; Schorderet-Slatkine, 1972; Schorderet-Slatkine and Drury, 1973; Reynhout and Smith, 1974).

In oocytes maturing with and without a GV protein, synthesis changes in the same way, and these changes involve not only the intensity of the synthesis, but also the range of the proteins undergoing synthesis (Ecker et al., 1968; Ecker and Smith, 1968; Smith and Ecker, 1969a; Smith et al., 1966). Thus, both the intensity and the pattern of protein synthesis during the maturation period, and in all probability also in the period of cleavage, is controlled by the cytoplasm (Ecker and Smith, 1971a).

The cytoplasm of enucleated oocytes not only acquires all the above properties under the influence of progesterone but, when injected into the intact oocytes, induces their complete maturation (Masui and Markert, 1971). The roles of GV and cytoplasm in oocyte maturation will be dealt with in Section 3.4.2.1.

3.4.2. Egg Properties Emerging with GV Nucleoplasm Participation

It has been demonstrated that oocytes which mature after the removal of the GV lack the capacity for cytokinesis, and phenomena essential for development, such as chromosome condensation, conversion of sperm nuclei into pronuclei, and DNA synthesis, do not take place in the nuclei of somatic cells and sperm nuclei transplanted into oocytes lacking a GV. The structure of the cytoplasm of such oocytes shows some deviation from normal. Absent in such oocytes is about 90% of the DNA- and RNA-polymerase and other substances which have accumulated in the GV nucleoplasm during the growth period.

3.4.2.1. Capacity of Nucleoplasm to Induce the Maturation of Intact Full-Grown Oocytes

Convincing evidence has recently been obtained that the nucleoplasm of the GV of oocytes matured under the influence of hormones acquires the capacity, when injected into full-grown oocytes, to stimulate their maturation in the absence of hormones; this capacity arises in the nucleoplasm even earlier than in the cytoplasm. According to Kishimoto et al. (1981), enucleated oocytes of *Asterina pectinifera* produce much less MPF than intact oocytes and are unable to amplify MPF production. Injection of GV material into such oocytes restored the MPF activity to a considerable degree. They concluded that some GV material is required for MPF amplification. Picard and Dorée (1984) performed removal and transplantation of nuclear material into full-grown oocytes of *Marthasterias glacialis* and *Asterias rubens*. Less than 5 min after 1-Me-Ade addition, and about 10 min before appearance of the cytoplasmic MPF, a factor appeared in the GV which triggered the production of ooplasmic MPF and the subsequent events of meiotic maturation when injected in any full-grown oocytes. Transfer of either "nuclear MPF" or cytoplasmic MPF, initially produced in hormone-treated maturing oocytes, triggers the production of both cytoplasmic and "nuclear" MPF in the intact recipient oocytes. Before hormone action the GV also contains "nuclear meiosis-inducing factor" capable of inducing meiosis reinitiation, but only in advanced "competent" oocytes. These latter are capable of producing MPF to some extent after enucleation upon hormonal stimulation.

Picard and Dorée (1984) concluded that the GV is involved in MPF production. Guerrier et al. (1983) succeeded in inducing the hormone-independent maturation of the *A. forbesi*, *A. rubens*, and *M. glacialis* oocytes by mechanical disruption (vigorous shaking) of the GV. They believe that the dispersion of the nucleoplasm into the cytoplasm appears to be a sufficient stimulus for sperm aster differentiation, meiotic spindle formation, polar body extrusion, and female pronuclear development. But the dispersion of the nucleoplasm into the cytoplasm suggests that in their experiments the GV membrane was disrupted in advanced oocytes.

In amphibian oocytes the reinitiation of meiosis was first obtained by injection of the nucleoplasm with traces of cytoplasm into the *B. viridis* full-grown oocytes either after the onset of maturation inertia, or at the beginning of GV breakdown (Dettlaff et al., 1964). Later, Masui and Markert (1971) published the results of injecting cytoplasm from the progesterone-treated oocytes into *R. pipiens* intact full-grown oocytes. The injected cytoplasm becomes effective in inducing maturation within a few hours after progesterone administration, reaching a maximum close to GV breakdown time. They have also established that enucleated donor oocytes produce MPF after progesterone treatment just as readily as do the GV-containing oocytes. In their paper they state: "Thus, the production of the maturation-promoting factor can be attributed entirely to a cytoplasmic process independent of immediate nuclear function" (Masui and Markert, 1971, p. 135). The MPF activity of nucleoplasm was not tested in this and many other investigations. The injection of nucleoplasm of the maturing oocytes into *B. bufo* full-grown oocytes failed to induce maturation (Wang Yu-lan et al., 1982a). The results of Masui and Markert were confirmed by the other investigators of *R. pipiens* (Smith and Ecker, 1971), *X. laevis* (Schorderet-Slatkine and Drury, 1973), *B. bufo* (Wang Yu-lan et al., 1982a, b), *R. temporaria*, and *A. stellatus* (Skoblina, 1969). Only in *Pleurodeles waltlii* did the enucleation of oocytes prior to maturation induction result in a significantly lower incidence of maturation than in the control oocytes under the influence of different steroids (Skoblina et al., 1984b). Skoblina concluded that the nucleoplasm appears to be essential for MPF production in the cytoplasm of the *P. waltlii* oocytes.

In order to elucidate the capacity of the GV nucleoplasm to induce oocyte maturation, the experiments which had been performed on *B. viridis* oocytes (Dettlaff et al., 1964) were repeated on *B. viridis*, *X. laevis*, and *A. stellatus* oocytes (Nikitina and Dettlaff, 1986). GVs were isolated from the oocytes in a hormone-free Ringer solution within different times after progesterone administration. The intact GVs or the membrane-free GV contents were sucked into micropipettes and injected into full-grown oocytes. At the same time, the cytoplasm was taken from the enucleated oocytes and injected into the other full-grown oocytes. The recipient oocytes were incubated in a hormone-free Ringer solution until the maturation of the progesterone-treated control oocytes was complete, and were then checked for GV breakdown. The experiments have shown that, in all studied species, the nucleoplasm of the donor oocytes acquires the capacity to induce maturation of the recipient oocytes (GV breakdown) and, moreover, at a much earlier time than does the cytoplasm. These facts confirm earlier observations (Dettlaff et al., 1964), and are consisted with data obtained on the starfish oocytes (Picard and Dorée, 1984). Thus the nucleoplasmic factor does not differ by its final effect from the MPF discovered by Masui and Markert (1971) in the cytoplasm of the enucleated *R. pipiens* oocytes.

The nature of the nucleoplasmic maturation factor is still obscure. Since it is known that progesterone induces the maturation of amphibian oocytes despite the inhibition of transcription (see Dettlaff, 1977b; Masui and Clarke, 1979; Skoblina et al., 1984b), it is evident that the karyoplasmic factor is not *de novo* synthesized mRNA.

A number of questions arise with respect to these data: What is the relationship between the karyoplasmic and cytoplasmic maturation factors? What is the nature of the karyoplasmic factor? What is the chain of events leading from the induction of oocyte maturation by progesterone to the appearance of the karyoplasmic factor? How complete is the maturation of oocytes stimulated by the karyoplasmic factor?, etc.

Despite many unanswered questions, it is evident that in various animal species a previously unknown maturation factor arises in the GV nucleoplasm at the early stages of progesterone-induced maturation. It is quite natural that this factor could not be taken into account in explaining the experimental results. For example, the appearance of MPF in the cytoplasm of the maturing oocytes shortly before GV breakdown and its maximum activity at the stage when the entire nucleoplasm passes into the cytoplasm were explained only by the existence of a GV-independent cytoplasmic MPF (see Masui and Clarke, 1979). Furthermore, the biochemical studies of MPF are usually performed on the mature egg cytoplasm, which contains the nucleoplasm of the disrupted GV (see Wu and Gerhart, 1980). Finally, the MPF-like activity in the GV should also be taken into account when estimating the role of MPF in cell cycle regulation (Gerhart et al., 1984; Newport and Kirschner, 1984; Ford, 1985; Masui, 1985b); see Section 3.5.1.2.

3.4.2.2. Capacity for Cytokinesis

It was shown (Dettlaff et al., 1964) that hormone-treated oocytes of the *B. bufo* and *B. viridis* from which the GV had been removed are incapable of cleavage after transplanting blastula cell nuclei into them, whereas the control oocytes from which the GV had not been removed did cleave under the same conditions.

The necessity of a GV for the oocytes of lower vertebrates to acquire the capacity for cleavage was confirmed in experiments involving the removal of the GV from oocytes of different amphibian species (Skoblina, 1968, 1969; Smith and Ecker, 1969a), and those of sturgeons (Skoblina, 1968, 1969; Nikitina, 1972). These tests also demonstrated that the injection of GV nucleoplasm from full-grown oocytes not exposed to hormones into oocytes in which the GV was removed (incubated with progesterone or with gonadotropic hormones) allows cleavage. Such oocytes of *R. pipiens*, after transplantation of somatic cell nuclei into them, cleaved and reached the blastula stage (Smith and Ecker, 1968, 1969a), while *A. stellatus* oocytes after insemination even underwent gastrulation (Nikitina, 1972). It was further demonstrated that in oocytes lacking a GV the capacity for cytoplasmic cleavage can be restored not only by injecting the nucleoplasm from the GVs of intact oocytes of a definite size, but also by transplanting GVs taken from oocytes at the beginning of the vitellogenic growth period (Nikitina, 1972, 1978). This shows that the factor essential for the cytoplasm to acquire the capacity for cleavage is already present in the oocyte nucleus at an early stage of vitellogenesis, and undergoes no qualitative changes during the maturation period. Supporting the latter is also the fact that the nucleoplasm taken from GVs both prior to and following the effect of hormones is equally effective (Smith and Ecker, 1968; 1969a).

For the cytoplasm to acquire the capacity for cleavage, not only is the transition of the nucleoplasm into the cytoplasm needed, but also the changes during this period in the properties of the cytoplasm itself, since the introduction of nucleoplasm into the cytoplasm of full-grown oocytes not exposed to the effect of hormones does not permit cleavage (Dettlaff et al., 1964).

At the same time, the cytoplasm of the growing *P. waltlii* (0.75-1.35 mm in diameter) oocytes can acquire the capacity for cytokinesis if their maturation is stimulated first by the injection of very small amounts of cytoplasm from the maturing oocytes (0.01-0.04% of the volume of recipient oocyte), followed by the transplants of somatic cell nuclei (Skoblina et al., 1984a). In starfish it has been shown that the volume of cytoplasm affects the number of cleavage divisions. It was shown with *M. glacialis* oocytes (Buznikov et al., 1985) that in karyoplasts surrounded by a hardly discernible layer of cytoplasm and matured under the influence of 1-Me-Ade the number of cleavage divisions is limited to 1-2 and, for a larger cytoplasm layer, to 3-4.

The only data which at first sight contradict these conclusions of the effect of nucleoplasm on the capacity for cytokinesis were reported by Iwamatsu (1966, 1967). By centrifuging *Oryzias latipes* oocytes at the stage of maturation inertia, not long before GV breakdown, Iwamatsu shifted the GV from the animal region into the yolk, or removed the GV by means of pricking. In many oocytes the GV remained intact in the yolk and, nevertheless, the oocytes were activated and cleaved, and there were no nuclei in their blastomeres although cytasters were present. This result, however, does not contradict the conclusion that the transition of the nucleoplasm into the cytoplasm is needed for the acquisition of the capacity for cytokinesis. It is known that part of the nucleoplasm passes into the cytoplasm before GV breakdown (Iwamatsu, 1967) and the later the stage at which the GV is removed, the greater is the amount of nucleoplasm that remains in the oocyte cytoplasm. In the oocytes of sturgeons, in which a network of anilinophilic nucleoplasmic lacunae is particularly well distinguished in the cytoplasm, the emergence of part of the nucleoplasm into the cytoplasm before GV breakdown is perfectly visible in sections. Accordingly, the oocytes of amphibians and sturgeons, just like the oocytes of *Oryzias latipes*, are capable, after the transplantation into them of somatic cell nuclei, of atypical cleavage provided the GVs are removed from them not long before their complete breakdown (Dettlaff et al., 1964).

3.4.2.3. Rhythm of Cleavage Divisions

It has long been known that the rhythm of synchronous cleavage divisions is controlled by the cytoplasm (Moore, 1933); however, only recently has the nature of this control become the subject of special study. Experiments involving reciprocal replacement of 40-50% of the cytoplasm between the embryos of *X. laevis* and *Pleurodeles waltlii* at the stage of 2-4 blastomers with cytoplasm from activated eggs resulted in a temporary acceleration of cleavage of *P. waltlii* cytoplasm (Aimar et al., 1981, 1983). It was found that the cytoplasm was effective only at the GV breakdown stage. Neither the GV nucleoplasm nor the cytoplasm of the full-grown oocyte nor the oocyte after 2-h exposure to progesterone, possess this property. It is not yet clear whether the capacity to regulate the cleavage rhythm (called by the authors "cleavage timing system," or CTS) appears in the cytoplasm of oocytes from which the GV had been recently removed, or if this requires the passage of nucleoplasm into the cytoplasm.

A comparison of the above observations with what is known about the appearance of the capacity for cytokinesis in the cytoplasm suggests that the capacity for cytokinesis and the cleavage rhythm are determined by different, or only partially coinciding, changes in oocytes. CTS is not identical with the MPF, although they appear simultaneously in the oocyte during the GV breakdown period. A supernatant fraction from the oocytes of *X. laevis* isolated by Drury's method (1978) and having MPF activity does not influence the cleavage rate upon its injection into *Pleurodeles* embryos. (The role of MPF in the control of the cell cycle is also discussed in Section 3.4.1.2.)

3.4.2.4. Chromosome Condensation

Chromosome condensation is observed in oocytes at the stage of GV breakdown during maturation both in vivo and in vitro, and has been described in a number of species, particularly in the oocytes of the sturgeon (see Dettlaff and Skoblina, 1969) and *R. pipiens* (Ziegler and Masui, 1973). Gurdon (1968) described the breakdown of the nuclear membrane and chromosome condensation in the nuclei of brain cells of an adult animal when they were transplanted into the oocytes of *X. laevis* at a stage following GV breakdown. Ziegler and Masui (1973) followed Gurdon's experimental approach for obtaining the chromosome condensation. They injected a suspension of brain cell nuclei of *R. pipiens* into oocytes at different stages of maturation, and fixed them 2 and 4 h later. In this case, chromosome condensation in the transplanted nuclei only took place some time after GV breakdown and only in the cytoplasm of oocytes that matured with GVs. At the same maturation stages, chromosome condensation does not take place in the transplanted nuclei in recently enucleated oocytes. However, if the nucleoplasm from another maturing oocyte is injected into such enucleated oocytes, chromosome condensation takes place. Thus, as in the case of the capacity for cytokinesis, the passage of nucleoplasm into the cytoplasm is not sufficient for the cytoplasm to cause chromosome condensation, but in addition, a certain stage of cytoplasmic maturity is required. In the experiments of Ziegler and Masui (1973), injection of nucleoplasm into oocytes containing transplanted nuclei, but untreated with progesterone, had no effect on the chromosomes.

In normally maturing oocytes, the nuclear membranes do break down, as in similar tests by Gurdon (1968), although it is not ruled out that they also break down in oocytes that have matured without a GV, since the disintegration of annulate lamellae takes place in such oocytes. At the stage of the mature activated egg, the capacity to cause chromosome condensation in the cytoplasm disappears; somatic nuclei transplanted into it, as well as sperm nuclei, swell and despiralize.

Thus, in the period between GV breakdown and egg activation, there are in the cytoplasm certain conditions [or an activity, which Masui calls "chromosome condensation activity" (CCA)] which cause chromosome condensation. Both nucleoplasm and cytoplasm are required for this activity. The effect of CCA is not species specific; the CCA of *X. laevis* cytoplasm may cause the condensation of chromosomes even in the somatic cells of plants (von der Haar et al., 1981).

G_0 noncycling erythrocyte nuclei can also be induced to undergo chromosome condensation by maturing oocyte cytoplasm (Leonard et al., 1982).

3.4.2.5. Capacity to Convert Sperm Nuclei into Pronuclei and Induce DNA Synthesis in Them

During the process of fertilization in the activated egg cytoplasm, the nucleus of a spermatozoon turns into a pronucleus, while DNA synthesis starts in the male and female pronuclei (see Ginsburg, 1968). Although these changes do take place during fertilization, the oocyte cytoplasm requires the capacity to cause them in the process of maturation, but only on condition that nucleoplasm enters the oocyte cytoplasm (Skoblina, 1974, 1976; Katagiri and Moriya, 1976; Katagiri, 1980). These authors have demonstrated that the nuclei of spermatozoa of *R. temporaria* and *B. bufo* obtained by treating spermatozoa with pronase and Triton X-100, when injected into a mature intact oocyte, soon swell, turn into pronuclei, and incorporate ^{3}H-thymidine. When pronase-treated spermatozoa are injected into the oocyte at earlier maturation stages, their transformation is delayed until the oocytes reach metaphase II. When similar nuclei are transplanted into the cytoplasm of enucleated oocytes which had matured without GVs and were activated, neither the transformation of sperm nuclei into pronuclei, nor DNA synthesis, takes place in them. The nuclei of spermatozoa which penetrate cytoplasm of such enucleated oocytes following normal insemination behave in the same manner.

The appearance of the capacity to induce DNA synthesis in the nuclei in the process of oocyte maturation has been well studied. It has been shown (Gurdon, 1968; Gurdon and Speight, 1969) that in the full-grown oocyte neither the GV nucleoplasm nor the cytoplasm possess this property separately. The nuclei of brain cells injected together with ^{3}H-thymidine into the GV or into the cytoplasm do not incorporate the label. DNA synthesis occurs in transplanted nuclei only after the nucleoplasm passes into the cytoplasm. This capacity to induce DNA synthesis, according to Gurdon, is preserved by the cytoplasm for a fairly long time after fertilization.

3.4.2.6. Special Structural Features of the Cytoplasm of Oocytes Matured without GV

If the GV is removed from oocytes of *A. güldenstädti* and *A. stellatus* with a long hormone-dependent period before they have proceeded to maturation, one can observe that their cytoplasm contains practically none or very little nucleoplasm stainable by aniline blue (Dettlaff et al., 1964). In the absence of a GV in the cytoplasm of such oocytes (following hormone treatment), the network of anilinophilic lacunae containing globules of hydrophilic colloid, cytasters, and glycogen either fails to appear or is developed weakly. It should be added that, despite the presence of much tubulin in the cytoplasm of oocytes of *X. laevis* (Pestell, 1975), cytasters do not form in enucleated oocytes even after treatment with D_2O, which promotes the polymerization of microtubules in normally maturing oocytes (Hanocq-Quertier et al., 1978). After the activation by pricking of the enucleated oocytes of *A. stellatus* and *R. temporaria*, lacunae with hydrophilic colloid do not appear on the animal surface, no hyaline layer forms at the egg surface, and no colloid, or very little of it, is excreted under the detaching membrane (Dettlaff et al., 1964; Skoblina, 1970).

Lastly, there is less hydration of the cytoplasm of enucleated oocytes than in normally matured eggs. This phenomenon requires closer study, however, since cytoplasm hydration is influenced by many conditions, including the physiological state of the oocyte, the permeability of the surface membrane, and changes in the ionic composition of the cytoplasm during the passage of nucleoplasm into it (Morrill, 1965).

The fine structure of the cytoplasm of enucleated oocytes differs slightly from that of normal oocytes.

3.5. MOLECULAR BIAS OF OOCYTE CHANGES DURING THE MATURATION PERIOD

Toward the end of the growth period an oocyte contains a tremendous store of structural and biochemical components, most of which do not take part in the transformation of the oocyte during maturation and are merely stored in it until the beginning of embryogenesis (see Chapters 1, 2, 6, and 7). Thus, apart from mRNA and protein-synthesizing systems which take part in protein synthesis during the maturation period, an oocyte also contains long-lived masked mRNA (Davidson, 1969; Denis, 1974; Bienz and Gurdon, 1982; Bienz, 1984). Oocytes also contain precursors of nucleic acids, DNA and RNA polymerases, and histones. These components not only ensure specific syntheses at early stages of embryogenesis, but may also begin functioning earlier if mRNA, nuclei, DNA, or individual genes are injected into the oocytes (Gurdon, 1980, 1982; Gurdon and Melton, 1981).

Amphibian oocytes also contain large reserves of tubulin (Pestell, 1975; Moreau and Gounon, 1977; Raff and Raff, 1978) and actin. Nonpolymerized actin is contained not only in the cytoplasm, but also in the GV, where it influences nucleoplasm viscosity (Clark and Rosenbaum, 1979). Thus, in the more viscous nucleoplasm of *X. laevis*, its concentration is 50% higher than in the nucleoplasm of *P. waltlii* (Moreau and Boucher, 1981; Gounon and Karsenti, 1981).

The GV contains tissue-specific antibodies (Dreyer et al., 1981) and other proteins (Merriam and Hill, 1976; Hausen et al., 1985) which are distributed in the cytoplasm after GV breakdown. The substances stored in the cytoplasm of the oocytes include lamin III, a protein specific for the GV membrane, pronuclei, and cleavage nuclei. Upon GV breakdown it is solubilized and serves as a storage pool for the formation of nuclear membranes in the early embryos of *X. laevis* until stage 13 (Nieuwkoop and Faber, 1956). This reserve of lamin III in the *X. laevis* oocyte is enough to supply the membranes of about 14,000 somatic nuclei (Benavente et al., 1985; Stick and Hanson, 1985). The *Xenopus* egg also stores enough histones to assemble 20,000-30,000 nuclear equivalents of DNA into chromatin (Laskey et al., 1977; Woodland and Adamson, 1977).

There are many studies on the molecular nature of the processes underlying the structure and functional transformations in the oocytes during the maturation period. The following facts serve as starting points. During the induction of maturation by gonadotropic hormones and progesterone there is a considerable activation of protein synthesis and an alteration of its pattern in oocytes (Smith et al., 1966; Ecker et al., 1968; Ecker and Smith, 1968, 1971a; Smith and Ecker, 1969a). It is also established that inhibitors of protein synthesis (puromycin and cycloheximide) block the gonadotropic hormone induction of maturation (Dettlaff, 1966; Brachet, 1967; Dettlaff and Shmerling, 1968) and also its induction by progesterone (Schuetz, 1967c; Wasserman and Masui, 1975a; Dettlaff et al., 1977). These observations indicate that the synthesis of some proteins in the oocyte itself is a necessary condition for the induction of oocyte maturation. Protein synthesis is also required for other events taking place during oocyte maturation, such as chromosome condensation (Ziegler and Masui, 1976) and pseudo-gastrulation (Smith and Ecker, 1969b, 1970a; Dettlaff et al., 1977).

The discovery that the nucleoplasm and cytoplasm of maturing and mature oocytes, when injected into intact oocytes not treated with hormones, causes their maturation is of great importance for studying the molecular nature of the substances appearing in the oocyte during maturation (Dettlaff et al., 1964; Masui and Markert, 1971). This led to attempts to separate the active fractions from the cytoplasm of maturing and mature oocytes, and to test their effect on intact oocytes not exposed to the effect of hormones. The capacity of the nucleoplasm (Nikitina and Dettlaff, 1986) has not yet been studied in this respect and presents a very important task for future investigations.

Similarly, the nature of the active fractions of the nucleoplasm can also be studied, since injection of the latter into enucleated oocytes imparts to them the capacity for cytokinesis (Smith and Ecker, 1969a, b; Nikitina, 1972), for chromosome condensation (Ziegler and Masui, 1973), for nuclear transformation, and for DNA synthesis (Gurdon, 1967; Skoblina, 1974, 1976; Katagiri and Moriya, 1976). Methods allowing mass separation of GVs from oocytes (Burzio and Koide, 1976; Scalenghe et al., 1978; Moreau and Boucher, 1981; Vassetzky et al., 1984; Chiba and Hoshi, 1985) may prove very useful for these purposes.

Before considering the results of these investigations, we should dwell on questions of terminology. A number of tentative designations have already been introduced for hypothetical maturation factors of a presently unknown nature. Most of them have been contributed by Masui. In addition to the factors already mentioned above (MPF, CCA, cytokinesis factor), there is also a cytostatic factor (Masui and Markert, 1971), a maturation initiator (Wasserman and Masui, 1975a), and a factor inducing pseudomaturation (Baltus et al., 1973).

One can see that some of these hypothetical factors have been designated according to structural or functional features, whose emergence they presumably determine. Many other factors of this kind can be named, e.g., factors for the appearance of contractility of the cytoplasm's surface layer, for the formation of cytasters, for cytoplasm hydration, for the appearance of a capacity for cortical reaction, for the transformation of sperm nuclei into pronuclei, for DNA synthesis, and factors determining the fate of germ cells (sex determinants) (Fig. 35). The designation of such factors is justified in order to specify the subjects of investigation. After their nature is clarified, the "factors" assume a reality of their own.

Other factors, however, are determined more broadly and less precisely. This refers, particularly, to the MPF and the factor inducing pseudomaturation. Let us now consider what is known at present about the individual factors.

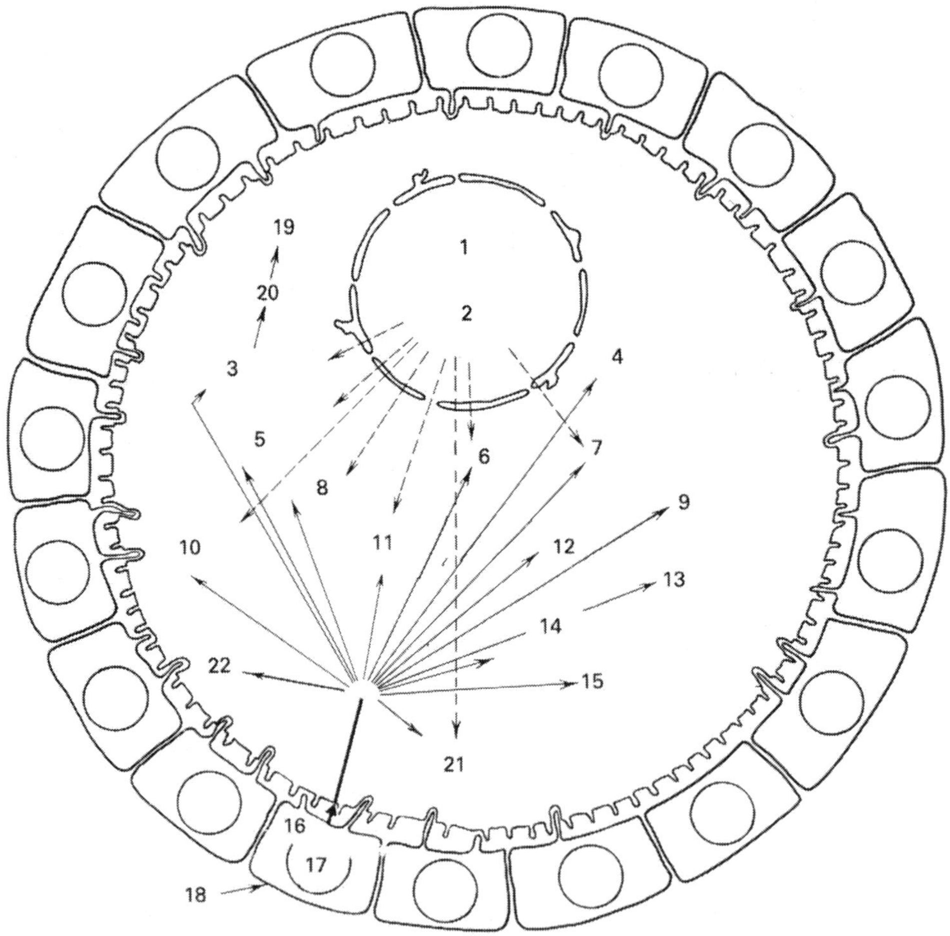

3.5.1. Maturation Promoting Factor (MPF)

Since Masui and Markert (1971) first introduced the concept of the "maturation promoting factor" (MPF), its meaning has changed (Wasserman and Masui, 1975a). Initially this term referred to some factor formed during oocyte maturation not only in situ, but after the injection of this factor with the cytoplasm into another oocyte that had not been exposed to the effect of hormones. Here the basic criterion for assessing its influence was at first GV breakdown.

However, the injection of cytoplasm taken from a maturing oocyte into an intact one causes in the latter not only GV breakdown, but other changes as well (Masui and Markert, 1971; Smith and Ecker, 1971; see Masui and Clarke, 1979; Hirai et al., 1981). The cytoplasm of a maturing oocyte apparently contains a complex of factors rather than a single factor, actuating the entire chain of maturation events; the efficiency can be assessed according to different criteria. In addition, Wasserman and Masui (1975a) established that the oocyte's cytoplasm acquires the capacity to cause GV breakdown not as a primary reaction of the oocyte to the effect of progesterone, and that the formation of the MPF is preceded by the formation of a still earlier protein factor, called the "initiator." Wasserman and Masui now define the MPF as a factor responsible for GV breakdown (see Fig. 2 in their paper), although they retain its previous name.

Fig. 35. Schematic section through the oocyte and the surrounding follicle epithelium in which the properties acquired by the cytoplasm during maturation are indicated. Some of these properties are revealed during the maturation period itself, others only at the beginning of development (although the potential for these is already in the oocyte). Also indicated are properties in the development of which the karyoplasm takes no part (continuous arrows), and those properties dependent on the participation of components of the karyoplasm (dotted-line arrows). The bold line indicates the puromycin-sensitive stage (emergence of the initiator). 1) GV; 2) karyoplasm; 3) chromosomal condensation; 4) GV breakdown and disintegration of annulate lamellae; 5) capacity for DNA synthesis; 6) hydration; 7) structural changes of cytoplasmic organelles; 8) cytasters; 9) permeability; 10) capacity for cytotomy; 11) capacity for the conversion of sperm nuclei into pronuclei; 12) cortical layer contractility; 13) capacity for fertilization; 14) capacity for cortical reaction; 15) capacity for activation impulse conduction; 16) progesterone; 17) RNA, DNA; 18) gonadotropins; 19) arrest of meiosis; 20) metaphase II; 21) capacity for sperm aster formation; 22) cell clock.

A more exact definition is particularly important in connection with the extensive studies of the properties, species-specificity, and chemical nature of the MPF. It is most likely that the cytoplasmic factors involved in the different changes of the oocytes during their maturation, and ascribed to the effect of the MPF, in particular protein synthesis and GV breakdown stimulation, and the acquisition by the cortex of the capacity to transmit the activation impulse and cortical reaction, will turn out to be separated in the cytoplasm of the maturing oocyte. It is therefore necessary to clearly differentiate the criteria whereby the MPF is identified, and take into account the greatest possible number of criteria indicated by various observations.

As to this question, let us cite Wu and Gerhart (1980, p. 475): "...We have no indication at this time of a separation of factors for surface maturation or for cytoplasmic maturation. Perhaps it is the situation that the maturing oocyte contains a battery of such secondary factors, each specific for one or more elemental biochemical changes in maturation, but that MPF serves as single common trigger of their activation... On the other hand, there are reasons to anticipate that our MPF, even as a single factor, may depend on the copurification of two or more components for the preservation of its activity."

Another circumstance which must be borne in mind is that data concerning the MPF cannot necessarily be extended to different species even within the boundaries of the single class of amphibians. For instance, *X. laevis*, which is the main subject of these investigations, differs from other *Anura* over a whole range of MPF properties.

Since the GV breakdown has been, and remains, a criterion of MPF activity in the majority of investigations, let us begin with the effect of the MPF on GV breakdown, and call it the GV breakdown factor (GVBDF, or GV breakdown activity). In papers published earlier in Russian (Dettlaff, 1977b; Dettlaff et al., 1977), it was called the GV membrane disintegration factor (MDF). The cytoplasm from maturing and mature oocytes capable of causing the appearance of various maturation characteristics in full-grown oocytes was called active cytoplasm, and it is in this sense that the symbol AC is used in this review. Halleck et al. (1984) gave a better name to this capacity of the cytoplasm: maturation-promoting activity (MPA).

3.5.1.1. Germinal Vesicle Breakdown Factor (GVBDF)

GV breakdown factor is revealed in the oocyte of amphibians not long before the GV breakdown, and attains its maximum activity in the period of GV disintegration; after this, the activity remains at a lower level, yet it is still detectable in the mature egg even after fertilization (Masui and Markert, 1971; Wasserman and Masui, 1976). New and more complete data about the dynamics of its activity in *Xenopus* oocytes and embryos were obtained by Gerhart et al. (1984).

In the cytoplasm of maturing oocytes of the starfish *Asterina pectinifera* (Kishimoto and Kanatani, 1976), GVBDF appears shortly before GV breakdown; the peak of its activity coincides with the period of GV breakdown, whereas toward the stage of the extrusion of the I polar body GVBDF activity drops steeply. In sturgeons it is found in the cytoplasm not only of maturing oocytes (Davydova, 1968), but also in mature eggs and cleaving embryos at the stage of synchronous cleavage divisions without any noticeable drop in activity (after Felgenhauer; see Detlaff, 1977b).

It has been demonstrated that the cytoplasm of a maturing oocyte causes nuclear membrane breakdown not only in the full-grown oocytes, but also in small oocytes still in the growth period (Nikitina, 1974; Hanocq-Quertier et al., 1976; Felgenhauer and Chulitskaya, 1977; Skoblina et al., 1984a), as well as in somatic cell nuclei transplanted into the cytoplasm (Gurdon, 1967; Ziegler and Masui, 1973) and spermatozoa (Graham, 1966; Graham et al., 1966; Skoblina, 1976; Katagiri and Moriya, 1976).

The GVBDF is not species-specific. GV breakdown may be induced not only by injecting the cytoplasm from a mature oocyte of one amphibian species into another (Reynhout and Smith, 1974; Wasserman and Masui, 1976), but also from one class to another. Thus, the cytoplasm from a maturing oocyte of *X. laevis* causes GV breakdown in an intact oocyte of *A. stellatus*, and vice versa (Dettlaff et al., 1977). Moreover, the cytoplasm of maturing *B. bufo* and *X. laevis* oocytes causes GV breakdown in starfish oocytes (Kishimoto, 1982; Kishimoto et al., 1982b). Many such data have been summarized by Meijer and Guerrier (1984, Fig. 19).

The efficiency of injection depends on the quantity of cytoplasm introduced (Masui and Markert, 1971; Wasserman and Masui, 1975a), and also from which region of the oocyte it is taken. It has been established that GVBDF is contained in the hyaloplasm (Masui, 1972), more of which is concentrated in the animal region, with none at all in the yolk (Iwamatsu, 1966; Masui, 1972). Masui and Markert (1971) assess the activity of GVBDF, all other conditions being equal, by the percentage of oocytes in which GV breakdown had taken place.

When active cytoplasm from an oocyte at the stage of GV breakdown or later is injected into an amphibian oocyte that has not been exposed to the action of progesterone, the GV membrane of this oocyte breaks down 3-4 h earlier than in an oocyte maturing under the action of progesterone (Masui and Markert, 1971; Schorderet-Slatkine and Drury, 1973).

The amplification of GVBDF is of particular interest. It has been shown repeatedly that when the cytoplasm of a maturing oocyte is injected into an intact one (not treated with hormones), the latter's GV membrane undergoes breakdown. Moreover, the cytoplasm of an oocyte which matured without hormones may be injected with the same effect into the second intact oocyte, from the second into a third one, and so on. Even after ten such passages the activity of the cytoplasm being injected does not diminish, indicating that there is no dilution of the cytoplasmic component causing GV breakdown (Masui and Markert, 1971; Reynhout and Smith, 1974; Drury and Schorderet-Slatkine, 1975; Wasserman and Masui, 1975a). Opinions differ as to the mechanism of GVBDF amplification and the necessity of protein synthesis for it. Wasserman and Masui (1975a), when repeatedly transferring the cytoplasm from a maturing *X. laevis* oocyte through oocytes treated with cycloheximide, observed GV breakdown in the latter. However, it occurred in a lower percentage of cases and was slower than in control oocytes not treated with cycloheximide. Wasserman and Masui concluded that neither GVBDF amplification nor its action on GV breakdown requires protein synthesis. They suppose either that GVBDF is not a protein, or, which is in their opinion more likely, that the oocyte cytoplasm contains inactive polypeptide precursors. Already in this paper Wasserman and Masui were inclined toward the second supposition. They suggest that the autocatalytic reaction involving the transformation of the preformed GVBDF precursors into an active form spreads from the inside of the oocyte surface and constitutes the mechanism by which the hormonal signal is carried to the GV.

Contrary to the above, Drury and Schorderet-Slatkine (1975) believe that protein synthesis is necessary for GVBDF amplification since, in their experiments with *X. laevis* oocytes, the GV failed to undergo breakdown during the third passage of the cytoplasm from cycloheximide-treated oocytes to initial oocytes in the presence of cycloheximide. However, in another experiment also carried out on *X. laevis* oocytes (Dettlaff et al., 1977), the cytoplasm of the oocytes preserved the capacity to cause GV breakdown even after five passages through cycloheximide-treated oocytes. In this case, the amplification of the GVBDF and GV breakdown itself took place when the suppression of protein synthesis was almost 95% complete. However, in *A. stellatus*, the preincubation of recipient oocytes in a cycloheximide solution prevents GV breakdown after the injection of cytoplasm from mature *A. stellatus* eggs not treated by an inhibitor. Thus, in *A. stellatus* oocytes, unlike *X. laevis* oocytes, protein synthesis is necessary for increasing the quantity of GVBDF and for GB breakdown even during the first passage. A similar result was obtained in *R.*

temporaria oocytes (Felgenhauer; see Dettlaff et al., 1977) and in *R. pipiens* oocytes (Schuetz and Samson, 1979).

These observations favor the assumption by Wasserman and Masui of the protein nature of GVBDF, but suggest that synthesis of GVBDF or its precursors in the oocytes of sturgeon and *Ranidae* takes place later during maturation than in *X. laevis* oocytes.

The protein nature of the GVBDF is confirmed by tests involving its extraction from mature eggs of *R. pipiens* (Masui, 1974; Wasserman and Masui, 1976). GVBDF-containing fractions were separated from the yolk, cortex, and partially from lipid by twice centrifuging intact eggs (25,000g for 15 min at 2°C and 150,000g for 120 min at 2°C). Preliminary homogenization of eggs, accompanied by the destruction of the cortex and cytoplasmic inclusions, causes rapid inactivation of GVBDF. Tris and Ca^{++} ions also possess an inactivating effect. The addition of EGTA, which selectively binds Ca^{++} ions, preserves GV breakdown activity in the extract for 24 h, while the presence of Mg^{++} ions prolongs its activity for up to 48 h (at 2°C) and longer. An optimal medium for extracting the GBVDF is a solution containing 0.2 M NaCl, 0.25 M sucrose, 0.01 M $MgSO_4$, 0.02 M EGTA, and 0.01 M Na_2HPO_4/NaH_2PO_4 at pH 6.5. When kept on ice, GVBDF activity in such an extract is preserved for at least 48 h and sometimes for up to 72 h. At 25°C it is inactivated within 4 h, and at 37°C within 15 min. Its activity is completely lost under the effect of proteases and it is resistant to the effect of RNase. Eighteen fractions were obtained by centrifuging the extract using a sucrose density gradient. Each fraction was tested for GVBDF activity by injection into intact oocytes. This revealed three peaks of activity with sedimentation coefficients close to 30S, 13S, and 4S (Masui and Clarke, 1979). Wasserman and Masui suggest that in vitro MPF (GVBDF-TD) is probably connected with thermolabile polypeptides, which have three different molecular weights, and this might indicate either the binding of GVBDF with different proteins, or a different degree of aggregation of GVBDF molecules.

The method of isolating MPF (GVBDF) from *X. laevis* oocytes was further improved by Drury (1978), who was able to stabilize it for an extended time with a new extraction medium consisting of a glycerophosphate-buffered sucrose solution with NaF or ammonium molybdate. By diluting this solution in the absence of ATP, Drury obtained the destabilization of the MPF (GVBDF). Since NaF and ammonium molybdate are phosphatase inhibitors and may prevent the dephosphorylation of MPF (GVBDF) precursors, it has been suggested that MPF may be a phosphoprotein (Masui and Clarke, 1979). According to Drury (1978), the MPF (GVBDF) in *X. laevis* oocytes has a molecular weight from $(0.6 \text{ to } 1.0)\cdot 10^6$ daltons and is extractable with a 30% $(NH_4)_2SO_4$ solution.

Wu and Gerhart (1980) have purified MPF (AC) from dejellied *X. laevis* eggs by 30-fold, using an extraction buffer containing 80 mM Na-β-glycerophosphate, 20 mM Na EDTA, 15 mM $MgCl_2$, and 1 mM dithiothreitol (pH 7.3). After the removal of more than 95% of the protein contaminants, the MPF appears as a single component possessing the properties of the "complete" MPF, which induces not only GV breakdown, but also complete maturation. The successful extraction of MPF requires the presence of small phosphorylated molecules such as β-glycerophosphate, phosphoserine, or phosphoacetate or fluoride (Drury, 1978), which may act to inhibit phosphoprotein phosphatases of the egg. Wu and Gerhart conclude that the purified MPF preparations contain active kinases, phosphatases, and protein substrates for phosphorylation, but they do not have direct evidence as yet that MPF is itself a kinase, phosphatase, or substrate. Gerhart et al. (1984) have elaborated on the method of small-scale extraction of eggs and oocytes, and measured the GV breakdown activity in the extracts by endpoint dilution assay (see Section 3.5.2). More recently, the MPF (GVBDF) was further purified (Cyert et al., 1985) by immunoaffinity chromatography using monoclonal antibodies that immunoprecipitate the GV breakdown activity. These antibodies also inactivate the GV breakdown activity in a solution of the extract.

At the same time, the results obtained by Hermann et al. (1984) indicate that phosphate- or phosphoester-containing buffer should be avoided in the course of MPF (GVBDF) purification, since 2-glycerophosphate, as well as some other phosphate compounds, facilitate the first meiotic division and, in some cases, induce it in the absence of hormonal stimuli.

Finally, new candidates for mediating the maturation-promoting activity (MPA) were found by Halleck et al. (1984). They have separated and identified a number of protein kinases from the mitotic and interphase culture cells and from the mature and immature amphibian oocytes. Some of these protein kinase activities appear to correlate with the biological activity of extracts, assayed by their ability to induce meiotic maturation following injection into the *Xenopus* oocytes.

Close correlation between the time of protein phosphorylation stimulation and the appearance of GVBDF was observed both during oocyte maturation under the effect of a hormone and following the injection of cytoplasm taken from the oocytes of *X. laevis* and *R. pipiens* and starfish that had matured with and/or without a GV (Masui and Clarke, 1979; Wu and Gerhart, 1980). Despite considerable success in investigating the nature of GVBDF, the manner in which it causes GV breakdown is still unknown (Masui and Clarke, 1979; Meijer and Guerrier, 1984; Maller and Smith, 1985).

By incubating somatic cell nuclei in a cell-free system, Miake-Lye and Kirschner (1985) have shown that the major structural proteins of nuclear membrane are hyperphosphorylated within 15 min after addition of the ripe egg cytoplasm fraction free of lipid and yolk granules, and this is followed by a gradual depolymerization of somatic cell nuclear membrane until the nuclear membrane breaks down 30 min later. It remains to determine which component of the purified active cytoplasm fraction induces the hyperphosphorylation of the membrane proteins and their depolymerization.

The supposed conversion of inactive GVBDF precursors into the active form may take place not only after the effect of progesterone, but also under the action of bivalent cations in the presence of ionophore A23187, while the effect of the latter, just like that of progesterone, is inhibited by cycloheximide (Wasserman and Masui, 1975b). It should be noted that stimulation of oocyte maturation by Ca^{++} and Mg^{++} in the presence of an ionophore was obtained in *X. laevis*; in *R. pipiens* oocytes this effect was not obtained. Regarding the possible mechanism of the action of Ca^{++} and Mg^{++}, these ions, in addition to their participation in the regulation of translational processes leading to the appearance of the GVBDF, also stimulate the ATPase activity of the pigment granules. For a detailed review of the role of Ca ions in the initiation of oocyte maturation in different species, see Masui and Clarke (1979) and Meijer and Guerrier (1984).

The necessity of Ca^{++} for GV breakdown was established long ago in tests with starfish oocytes (Dalcq, 1924; Kanatani, 1975), *Nereis* (Heilbrunn and Wilbur, 1937), *Chaetopterus* (Goldstein, 1953; Ikegami et al., 1976), and *Spisula* (Sawada, 1952; Schuetz, 1975) both during "spontaneous"* oocyte maturation in sea water, and during maturation under the effect of trypsin, KCl, and other agents. It was demonstrated in *Chaetopterus* oocytes that GV breakdown, requiring the presence of Ca^{++} in the solution or surface structures of the cytoplasm for only 3 min or even 1 min, does not take place when the oocytes are subsequently placed in conditions of anaerobiosis (Goldstein, 1953), or treated with cycloheximide (reference to unpublished data of Ikegami et al., 1976). Thus, one gets the impression that in *Chaetopterus* oo-

*According to a number of authors (Cloud and Shuetz, 1973; Shuetz, 1975), the so-called spontaneous oocyte maturation in sea water is due to the fact that the bivalent ions of the water stimulate the release of 1-Me-Ade or even its synthesis (Shirai, 1974) by the cells of the follicular epithelium.

cytes, as in *R. pipiens* oocytes (Wasserman and Masui, 1975b), the activating effect of Ca^{++} is inappropriate in the absence of protein synthesis.

Goldstein (1953) noted that the effect of Ca^{++} is suppressed by a protease inhibitor. In the same paper he records the proteolytic activity of homogenates from oocytes at different times after placing them in a solution containing Ca^{++}: the activity appears after the effect of Ca^{++}, and its peak coincides with the period of GV breakdown, after which it disappears.

However, it was shown in *X. laevis* (Guerrier et al., 1977b) and starfish oocytes (Kishimoto et al., 1982a) that leupeptin and antipain, inhibitors of proteolytic enzymes, suppress oocyte maturation induced either by progesterone or 1-Me-Ade and dithiothreitol, but do not prevent their maturation induced by GVBDF (AC). Nor do they influence MPF (GVBDF) amplification. Thus, although proteolytic enzymes participate in the molecular processes going on during oocyte maturation, they are not involved in processes leading from the GVBDF to GV breakdown.

During the stimulation of starfish oocyte maturation with 1-Me-Ade, the phosphorylation of cortex proteins is strongly enhanced (Guerrier et al., 1977a). It has been demonstrated that levels of phosphorylation depend on the concentration of 1-Me-Ade and on the effect of inhibitors, as does the biological reaction of oocytes. Guerrier et al. suggest that the phosphorylation of some proteins in the oocyte's cortical layer constitutes not only a very early, but even the main event in the reinitiation of meiosis (see also Masui and Clarke, 1979). However, Hanocq-Quertier et al. (1978) regard the release of intracellular Ca and the opening of Ca channels in the oocyte membrane to be the primary trigger of meiotic reinitiation.

3.5.1.2. GVBDF in the Mitotic Cycle in the Period of Synchronous Cleavage Divisions, Desynchronization, Blastulation, and during Divisions of Synchronized Mammal Cells

In contrast with the preliminary data for the oocytes of *X. laevis* and *R. pipiens* (Masui and Markert, 1971; Wasserman and Masui, 1976), GV breakdown activity in *A. stellatus* oocytes does not drop following GV breakdown and remains at a high level not only in the mature egg but also in cleaving eggs, including the 32-blastomere stage (Dettlaff, 1977b). Moreover, the cytoplasm from the animal region of *A. stellatus* eggs, collected throughout the interval between the appearance of the first and second cleavage divisions, caused GV breakdown in full-grown oocytes at an equally high frequency. This gave grounds for believing (Dettlaff, 1977b) that GVBDF is active not only during oocyte maturation, but also during the period of synchronous cleavage divisions and, in the latter period, probably appears anew in each cell cycle.

The periodic appearance of GVBDF in the eggs of *R. pipiens* and *X. laevis* was, however, described during the transition from G_2 to M during the first cleavage divisions (Wasserman and Smith, 1978b). Shortly after fertilization in *R. pipiens*, GV breakdown activity is not detected at all, while in *X. laevis* it is very low. Wasserman and Smith observe that enucleated activated eggs of these species preserve the same periodicity in the appearance of the GVBDF, i.e., it is controlled by the cytoplasm with a rhythm equal to the duration of a single mitotic cycle of synchronous cleavage divisions. According to their data, the periodical appearance of GVBDF requires protein synthesis, although for its reduction or disappearance such synthesis is unnecessary. Wasserman and Smith conclude that MPF (GVBDF) controls nuclear membrane disintegration not only in meiosis, but also in mitosis. In order to reveal the causes of such marked differences in GV breakdown activity at different phases of the cell cycle of synchronous cleavage divisions in *A. stellatus* eggs, on the one hand, and of *R. pipiens* and *X. laevis*, on the other, additional tests were undertaken involving the injection of cytoplasm taken from the eggs of *A. stellatus*, *R. temporaria*, and *X. laevis* at different phases of the first two cleavage divisions into full-grown oocytes of the same species (Ryabova and Dettlaff, 1981), and, after that, at different stages of synchronous cleavage divisions, desynchronization, and blastulation (after the mitotic index fell) (Ryabova and Dettlaff, 1984; Dettlaff and Ryabova, 1986). The test results are given in Tables 3.1A and 3.1B. They indicate that GV breakdown activity in *A. stellatus* and *R. temporaria* eggs is similar during all phases of the cell cycle. Nor were any significant differences in its activity revealed in the eggs of *X. laevis*. The GV breakdown activity does not diminish in the activated *R. temporaria* eggs. In all experiments the GV breakdown activity in the cytoplasm during synchronous cleavage divisions remained at the same level as in the nonactivated eggs, but in the period of desynchronization it decreased by almost half. The same results were obtained when the homogenate (supernatant) of the embryos at the same cleavage stages was injected, although its activity was lower than that of the cytoplasm. The GV breakdown activity in the cytoplasm of animal blastomeres at the blastula stage dropped steeply to 15-17% of that in the egg. Since the protein content in equal volumes of the cytoplasm taken from the nonactivated egg and the embryos during synchronous cleavage and blastulation was practically the same, it may be supposed that the differences actually reflect existing changes in GV breakdown activity. The same results were obtained with *A. stellatus* embryos.

Table 3.1A. GV Breakdown in Full-Grown A. stellatus and Anura Oocyte Drawn from Cleaving Embryos at Different Mitotic Phases during the F

Donor	Recipient	Dose, nl	Number of oocytes given GVBDF from donors at th oocytes with GVBD)		
			interphase–prophase	interphase–prophase + prometaphase	met
A. stellatus	A. stellatus	200	148 (56)	77 (43)	7
R. temporaria	X. laevis	80	49 (82)	17 (94)	2(
R. temporaria	X. laevis	30	120 (56)	54 (66)	5(
R. pipiens[b]	X. laevis	80	10 (0)	10 (20)	1(
X. laevis[b]	X. laevis	80	10 (60)	-	1(

[a]Shown in parentheses is the percentage of oocytes with GV breakdown.
[b]Data by Wasserman and Smith (1978).

Table 3.1B. GVBDF Activity in the Cytoplasm of Rana temporaria Embryos at Different Developmental Stages: In the Egg, after Its Activation, and during Cleavage and Blastulation

Experiment number	Cytoplasm donors: age (in τ_0)	Number of blastomeres	Period	Native cytoplasm: Number of oocytes	Native cytoplasm: GV breakdown, %	Homogenate: Number of oocytes	Homogenate: GV breakdown, %
1	0	-	Mature egg	50	48	-	-
	5-6	8-16	Synchronous divisions	51	51	52	23
	9	128		48	44	51	21.6
	12-13	~ 1024	Desynchronization	51	21.6	47	12.6
2	0	-	Mature egg	62	79		
	4-6	4-16	Synchronous divisions	57	77		
	12	~ 1024	Desynchronization	61	39		
	17-18		Blastulation	62	13		
3	0	-	Mature egg activated by pricking	29	73		
	1	-		28	66		
	2	-		39	69		
	3	2-4	Synchronous divisions	40	70		
	13-14		Desynchronization	40	35		
	15-16			42	31		
	18-19		Blastulation	70	12.5		
	19-20		Blastulation	72	11		

Thus, the correlation between the GV breakdown activity of the egg cytoplasm and the rearrangement of the cell cycle during the transition from synchronous divisions to desynchronization, and from desynchronization to blastulation, is evident. A relationship probably exists between the drop in the GV breakdown activity and the onset of desynchronization of nuclear divisions. This is supported by the results of experiments involving injections of the cytoplasm and homogenate of the eggs into the cleaving *A. stellatus* embryos (Chulitskaya, 1970), in which a greater amount of cytoplasm, and consequently of GVBDF, delayed the onset of desynchronization by one or two cyles. These injections had no effect on the transition to blastulation. A low GV breakdown activity during blastulation coincides with the low mitotic activity of the cells.

The small number of oocytes used in Wasserman and Smith's experiments is insufficient to explain the divergence between our results and theirs by the species-specific differences in GVBDF activity during different phases of the cell cycle.

Thus, we find the conclusion of Wasserman and Smith (1978b) that "the peak of embryonic MPF activity (GV breakdown activity of cytoplasm or GVBDF-T.D.) during each cycle coincided with the time the embryonic nuclei were entering the G_2–M transition, i.e., mitosis" (p. 15), despite its wide recognition, unjustified for the early cleavage period. We cannot agree with this conclusion, not only because it is not consistent with our results, but also because it is not supported by sufficient data (see Dettlaff and Ryabova, 1986). In addition, it should be noted that the presence of the G_2 phase in cell cycles of the first synchronous cleavage divisions has not been proved (Graham and Morgan, 1966; Newport and Kirschner, 1982a, b; for review see Rott, 1980).

Miake-Lye et al. (1983) have performed the reverse experiment. They have studied the effect of partly purified fraction obtained according to Wu and Gerhart (1980) from the cytoplasm of the *X. laevis* eggs on the embryonic nuclei during the early cleavage stages. In the recipient eggs cytokinesis was inhibited by cytochalasin B and, after 6-7 cell cycles, about 100 nuclei were arrested at a G_2-like stage by cycloheximide in common nondivided cytoplasm. Within 5 min after injection of AC into such embryos, the nuclear envelopes disintegrate. No amplification of GV breakdown activity is observed in such arrested oocytes with inhibited protein synthesis, unlike in the *X. laevis* oocytes. The authors suggest that MPF can advance the interphase nuclei to the first events of mitosis and that the nuclear envelope responds rapidly to MPF (GCBDF).

Gerhart et al. (1984) have studied in detail the MPF (GVBDF)-dependent cell cycle regulation. With this in view, they developed the method for small-scale extraction of maturing oocytes and activated fertilized eggs of *X. laevis*, and measured the GV breakdown activity in the extracts. The time scale was normalized to the interval from fertilization to the first cleavage. Extraction was performed two or three times at every tenth of this interval. Five full-grown oocytes were injected each with 40 nl of the extract for a single assay point and, after 2-2.5 h, were scored for GV breakdown. This method allowed them to quantify the GV breakdown activity during meiosis and the first cleavage divisions. They define a unit of MPF (GV breakdown) activity as the amount causing a 50% incidence of GV breakdown in the oocytes injected each with 40 nl of test material. They write: "...100 units per egg; that is, if a single egg is extracted in a volume of 4 μl, the extract will contain enough MPF (GVBDF-T.D.) to inject 100 oocytes with 40 nl each and obtain 50% frequency of maturation" (p. 1249). The lower limit of detection in their assay was 15 units per egg, due to the inherent dilution of the homogenate itself. At the same time, one unit is the minimum dose per oocyte to obtain GV breakdown in 50% of the injected oocytes. After injection of full-grown oocytes with 5 units of MPF (GVBDF) (20 ng protein), they found endogenously produced GV breakdown activity within 30 min. This activity was then raised to 110-120 units (GV breakdown takes place within 70-80 min), and disappeared rapidly by the end of the first meiotic division. During the second meiotic division GV breakdown activity reappeared before metaphase II, when meiosis is arrested.

When oocyte maturation was stimulated with progesterone, the dynamics of GV breakdown activity were the same as when MPF (AC) was injected into the oocytes. After insemination or artificial activation the GV breakdown activity dropped from 100-120 to 15 units within a few minutes, remained practically undetectable for a certain time, and then rapidly attained the maximum value. During the first cleavage divisions, the minimum GV breakdown activity was recorded in the middle of the interval between the appearance of successive cleavage furrows, at prophase. In the periods of minimum activity it did not exceed 15 units. On the basis of these data the authors postulate the cyclic "appearance" and "disappearance" of GV breakdown activity. However, taking into account that 50% of GV breakdown is attained if one unit is introduced into the oocytes, 15 units cannot be considered as evidence of the absence of GVBDF activity. It will therefore be more precise to speak of periodic oscillations of the GV breakdown activity during the cell cycle in the cytoplasm of the embryos.

Gerhart et al. (1984) have also shown that the protein synthesis inhibition in the activated eggs by cycloheximide prevented the cyclic changes in GV breakdown activity and preserved it at a level of 15 units. The authors suggested that partly purified cytoplasm of the cleaving embryos also contains an agent for MPF (GVBDF) inactivation which, just like MPF, is activated at the posttranslational level and can inactivate not only endogenous MPF but also large amounts of injected MPF. They explain these data by the repeated reciprocal cycling of MPF and the inactivating agent during cleavage. They assume also that MPF, being a trigger of M-phase, may be regulated by a still more fundamental and autonomous cell-cycle oscillator.

Newport and Kirschner (1984) also concluded that cell-cycle events of the nucleus are controlled by the level of MPF (GVBDF) activity, which is driven by or may be part of an autonomous cell-cycle oscillator.

The problem of cytoplasmic GV breakdown activity at different phases of the cell cycle has thus become important for understanding the ways of its regulation. It was even proposed to rename MPF and call it "metaphase factor" (MF) (Gerhart et al., 1984).

However, it can be concluded from the available data concerning the GV breakdown activity at different phases of the cell cycle during synchronous cleavage divisions in the intact cytoplasm (Wasserman and Smith, 1978b; Ryabova and Dettlaff, 1981; Dettlaff and Ryabova, 1986) and in the cytoplasm extracts (Kishimoto et al., 1982b; Gerhart et al., 1984) that a certain amount (at the minimum 15 units) of GV breakdown activity is present in the cytoplasm throughout the whole cell cycle. Therefore, it would be more precise to speak of fluctuation of the GV breakdown activity rather than of its appearance and disappearance. As for cyclic changes in the GV breakdown activity, they were clearly proved in the experiments with cytoplasm extracts but were not reliably shown in experiments with intact cytoplasm. The causes of these differences require further investigation. It is certain, however, that the GVBDF is a factor of nuclear membrane disintegration not only in meiosis, but also in mitosis. This is corroborated by numerous facts. GVBDF has been discovered not only in cleaving eggs of different species of amphibians, *A. stellatus*, and starfish (Kishimoto et al., 1982b), but also in the synchronized cell cultures of mammals (Sunkara et al., 1979; Nelkin et al., 1980). Here, an extract from synchronized HeLa cells causes GV breakdown (and chromosome condensation) only when it is obtained from cells during the G_2–M phases. It is ineffective during the other phases of the cycle. Thus, the nuclear membrane disintegration factor appears in these cells periodically in every cycle. New data relevant to this problem were published by Swenson et al. (1986). They found that the mRNA of the clam embryo cyclin A, which induces the entry of the blastomeres into M phase, is translated after its injection into *Xenopus* oocytes and induces GV breakdown.

At present, however, the factors regulating oocyte maturation and cell cycle should include, in addition to the factor arising in the cytoplasm, a factor arising in the nucleoplasm (see Section 3.4.2.1).

3.5.2. MPF and the Appearance in the Cortex of the Ability to Transmit the Activation Impulse and to Produce a Cortical Reaction and Cortical Granule Breakdown

The fact that injection of cytoplasm of maturing oocytes (AC) into full-grown oocytes causes not only GV breakdown but the acquisition by the oocyte surface layer of the ability to transmit an activation impulse (AI) and to produce cortical granule breakdown is not grounds for believing that all these changes are determined by the action of one and the same cytoplasmic factor – MPF. A priori it seems more probable that such obviously different processes should involve different factors. Consequently, given certain experimental conditions, GV breakdown should be obtained without the appearance of the capacity for cortical reaction (CRC), and vice versa. For the sake of brevity we will refer to these capacities of the cytoplasm as GV breakdown activity, or GVBDF and CRCF.

Attempts to produce such separation were undertaken in *A. stellatus* oocytes (Dettlaff et al., 1977; Dettlaff and Felgenhauer, 1980; Detlaff and Ryabova, 1986). For this purpose, the oocytes were maintained under different experimental conditions until the time when the control oocytes reached maturity, whereupon both the test and control oocytes were activated by pricking. The state of the nucleus and the state of the cortical granules were determined from histological sections. In these experiments, the injections of different volumes of mature egg cytoplasm revealed no dose-dependent differences between the appearance of CRC and GV breakdown. In all cases either GV and cortical granules remained intact, or GV breakdown coincided with the appearance of CRC. Separation could not be produced even with AC injection into *A. stellatus* oocytes against a background of suppressed protein synthesis under these conditions. Changes in the cortical granule layer occurring under these conditions were at first mistaken by us for an atypical cortical reaction (Dettlaff et al., 1977). These changes consist in the expansion of the yolk-free cytoplasm surface layer, a shift of cortical granules from the plasma membrane deep under the layer of pigment granules, and partial dispersion of the granules. These changes resemble changes in the cortex during pseudomaturation (Dettlaff and Felgenhauer, 1980).

Another experiment staged for the same purpose (Dettlaff and Ryabova, 1986) consisted of the injection of cytoplasm from the animal region of a cleaving *A. stellatus* embryo, collected during different periods of cleavage and blastulation, into intact *A. stellatus* oocytes and the fixation of the recipient oocytes following their activation by pricking.

Since the presence of the GVBDF in the cytoplasm of dividing cells is biologically essential, while the presence of the CRCF has no biological rationale, it was apparent that the above experiment is most likely to produce the separation of these two effects on maturation. But the result of this experiment was negative.

The capacity for inducing not only GV breakdown in the oocytes but also maturation of the cortex and appearance of CRC was found in the cytoplasm of cleaving *A. stellatus* embryos. This fact can be explained in different ways: either the cytoplasm of cleaving embryos contains the same complex of factors as does the cytoplasm of maturing oocytes, or the cytoplasm of cleaving embryos contains a certain initiator which, when introduced into full-grown oocytes, activates the entire complex of factors needed for oocyte maturation, but in the cytoplasm of cleaving embryos, activates only those which contribute to the movement of cells through the cell cycle, in particular the nuclear envelope disintegration (GVBDF). This problem requires further studies.

3.5.3. Stimulation of Protein Synthesis and MPF (AC) Action

The information on protein synthesis in oocytes during the maturation period, obtained by different authors using a variety of species, is not easily compared. Different methods are used to induce maturation, and systematic differences exist in the dynamics and volume of the protein being synthesized during this period (see Masui and Clarke, 1979). Nevertheless, a substantial (two- to fourfold) increase in protein synthesis commonly occurs, both under the influence of progesterone or, in starfishes, of 1-Me-Ade, and following MPF (AC) injection in all animals studied in this respect. The volume of the protein being synthesized is the same in both cases, but after exposure to progesterone (or 1-Me-Ade) the intensification of synthesis begins relatively later than after MPF (AC) injection into full-grown oocytes. MPF (AC) injection causes an almost immediate increase in protein synthesis and then, earlier than under the influence of progesterones, induces GV breakdown.

On the other hand, the use of protein synthesis inhibitors demonstrates that for oocyte maturation under the influence of hormones, protein synthesis in the oocytes of *R. temporaria* (Dettlaff, 1966; Brachet, 1967), *A. stellatus* (Dettlaff, 1966; Dettlaff et al., 1980), and *R. pipiens* (Smith and Ecker, 1970b; Ecker and Smith, 1971a) is needed for implementing maturation within about the first two thirds of the period from placing the oocytes in the hormonal solution to GV breakdown, and in *X. laevis* oocytes, for a very short time interval (Wasserman and Masui, 1975a). After that, suppression of protein synthesis no longer prevents GV breakdown (Morrill et al., 1981). Masui and Clarke (1979, p. 217) think it is probable that the synthesis necessary for the maturation of *X. laevis* oocytes is not "new protein synthesis...since it is possible that the inhibitors also interfere with ongoing synthesis of pre-existing, but short-lived, proteins in the oocytes to deprive them of the ability to respond to maturation-inducing agents."

In oocyte maturation induced by MPF (AC) GV breakdown depends on protein synthesis differently in different amphibian species. The *X. laevis* oocytes require very little synthesis of new polypeptides to act as a trigger or starting mechanism and convert the nonactive form of MPF (GVBDF) precursors into the active form by their phosphorylation (Wasserman and Masui, 1975a; Masui and Clarke, 1979). This hypothetical protein has been called an "initiator" (Wasserman and Masui). In *X. laevis* it is distinguished from the MPF (GVBDF) by its dependence on protein synthesis inhibitors, while MPF (GVBDF) amplification is independent of them. However, the situation revealed in *X. laevis* is not universal, but rather exceptional, since in the majority of animals studied, *A. stellatus* (Dettlaff et al., 1977), *R. temporaria* (Felgenhauer, see Dettlaff, 1977), and *R. pipiens* (Schuetz and Samson, 1979; Samson and Schuetz, 1979), MPF (GVBDF) amplification does not take place when protein synthesis is inhibited, and there are no grounds for distinguishing the initiator and the GVBDF, nor, apparently, its nonactive and active forms.

Another important question is whether protein synthesis stimulation and the appearance of the GV breakdown activity are the successive elements in the operation of a common factor (MPF) contained in the cytoplasm of maturing oocytes (AC). An answer to this question is given in a series of papers (Wasserman et al., 1982; Richter et al., 1982; Wasserman, 1982). In these works, a detailed quantitative and qualitative study of protein synthesis and the influence of progesterone and the AC has been carried out on *X. laevis* oocytes at different stages of maturation. The experiments showed that the synthesis and phosphorylation of proteins increased 10 min after the injection of AC and hours after exposure to progesterone. Two-dimensional electrophoresis revealed the synthesis of 300 different polypeptides, only one of which was absent during AC-stimulated maturation. Wasserman et al. (1982) assume that the cytoplasm of the mature egg (MPF) contains an active protein kinase and, possibly, also a substrate necessary for phosphorylation. Richter et al. (1982) demonstrated that during maturation the increase of protein synthesis depends, above all, on the recruitment of mRNA into polysomes without a change in translation efficiency; i.e., it proceeds at the expense of accumulated masked mRNA. Further, they discovered conditions which inactivated MPF (GVBDF) activity in cytosol, but did not influence its stimulating effect on protein synthesis. Cytosole, which did not induce GV breakdown in their experiments, increased protein synthesis 1.5-2 times (Wasserman et al., 1982). On these grounds it was assumed that the substance or substances responsible for inducing GV breakdown could be isolated from the substances determining an increase of protein synthesis, and they were, in fact, separated (Wasserman, 1982).

When injected into full-grown oocytes, cytosol from mature eggs of *X. laevis*, in a glycerophosphate buffer with NaF and EGTA-induced protein synthesis, acted as a protein synthesis-stimulating factor (PSSF). After triple freezing and defrosting, or after 10-day storage at 4°C, or after the addition to it of 100 μM Ca^{++}, the same cytosol in similar experiments stimulated protein synthesis, but did not induce GV breakdown. By using ammonium sulfate solutions at different levels of saturation, the GVBDF and PSSF were separated. The GVBDF was contained predominantly in a fraction of 20-30% saturation, while the PSSF was found mainly in a fraction of 30-40% saturation. Wasserman concluded that the PSSF must reside with a substance or group of substances that are physically distinct from those responsible for inducing GV breakdown in oocytes.

In this connection, we would like to stress once again that one and the same symbol (MPF) cannot be used for designating different effects of the active cytoplasm on oocyte maturation.

3.5.4. Cytoplasmic Control of Chromosome Condensation Activity (CCA)

Inhibitors of both nucleic acid and protein synthesis do not prevent chromosome condensation in nuclei transplanted into mature nonactivated eggs, but prevent chromosome condensation when similar nuclei are transplanted into oocytes at metaphase I (Ziegler and Masui, 1976). However, the cytoplasm of oocytes at this stage already contains the conditions necessary for their condensation (Ziegler and Masui, 1973). These data suggested to Ziegler and Masui that cytoplasmic factors do not directly cause chromosome condensation, but, in nuclei transplanted into the oocyte, they induce synthesis of mRNA molecules on which the corresponding protein is translated and which, in turn, causes chromosome condensation. In their 1976 paper, Ziegler and Masui established that GV substances are needed, not for chromosome condensation, but for their long-term preservation in the cytoplasm in the condensed state.

Further, Masui et al. (1979, 1980) demonstrated that labeled GV nuclear proteins become included in the condensed chromosomes of brain cell nuclei following the injection of both, together with cycloheximide, into enucleated oocytes that had matured under the effect of progesterone and reached the stage equivalent to metaphase II.

As noted by Morrill (1965), it may be significant that during GV breakdown the ionic environment of the chromosomes undergoes considerable change consequent upon their entry into the cytoplasm. The Na^+ content per unit of volume is 3.2, and K^+ content is 2.4 times higher in the GV than in the cytoplasm (Naora et al., 1962).

But the most important role in condensation and decondensation of chromatin is played by Ca^{++} (Lohka and Masui, 1984a). The chromosome condensation activity of nonactivated egg cytoplasm in the experiments of Lohka and Masui was sustained in cell-free preparations when the Ca^{++} level was kept low by addition of EGTA; when Ca^{++} level was raised by addition of Ca^{++}, this activity was lost and replaced by new activity which decondensed chromatin of the demembraned sperm nuclei. Lohka and Masui (1984b) also studied the formation of the nuclear envelope around the dispersed chromatin and nuclei enlargement. Ultrastructural studies showed that membrane vesicles, 200 nm in diameter, found in the particulate fraction of the cytoplasmic preparation obtained at 150,000g for 2 h, flattened and fused together to contribute to the membranous components of the nuclear envelope.

In a review concerning oocyte maturation and cyclic changes in chromosomes during meiosis and mitosis, Masui (1985) concluded that the chromosome cycle appears to be controlled in some way by common cytoplasmic factors. Among them, the MPF plays an especially important role, since its cyclic changes, according to Masui, control the other parameters of the cell cycle, including the nuclear envelope disintegration, chromosome condensation, and microtubule assembly.

3.5.5. Cytostatic Factor (CSF)

Masui and Markert (1971) used the term cytostatic factor (CSF) to designate a hypothetical substance appearing in the oocyte cytoplasm soon after GV breakdown. This substance causes the arrest of oocyte nucleus division at metaphase II and maintains this nuclear state immediately up to the moment of fertilization or artificial egg activation, when the cytostatic factor is reversibly inactivated (Masui, 1974). The CSF was discovered by injecting the cytoplasm from a mature nonactivated egg into one of two blastomeres of a cleaving egg in *R. pipiens*. In the injected blastomere nuclear division was usually arrested at metaphase, and cleavage was stopped. This effect was not observed with cytoplasm from an activated egg. The appearance of this factor does not require the participation of the GV nucleoplasm, since it is also formed in an enucleated oocyte.

Meyerhof and Masui (1977) have isolated the primary cytostatic factor (CSF), which is extractable only from mature nonfertilized eggs, inactivated by endogenous Ca^{++} and EDTA, and sensitive to Mg^{++} and EGTA. Following injection into the zygote, the CSF is inactivated within 45 min after insemination. The authors presumed the CSF is a factor which blocks meiosis at metaphase II, while the liberation of Ca^{++} ions during fertilization removes its blocking effect and thereby causes meiotic reinitiation. There are a number of objections to this very important conclusion (Chulitzkaya and Felgenhauer, 1977; Dettlaff, 1977b; Ryabova, 1983).

There is no doubt whatsoever that a cytostatic effect arises toward the end of the maturation period, but it is not certain that it can be ascribed to the isolated extract. It is somewhat puzzling that the CSF is revealed, beginning from the stage of GV breakdown, just when the oocyte nucleus is going over to the I maturation division.

One may suppose that the cytostatic effect of the cytoplasm of mature nonactivated eggs is not universal. For instance, the injection of the cytoplasm from mature nonactivated eggs and the supernatant from homogenized eggs of the sturgeon into their cleaving eggs resulted not in cleavage arrest, but in the prolongation of the period of synchronous cleavage divisions (Chulitskaya, 1968, 1970).

Using the method of Masui and Markert, Chulitskaya and Felgenhauer (1977) injected the cytoplasm from maturing and mature *R. temporaria* oocytes into one of two blastomeres of *R. temporaria* embryos at different stages from the start of the cleavage furrowing, yet in none of 536 cases did they obtain a cytostatic effect. Equally unsuccessful were tests with *A. stellatus* eggs. However, Meyerhof and Masui (1979a) obtained a cytostatic effect upon injecting the cytoplasm from mature eggs into one of two blastomeres not only in the embryos of *R. pipiens*, but, in single cases, also of *X. laevis*. The authors explained the low percentage of division arrest in this case by the inactivation of CSF by Ca^{++} ions at the moment of cytoplasm transfer from the oocyte into the embryo. To stabilize CSF, Meyerhof and Masui introduced EGTA into each donor oocyte before withdrawing the cytoplasm from it, so that the approximate EGTA concentration within the embryo donor of cytoplasm would be 5 mM. After pipetting EGTA into the cytoplasm of the animal region, 60 nl of cytoplasm was taken for injection into *R. pipiens* embryos and 30 nl for *Xenopus* embryos. In this experiment, blastomere division arrest was effected in 94% of both *R. pipiens* and *X. laevis* embryos. Xenoplastic transfer of cytoplasm from a *Xenopus* oocyte into a *R. pipiens* embryo gave 100% arrest. In the reverse combination, the transfer of EGTA-containing cytoplasm from a *R. pipiens* oocyte to a *Xenopus* embryo, division arrest occurred in only one out of 27 embryos. On the basis of these and previous data, Meyerhof and Masui concluded: "It may be predicted that CSF represents the general factor for establishing meiotic arrest in eggs which are normally blocked at metaphase before fertilization, and that the properties of this factor would be similar in diverse species" (Meyerhof and Masui, 1979a, p. 186). The termination of the meiotic block takes place at the moment of fertilization or activation due to CSF inactivation by Ca ions liberated in the eggs at this time. The authors then write that since CSF inactivation by Ca ions determines the termination of meiotic blocking, the absence of a detectable quantity of CSF in the oocyte cytoplasm may serve as a precise criterion of egg activation.

The negative results of Chulitskaya (1968, 1970) and Chulitskaya and Felgenhauer (1977) were explained by Meyerhof and Masui (1979b) by the fact that the cytoplasm donors used by Chulitskaya and Felgenhauer might have been activated eggs and, later (Masui and Clarke, 1979), by the fact that their CSF inactivation might have taken place during the transfer of the cytoplasm from donor to recipient under the action of Ca^{++}, just as in their own experiments with *Xenopus*. Thus, Masui and collaborators remained convinced that their attractive explanation of the meiotic block emergence and termination was correct.

The experiments with *A. stellatus* and *R. temporaria* oocytes have been repeated with the use of EGTA for CSF stabilization after the method of Meyerhof and Masui (1979b). In these experiments (Ryabova, 1983), the cytoplasm + EGTA was taken in one series from mature nonactivated eggs and, in another, from the embryos at the 2-4 blastomere stage. In each series three different EGTA doses were introduced into the cytoplasm donors. EGTA-free cytoplasm from mature and cleaving eggs was also injected into one of the two blastomeres of *A. stellatus* and *R. temporaria* embryos. Xenoplastic transfers of EGTA-free cytoplasm from mature nonactivated *X. laevis* eggs into one of two *R. temporaria* blastomeres, and from a *R. temporaria* egg into a *X. laevis* blastomere, were also made. Just as in the experiments by Chulitskaya and Felgenhauer, the EGTA-free cytoplasm from the mature eggs in the *A. stellatus* and *X. laevis* embryos had no cytostatic effect in these experiments. However, the EGTA-containing cytoplasm induced division arrest in one out of two or one out of four blastomeres, or even in the entire embryo. The same cytostatic effect was obtained upon the injection of EGTA-containing cytoplasm from cleaving embryos. A clear dependence of the percentage of blocked embryos at one or another division stage on the dose of EGTA injected into the donors was established. Sections of the arrested blastomeres revealed karyomeric vesicle-like nuclei: usually one or two, seldom more; no nuclei arrested at metaphase were detected.

The cytostatic effect of EGTA-free cytoplasm taken from a mature oocyte was revealed in Ryabova's experiments only when the cytoplasm from a *Xenopus* oocyte was injected into a *R. temporaria* embryo, but in this experiment, out of the 32 arrested blastomeres studied the nuclei were at metaphase only in 4 cases, while 4 others contained 3-pole mitoses and the rest contained vesicle-like karyomeric nuclei surrounded by a pigmented ring. Meiotic arrest was also obtained in a low percentage of cases, when the injected cytoplasm was taken from cleaving *Xenopus* embryos. In a reciprocal variant of the experiment, injection of cytoplasm from mature *R. temporaria* eggs into *Xenopus* blastomeres did not cause meiotic arrest.

Table 3.2. Cytostatic Effect of Cytoplasm from Mature Nonactivated Eggs and Cleaving Embryos of Various Amphibian Species under Different Experimental Conditions

Cytoplasm drawn from	Quantity, nl	EGTA in donor, mM	Donor	Recipient	Total recipients	Cleavage arrested, %*	Perished†	References
Nonactivated eggs	48-60	0	*R. pipiens*	*R. pipiens*	93	91	-	Masui and Markert, Table 4
	60	0	*R. pipiens*	*R. pipiens*	70	93	-	Meyerhof and Masui, 1977,. Table 1
	60	0	*R. pipiens*	*R. pipiens*	186	46	62	Mayerhof and Masui, 1979, Table 1
	30	0	*X. laevis*	*X. laevis*	39	35	56	Meyerhof and Masui, 1979, Table 1
	60	0	*R. temporaria*	*R. temporaria*	161	0	13	Chulitskaya and Felgenhauer, 1977
	60	0	*R. temporaria*	*R. temporaria*	64	0	0	Ryabova, 1983
	200	0	*A. stellatus*	*A. stellatus*	93	0	0	Chulitskaya and Felgenhauer, 1977
Embryos at stage of 2 blastomeres	200	0	*A. stellatus*	*A. stellatus*	65	0	6	Ryabova, 1983
	?	0	*A. güldenstädti*	*A. güldenstädti*	?	0	?	Chulitskaya, 1968
	60	0	*X. laevis*	*R. temporaria*	90	86	0	Ryabova, 1983
	30	0	*R. temporaria*	*X. laevis*	73	24	55	Ryabova, 1983
	48-60	0	*R. pipiens*	*R. pipiens*	101	8	-	Masui and Markert, 1971
	60	0	*R. pipiens*	*R. pipiens*	66	2	-	Meyerhof and Masui, 1977
	?	0	*X. laevis*	*R. pipiens*	40	0	-	Mayerhof, 1978 (Masui and Clarke, 1979)
	30	0	*X. laevis*	*R. temporaria*	82	23.5	9	Ryabova, 1983
	60	0	*R. temporaria*	*R. temporaria*	83	0	0	Chulitskaya and Felgenhauer, 1977
	200	0	*A. stellatus*	*A. stellatus*	25	0	-	Chulitskaya and Felgenhauer, 1977
Activated eggs	500	0	*X. laevis*	*Pl. waltlii*	?	57	-	Aimar et al., 1981
	?	0	*X. laevis*	*R. pipiens*	66	0	-	Meyerhof, 1978
Nonactivated eggs	60	0	*R. pipiens*	*R. pipiens*	221	97	22	Meyerhof and Masui, 1979, Table 1
	30	5	*X. laevis*	*X. laevis*	53	98	6	Meyerhof, and Masui, 1979, Table 1

	60?	5	*X. laevis*	*R. pipiens*	71	100	1.5	Meyerhof and Masui, 1979, Table 1
	30?	5	*R. pipiens*	*X. laevis*	61	4	56	Mayerhof and Masui, 1979, Table 1
	60	50	*R. temporaria*	*R. temporaria*	83	100	24	Ryabova, 1983
	60	10	*R. temporaria*	*R. temporaria*	134	87	12	Ryabova, 1983
	60	1	*R. temporaria*	*R. temporaria*	40	42	7	Ryabova, 1983
	200	100	*A. stellatus*	*A. stellatus*	57	67	47	Ryabova, 1983
	2000	50	*A. stellatus*	*A. stellatus*	52	46	28	Ryabova, 1983
	200	10	*A. stellatus*	*A. stellatus*	48	17	24	Ryabove, 1983
Activated eggs	30	5	*X. laevis*	*X. laevis*	14	7	-	Meyerhof, 1978
	30	5	*X. laevis*	*X. laevis*	12	0	-	Meyerhof, 1978 (Masui and Clarke, 1979)
	60	5	*R. pipiens*	*R. pipiens*	67	6	-	Masui et al., 1980, Table 3
Embryos at stage of 2-4 blasto-meres	60	50	*R. temporaria*	*R. temporaria*	80	100	31	Ryabova, 1983
	60	10	*R. temporaria*	*R. temporaria*	130	82	15	Ryabova, 1983
	60	1	*R. temporaria*	*R. temporaria*	47	36	2	Ryabova, 1983
	200	100	*A. stellatus*	*A. stellatus*	41	58	36	Ryabova, 1983
	200	50	*A. stellatus*	*A. stellatus*	48	39	21	Ryabova, 1983
	200	10	*A. stellatus*	*A. stellatus*	49	24	14	Ryabova, 1983

*Percentage of developing embryos.
†Percentage of injected embryos.

The rather contradictory data above have been summarized in Table 3.2. The existence of CSF in mature nonactivated eggs remains uncertain, but some conclusions can be drawn. It is apparent that EGTA cannot be used as a common CSF stabilizer for different species, since its presence imparts a cytostatic effect also to the cytoplasm of activated and cleaving *A. stellatus* and *R. temporaria* embryos. So far as *Xenopus* and *R. pipiens* are concerned, the papers by Meyerhof and Masui (1977, 1979a, b) contain no such data. Masui and Clarke (1979) and Masui et al. (1980) present the data from Meyerhof's dissertation, where it was demonstrated that the cytoplasm of *Xenopus* and *R. pipiens* embryos at the 2-4-blastomere stage, initially injected with EGTA, has no cytostatic effect. As to the cytostatic effect of EGTA-free cytoplasm from mature nonactivated eggs, of all five animal species studied such an effect is obtained only for the cytoplasm of *R. pipiens*. In *Xenopus* it manifests a distinct but very low activity (Meyerhof and Masui, 1979a, b; Masui and Clarke, 1979; Masui et al., 1980). However, a high cytostastic effect (irreversible or reversible) is revealed when EGTA-free cytoplasm from activated *Xenopus* eggs is injected into *P. waltlii* embryos (Aimar et al., 1981).

It is interesting that the special feature revealed by Meyerhof and Masui (1979a) in reciprocal transfer of cytoplasm with EGTA between *Xenopus* and *R. pipiens* has also been confirmed by a cross-injection of cytoplasm between *Xenopus* and *R. temporaria*. The cytoplasm from nonactivated *Xenopus* eggs, both with and without EGTA, has a cytostatic effect on *R. pipiens* and *R. temporaria* blastomeres, while the cytoplasm from nonactivated *R. pipiens* and *R. temporaria* eggs, either with EGTA or without it, induces no meiotic arrest in *Xenopus* embryos. If one assumes that the cytostatic effect of cytoplasm "stabilized" with EGTA is due to its damaging effect, then one cannot rule out the probability of a damaging effect of *Xenopus* cytoplasm upon its transfer into the embryos of *R. pipiens*, *R. temporaria*, and *P. waltlii*. Thus, the cytostatic effect of the cytoplasm of nonactivated *R. pipiens* eggs is, for the time being, an exception, and requires explanation.

Recently we obtained adult frogs *R. pipiens* (courtesy of Prof. B. M. Carlson). The cytoplasm from the nonactivated mature eggs of *R. pipiens* was injected without EGTA into one of two or one of four blastomeres of *R. pipiens* and *R. temporaria* embryos. A cytostatic effect was obtained in 15 out of 19 injected *R. pipiens* embryos, while all 22 *R. temporaria* embryos continued their normal development. The metaphase arrest of nuclei was found only in 4 out of 15 arrested blastomeres.

Karsenti et al. (1984) used EGTA for stabilizing the unfertilized eggs in the nonactivated state. They postulate that, since Ca^{++} ions are an important regulator of many processes, it is difficult to determine which process is affected by Ca^{++} binding with EGTA in every case. Similar doubts might be cast with respect to the EGTA influence on the mitosis blockade. It is not excluded that the Ca^{++} binding directly affects the spindle microtubules rather than the stabilization of a hypothetical CSF.

New data concerning the CSF were obtained by Newport and Kirschner (1984) on the *X. laevis* embryos. Nonactivated eggs used as the CSF cytoplasmic donors were injected with 50 nl of 10 mM EGTA before cytoplasm was removed for injection. Recipient eggs were arrested at metaphase by injecting each of them with 100 nl of the donor cytoplasm. The CSF effect was tested in two experimental variants. In the first one, the eggs were subjected to centrifugation at 500g for 40-60 min soon after fertilization. This treatment suppressed cytokinesis but not karyokinesis. Within 7 h after fertilization, when the total number of nuclei in the noncleaved cytoplasm attained 2000-4000, the eggs were injected with CSF which blocked the nuclear division at metaphase. The same result was obtained in the second experimental variant on exogenous nuclei from the cells of various tissues injected into the fertilized eggs, which were previously arrested at metaphase by the CSF. According to these authors, the CSF does not cause acceleration of the cell cycle or immediate onset of metaphase, but appears to arrest the cell cycle at the next metaphase and can affect the embryonic cell cycle at much later stages than the first two cycles. The arrest at metaphase was determined by the nuclear envelope disintegration, chromosome condensation, and spindle formation, i.e., only by part of the features which characterize metaphase. For this aim they used the fluorescent dye bisbenzimide for staining and dropped a cover slip on the top to lyse the eggs. The paper contains photographs which illustrate individual blocked nuclei, but presents no data as to what fraction of endogenous or exogenous nuclei are at metaphase in what number of operated embryos. On the other hand, the estimation of the cell cycle phase by the absence of nuclear envelope and the presence of condensed chromosomes is insufficient for the firm conviction that the nuclei were arrested at metaphase. In any case, Ryabova (in press) has found in the cytoplasm of parts of the arrested *X. laevis* embryo, in which the cleavage was arrested by the injection of CSF (cytoplasm of nonactivated egg + EGTA, according to Gerhart et al., 1984), nuclei at different phases: interphase karyomeric nucleus, karyomeres of the telophase nucleus, large metaphase spindle, fragments of fibrillar structures, and isolated asters.

The CSF-arrested nuclei can be reversed by injection of sperm (Shibuya and Masui, 1982) or of Ca^{++} (Newport and Kirschner, 1984). When the arrested eggs were injected with 50 nl of 5 mM $CaCl_2$, they were released from metaphase arrest and entered S-phase, as determined by decondensation of their chromosomes, reformation of nuclear envelope, and incorporation of ^{32}P-α-dCTP into DNA. Unfortunately, the materials do not permit the discrimination of the roles of cytoplasm and EGTA in the complex of the nonactivated egg cytoplasm and EGTA (CSF). The fact that Ca^{++} excess eliminates the CSF effect suggests that the cytostatic effect, as well as chromosome condensation, is due to the EGTA effect, which does not stabilize a hypothetic CSF, but blocks mitosis by binding Ca^{++}. A similar idea was put forward by Warner (see Ford, 1985).

The presence of CSF in the cytoplasm of mature nonactivated eggs, which is transmitted with the cytoplasm into cleaving embryos, cannot be considered as proved, even taking into account the data by Newport and Kirschner (1984).

The factors involved in meiotic arrest at metaphase II during oocyte maturation in vertebrates remain unclear.

3.5.6. Factors of DNA Synthesis

Three types of DNA polymerases, α, β, and γ, have been discovered in animal oocytes, just as in somatic cells. DNA polymerases specific for the oogenesis of animals have not been reported. Increasing DNA polymerase activity of the cytoplasm occurs during oocyte maturation (Grippo and Lo Scavo, 1972; Benbow et al., 1975; Ford et al., 1975; Ford and Woodland, 1975; Grippo et al., 1975). This activity apparently reflects the accumulation in the oocyte of high-molecular-weight DNA polymerase-α, which provides for the intensive replication of nuclear DNA in the embryos following fertilization (Ikegami et al., 1978; Mikhailov and Gulyamov, 1980). Further investigations have, however, failed to confirm the predominantly cytoplasmic location of this enzyme. Microsurgical separation of GVs convincingly demonstrated that in the *X. laevis* practically all the DNA polymerase-α is located in the GV of maturing oocytes (Grippo et al., 1977; Fox et al., 1980). The same DNA polymerase-α location was also found in the oocytes of the loach *Misgurnus fossilis* (Gulyamov et al., 1982; Mikhailov et al., 1983a) and *B. bufo* (Nagano et al., 1982). The cytoplasmic location of DNA polymerase-α indicated by early investigations is due, apparently, to the fact that the conditions used for extracting oocytes, aqueous solutions, extract DNA polymerase-α from the nuclei (Lynch et al., 1975). Upon transfer of DNA polymerase during GV breakdown from the GV into the cytoplasm, DNA polymerase-α apparently becomes bound to the ER (Nagano et al., 1982).

At present it is still not clear whether the accumulation of DNA polymerase-α during oocyte growth and maturation is accompanied also by the accumulation of DNA polymerase-β. This enzyme is detectable both in amphibian eggs (Fox et al., 1980) and in those of echinoderms (Hobart and Infante, 1978), whereas it is practically absent in the eggs of teleosts (Mikhailov et al., 1980; Mikhailov and Gulyamov, 1983). In the maturing oocytes of *X. laevis* the bulk of polymerase-β is located in the cytoplasmic fraction (Fox et al., 1980). In the unfertilized eggs of *H. pulcherrimus* DNA polymerase-β is also detected mainly in the cytoplasm (Shioda et al., 1977).

DNA polymerase-γ in amphibians and teleosts appears to be located in the cytoplasm of maturing oocytes (Fox et al., 1980; Gulyamov et al., 1982; Mikhailov et al., 1983b). In all probability this enzyme is contained only in the oocyte mitochondria (Bazzicalupo, 1979; Mikhailov et al., 1983b). The mitochondrial location of this enzyme agrees with the known involvement of polymerase-γ in mitochondrial DNA synthesis (Weissbach, 1979).

It has been demonstrated that endogenous DNA, upon injection into the GV, remains intact for an extended time, whereas it quickly decomposes upon being injected into the cytoplasm (Wylie et al., 1977). The capacity of the GV contents to preserve the native state of endogenous DNA has been used by Gurdon and his colleagues to study the replication and translation of genes injected into the GV (see Gurdon and Melton, 1981; Etkin, 1982; Gurdon, 1982).

Lastly, one should mention that the initiation of DNA synthesis in nuclei also requires, apart from DNA polymerases, the activation of the egg (Grippo et al., 1976). However, the molecular nature of the role of egg activation on the initiation of DNA synthesis is not clear.

Now that it is known that DNA polymerase-γ resides in the GV of full-grown oocytes rather than in the cytoplasm, it is clear that the transfer of this enzyme into the cytoplasm during GV breakdown enables transplanted nuclei to acquire the capacity for DNA synthesis.

3.5.7. Factors of RNA (RNA Polymerase) Synthesis

At the beginning of the maturation period, the GV of the oocytes of *R. pipiens* and *X. laevis* contain a great amount of RNA polymerase I, IIa, IIb, and III (Wasserman et al., 1972; Roeder, 1974; Hollinger and Smith, 1976). Isolated GVs contain about 90% of the oocyte's total RNA polymerase activity. The RNase content of an enucleated oocyte, both at the initial stage and after progesterone treatment, remains unchanged and is less than 10% of the total of an intact oocyte. After GV breakdown and the transfer of its contents into the cytoplasm, RNA polymerases remain in the cytoplasm in the inactive state, and during early embryogenesis they appear to migrate into the nuclei (Hollinger and Smith, 1976).

Data obtained by the injection of genes into oocytes demonstrate that RNA polymerases are present in the oocyte in a functionally active state, and that the amount of RNA polymerase in a single *Xenopus* oocyte equals the amount contained in several thousand somatic cells (see reviews by Laskey, 1979; Gurdon and Melton, 1981).

3.5.8. Factors of Cytokinesis

Little is known about factors involved in cytokinesis following egg activation. The quality of tubulin bound with labeled colchicine is the same in intact oocytes and enucleated oocytes, and also before and after the effect of hormone (Smith and Ecker, 1968, 1969b; Heidemann and Kirschner, 1975, 1978). On this basis the conclusion may be drawn that the nucleoplasm contains no structural elements of the mitotic apparatus. There is strong evidence suggesting that a karyoplasmic cytokinesis factor is necessary for the emergence of the mitotic apparatus and cytasters in the cytoplasm, and for the propagation of centrioles introduced in the egg by the spermatozoa during fertilization. Transplantation of an isolated fraction of centrioles provides a new method for studying this matter (Heidemann and Kirschner, 1975, 1978; Maller et al., 1976; Karsenti et al., 1984). Such experiments suggest that the inability of full-grown enucleated oocytes to form asters may be due to the inability of tubulin to polymerize microtubules present in the ooplasm (Heidemann and Kirschner, 1975).

Karsenti et al. (1984) found that in nonactivated eggs centrosomes nucleate microtubules exclusively in the proximity of condensed chromatin of the injected sperm nuclei. Individual centrosomes are unable to nucleate microtubule assembly in nonactivated eggs, but serve as efficient sites of assembly in the activated eggs. These facts suggest that the mitotic and interphase states of cytoplasm may differ in their capacity for microtubule polymerization. Special studies are needed to learn how these changes are induced by egg activation. But the problem still remains open whether an unknown component of the GV nucleoplasm essential for the oocyte to acquire the capacity for karyo- and cytokinesis relates somehow to changes in the cytoplasm of the egg upon its activation.

Cytokinesis is deeply connected with changes in the cytoplasm tension, which correlate with the cell cycle phases and are timed to the extrusion of the I and II polar bodies. In starfish oocytes these tension changes arise in the enucleated oocytes only in those cases when the oocytes were enucleated some time after GV breakdown. When the nucleoplasm was injected into the oocytes, there arose in them, under the influence of 1-Me-Ade, cyclic activity of cytoplasm tension (Yamamoto, 1985). One can suggest that these phenomena have something in common with the "surface contraction wave" described by Hara et al. (1980) for the *Xenopus* embryos.

3.5.9. Morphogenetically Active Substances

In addition to the factors involved in the transition of the full-grown oocyte into a mature egg, and factors ensuring the initial stages of embryonic development (changes in the egg during fertilization and cleavage), the oocyte also contains substances which influence the morphogenetic processes taking place at still later stages of development by switching off and regulating the gene activity (Gurdon, 1982; Gurdon et al., 1985). Thus, one can confidently speak, in the case of amphibians, about substances necessary for gastrulation and organogenesis (Briggs and Cassens, 1966; Nikitina, 1974, 1982, 1983; Leonard et al., 1982) and histogenesis (Mohun et al., 1984; Gurdon et al., 1984), and about sex determinants (see Smith, 1966).

It has been demonstrated that the GV nucleoplasm of an axolotl oocyte at the end of the growth stage contains a protein factor which in some way participates in gastrulation. The females of the axolotl, homozygotic for gene O, lay eggs in which this factor is absent, and their development does not proceed beyond the stage of an atypical gastrula. When the nucleoplasm from an oocyte (or cytoplasm from an egg) of an axolotl, normal with respect to the O gene, is injected into an unfertilized egg of an O/O mutant, development becomes normal (Briggs and Cassens, 1966; Briggs and Justus, 1968; Briggs, 1972). According to Brothers (1976), this normalization is due to the fact that the O substance becomes bound to cell nuclei during the formation of the blastula and causes the activation of genes necessary for gastrulation and organogenesis. (When a nucleus from an O+ cell strain, taken at the stage of late blastula, is transplanted into an enucleated egg of the O/O strain, development proceeds normally, but is arrested beginning from the stage of early blastula.) Questions about the nature, properties, and modes of action of the O+ substance are as yet unanswered.

The influence of the cytoplasm of maturing oocytes and eggs on the morphogenetic function of nuclei was demonstrated by transplanting small GVs with lampbrush chromosomes into the cytoplasm of enucleated *A. stellatus* eggs (Nikitina, 1974, 1982, 1983). Increasing the period of action of the egg cytoplasm on such nuclei by means of serial cloning, Nikitina obtained embryos which not only began gastrulation (Nikitina, 1974), but proceeded through neurulation as well (Nikitina, 1982).

Complete reactivation of the erythrocyte nuclei was observed after their transplantation into the cytoplasm of maturing *X. laevis* oocytes taken after their ovulation from the body cavity. Under the influence of the cytoplasm, these terminally differentiated nuclei acquired the capacity to provide for development up to the stage of swimming larva (Leonard et al., 1982; DiBerardino and Hoffner, 1983; DiBerardino et al., 1984).

Of great interest are the substances localized in the cytoplasm of the egg's vegetal region which determine the fate of the embryo germ cells forming in these regions the so-called germ-cell determinants. They can be transferred together with the cytoplasm from one egg to another (Smith, 1966). Their localization is an example of early ooplasmic segregation. The questions of the time of emergence and the nature of the sex determinants, and also of how they affect the nuclei of future germinal cells, are of great interest (see reviews by Beams and Kessel, 1974; Aizenshtadt, 1975; Eddy, 1975).

Although there is a clear demonstration of the presence in the cytoplasm of maturing amphibian oocytes and mature eggs of substances responsible for gene activation and control of gene activity, the molecular mechanisms of these processes remain unclear. Interesting and promising model systems for purification of oocyte substances which control gene activity to the extent of removing molecules responsible for the developmental repression were developed and used in experimental studies by Gurdon and his co-workers (Gurdon, 1982).

3.5.10. Pseudomaturation-Inducing Factor (PIF)

"Pseudomaturation" is a term used by Brachet and his co-workers (Baltus et al., 1973; Hanocq et al., 1974; Brachet et al., 1974; Brachet, 1974b; Steinert et al., 1974) to describe the incomplete maturation process in which the GV membrane breaks down, yet chromosome condensation does not take place, the nucleoli do not completely disappear, and atypical changes of membranous structures of the cytoplasm are observed, while protein synthesis is steeply reduced. Electron microscopy of this phenomenon established (Steinert et al., 1974) that the oocyte plasmalemma forms deep depressions. The breakdown of nuclear membranes frequently begins not in the basal, but in the apical part of the GV. At the moment of nuclear membrane breakdown the nucleoli break down to granular and fibrillar parts. During pseudomaturation, as distinct from normal maturation, the cytoplasm of the oocytes has no microtubules, but, in all probability, microfilaments are present (Steinert et al., 1974, Fig. 16). During pseudomaturation an increasing concentration of cytoplasmic membranes occurs in the nearly complete absence of protein synthesis, and it is assumed that part of the membranes arise from preformed elements and, in particular, from annulate lamellae.

Defective maturation takes place during lasting oocyte incubation in a mixture of progesterone and cycloheximide, and also when the supernatant of a homogenate prepared from mature intact eggs of *X. laevis* is injected into the full-grown oocytes not exposed to the effect of progesterone (homogenate from decapsulated eggs causes normal maturation). An extract from the egg capsules and cortical granules of mature *X. laevis* eggs causes a 50-90% reduction in protein synthesis. An extract from immature oocytes does not possess this action (Brachet et al., 1974). Experiments with oocytes of some females showed pseudomaturation under the effect of progesterone in vitro in the absence of inhibitors; this was observed in *X. laevis* much more frequently than in *R. pipiens* (Baltus et al., 1973; Hanocq et al., 1974). It has been shown that the suppression of protein synthesis precedes morphological manifestations of pseudomaturation and, apparently, is the cause of them.

It is still unclear, however, whether the substance (PIF) extracted from homogenized membranes and the cortex normally participates in the oocyte maturation process. If not, it is hardly advisable to put PIF in the same category with factors determining the changes of the oocyte in the process of maturation. However, the phenomenon of pseudomaturation *per se* is of interest, since it shows that, under conditions deviating from the normal, individual processes in the oocyte may be separated, and some of them may be produced in isolation from the others. This may be important for shedding light on their interrelationships.

Among such deviations from the normal maturation process caused by the suppression of protein synthesis, one should also list the formation of giant mitochondria in frog oocytes placed at the period of GV breakdown in a Ringer solution with puromycin (Aizenshtadt, 1969). In such oocytes the giant mitochondria frequently assumed the shape of twisted plates up to 20-25 μm in diameter, their sections appearing in the form of a spiral. Their formation indicates that during the maturation period part of the mitochondria in the oocytes are still capable of division and that, by the inhibition of protein synthesis, it is possible to separate the growth of mitochondria from their division.

3.6. CONCLUDING REMARKS

The above review makes it clear that the development of mature egg organization is a complicated process, made up of interrelated and relatively autonomously developing events which are determined by the genetic information supplied by the nucleus during the growth period of the oocyte. Some of these events take place in the oocyte cytoplasm without the participation of the GV nucleoplasm, whereas others require the passage of the nucleoplasm into the cytoplasm. It has been shown that regulation of the successive changes of the oocyte during maturation takes place at the level of translation processes, without the participation of the genetic apparatus of the nucleus, and requires, during the initial stages, the synthesis of some so far unidentified proteins. In recent years, wide-ranging investigations have taken place into the molecular basis underlying the individual features of the mature egg. Here one must bear in mind that during maturation changes occur inherent in the oocyte alone (for example, the development of the organization and properties of the cortical layer of the cytoplasm and the appearance of contractility, as well as the appearance of the capacity for fusing with spermatozoa, for cortical reaction, for conducting the activation impulse, and also for blocking meiotic divisions). Changes also occur which are common to the oocyte and somatic cells as they pass through the cellular cycle. Such changes induce nuclear membrane breakdown, chromosome condensation, the appearance of conditions necessary for the switchover of nuclear synthesis from RNA to DNA synthesis, and the appearance of a capacity for cytokinesis.

Since during the maturation period the oocyte undergoes profound structural transformations within a short period of time, and these take place in a unicellular system amenable to comprehensive examination by the methods of molecular biology and cytology, it is to be hoped that investigations into oocyte maturation will make it possible to unravel further the causative relationships between the processes proceeding at the molecular and cytological levels.

Finally, a few words on the significance of the data cited for an understanding of the evolutionary aspects of maturation and ovulation processes in oocytes among different animals. There is a great similarity in the basic changes in the oocytes during their maturation; they are characterized by a number of specific features and, above all, differences in the successive onset of individual events (heterochronies).

An analysis of these heterochronies reveals some regularities. Historically fixed in different groups of animals are relative shifts only in the time of onset of those changes in the oocytes, between which there is no causative relation during maturation, whereas the sequence of events between which there are causative relationships during oocyte maturation is preserved intact.

Thus, it has been shown that oocyte maturation and ovulation are relatively independent processes which are correlated only in time. Similarly, the maturation of the cortical layer of the oocyte cytoplasm and GV breakdown do not share a causative relationship, and so there is evolutionary variability regarding these features. Thus, for example, starfish and annelid oocytes ovulate even before GV breakdown; in amphibians, at the time of ovulation, they are at metaphase I; the oocytes of sturgeon ovulate at metaphase II. In the oocytes of starfish and some annelid worms, the cortical layer of the cytoplasm matures first, while GV breakdown occurs later; in amphibians and sturgeon, GV breakdown takes place shortly before the maturation of the cortical layer of the cytoplasm. Correlated with the differences in the time of cortical-layer maturation are also differences in the stages at which it becomes possible for the spermatozoa to penetrate the cytoplasm. In starfish and some annelids this may take place before, and in amphibians and sturgeons only after, GV breakdown.

Whereas cortical-layer maturation and the possibility of penetration by spermatozoa into the cytoplasm do not depend on the passage of the nucleoplasm into it, the conversion of the sperm nucleus into a pronucleus requires the presence of the nucleoplasm. Accordingly, whatever the stage at which the spermatozoon penetrates the oocyte in different animal species, its transformation into a pronucleus takes place in all of them only after GV breakdown.

Thus, a comparative investigation into the regularities of the maturation process, in addition to expanding our knowledge about these processes, helps to clarify their evolutional relationships. On the other hand, and this is of special importance, differences in the succession of similar events in the oocytes of different animal species may facilitate the investigation of the molecular processes underlying them.

REFERENCES

Aimar, Ch., Delarue, M., and Vilain, C., 1981, "Cytoplasmic regulation of the duration of cleavage in Amphibian eggs," J. Embryol. Exp. Morphol., *64*, 259-274.

Aimar, Ch., Vilain, C., and Delarue, M., 1983, "Regulation of cell cycle by cytoplasmic components in the amphibian eggs," Cell Differ., *13*, 293-300.

Aizenshtadt, T. B., 1969, "Gigantic mitochondria in frog oocytes," Dokl. Akad. Nauk SSSR, *189*, 1367-1370.

Aizenshtadt, T. B., 1975, "Modern ideas about determinants of cells of the germ path," Ontogenez, *6*, 379-386.

Aizenshtadt, T. B., and Dettlaff, T. A., 1968, "Fine structure of the frog oocyte at the maturation period," Tsitologiya, *10*, 783-797.

Aizenshtadt, T. B., and Dettlaff, T. A., 1972, "Ultrastructure of stellate sturgeon oocyte during maturation. I. Annulate lamellae and Golgi complex," Ontogenez, *3*, 220-229.

Aizenshtadt, T. B., and Skoblina, M. N., 1970, "Fine structure of frog oocytes maturing in vitro under the influence of hypophyseal hormones after the removal of their germinal vesicles," Tsitologiya, *12*, 713-717.

Aizenshtadt, T. B., and Vassetzky, S. G., 1986, "The fine structure of oocytes and accessory cells of the ovary in a starfish *Patiria* (*Asterina*) pectinifera at different stages of oogenesis and after 1-methyladenine-induced maturation," Dev., Growth, Differ., *28*, 449-460.

Anderson, D. M., and Smith, L. D., 1978, "Synthesis and accumulation of heterogeneous RNA in lampbrush stage of *Xenopus laevis* (Daudin)," Dev. Biol., *69*, 274-285.

Anderson, J. W., and Yatvin, M. B., 1970, "Metabolic and ultrastructural changes in the frog ovarian follicles in response to pituitary stimulation," J. Cell Biol., *46*, 491-504.

Balinsky, B. I., 1966, "Changes in the ultrastructure of amphibian eggs following fertilization," Acta Morphol. Embryol. Exp., *9*, 132-154.

Balinsky, B. I., and Devis, R. J., 1963, "Origin and differentiation of cytoplasmic structures in the oocytes of *Xenopus laevis*," Acta Morphol. Embryol. Exp., *6*, 55-108.

Baltus, E. J., Brachet, J., Hanocq-Quertier, J., and Hubert, E., 1973, "Cytochemical and biochemical studies on progesterone-induced maturation in amphibian oocytes. I. Ribonucleic acid and protein synthesis (effects of inhibitors and of a 'maturation promoting factor')," Differentiation, *1*, 127-143.

Barsacchi, G., 1971, "Observations on *Triturus viridescens* oocytes during maturation induced in vitro by progesterone," Boll. Zool., *38*.

Batistoni, R., and Barsacchi, G., 1974, "Induction of in vitro maturation in oocytes of *Triturus* (*Amphibia*, *Urodela*)," Experientia, *30*, 77-79.

Baulieu, E. E., Godeau, F., Schorderet, M., and Schorderet-Slatkine, S., 1978, "Steroid-induced meiotic division in *Xenopus laevis* oocytes: surface and calcium," Nature, *275*, 593-598.

Bazzicalupo, P., 1979, "Mitochondrial DNA polymerase from *Xenopus laevis* oocytes," Biochem. Biophys. Res. Commun., *87*, 1218-1225.

Beams, H. W., and Kessel, R., 1974, "The problem of germ cell determinants," Int. Rev. Cytol., *39*, 413-479.

Belanger, A. M., and Shuetz, A. W., 1975, "Precocious induction of activation responses in amphibian oocytes by divalent ionophore A23187," Dev. Biol., *45*, 378-381.

Bellé, R., Marot, J., and Ozon, R., 1976, "Nature of progesterone action on amino-acid uptake by isolated full-grown oocytes of *Xenopus laevis*," Biochim. Biophys. Acta, *419*, 342.

Bellé, R., Boyer, J., and Ozon, R., 1979, "Endogenous phosphorylation proteins during maturation of *Xenopus laevis* oocytes," Gamete Res., *2*, 137-145.

Belyaeva, V. N., and Cherfas, N. B., 1965, "On the processes of maturation and fertilization in the eggs of the loach *Misgurnus fossilis*," Vopr. Ikhtiol., *5*, 82-90.
Benavente, R., Krohne, G., and Franke, W. W., 1985, "Cell type-specific expression of nuclear lamina proteins during development of *Xenopus laevis*," Cell, *41*, 177-190.
Benbow, R. M., Pestell, R. Q. W., and Ford, Ch. C., 1975, "Appearance of DNA polymerase activities during early development of *Xenopus laevis*," Dev. Biol., *43*, 159-174.
Bienz, M., 1984, "Developmental control of the heat-shock response in *Xenopus*," Proc. Natl. Acad. Sci. USA, *81*, 3138-3142.
Bienz, M., and Gurdon, I. B., 1982, "The heat-shock response at the translational level," Cell, *29*, 811-819.
Brachet, J., 1967, "Effects of actinomycin, puromycin, and cycloheximide upon the maturation of amphibian oocytes," Exp. Cell Res., *48*, 233-235.
Brachet, J., 1974a, "Observations cytologiques et cytochimiques sur la maturation de l'oocyte chez les Urodèles," Année Biol., *13*, 271-284.
Brachet, J., 1974b, "Le contrôle de la maturation chez les oocytes d'Amphibien," Année Biol., *13*, 403-434.
Brachet, J., Hanocq, F., and Van Gansen, P., 1970, "A cytochemical and ultrastructural analysis of in vitro maturation in amphibian oocytes," Dev. Biol., *21*, 157-195.
Brachet, J., Baltus, E., De Schutter, A., Hanocq-Quertier, J., Hubert, E., Iacobelli, S., and Steinert, G., 1974, "Biochemical changes during progesterone-induced maturation in *Xenopus laevis* oocytes," Mol. Cell Biochem., *3*, 189-205.
Briggs, R., 1972, "Further studies on the maternal effect of the O gene in the Mexican axolotl," J. Exp. Zool., *181*, 271-280.
Briggs, R., and Cassens, G., 1966, "Accumulation in oocyte nucleus of a gene product essential for embryonic development beyond gastrulation," Proc. Natl. Acad. Sci. USA, *55*, 1103-1109.
Briggs, R., and Justus, G. T., 1968, "Partial characterization of the component from normal eggs which corrects the maternal effect of O gene in the Mexican axolotl (*Ambystoma mexicanum*)," J. Exp. Zool., *167*, 105-116.
Brothers, A. J., 1976, "Stable nuclear activation dependent on a protein synthesized during oogenesis," Nature, *260*, 112-115.
Brown, D. D., and Littna, E., 1964, "RNA synthesis during the development of *Xenopus laevis*, the South African clawed toad," J. Mol. Biol., *8*, 669-687.
Burzio, L. O., and Koide, S. E., 1976, "Isolation and characterization of *Xenopus laevis* oocyte nucleus," Arch. Biol. Med. Exp., *10*, 22-27.
Buznikov, G. A., Malchenko, L. A., Delone, G. V., and Brodsky, V. Ya., 1985, "The effect of cytostatic neurochemicals on the development of embryos derived from the karyoplasts of starfish oocytes," Dokl. Akad. Nauk SSSR, *284*, 1477-1480.
Byskov, A. G. S., 1969, "Ultrastructural studies on the pre-ovu-

latory follicle in the mouse ovary," Z. Zellforsch. Mikrosk. Anat., *100*, 285-299.

Cajander, S., and Bjersing, L., 1976, "Further studies of the surface epithelium covering rabbit follicles with special reference to lysosomal alterations," Cell Tissue Res., *169*, 129-141.

Cartaud, A., Huchon, D., Marot, J., Ozon, R., and Demaille, J. G., 1981, "Calmodulin is involved in the first step of maturation: effects of the antipsychotic drug fluphenazine and of anticalmodulin antibodies on the progesterone-induced maturation of *Xenopus laevis* oocytes," Cell Differ., *10*, 357-363.

Chambers, R., 1921, "Microdissection studies. III. Some problems in the maturation and the fertilization of the echinoderm egg," Biol. Bull., *41*, 318-322.

Charbonneau, M., and Grey, R. D., 1984, "The onset of activation responsiveness during maturation coincides with the formation of the cortical endoplasmic reticulum in oocytes of *Xenopus laevis*," Dev. Biol., *102*, 90-97.

Chiba, K., and Hoshi, M., 1985, "Mass isolation of germinal vesicles from starfish oocytes," Dev., Growth, Differ., *27*, 277-282.

Chulitskaya, E. V., 1968, "Effect of cytoplasm on synchronization and desynchronization of nuclear division during cleavage in the sturgeon embryo," Dokl. Akad. Nauk SSSR, *178*, 496-499.

Chulitskaya, E. V., 1970, "Desynchronization of cell divisions in the course of egg cleavage and an attempt at experimental shift of its onset," J. Embryol. Exp. Morphol., *23*, 359-374.

Chulitskaya, E. V., and Felgenhauer, P. E., 1977, "Noncytostatic effect of cytoplasm from oocytes and mature eggs of the grass frog and sturgeon," Ontogenez, *8*, 258-260.

Clark, T. G., and Merriam, R. W., 1977, "Diffusible and bound actin in nuclei of *Xenopus laevis* oocytes," Cell, *18*, 1101-1108.

Clark, T. G., and Rosenbaum, J. L., 1979, "An actin filament matrix in hand-isolated nuclei of *Xenopus laevis* oocytes," Cell, *18*, 1101-1108.

Cloud, Y., and Schuetz, A. W., 1973, "Spontaneous maturation of starfish oocytes. Role of follicle cells and calcium ions," Exp. Cell Res., *79*, 447-450.

Costello, D. P., 1940, "The fertilizability of nucleated and nonnucleated fragments of centrifuged *Nereis* eggs," J. Morphol., *66*, 99-114.

Cyert, M., Wu, M., Gerhart, J., and Kirschner, M., 1985, "Isolation of monoclonal antibodies that precipitate maturation-promoting factor," Cell Differ., *16*, Suppl. N442, 180S.

Dalcq, A., 1924, "Récherches expérimentales et cytologiques sur la maturation et l'activation de l'oeuf d *Asterias glacialis*," Arch. Biol., *34*, 507-674.

Davidson, E. H., 1969, Gene Activity in Early Development, Academic Press, New York—London.

Davydov, P. V., Shubravyi, O. I., and Vassetzky, S. G., 1986, "Methods for maintaining starfish and obtaining gametes and embryos year-round," Methods Cell Biol., *27*, 375-378.

Davydova, S. I., 1968, "Maturation of oocytes of *Acipenseridae* outside the body of the female," Dokl. Akad. Nauk SSSR, *179*, 750-752.

Davydova, S. I., 1973, "Effect of temperature and keeping time of female sturgeons in captivity on maturation of oocytes under the influence of hormones in vitro," Ontogenez, *4*, 415-420.

Delage, Y., 1899, "Études sur la mérogonie," Arch. Zool. Exp. Gen., Sér. 3, *7*, 383-417.

Delage, Y., 1901, "Études expérimentales sur la maturation cytoplasmique et sur la parthénogenèse artificielle chez les Echinodermes," Arch. Zool. Exp. Gen., Sér. 3, *2*, 284-336.

Denis, H., 1974, Precis d'Embryologie Moleculaire, Presses Univ., France.

Dettlaff, T. A., 1961, "The ovulation and activation of oocytes of acipenserid fishes in vitro," Sympos. Germ. Cells and Dev., Int. Inst. Embryol. Fond., Milano, 141-144.

Dettlaff, T. A., 1962, "Cortical changes in acipenserid eggs during fertilization and artificial activation," J. Embryol. Exp. Morphol., *10*, 1-16.

Dettlaff, T. A., 1966, "Action of actinomycin and puromycin upon frog oocyte maturation," J. Embryol. Exp. Morphol., *16*, 183-195.

Dettlaff, T. A., 1970a, "Influence of environmental temperature during oocyte maturation and ovulation on the quality of sturgeon eggs (on thermal regime of keeping sturgeons in captivity during the period of obtaining eggs)," Tr. TZNI-ORKH, *2*, 112-126.

Dettlaff, T. A., 1970b, "Intercellular influences in the process of oocyte maturation in sturgeons," in: Intercellular Interactions in Differentiation and Growth [in Russian], Nauka, Moscow, pp. 241-243.

Dettlaff, T. A. (ed.), 1975, Objects of Developmental Biology [in Russian], Nauka, Moscow, p. 84.

Dettlaff, T. A., 1977a, "Some thermal temporal patterns of embryogenesis," in: Problems of Experimental Biology [in Russian], Nauka, Moscow, pp. 269-287.

Dettlaff, T. A., 1977b, "Development of the mature egg organization in amphibians and fishes during the concluding stages of oogenesis, in the period of maturation," in: Modern Problems of Oogenesis [in Russian], Nauka, Moscow, pp. 99-144.

Dettlaff, T. A., and Davydova, S. I., 1974, "Effect of triiodothyronine on ripening of oocytes in stellate sturgeon after the action of low temperatures and reservation of females," Ontogenez, *5*, 402-410.

Dettlaff, T. A., and Davydova, S. I., 1979, "Differential sensitivity of cells of follicular epithelium and oocytes in the stellate sturgeon to unfavorable conditions and the correlating influence of triiodothyronine," J. Gen. Comp. Endocrinol., *39*, 236-243.

Dettlaff, T. A., and Dettlaff, A. A., 1961, "On relative dimensionless characteristics of the development duration in embryology," Arch. Biol., *72*, 1-16.

Dettlaff, T. A., and Felgenhauer, P. E., 1980, "Maturation of stellate sturgeon oocytes following the injection of cytoplasm of mature stellate sturgeon eggs and maturing *Xenopus laevis* oocytes and the effect of cycloheximide on this process," Ontogenez, *11*, 223-228.

Dettlaff, T. A., and Ginsburg, A. S., 1950, "A case of parthenogenetic cleavage of nonovulated oocytes in the ovary of *Acipenser stellatus*," Dokl. Akad. Nauk SSSR, *72*, 429-432.

Dettlaff, T. A., and Ginsburg, A. S., 1951, "Early parthenogenetic activation as a cause of nonfertilizability of the eggs in *Acipenser stellatus*," Dokl. Akad. Nauk SSSR, *77*, 537-540.

Dettlaff, T. A., and Ginsburg, A. S., 1954, The Embryonic Development of Sturgeons in Connection with the Problems of Their Breeding [in Russian], Izd. Akad. Nauk SSR, Moscow.

Dettlaff, T. A., and Ryabova, L. V., 1986, "Maturation of *Rana temporaria* and *Acipenser stellatus* oocytes induced by the cytoplasm of embryos at different cell cycle phases and at different stages of cleavage and blastulation," Cell Differ., *18*, 9-16.

Dettlaff, T.A., and Schmerling, Zh. G., 1968, "Puromycin-sensitive phases and protein synthesis in the period of oocyte maturation in *Rana temporaria*," Dokl. Akad. Nauk SSSR, *181*, 501-504.

Dettlaff, T. A., and Skoblina, M. N., 1969, "The role of germinal vesicle in the process of oocyte maturation in *Anura* and *Acipenseridae*," Ann. Embryol. Exp. Morphol., Suppl., *1*, 133-151.

Dettlaff, T. A., and Vassetzky, S. G. (eds.), 1988, Animal Species for Developmental Studies, Consultants Bureau (in press).

Dettlaff, T. A., and Zubova, S. E., 1962, "A correlation between the periods of maturation and embryonic development of *Acipenser güldenstädti* and *A. stellatus*," Dokl. Akad. Nauk SSSR, *143*, 746-748.

Dettlaff, T. A., Nikitina, L. A., and Stroeva, O. G., 1964, "The role of the germinal vesicle in oocyte maturation in anurans as revealed by the removal and transplantation of nuclei," J. Embryol. Exp. Morphol., *12*, 851-873.

Dettlaff, T. A., Skoblina, M. N., and Davydova, S. I., 1968a, "Intercellular influences in the process of oocyte maturation in the sturgeon," in: Intercellular Interactions in the Process of Differentiation and Growth [in Russian], Tbilisi, pp. 5-6.

Dettlaff, T. A., Vassetzky, S. G., and Davydova, S. I., 1968b, "Maturation of sturgeon oocytes under different temperatures and time of obtaining eggs after the hypophyseal injection," VIe Congrès Int. Reprod. Anim., Paris, Vol. 1, pp. 129-131.

Dettlaff, T. A., Felgenhauer, P. E., and Chulitskaya, E. V., 1977, "Effect of cycloheximide on germinal vesicle breakdown and changes in the cortical layer in clawed toad and sturgeon oocytes during their maturation under the influence of active cytoplasm," Ontogenez, *8*, 417-424.

Dettlaff, T. A., Chulitskaya, E. V., Stepanov, V. S., and Felgenhauer, P. E., 1980, "Protein synthesis in the *Acipenser stellatus* oocytes during the period of maturation and influence of protein synthesis inhibition on changes in oocytes," Ontogenez, *11*, 24-30.

Dettlaff, T. A., Ginsburg, A. S., and Schmalhausen, O. I., 1981, Development of the Sturgeon – Maturation, Fertilization, and Development of Embryos and Prelarvae [in Russian], Nauka, Moscow.

Dettlaff, T. A., Ignatieva, G. M., and Vassetzky, S. G., 1986, "The problem of time in developmental biology: its study by the method of relative characteristics of development duration," Soviet Scientific Reviews, Ser. F, General Biology and Physiology, Vol. 1, Developmental Biology (in press).

DiBerardino, M. A., and Hoffner, N. J., 1983, "Gene reactivation in erythrocytes: nuclear transplantation in oocytes and eggs of *Rana*," Science, *219*, 862-864.

DiBerardino, M. A., Hoffner, N. J., and Etkin, L. D., 1984, "Activation of dormant genes in specialized cells," Science, *224*, 946-952.

Donaldson, E. M., and Hunter, G. A., 1983, "Induced final maturation, ovulation, and spermiation in cultured fish," in: Fish Physiology, Vol. IX, Part 13, W. S. Hoar, D. J. Randall, and E. M. Donaldson (eds.), Academic Press, New York–London, pp. 351-403.

Dorée, M., 1980, "Calmodulin content does not change following hormone-induced meiosis reinitiation in starfish oocytes," Experientia, *36*, 932-933.

Dorée, M., 1982, "Protein synthesis is not involved in the initiation of amplification of the maturation-promoting factor (MPF) in starfish oocytes," Exp. Cell Res., *139*, 127-134.

Dorée, M., and Guerrier, P., 1975, "Site of action of 1-methyladenine in inducing oocyte maturation in starfish. Kinetic evidence for receptors localized on the cell membrane," Exp. Cell Res., *96*, 296-300.

Dorée, M., Moreau, M., and Guerrier, P., 1978, "Hormonal control of meiosis. In vitro induced release of calcium ions from the plasma membrane in starfish oocytes," Exp. Cell Res., *115*, 251-260.

Dorée, M., Cavadore, J.-C., Le Peuch, C. J., and Demaille, J. G., 1981a, "Hormone-induced phosphorylation of a 16,000-dalton polypeptide following meiosis reinitiation in starfish oocytes," Exp. Cell Res., *135*, 251-257.

Dorée, M., Kishimoto, T., Le Peuch, C. J., Demaille, J. G., and Kanatani, H., 1981b, "Calcium-mediated transduction of the hormonal message in meiosis reinitiation of starfish oocytes. Modulation following injection of cholera toxin and cAMP-dependent protein kinase," Exp. Cell Res., *135*, 237-249.

Dreyer, C., Singer, H., and Hansen, P., 1981, "Tissue-specific nuclear antigens in the germinal vesicle of *Xenopus laevis* oocytes," W. Roux's Arch. Dev. Biol., *190*, 197-207.

Drury, K., 1978, "Method for preparation of active maturation-promoting factor (MPF) from in vitro matured oocytes of *Xenopus laevis*," Differentiation, *10*, 181-186.

Drury, K. C., and Schorderet-Slatkine, S., 1975, "Effects of cycloheximide on the 'autocatalytic' nature of the maturation-promoting factor (MPF) in oocytes of *Xenopus laevis*," Cell, *4*, 269-274.

Dumont, J. N., 1972, "Oogenesis in *Xenopus laevis* (Daudin). I. Stages of oocyte development in laboratory-maintained animals," J. Morphol., *136*, 153-180.

Dumont, J. N., and Brummett, A. R., 1978, "Oogenesis in *Xenopus laevis* (Daudin). V. Relationships between developing oocytes and their investing follicular tissues," J. Morphol., *155*, 73-77.

Duryee, W. R., 1950, "Chromosomal physiology in relation to nuclear structure," Ann. N. Y. Acad. Sci., *50*, 920-953.

Ecker, R. E., and Smith, L. D., 1968, "Protein synthesis in amphibian oocytes and early embryos," Dev. Biol., *18*, 232-249.

Ecker, R. E., and Smith, L. D., 1971a, "The nature and fate of *Rana pipiens* proteins synthesized during maturation and early cleavage," Dev. Biol., *24*, 559-576.

Ecker, R. E., and Smith, L. D., 1971b, "Influence of exogenous ions on the events of maturation in *Rana pipiens* oocytes," J. Cell Physiol., *77*, 61-70.

Ecker, R. E., Smith, L. D., and Subtelny, S., 1968, "Kinetics of protein synthesis in enucleate frog oocytes," Science, *160*, 1115-1117.

Eddy, E. M., 1975, "Germ plasm and the differentiaton of the germ cell line," Int. Rev. Cytol., *43*, 229-280.

Elinson, R. P., 1971, "Fertilization of partially jellied and jellyless oocytes of the frog *Rana pipiens*," J. Exp. Zool., *176*, 415-428.

Elinson, R. P., 1973a, "Fertilization of frog body cavity eggs enhanced by treatments affecting the vitelline coat," J. Exp. Zool., *189*, 291-301.

Elinson, R. P., 1973b, "Fertilization of frog body cavity eggs: *Rana pipiens* eggs and *Rana clamitans* sperm," Biol. Reprod., *8*, 362-368.

Elinson, R. P., 1980, "The amphibian egg cortex in fertilization and early development," in: The Cell Surface: Mediator of Developmental Processes, S. Subtelny and N. K. Wessells (eds.), Academic Press, New York.

Etkin, L. D., 1982, "Analysis of the mechanism involved in gene regulation and cell differentiation by microinjection of purified genes and somatic cell nuclei into amphibian oocytes and eggs," Differentiation, *21*, 149-159.

Felgenhauer, P. E., 1985, "RNA synthesis in maturing amphibian oocytes," Ontogenez, *16*, 290-294.

Felgenhauer, P. E., and Chulitskaya, E. V., 1977, "Induced maturation of oocytes in the clawed frog and the starred sturgeon at stages of incomplete vitellogenesis," Ontogenez, *8*, 167-169.

Ford, C. C., 1985, "Maturation-promoting factor and cell cycle regulation," J. Embryol. Exp. Morphol., *89*, Suppl., 271-284.

Ford, C. C., and Woodland, H. P., 1975, "DNA synthesis in oocytes and eggs of *Xenopus laevis* injected with DNA," Dev. Biol., *43*, 189-199.

Ford, C. C., Pestell, R. Q. W., and Benbow, R. M., 1975, "Template preferences of DNA polymerase and nuclease activities appearing during early development of *Xenopus laevis*," Dev. Biol., *43*, 175-188.

Fox, A. M., Breaux, C. B., and Benbow, R. M., 1980, "Intracellular localization of DNA polymerase activities within large oocytes of the frog *Xenopus laevis*," Dev. Biol., *80*, 79-95.

Franke, W. W., Rathke, P. C., Trendelenburg, M. F., Seib, E., Osborn, M., and Weber, K., 1976, "Distribution and mode of arrangement of microfilamentous structures and action in the cortex of amphibian oocytes," Cytobiologiya, *14*, 111-130.

Georgia, T., 1980, "The in vitro effects of protein synthesis inhibitors on steroid-induced final maturation and ovulation of yellow perch (*Perca flavescens*) oocytes," Biol. Reprod., *22*, Suppl. 1, 124A.

Gerhart, J., Wu, M., and Kirschner, M., 1984, "Cell cycle dynamics of an M-phase specific cytoplasmic factor in *Xenopus laevis* oocytes and eggs," J. Cell Biol., *98*, 1247-1255.

Ginsburg, A. S., 1968, Fertilization in Fishes and the Problem of Polyspermy, Israel Program for Scientific Translation, Jerusalem.

Ginsburg, A. S., 1987, "Egg cortical reaction during fertilization and its role in blocking polyspermy," in: Soviet Scientific Reviews, Ser. F. General Biology and Physiology, Vol. 1, Developmental Biology (in press).

Ginsburg, A. S., and Dettlaff, T. A., 1969, Development of Acipenserid Fishes. Oocyte Maturation, Fertilization, and Embryogenesis [in Russian], Nauka, Moscow.

Ginsburg, A. S., and Nikiforova, G. P., 1979, "Development of cortical reaction during oocyte maturation in *Acipenser stellatus*," Ontogenez, *10*, 385-392.

Godeau, F., Schorderet-Slatkine, S., and Baulieu, E. E., 1978a, "Role of cell surface in steroid-induced oocyte maturation in *Xenopus laevis*," Biol. Cell., *32*, 83-87.

Godeau, F., Francois, J., Schorderet-Slatkine, S., Hubert, P., and Baulieu, E. E., 1978b, "Induction of maturation in *Xenopus laevis* oocytes by a steroid linked to a polymer," Proc. Natl. Acad. Sci. USA, *75*, 2353-2357.

Goetz, F. W., 1979, "The role of steroids in the control of oocyte final maturation and ovulation in yellow perch (*Perca flavescens*)," Proc. Indian Natl. Sci. Acad., *B45*, 497-504.

Goetz, F. W., 1980, "Prostaglandin levels during steroid-induced in vitro final maturation of brook trout (*Salvelinus fontinalis*)," Am. Zool., *20*, 730.

Goldstein, L., 1953, "A study of the mechanism of activation and nuclear breakdown in the *Chaetopterus* egg," Biol. Bull., *105*, 87-102.

Goncharov, B. F., 1969a, "On the regularities of emergence of gonad sensitivity to the pituitary gonadotropic hormones," in: A Contemporary Method of Pituitary Injections [in Russian], Astrakhan, pp. 39-44.

Goncharov, B. F., 1969b, "Inhibitory effect of hypophysis on progesterone-induced oocyte maturation in vitro in frogs and sturgeon," in: Demonstrations Presented at the 9th Int. Embryological Conference, pp. 31-32.

Goncharov, B. F.,1971a, "Functional relationship between length of the hormone-dependent period of follicle maturation in common frog and the concentration of a pituitary suspension. A new method for testing pituitaries," Ontogenez, *2*, 50-55.

Goncharov, B. F., 1971b, "Studying the regularities of the transition of amphibian and sturgeon fish oocytes from growth to maturation," Author's Abstract of Candidate's Dissertation, Moscow.

Goncharov, B. F., 1976, "The physiological state of the elder generation of sevryuga follicles during their spawn migration," in: Ecological Physiology of Fishes, Pt. 2 [in Russian], Naukova Dumka, Kiev, pp. 139-140.

Goncharov, B. F., 1984a, "A synthetic analog of luteinizing-hormone-releasing hormone as a new promising stimulator of gamete maturation in *Acipenseridae*," Dokl. Akad. Nauk SSSR, *276*, 1002-1006.

Goncharov, B. F., 1984b, "Application of LHRH synthetic analogs for induction of gamete maturation in the *Acipenseridae*," in: Sturgeon Farming in Ponds in the USSR [in Russian], Astrakhan, pp. 94-95.

Goncharov, B. F., 1985, "Prospect of application of LHRH synthetic analogs for obtaining mature gametes of the sturgeon," in: Ecological Physiology and Biochemistry of Fishes [in Russian], Vilnius, pp. 384-385.

Good, G. M., and Daniel, J., 1943, "Fertilization of coelomic eggs of *Triturus torosus*," Univ. California Publ. Zool., *51*, 149-158.

Gounon, P., and Karsenti, E., 1981, "Involvement of contractile proteins in the changes in consistency of oocyte nucleoplasm of the newt *Pleurodeles waltlii*," J. Cell Biol., *88*, 410-421.

Graham, C. F., 1966, "The regulation of DNA synthesis and mitosis in multinucleate frog eggs," J. Cell. Sci., *1*, 363-374.

Graham, C. F., and Morgan, R. W., 1966, "Changes in the cell cycle during early embryonic development," Dev. Biol., *14*, 439-460.

Graham, C. F., Arms, K., and Gurdon, J. B., 1966, "The induction of DNA synthesis by frog egg cytoplasm," Dev. Biol., *14*, 349-381.

Grippo, P., and Lo Scavo, A., 1972, "DNA polymerase activity during maturation in *Xenopus laevis* oocytes," Biochem. Biophys. Res. Commun., *48*, 280-285.

Grippo, P., Locorotondo, G., and Caruso, A., 1975, "Characterization of the two major DNA polymerase activities in oocytes and eggs of *Xenopus laevis*," Fed. Eur. Biochem. Soc. Letters, *51*, 137.
Grippo, P., Caruso, A., Locorotondo, G., and Labella, T., 1976, "Multiple forms of DNA polymerase during early development and in somatic cells of *Xenopus laevis*," Cell Differ., *5*, 121-128.
Grippo, P. Taddei, C., Locorotondo, G., and Gambino-Giuffrida, A., 1977, "Cellular localization of DNA polymerase activities in full-grown oocytes and embryos of *Xenopus laevis*," Exp. Cell Res., *109*, 247-252.
Grusova, M. N., 1975, "The karyosphere during oogenesis," Tsitologiya, *17*, 219-237.
Guerrier, P., and Dorée, M., 1975, "Hormonal control of reinitiation of meiosis in starfish. The requirement of 1-methyladenine during nuclear maturation," Dev. Biol., *47*, 341-348.
Guerrier, P., Moreau, M., and Dorée, M., 1977a, "Hormonal control of meiosis in starfish: stimulation of protein phosphorylation induced by 1-methyladenine," Mol. Cell Endocrinol., *7*, 137-150.
Guerrier, P., Moreau, M., and Dorée, M., 1977b, "Inhibition de la réinitiation de la méiose des ovocytes de *Xenopus laevis* par trois antiprotéases naturelles, l'antipaine, la chymostatine, et la leupeptine," C. R. Acad. Sci. Paris, *284*, 317-319.
Guerrier, P., Meijer, L., Moreau, M., and Longo, F., 1983, "Hormone-independent GVBD induces cytoplasmic maturity in the starfish oocyte," J. Exp. Zool., *226*, 303-309.
Gulyamov, D. B., Knyazeva, E. F., Kostomarova, A. A., and Mikhailov, V. S., 1982, "Localization of DNA polymerase α in the nucleus and DNA γ polymerase in the cytoplasm of maturing oocytes in the loach," Ontogenez, *13*, 641-644.
Gurdon, J. B., 1967, "On the origin and persistence of a cytoplasmic state inducing nuclear DNA synthesis in frog's eggs," Proc. Natl. Acad. Sci. USA, *58*, 545-552.
Gurdon, J. B., 1968, "Changes in somatic cell nuclei inserted into growing and maturing amphibian oocytes," J. Embryol. Exp. Morphol., *20*, 401-414.
Gurdon, G. B., 1980, "Molecular mechanism in the control of gene expression," 15th Ciba Medical Lecture, Biochemical Society Meeting, Sheffield, *9*, 13-21.
Gurdon, J. B., 1982, "Amphibian oocytes and gene control in development," in: Embryonic Development, Part A: General Aspects, Alan R. Liss, New York, pp. 211-225.
Gurdon, J. B., and Melton, D. A., 1981, "Gene transfer in amphibian eggs and oocytes," Annu. Rev. Genet., *15*, 189-218.
Gurdon, J. B., and Speight, V. A., 1969, "The appearance of a cytoplasmic DNA polymerase activity during the maturation of amphibian oocytes into eggs," Exp. Cell Res., *55*, 253-256.
Gurdon, J. B., Brennan, S., Fairman, S., and Mohun, T. J., 1984, "Transcription of muscle-specific actin genes in early *Xenopus* development: nuclear transplantation and cell dissociation," Cell, *38*, 691-700.

Gurdon, J. B., Mohun, T. J., Fairman, S., and Brennan, S., 1985, "All components required for the eventual activation of muscle-specific actin genes are localized in the subequatorial region of an uncleaved amphibian egg," Proc. Natl. Acad. Sci. USA, *82*, 139-141.

Halleck, M. S., Lumley-Sapanski, K., Reed, J. A., Iyer, A. P., Mastro, A. M., and Schlegel, R. A., 1984, "Characterization of protein kinases in mitotic and meiotic cell extracts," Fed. Eur. Biochem. Soc.Lett., *167*, No. 2, 193-198.

Hamaguichi, M. S., and Hiramoto, Y., 1978, "Protoplasmic movement during polar body formation in starfish oocytes," Exp. Cell Res., *112*, 55-62.

Hanocq, F., De Schutter, A., Hubert, E., and Brachet, J., 1974, "Cytochemical and biochemical studies on progesterone-induced maturation in amphibian oocytes. II. DNA synthesis," Differentiation, *2*, 75-90.

Hanocq-Quertier, J., and Baltus, E., 1981, "Phosphorylation of ribosomal proteins during maturation of *Xenopus laevis* oocytes," Eur. J. Biochem., *120*, 351-355.

Hanocq-Quertier, J., and Baltus, E., 1982, "Phosphorylation de la proteine ribosomale S6 au cours de la maturation de l'oocyte de Xenope: identification des acides aminés phosphorylés," C. R. Acad. Sci. Paris, *294*, 699-702.

Hanocq-Quertier, J., Baltus, E., Ficq, A., and Brachet, J., 1968, "Studies on the DNA of *Xenopus laevis* oocytes," J. Embryol. Exp. Morphol., *19*, 273-282.

Hanocq-Quertier, J., Baltus, E., and Brachet, J., 1976, "Induction of maturation (meiosis) in small *Xenopus laevis* oocytes by injection of maturation-promoting factor," Proc. Natl. Acad. Sci. USA, *73*, 2028-2032.

Hanocq-Quertier, J., Baltus, E., and Brachet, J., 1978, "A study of the induction of spindle and astral fibers in maturing *Xenopus laevis* oocytes," Biol. Cell., *32*, 103-108.

Hara, K., Tydeman, P., and Kirschner, M., 1980, "A cytoplasmic clock with the same period as the division cycle in *Xenopus* eggs," Proc. Natl. Acad. Sci. USA, *77*, 462-466.

Hausen, P., Ya Hui Wang, Dreyer, C., and Stick, R., 1985, "Distribution of nuclear proteins during maturation of *Xenopus* oocytes," J. Embryol. Exp. Morphol., *89*, Suppl., 17-34.

Heidemann, S. R., and Kirschner, M. W., 1975, "Aster formation in eggs of *Xenopus laevis*. Induction by isolated basal bodies," J. Cell Biol., *67*, 105-117.

Heidemann, S. R., and Kirschner, M. W., 1978, "Induced formation of asters and cleavage furrows in oocytes of *Xenopus laevis* during in vitro maturation," J. Exp. Zool., *204*, 431-444.

Heilbrunn, L. V., 1952, An Outline of General Physiology, 3rd edn., W. B. Saunders, Philadelphia.

Heilbrunn, L. V., and Wilbur, K. M., 1937, "Stimulation and nuclear breakdown in the *Nereis* egg," Biol. Bull., *73*, 557-564.

Heilbrunn, L. V., and Wilson, W. G., 1955, "Changes in the protoplasm maturation," Biol. Bull., *109*, 271-275.

Heilbrunn, L. V., Daugherty, K., and Wilbur, K. M., 1939, "Initiation of maturation in the frog egg," Physiol. Zool., *12*, 97-100.

Hermann, J., Malner, O., Bellé, R., Marot, J., Tso, J., and Ozon, R., 1984, "In vivo effects of microinjected alkaline phosphatase and its low-molecular-weight substrates on the first meiotic cell division in *Xenopus laevis* oocytes," Proc. Natl. Acad. Sci. USA, *81*, 5159-5154.

Hirai, S., Kubota, I., and Kanatani, H., 1971, "Induction of cytoplasmic maturation by 1-methyladenine in starfish oocytes after removal of the germinal vesicle," Exp. Cell Res., *68*, 137-143.

Hirai, S., Nagahama, Y., Kishimoto, T., and Kanatani, H., 1981, "Cytoplasmic maturity revealed by the structural changes in incorporated spermatozoon during the course of starfish oocyte maturation," Dev., Growth, Differ., *23*, 465-478.

Hirose, K., 1976, "Endocrine control of ovulation in medaka (*Oryzias latipes*) and ayu (*Plecoglossus altivelis*)," J. Fish Res. Board Can., *33*, 989-994.

Hobart, P. M., and Infante, A. A., 1978, "A low-molecular-weight DNA polymerase β in the sea urchin *Strongylocentrotus purpuratus*. Partial purification, properties, and changes in development," J. Biol. Chem., *253*, 8229-8238.

Hollinger, T. G., and Schuetz, A. W., 1976, "'Cleavage' and cortical granule breakdown in *Rana pipiens* oocytes by direct microinjection of calcium," J. Cell Biol., *71*, 395-401.

Hollinger, T. G., and Smith, L. D., 1976, "Conservation of RNA polymerase during maturation of the *Rana pipiens* oocyte," Dev. Biol., *51*, 86-97.

Holtfreter, J., 1943, "A study of the mechanics of gastrulation," Part 1," J. Exp. Zool., *94*, 261-318.

Houle, J. G., and Wasserman, W. J., 1983, "Intracellular pH plays a role in regulating protein synthesis in *Xenopus* oocytes," Dev. Biol., *97*, 302-312.

Huchon, D., Crozet, N., Cantenot, N., and Ozon, R., 1981a, "Germinal vesicle breakdown in the *Xenopus laevis* oocyte: description of a transient microtubular structure," Reprod. Nutr. Dev., *21*, 135-148.

Huchon, D., Ozon, R., Fisher, E. H., and Demaille, J. G., 1981b, "La maturation méiotique de l'ovocyte de *Xenopus laevis*: un méchanisme en quatre étapes," C. R. Acad. Sci. Paris, *292*, Ser. III, 65-67.

Iacobelli, S. J., Hanocq, F., Baltus, E. J., and Brachet, J., 1974, "Hormone-induced maturation of *Xenopus laevis* oocytes: effect of different steroids and study of the properties of a progesterone receptor," Differentiation, *2*, 129.

Igumnova, E. V., 1974, "A characteristic of the duration of maturation of females of the great sturgeon (*Huso huso*) following hypophysial injection at different temperatures," Vopr. Ikhtiol., *14*, 1029-1035.

Ikegami, S., Okada, T. S., and Koide, S. S., 1976, "On the role of

calcium ions in oocyte maturation in the polychaete *Chaetopterus pergamentaceus*," Dev., Growth, Differ., *18*, 33-43.
Ikegami, S., Taguchi, T., and Ohashi, M., 1978, "Aphidicolin prevents mitotic cell division by interfering with the activity of DNA polymerase α," Nature, *275*, 458-460.
Imoh, H., 1981, "Changes in nucleoli at maturation of newt oocyte," Dev., Growth, Differ., *23*, 33-39.
Ishikawa, K., Hanaoka, Y., Kondo, Y., and Imai, K., 1977, "Primary action of steroid hormone at the surface of amphibian oocyte in the induction of germinal vesicle breakdown," Mol. Cell Endocrin., *9*, 91-100.
Iwamatsu, T., 1965a, "On fertilizability of preovulation eggs in the medaka *Oryzias latipes*," Embryologia, *8*, 327-336.
Iwamatsu, T., 1965b, "A further study of fertilizability of fish oocytes," Annot. Zool. Jpn., *38*, 90-97.
Iwamatsu, T., 1966, "Role of germinal vesicle materials in the acquisition of developmental capacity of the fish oocyte," Embryologia, *9*, 205, 221.
Iwamatsu, T., 1967, "On acquisition of developmental capacity in oocytes of the medaka *Oryzias latipes*," Annot. Zool. Jpn., *40*, 6-19.
Iwamatsu, T., 1974, "Studies on oocyte maturation of the medaka *Oryzias latipes*. I. Effect of several steroids and calcium ions and the role of follicle cells on in vitro maturation," Annot. Zool. Jpn., *47*, 30-42.
Iwamatsu, T., and Ohta, T., 1981, "On a relationship between oocyte and follicle cells around the time of ovulation in the medaka *Oryzias latipes*," Annot. Zool. Jpn., *54*, 17-29.
Iwamatsu, T., Ohta, T., Nakayana, N., and Shoji, H., 1976, "Studies of oocyte maturation of the medaka *Oryzias latipes*. III. Cytoplasmic and nuclear changes of oocyte during in vitro maturation," Annot. Zool. Jpn., *49*, 28-37.
Iwao, Y., 1982, "Differential emergence of cortical granule breakdown and electrophysiological responses during meiotic maturation of toad oocytes," Dev., Growth, Differ., *24*, 464-477.
Jalabert, B., 1976, "In vitro maturation and ovulation in rainbow trout (*Salmo gairdnerii*), northern pike (*Esox lucius*), and goldfish (*Carassius auratus*)," J. Fish Res. Board Can., *33*, 974-988.
Jalabert, B., and Szöllösi, D., 1975, "In vitro ovulation of trout oocytes: effect of prostaglandins on smooth muscle-like cells of the theca," Prostaglandins, *9*, 765-778.
Jalabert, B., Breton, B., and Bry, C., 1972, "Maturation et ovulation in vitro des ovocytes de la truite arc-en-ciel *Salmo gairdnerii*," C. R. Acad. Sci. Paris, *275*, 1139-1142.
Jalabert, B., Bry, C., Szöllösi, D., and Fostier, A., 1973, "Étude comparée de l'action des hormones hypophysaires et steroides sur la maturation in vitro des ovocytes de la truite et du carassin," Ann. Biol. Anim., Biochem., Biophys., *13*, 59-72.

Jalabert, B., Breton, B., Brzuska, E., Fostier, A., and Wieniawski, J., 1977, "A new tool for induced spawning: the use of 17α-hydroxy-20β-dihydroprogesterone to spawn carp at low temperature," Aquaculture, *10*, 353-364.

Jalabert, B., Breton, B., and Fostier, A., 1978a, "Precocious induction of oocyte maturation and ovulation in rainbow trout (*Salmo gairdnerii*): problems when using 17α-hydroxy-20β-dihydroprogesterone," Ann. Biol. Anim., Biochem., Biophys., *18*, 977-984.

Jalabert, B., Goetz, F. W., Breton, B., Fostier, A., and Donaldson, E. M., 1978b, "Precocious induction of oocyte maturation and ovulation in coho salmon *Oncorhynchus kisutch*," J. Fish Res. Board Can., *35*, 1423-1429.

Johnson, C. H., and Epel, D., 1980, "Intracellular pH does not regulate protein synthesis of starfish oocytes," J. Cell Biol., *87* (2, Pt. 2), 142a.

Just, E. E., 1915, "Initiation of development in *Nereis*," Biol. Bull., *28*, 1-17.

Kalimi, M., Ziegler, D., and Morrill, G. A., 1979, "Characterization of progestin-binding macromolecule in the amphibian oocyte cytosol," Biochem. Biophys. Res. Commun., *86*, 560-569.

Kambara, S., 1953, "Role of jelly envelope in toad eggs in fertilization," Annot. Zool. Jpn., *36*, 78-89.

Kanatani, H., 1964, "Spawning of starfish: action of gamete-shedding substance obtained from radial nerves," Science, *140*, 1177-1179.

Kanatani, H., 1969, "Induction of spawning and oocyte maturation by 1-methyladenine in starfish," Exp. Cell Res., *57*, 333-337.

Kanatani, H., 1972, "Adenine derivatives and oocyte maturation in starfish," in: Oogenesis, J. D. Biggers and A. W. S. Schuetz (eds.), Univ. Park Press, Baltimore; Butterworths, London, pp. 450-478.

Kanatani, H., 1973, "Maturation-inducing substance in starfish," Int. Rev. Cytol., *35*, 253-298.

Kanatani, H., 1975, "Maturation-inducing substances in asteroid and echinoid oocytes," Am. Zool., *15*, 493-505.

Kanatani, H., 1979, "Hormones in echinoderms," in: Hormones and Evolution, E. I. W. Barrington (ed.), Academic Press, London, pp. 273-307.

Kanatani, H., and Hiramoto, H., 1970, "Site of action of 1-methyladenine in inducing oocyte maturation in starfish," Exp. Cell Res., *61*, 280-284.

Kanatani, H., and Shirai, H., 1967, "In vitro production of meiosis-inducing substance by nerve extract in ovary of starfish," Nature, *216*, 280-284.

Kanatani, H., and Shirai, H., 1972, "On the maturation-inducing substance produced in starfish gonad by neural substance," Gen. Comp. Endocrinol., Suppl., *3*, 571-579.

Kanatani, H., Shirai, H., Nakanishi, K., and Kurokawa, T., 1969, "Isolation and identification of meiosis-inducing substance in starfish," Nature, *221*, 273-274.

Karsenti, E., Newport, J., Hubble, R., and Kirschner, M., 1984, "Interconversion of metaphase and interphase microtubule arrays, as studied by the injection of centrosomes and nuclei into *Xenopus* eggs," J. Cell Biol., *98*, 1730-1745.

Kashchenko, L. A., 1962, "Ovulation mechanism in *Rana temporaria*," Dokl. Akad. Nauk SSSR, *145*, 693-700.

Kasyanov, V. L., 1969, "Studies of oocytes and eggs in *Rana temporaria* (some morphological observations)," Author's Abstract of Candidate's Dissertation, Leningrad.

Katagiri, C., 1966, "The fertilizing capacity of frog sperm in homologous and heterologous egg-jellies and polyvinylpyrrolidone (PVP)," J. Fac. Sci. Hokkaido Univ., Ser. 6, Zool., *16*, 77-84.

Katagiri, C., 1980, "Appearance of pronucleus-inducing activity during hormonally induced meiotic maturation in toad oocytes," Dev., Growth, Differ., *22*, 237-246.

Katagiri, C., and Moriya, N., 1976, "Spermatozoon response to the toad egg matured after removal of germinal vesicle," Dev. Biol., *50*, 235-241.

Kazansky, B. N., 1950, "Influence of hypophysis on nuclear process in fish oocytes," Dokl. Akad. Nauk SSSR, *75*, 311-314.

Kazansky, B. N., 1952, "Completion of ovulation outside the female body in sturgeons," Dokl. Akad. Nauk SSSR, *83*, 965-968.

Kazansky, B. N., 1956, "Oogenesis and adaptations related to reproduction in fishes," Author's Abstract of Doctoral Dissertation, Leningrad.

Kazansky, B. N., 1957, "Analysis of phenomena occurring in the sturgeon oocytes upon application of hypophyseal injections," in: Proc. of Conference on Fish Breeding [in Russian], Izd. Akad. Nauk SSSR, Moscow, pp. 130-138.

Kemp, N. E., and Ystock, N. Y., 1967, "Cortical changes in growing oocytes and in fertilized or pricked eggs of *Rana pipiens*," J. Cell Biol., *34*, 111-122.

Kessel, R. G., 1968, "Annulate lamellae," J. Ultrastruct. Res., Suppl., *10*, 1-82.

Kessel, R. G., and Panje, W. R., 1968, "Organization and activity in the pre- and postovulatory follicle of *Necturus maculosus*," J. Cell Biol., *39*, 1-34.

Kishimoto, T., 1982, "Maturation-promoting factor in starfish oocytes," Cell Differ., *11*, 255-257.

Kishimoto, T., and Kanatani, H., 1976, "Cytoplasmic factor responsible for germinal vesicle breakdown and meiotic maturation in starfish oocyte," Nature, *260*, 321-322.

Kishimoto, T., and Kanatani, H., 1980, "Induction of oocyte maturation by disulfide-reducing agent in the sea cucumber *Stichopus japonicus*," Dev., Growth, Differ., *22*, 163-167.

Kishimoto, T., Hirai, S., and Kanatani, H., 1981, "Role of germinal vesicle material in producing maturation-promoting factor in starfish oocyte (brief notes)," Dev. Biol., *81*, 177-181.

Kishimoto, T., Clark, T. G., Kondo, H. K., Shirai, H., and Kanatani, H., 1982a, "Inhibition of starfish oocyte maturation by some inhibitors of proteolytic enzymes," Gamete Res., *5*, 11-18.

Kishimoto, T., Kuriyama, R., Kondo, H., and Kanatani, H., 1982b, "Generality of the action of various maturation-promoting factors," Exp. Cell Res., *137*, 121-126.

Larsen, J. H., Schroeder, P. C., and Waldo, A. E., 1977, "Structure and function of the amphibian follicular epithelium during ovulation," Cell Tissue Res., *181*, 505-518.

Laskey, R. A., 1979, "Biochemical processes in early development," in: Companion to Biochemistry, Vol. 2, 137-160.

Laskey, R. A., Mills, A. D., and Morris, N. R., 1977, "Assembly of SV40 chromatin in a cell-free system from *Xenopus* eggs," Cell, *10*, 237-243.

Lau, V. T., Reynhout, T. K., and Horowitz, S. B., 1984, "Regional water changes during oocyte meiotic maturation: evidence of ooplasmic segregation," Dev. Biol., *104*, 106-110.

Lemaire, W. J., Clark, M. K., Chainy, E. B. N., and March, J. M., 1980, "The role of prostaglandins in the mechanism of ovulation," in: Endocrinol. Physiopathol. Ovary, Proc. Int. Symp., Rosario, 1979, Amsterdam, 207-217.

Leonard, R. A., Hoffner, N. J., and DiBerardino, M. A., 1982, "Induction of DNA synthesis in amphibian erythroid nuclei in *Rana* eggs following conditioning in meiotic oocytes," Dev. Biol., *92*, 343, 355.

Lillie, F. R., 1902, "Differentiation without cleavage in the egg of the annelid *Chaetopterus pergamentaceus*," W. Roux's Arch. Entwicklungsmech. Org., *14*, 477-499.

Lin, H. R., 1982, "Polycultural system of freshwater fish in China," Can. J. Fish Aquat. Sci., *39*, 143-150.

Lohka, M. J., and Masui, Y., 1984a, "Effects of Ca^{2+} ions on the formation of metaphase chromosomes and sperm pronuclei in cell-free preparations from unactivated *Rana pipiens* eggs," Dev. Biol., *103*, 434-442.

Lohka, M. J., and Masui, Y., 1984b, "Roles of cytosol and cytoplasmic particles in nuclear envelope assembly and sperm pronuclear formation in cell-free preparations from amphibian eggs," J. Cell Biol., *98*, 1222-1230.

Longo, F. J., and Schuetz, A. W., 1982, "Male pronuclear development in starfish oocytes treated with 1-methyladenine," Biol. Bull., *163*, 453-464.

Longo, F. J., So, F, and Schuetz, A. W., 1982, "Meiotic maturation and the cortical granule reaction in starfish eggs," Biol. Bull., *163*, 465-476.

Lynch, W. E., Surrey, S., and Lieberman, J., 1975, "Nuclear deoxyribonucleic acid polymerases of liver," J. Biol. Chem., *250*, 8179-8183.

Maller, J. L., 1985a, "Regulation of amphibian oocyte maturation," Cell Differ., *16*, 211-221.

Maller, J. L., 1985b, "Oocyte maturation in amphibians," in: Developmental Biology, Vol. 1, Oogenesis, pp. 289-312.

Maller, J. L., and Krebs, E. G., 1980, "Regulation of oocyte maturation," Curr. Top. Cell. Regul., *16*, 271-311.

Maller, J. L., and Smith, D. S., 1985, "Two-dimensional polyacrylamide gel analysis of changes in protein phosphorylation during maturation of *Xenopus* oocytes," Dev. Biol., *109*, 150-156.

Maller, J., Poccia, D., Nishioka, D., Kidd, P., Gerhart, J., and Hartman, H., 1976, "Spindle formation and cleavage in *Xenopus* eggs injected with centriole-containing fractions from sperm," Exp. Cell Res., *99*, 285-294.

Maller, J., Wu, M., and Gerhart, J. C., 1977, "Changes in protein phosphorylation accompanying maturation of *Xenopus laevis* oocytes," Dev. Biol., *58*, 295-312.

Marot, J., Belle, R., and Ozon, R., 1977, "In vitro facilitation of *Xenopus* oocyte maturation by subthreshold dose of progesterone," Dev. Biol., *59*, 91-95.

Martin, K., Osheim, J. N., Beyer, A. L., and Miller, O. L., 1980, "Visualization of transcriptional activity during *Xenopus laevis* oogenesis," in: Differentiation and Neoplasma, Berlin, pp. 37-44.

Maruyama, Y., 1980, "Artificial induction of oocyte maturation and development in the sea cucumbers *Holothuria leucospilota* and *Holothuria pardalis*," Biol. Bull., *158*, 339-348.

Masui, Y., 1967, "Relative roles of the pituitary, follicle cells, and progesterone in the induction of oocyte maturation in *Rana pipiens*," J. Exp. Zool., *166*, 365-376.

Masui, Y., 1972, "Distribution of the cytoplasmic activity inducing germinal vesicle breakdown in frog oocytes," J. Exp. Zool., *179*, 365-378.

Masui, Y., 1973a, "Effects of ionizing radiation on meiotic maturation of frog oocytes. I. In vivo studies," J. Embryol. Exp. Morphol., *20*, 87-104.

Masui, Y., 1973b, "Effects of ionizing radiation on meiotic maturation of frog oocytes. II. In vitro studies," J. Embryol. Exp. Morphol., *29*, 105-116.

Masui, Y., 1974, "A cytostatic factor in amphibian oocytes: its extraction and partial characterization," J. Exp. Zool., *187*, 141-147.

Masui, Y., 1985a, "Meiotic arrest in animal oocytes," in: Biology of Fertilization, Vol. 1, Academic Press, New York, pp. 189-219.

Masui, Y., 1985b, "Problems of oocyte maturation and the control of chromosome cycles," Dev., Growth, Differ., *27*, 295-309.

Masui, Y., and Clarke, N. J., 1979, "Oocyte maturation," Int. Rev. Cytol., *57*, 185-282.

Masui, Y., and Markert, C. L., 1971, "Cytoplasmic control of nuclear behavior during meiotic maturation of frog oocytes," J. Exp. Zool., *177*, 129-146.

Masui, Y., Meyerhof, P. G., and Ziegler, D. M., 1979, "Control of chromosome behavior during progesterone-induced maturation of amphibian oocytes," J. Steroid Biochem., *11*, 715-722.

Masui, Y., Meyerhof, P. G., and Miller, M. A., 1980, "Cytostatic factor and chromosome behavior in early development," in: The Cell Surface: Mediator of Developmental Processes, S. Subtelny and N. K. Wessels (eds.), Academic Press, New York, pp. 234-256.

Mazzei, G., Meijer, L., Moreau, M., and Guerrier, P., 1981, "Role of calcium and cyclic nucleotides during meiosis reinitiation in starfish oocytes," Cell Differ., *10*, 139-145.

Meijer, L., and Guerrier, P., 1985, "Maturation and fertilization in starfish oocytes," Int. Rev. Cytol., *86*, 129-196.

Meijer, L., Pendaven, P., Guerrier, P., and Moreau, M. V., 1984, "A starfish oocyte user's guide," Cah. Biol. Mar., *25*, 157-180.

Merriam, R. W., and Hill, R. J., 1976, "The germinal vesicle nucleus of *Xenopus laevis* oocytes as a selective storage receptacle for proteins," J. Cell Biol., *69*, 659-668.

Meyerhof, P. G., and Masui, Y., 1977, "Ca and Mg control of cytostatic factors from *Rana pipiens* oocytes which cause metaphase and cleavage arrest," Dev. Biol., *61*, 214-229.

Meyerhof, P. G., and Masui, Y., 1979a, "Chromosome condensation activity in *Rana pipiens* eggs matured in vivo and in blastomeres arrested by cytostatic factor (CSF)," Exp. Cell Res., *123*, 345-353.

Meyerhof, P. G., and Masui, Y., 1979b, "Properties of a cytostatic factor from *Xenopus laevis* eggs," Dev. Biol., *72*, 182-187.

Miake-Lye, R., and Kirschner, M. W., 1985, "Induction of early mitotic events in cell-free system," Cell, *41*, 165-175.

Miake-Lye, R., Newport, J., and Kirschner, M., 1983, "Maturation-promoting factor induces nuclear envelope breakdown in cycloheximide-arrested embryos of *Xenopus laevis*," J. Cell Biol., *97*, 81-91.

Mikhailov, V. S., and Gulyamov, D. B., 1980, "Two enzymatic systems of DNA synthesis in nuclei of loach *Misgurnis fossilis*. Involvement of DNA polymerase α in the DNA replication in embryo nuclei, and of DNA polymerase β in DNA synthesis in liver nuclei," Dokl. Akad. Nauk SSSR, *257*, 229-233.

Mikhailov, V. S., and Gulyamov, D. B., 1983, "Changes in DNA polymerase α, β, γ activities during early development of the teleost fish *Misgurnis fossilis* (loach)," Eur. J. Biochem., *135*, 303-306.

Mikhailov, V. S., Kostomarova, A. A., Gulyamov, D. B., and Knyazeva, E. F., 1983a, "Compartmentalization of DNA polymerases α and β in maturing oocytes of the loach (*Misgurnus fossilis* L.)," Cell Differ., *13*, 87-91.

Mikhailov, V. S., Gulyamov, D. B., Peskin, A. V., and Filatova, L. S., 1983b, "DNA polymerase γ in teleost fish embryos is localized in mitochondria," Biochemistry, *48*, 1012-1019.

Mohun, T. J., Brennan, S., Dathan, N. V., Fairman, S., and Gurdon, J. B., 1984, "Cell type-specific activation of actin genes in the early amphibian embryo," Nature, *311*, No. 5988, 716-721.

Montalembert, G. de, Jalabert, B., and Bry, C., 1978, "Precocious

induction of maturation and ovulation in the northern pike (*Esox lucius*)," Ann. Biol. Anim., Biochem., Biophys., *18*, 969-975.
Moore, A. R., 1933, "Is cleavage rate a function of the cytoplasm or of the nucleus?" J. Exp. Biol., *10*, 230-236.
Moreau, N., and Boucher, D., 1981, "Une méthode rapide d'extraction en masse des noyaux d'ovocytes de Pleurodèle (Amphibien, Urodele)," Biol. Cell, *42*, 185-188.
Moreau, N., and Gounon, P., 1977, "Synthèse de la tubuline au cours de l'ovogenèse de *Pleurodeles waltlii* (Amphibien, Urodele)," Biol. Cell, *28*, 19-22.
Moreau, N., Guerrier, P., and Dorée, M., 1978, "Hormonal control of meiosis reinitiation in starfish oocytes. New evidence for the absence of efficient intercellular receptors for 1-methyladenine recognition," Exp. Cell Res., *115*, 245-249.
Morrill, G. A., 1965, "Water and electrolyte changes in amphibian eggs at ovulation," Exp. Cell Res., *40*, 664-667.
Morrill, G. A., and Bloch, E., 1977, "Structure-function relationships of various steroids relative to induction of nuclear breakdown and ovulation in isolated amphibian oocytes," J. Steroid Biochem., *8*, 133-139.
Morrill, G. A., and Watson, D. E., 1966, "Transmembrane electropotential changes in amphibian eggs at ovulation, activation, and first cleavage," J. Cell Physiol., *67*, 85-92.
Morrill, G. A., Rosenthal, J., and Watson, D. E., 1966, "Membrane permeability changes in amphibian eggs at ovulation," J. Cell Physiol., *67*, 375-382.
Morrill, G. A., Schatz, F., and Zabranetzky, V. S., 1975, "RNA and protein synthesis during progesterone-induced germinal vesicle breakdown in *Rana pipiens* ovarian tissue," Differentiation, 4, 143-152.
Morrill, G. A., Ziegler, D., and Kostellow, A. B., 1981, "The role of Ca^{2+} and cyclic nucleotides in progesterone initiation of the meiotic divisions in amphibian oocytes," Life Sci., *29*, 1821-1835.
Nadamitsu, S., 1953, "Ovulation in vitro in several species of *Amphibia*," J. Sci. Hiroshima Univ., Ser. B, Div. 1, *14*, 1-7.
Nagano, H., Okano, K., Ikegami, S., and Katagiri, C., 1982, "Changes in intracellular location of DNA polymerase – during oocyte maturation of the toad *Bufo bufo japonicus*," Biochem. Biophys. Res. Commun., *106*, 683-690.
Nakamura, S., and Hiramoto, Y., 1978, "Mechanical properties of the cell surface in starfish eggs," Dev., Growth, Differ., *20*, 317-327.
Naora, H., Naora, H., Izawa, M., Alifrey, V. G., and Miskey, A. E., 1962, "Some observations on differences in composition between the nucleus and cytoplasm of the frog oocyte," Proc. Natl. Acad. Sci. USA, *48*, 853-859.
Negonovskaya, I. T., 1960, "Cytomorphological data on egg maturation process of the trout Gegarkuni from the Sevan lake," Tsitologiya, *2*, 536-546.

Nelkin, B., Nichols, C., and Vogelstein, B., 1980, "Protein factor(s) from mitotic CHO cells induce meiotic maturation in *Xenopus laevis* oocytes," Fed. Eur. Biochem. Soc. Lett., *109*, 233-238.

Newport, J., and Kirschner, M., 1982a, "A major developmental transition in early *Xenopus* embryos: I. Characterization and timing of cellular changes at the midblastula stage," Cell, *30*, 675-686.

Newport, J., and Kirschner, M., 1982b, "A major developmental transition in early *Xenopus* embryos: II. Control of the onset of transcription," Cell, *30*, 687-696.

Newport, J., and Kirschner, M., 1984, "Regulation of the cell cycle during early *Xenopus* development," Cell, *37*, 731-742.

Nielson, P. J., Thomas, G., and Maller, J. L., 1982, "Increased phosphorylation of *Xenopus* oocytes," Proc. Natl. Acad. Sci. USA, *79*, 2937-2941.

Nieuwkoop, P. D., and Faber, J., 1956, Normal Table of *Xenopus laevis* (Daudin), North-Holland, Amsterdam.

Nikitina, L. A., 1972, "Transplantation of germinal vesicle material in enucleated oocytes of *Acipenser stellatus*," Dokl. Akad. Nauk SSSR, *205*, 1487-1490.

Nikitina, L. A., 1974, "Transplantation of nuclei from growing oocytes into fully grown enucleated sturgeon oocytes," Ontogenez, *5*, 251-255.

Nikitina, L. A., 1978, "Presence of a cytotomy factor in the karyoplasm of *Acipenser* oocytes at the onset of the period of rapid growth," W. Roux's Arch. Dev. Biol., *184*, 109-113.

Nikitina, L. A., 1982, "A study of properties of nuclei of growing oocytes by means of serial cloning," Ontogenez, *13*, 40-45.

Nikitina, L. A., 1983, "Changes in the nuclei of the stellate sturgeon and axolotl oocytes produced by transplantation into the cytoplasm of mature activated eggs," Ontogenez, *14*, 261-266.

Nikitina, L. A., 1984, "The behavior of the nuclei from *Xenopus laevis* embryonic cells and vitellogenic oocytes in the cytoplasm of the axolotl maturing oocytes," Ontogenez, *15*, 535-538.

Nikitina, L. A., and Dettlaff, T. A., 1986, "Appearance of a maturation-stimulating factor in the karyoplasm of the *Bufo viridis*, *Xenopus laevis*, and *Acipenser stellatus* oocytes," Ontogenez, *17*, 252-256.

Osanai, K., Sato, R., and Hirai, S., 1972, "On the maturation of the oocytes in the chum salmon homing to the parent river," Bull. Jpn. Soc. Sci. Fisheries, *38*, 1251-1257.

Oshiro, T., and Hibiya, T., 1981, "Water absorption of oocytes in the plaice *Yomanda yokohamae* during meiotic maturation and its role in rupturing follicles," Bull. Jpn. Soc. Sci. Fisheries, *47*, 835-841.

Otto, J. J., and Schroeder, T. E., 1984a, "Assembly-disassembly of actin bundles in starfish oocytes: An analysis of actin-associated proteins in the isolated cortex," Dev. Biol., *101*, 263-273.

Otto, J. J., and Schroeder, T. E., 1984b, "Microtubule arrays in the cortex and near the germinal vesicle of immature starfish oocytes," Dev. Biol., *101*, 274-281.

Pendergrass, P. B., 1974, "Electron and light microscope study of the mechanism of oocyte extrusion during ovulation in the fish *Oryzias latipes*," Doctoral Dissertation, Washington State Univ., 1-66.

Pendergrass, P. B., and Schroeder, P. C., 1976, "The ultrastructure of the thecal cell of the teleost *Oryzias latipes* during ovulation in vitro," J. Reprod. Fertil., *47*, 229-233.

Pennequin, P., Schorderet-Slatkine, S., Drury, K. S., and Baulieu, E. E., 1975, "Decreased uptake of ^{3}H-leucine during progesterone-induced maturation of *Xenopus laevis* oocytes," Fed. Eur. Biochem. Soc. Lett., *51*, 156-160.

Pestell, R. O. W., 1975, "Microtubule protein synthesis during oogenesis and early embryogenesis in *Xenopus laevis*," Biochem. J., *145*, 527-534.

Picard, A., and Dorée, M., 1984, "The role of the germinal vesicle in producing maturation-promoting factor (MPF) as revealed by the removal and transplantation of nuclear material in starfish oocytes," Dev. Biol., *104*, 357-365.

Raff, E. C., and Raff, R. A., 1978, "Tubulin and microtubules in the early development of the axolotl and other *Amphibia*," Am. Zool., *18*, 237-254.

Reynhout, J. K., and Smith, L. D., 1973, "Evidence for steroid metabolism during the in vitro induction of maturation in oocytes of *Rana pipiens*," Dev. Biol., *30*, 392-402.

Reynhout, J. K., and Smith, L. D., 1974, "Studies on the appearance and nature of a maturation-inducing factor in the cytoplasm of amphibian oocytes exposed to progesterone," Dev. Biol., *38*, 394-400.

Reynhout, J. K., Smith, L. D., Taddei, C., and La Marca, M. J., 1975, "Response of large oocytes on *Xenopus laevis* to progesterone in vitro in relation to oocyte size and time after previous HCG-induced ovulation," Dev. Biol., *44*, 375-379.

Richter, T. P., Wasserman, W. F., and Smith, L. D., 1982, "The mechanism for increased protein synthesis during *Xenopus* oocyte maturation," Dev. Biol., *89*, 159-167.

Roeder, R. G., 1974, "Multiple forms of deoxyribonucleic acid-dependent ribonucleic acid polymerase in *Xenopus laevis*. Levels of activity during oocyte and embryonic development," J. Biol. Chem., *249*, 249-256.

Rott, N. N., 1980, "Cell divisions during pregastrulation period of development," Ontogenez, *11*, 3-23.

Rudneva, T. B., 1968, "The dynamics of oocyte maturation in *Rana temporaria*," Graduate Thesis, Moscow Univ.

Rugh, R., 1935, "Ovulation in the frog. II. Follicular rupture to fertilization," J. Exp. Zool., *71*, 163-193.

Ryabova, L. V., 1982, "An electron microscopic study of the development of cortex contractility in *Rana temporaria* oocytes," Ontogenez, *13*, 266-273.

Ryabova, L. V., 1983, "The cytostatic effect of the cytoplasm of mature nonactivated and cleaving eggs of *Rana temporaria*, *Acipenser stellatus*, and *Xenopus laevis*," Cell Differ., *13*, 171-175.

Ryabova, L. V., 1984, "Does Masiu's Ca-sensitive cytostatic factor exist in the cytoplasm of the mature nonactivated eggs of *Acipenser stellatus*, *Rana temporaria*, and *Xenopus laevis*? Ontogenez, *15*, 93-96.

Ryabova, L. V., and Dettlaff, T. A., 1981, "A study of activity of the factor of disintegration of germinal vesicle membrane at different phases of the two first cleavage divisions in the egg cytoplasm of *Acipenser stellatus*, *Rana temporaria*, and *Xenopus laevis*," Ontogenez, *12*, 517-520.

Ryabova, L. V., and Dettlaff, T. A., 1984, "The activity of the factor of germinal vesicle breakdown (GVBD) in the cytoplasm of embryos at different phases of cell cycle and at different stages of cleavage in some fishes and amphibians," Ontogenez, *15*, 438.

Ryabova, L. V., Betina, M. I., and Vassetzky, S. G., 1984, "Participation of microfilaments with maturation of *Xenopus laevis* oocytes," Dokl. Akad. Nauk SSSR, *277*, 704-706.

Ryabova, L. V., Betina, M. I., and Vassetzky, S. G., 1986, "Influence of cytochalasin B in oocyte maturation in *Xenopus laevis*," Cell Differ., *19*, 89-96.

Ryan, J. F., and Grant, R., 1940, "The stimulus for maturation and for ovulation of the frog's egg," Physiol. Zool., *13*, 383-389.

Saat, T. F., 1980, "Oocyte maturation and ovulation in the loach in different media and under different hormonal influences," Ontogenez, *11*, 545-554.

Saat, T. V., 1982, "The chronology of oocyte maturation in the loach and the acquisition of developmental ability," Ontogenez, *13*, 257-265.

Saat, T. V., 1985, "Influence of unfavorable factors on in vitro oocyte maturation and ovulation in *Carassius auratus* Gibelio," Ontogenez, *16*, 483-491.

Sakun, O. F., 1955, "Maturation of oocytes of *Vimba vimba* L.," Dokl. Akad. Nauk SSSR, *102*, 181-183.

Sakun, O. F., 1961, "The state of the nuclear apparatus of oocytes in *Teleostei* at the period preceding meiotic divisions," Dokl. Akad. Nauk SSSR, *137*, 749-751.

Sakun, O. F., 1966, "Transition to meiotic division of oocytes with incomplete vitellogeny under the influence of hormonal stimulation in *Salmo salar* L," Dokl. Akad. Nauk SSSR, *169*, 241-244.

Sakun, O. F., and Gureyeva-Preobrazhenskaya, E. V., 1975, "A study of in vitro maturation of teleostean oocytes. Technique of incubation and methods of analysis of the results," Vestn. Leningr. Univ., *15*, 15-22.

Samson, D. A., and Schuetz, A. W., 1979, "Progesterone induction of oocyte maturation in *Rana pipiens* and reversibility of cycloheximide inhibition," J. Exp. Zool., *208*, 213-220.

Sawada, N., 1952, "Effects of some cations on the breakdown of the germinal vesicle in the egg of clam *Mactra veneriformis*," Bull. Exp. Biol., *2*, 9-13.
Scalenghe, F., Buscaglia, M., Steinheil, Ch., and Crippa, M., 1978, "Large-scale isolation of nuclei and nucleoli from vitellogenic oocytes of *Xenopus laevis*," Chromosoma, *66*, 299-308.
Schlichter, L. C., 1983, "Spontaneous action potentials produced by Na and Ca channels in maturing *Rana pipiens* oocytes," Dev. Biol., *98*, 47-59.
Schorderet-Slatkine, S., 1972, "Action of progesterone and related steroids on oocyte maturation in *Xenopus laevis*. An in vitro study," Cell Differ., *1*, 179-189.
Schorderet-Slatkine, S., and Drury, K. C., 1973, "Progesterone-induced maturation in oocytes of *Xenopus laevis*. Appearance of a 'maturation-promoting factor' in enuclueated oocytes," Cell Differ., *2*, 247-254.
Schorderet-Slatkine, S., Schorderet, M., Godeau, M., Boquet, P., and Baulieu, M., 1980, "Role of cyclic AMP in process of meiosis reinitiation in *Xenopus laevis* oocytes: effects of progesterone, insulin, and cholera toxin," in: Steroids and Their Mechanism of Action in Nonmammalian Vertebrates, C. Delrio and J. Brachet (eds.), Raven Press, New York, pp. 93-101.
Schorderet-Slatkine, S., Schorderet, M., and Baulieu, E., 1982, "Cyclic AMP-mediated control of meiosis. Effects of progesterone, cholera toxin, and membrane-active drugs in *Xenopus laevis* oocytes," Proc. Natl. Acad. Sci. USA, *79*, 850-854.
Schroeder, P. C., 1973, "The inhibition of amphibian ovulation in vitro by cytochalasin B," Experientia, *29*, 617-618.
Schroeder, P. C., and Pendergrass, P. B., 1976, "The inhibition of in vitro ovulation from follicles of the teleost *Oryzias latipes* by cytochalasin B," J. Reprod. Fertil., *48*, 327-330.
Schroeder, T. E., 1985, "Cortical expressions of polarity in the starfish oocyte," Dev., Growth, Differ., *27*, 311-321.
Schroeder, T. E. (ed.), 1986, "Echinoderm gametes and embryos," in: Methods in Cell Biology, Vol. 27.
Schroeder, T. E., and Otto, J. J., 1984, "Cyclic assembly-disassembly of cortical microtubules during maturation and early development of starfish oocytes," Dev. Biol., *103*, 493-503.
Schroeder, T. E., and Stricker, S. A., 1983, "Morphological changes during maturation of starfish oocytes: surface ultrastructure and cortical actin," Dev. Biol., *98*, 373-384.
Schuetz, A. W., 1967a, "Mechanism of progesterone- and pituitary-induced germinal vesicle breakdown in oocytes of *Rana pipiens*," J. Cell Biol., *35*, 123A.
Schuetz, A. W., 1967b, "Effect of steroids on the germinal vesicle of oocytes of the frog (*Rana pipiens*) in vitro," Proc. Soc. Exp. Biol. Med., *124*, 1307-1310.
Schuetz, A. W., 1967c, "Action of hormones on germinal vesicle breakdown in frog (*Rana pipiens*) oocytes," J. Exp. Zool., *166*, 347-354.

Schuetz, A. W., 1969, "Chemical properties and physiological actions of a starfish radial nerve factor and ovarian factor," Gen. Comp. Endocrinol., *12*, 209-221.

Schuetz, A. W., 1972a, "Hormones and follicular functions," in: Oogenesis, J. D. Biggers and A. W. Schuetz (eds.), Univ. Park Press, Baltimore; Butterworths, London, pp. 479-512.

Schuetz, A. W., 1972b, "Induction of structural alterations in the preovulatory amphibian ovarian follicle by hormones," Biol. Reprod., *6*, 67-77.

Schuetz, A. W., 1974, "Role of hormones in oocyte maturation," Biol. Reprod., *10*, 150-178.

Schuetz, A. W., 1975, "Induction of nuclear breakdown and meiosis in *Spisula solidissima* oocytes by calcium ionophore," J. Exp. Zool., *191*, 433-440.

Schuetz, A. W., 1978, "The cell surface and its relationship to control of meiosis reinitiation," Biol. Cell., *32*, 89-95.

Schuetz, A. W., 1985, "Local control mechanisms during oogenesis and folliculogenesis," in: Developmental Biology. A Comprehensive Synthesis, Vol. I, L. W. Browder (ed.), Plenum Press, New York—London, pp. 3-83.

Schuetz, A. W., and Biggers, J. D., 1967, "Regulation of germinal vesicle breakdown in starfish oocytes," Exp. Cell Res., *46*, 624-628.

Schuetz, A. W., and Samson, D., 1979, "Protein synthesis requirement for maturation-promoting factor (MPF) initiation of meiotic maturation in *Rana* oocytes," Dev. Biol., *68*, 636-642.

Sekirina, G. G., Skoblina, M. N., Vassetzky, S. G., and Bilinkis, A. A., 1983, "The fusion of oocytes of the starfish *Aphelasterias japonica*. I. Formation of cell hybrids," Cell Differ., *12*, 67-71.

Selman, G. G., and Perry, M. M., 1970, "Ultrastructural changes in the surface layers of the newt's eggs in relation to the mechanism of its cleavage," J. Cell Sci., *6*, 207-227.

Shibuya, E. K., and Masui, Y., 1982, "Sperm-inducing cell cycle activities in blastomeres arrested by the cytostatic factor of unfertilized eggs in *Rana pipiens*," J. Exp. Zool., *220*, 381-385.

Shioda, N., Nagano, H., Mano, Y., 1977, "Cytoplasmic location of DNA polymerase-α and β of sea urchin eggs," Biochem. Biophys. Res. Commun., *78*, 1362-1368.

Shirai, H., 1974, "Effect of ^{3}H-phenylalanine on 1-methyadenine production and spontaneous oocyte maturation in starfish," Exp. Cell Res., *87*, 31-38.

Skoblina, M. N., 1968, "Maturation of the cortex of nonnucleate oocytes of frog and sevryuga induced by gonadotropic pituitary hormones," Dokl. Akad. Nauk SSSR, *183*, 643-645.

Skoblina, M. N., 1969, "Independence of the cortex maturation from germinal vesicle material during the maturation of amphibian and sturgeon oocytes," Exp. Cell Res., *55*, 142-144.

Skoblina, M. N., 1970, "An experimental study of the role of nucleus in the process of oocyte maturation in amphibians and sturgeons," Author's Abstract of Candidate's Dissertation, Moscow.

Skoblina, M. N., 1973, "The maturation of loach oocytes under the in vitro influence of chorionic gonadotropin," Ontogenez, 4, 284-288.

Skoblina, M. N., 1974, "Behavior of sperm nuclei in intact toad oocytes undergoing maturation and in oocytes matured after germinal vesicle removal," Ontogenez, *5*, 334-340.

Skoblina, M. N., 1976, "Role of karyoplasm in the emergence of capacity of egg cytoplasm to induce DNA synthesis in transplanted sperm nuclei," J. Embryol. Exp. Morphol., *36*, 67-72.

Skoblina, M. N., Kondratieva, O. T., and Nikitina, L. A., 1984a, "The capacity for maturation and cleavage of small-size oocytes of *Pleurodeles waltlii*," Ontogenez, *15*, 362-367.

Skoblina, M. N., Pivnitsky, K. K., and Kondratieva, O. T., 1984b, "The role of germinal vesicle in maturation of *Pleurodeles waltii* oocytes induced by steroids," Cell Differ., *14*, 153-157.

Smith, L. D., 1966, "The role of a 'germinal plasm' in the formation of primordial germ cells in *Rana pipiens*," Dev. Biol., *14*, 330-347.

Smith, L. D., 1972, "Protein synthesis during oocyte maturation," in: Oogenesis, J. D. Biggers and A. W. Schuetz (eds.), Univ. Park Press, Baltimore; Butterworths, London, pp. 227-240.

Smith, L. D., and Ecker, R. E., 1968, "Cytoplasmic regulation in early events of amphibian development," in: Proc. 8th Can. Cancer Conf., Honey Harbour, pp. 103-129.

Smith, L. D., and Ecker, R. E., 1969a, "Role of oocyte nucleus in physiological maturation in *Rana pipiens*,: Dev. Biol., *19*, 281-309.

Smith, L. D., and Ecker, R. E., 1969b, "Foundation for the expression of developmental potential," in: Problems in Biology: RNA in Development, E. W. Hanly (ed.), Univ. Utah Press, Salt Lake City, pp. 355-379.

Smith, L. D., and Ecker, R. E., 1970a, "Uterine suppression of biochemical and morphogenetic events in *Rana pipiens*," Dev. Biol., *22*, 622-637.

Smith, L. D., and Ecker, R. E., 1970b, "Regulatory processes in the maturation and early cleavage of amphibian eggs," in: Current Topics in Developmental Biology, Vol. 5, Academic Press, New York, pp. 1-38.

Smith, L. D., and Ecker, R. E., 1971, "The interaction of steroids with *Rana pipiens* oocytes in the induction of maturation," Dev. Biol., *25*, 232-247.

Smith, L. D., Ecker, R. E., and Subtelny, S., 1966, "The initiation of protein synthesis in eggs of *Rana pipiens*," Proc. Natl. Acad. Sci. USA, *56*, 1724.

Smith, L. D., Ecker, R. E., and Subtelny, S., 1968, "In vitro induction of physiological maturation in *Rana pipiens* oocytes removed from their ovarian follicles," Dev. Biol., *17*, 627-643.

Stacey, N. E., and Pandey, S., 1975, "Effects of indomethacin and prostaglandins on ovulation of goldfish," Prostaglandins, *9*, 597-608.

Stacey, N. E., and Peter, P. E., 1979, "Central action of prostaglandins in spawning behavior of female goldfish," Physiol. Behav., *22*, 1191-1196.

Steinert, G., Baltus, E., Hanocq-Quertier, J., and Brachet, J., 1974, "Ultrastructure of *Xenopus laevis* oocytes after injection of an extract from progesterone-treated oocytes," J. Ultrastruct. Res., *49*, 188-210.

Stick, R., and Hausen, P., 1985, "Changes in the nuclear lamina composition during early development of *Xenopus laevis*," Cell, *41*, 191-200.

Subtelny, S., and Bradt, G., 1959, "Transplantations of blastula nuclei into activated jellyless body cavity eggs in *Rana pipiens*," Anat. Rec., *134*, 644.

Subtelny, S., Smith, L. D., and Ecker, R. E., 1968, "Maturation of ovarian frog eggs without ovulation," J. Exp. Zool., *168*, 39-48.

Sunkara, P. S., Wright, P. A., and Rao, P. N., 1979, "Mitotic factors from mammalian cells induce germinal vesicle breakdown and chromosome condensation in amphibian oocytes," Proc. Natl. Acad. Sci. USA, *76*, 2799-2802.

Sunkara, P. S., Wright, P. A., and Nishioka, K., 1981, "An essential role for putrescine biosynthesis during meiotic maturation of amphibian oocytes," Dev. Biol., *87*, 351-355.

Swenson, K. J., Farrell, K. M., and Ruderman, T. W., 1986, "The clam embryo protein cyclin A induces entry in M phase and the resumption of meiosis in *Xenopus* oocytes," Cell, *47*, 861-870.

Tchou-Su, 1948, "Étude cytologique et expérimentale sur la maturation artificielle osmotique in vitro sur l'oeuf d'Anoures," Acta Zool. Taiwanica, *1*, 1-66.

Tchou-Su and Wang Yu-lan, 1956, "Étude éxperimentale sur la role du mucus des oviductes dans la fécondation chez le Crapaud, et la consideration générale sur le méchanisme de la pénétration spermatique," Acta Biol. Exp. Sin., *5*, 75-122.

Tchou-Su, and Wang Yu-lan, 1958, "Études comparatives sur l'ovulation de la maturation in vivo et in vitro chez le Crapaud asiatique (*Bufo bufo asiaticus*)," Acta Biol. Exp. Sin., *6*, 129-180.

Tchou-Su, and Wang Yu-lan, 1959, "La parthénogénèse traumatique sur l'oeuf maturé et le développement parthénogénétique," Acta Biol. Exp. Sin., *6*, 276-297.

Tchou-Su, and Wang Yu-lan, 1962, "Cytological study of egg maturation on two species of bony fishes: *Carassius auratus* L. and *Megalobrama terminalis*," Acta Biol. Exp. Sin., *8*, 1-133.

Tchou-Su, and Wang Yu-lan, 1963a, "L'hibernation, facteur déterminant de la maturation ovulaire chez le Crapaud (*Bufo bufo asiaticus*)," Scientia Sin., *12*, 1161-1164.

Tchou-Su, and Wang Yu-lan, 1963b, "La succession d'ovogénèse et l'impossibilité de maturation ovulaire chez le Crapaud fem-

elle, elevée dans le milieu à haute temperature pendant toute une année," Scientia Sin., *12*, 1165-1168.
Tchou-Su, and Wang Yu-lan, 1964, "The development of eggs of *Bufo bufo asiaticus* inseminated at different stages of maturity," Acta Biol. Exp. Sin., *9*, 101-118.
Tchou-Su, and Yen Pai-hu, 1950, "Rélation de l'augmentation d'eau dans l'oeuf et la maturation," Scientia Sin., *1*, 165-183.
Trubnikova, O. B., and Ryabova, L. V., 1988, "Preovulatory changes in the ovarian follicle of *Acipenser stellatus* Pall," Ontogenez, *19*, (in press).
Tso, I., Thibier, C., Mulner, O., and Ozon, R., 1982, "Microinjected progesterone reinitiates meiotic maturation of *Xenopus laevis* oocytes," Proc. Natl. Acad. Sci. USA, *79*, 5552-5556.
Tso Jia-ke, Zhu Guang, Hsu Kuo-giang, and Wang Yu-lan, 1984, "Experiments concerning the maturation resumption in *Bufo bufo gargarizans* oocyte: progesterone acts directly on the cytoplasm," Acta Biol. Exp. Sin., *17*, 103-113.
Vacquier, V. D., 1981, "Dynamic changes of the egg cortex," Dev. Biol., *84*, 1-26.
Van Assel, S., and Brachet, J., 1968, "Métabolisme des acides nucléiques et des proteines et formation de cytasters dans les oeufs d'amphibiens sous l'action de l'eau lourde," J. Embryol. Exp. Morphol., *19*, 261-272.
Van Gansen, P., 1966, "Ultrastructure comparée du cytoplasme périphérique des ovocytes mûrs et des oeufs vierges de *Xenopus laevis* (Batracien anoure)," J. Embryol. Exp. Morphol., *15*, 355-364.
Van Gansen, P., and Shram, A., 1968, "Ultrastructure et cytochimie ultrastructurale de la vésicule germinative et du cytoplasme périnucléaire de la vesicule germinative et du cytoplasme périnucléaire de l'oocyte mûr de *Xenopus laevis*," J. Embryol. Exp. Morphol., *20*, 375-389.
Vassetzky, S. G., 1966, "The thermosensitivity of sturgeon eggs at the stages of their maturation and cleavage," Zh. Obshch. Biol., *27*, 583-595.
Vassetzky, S. G., 1967, "Studies of oocyte maturation and early embryogenesis of Acipenserid fishes with respect to possible application of the method of thermal sex control in fishes," Author's Abstract of Candidate's Dissertation, Moscow.
Vassetzky, S. G., 1970, "Dynamics of the first maturation division in sturgeon oocytes," Zh. Obshch. Biol., *31*, 84-93.
Vassetzky, S. G., 1988, "Meiotic divisions," in: Oocyte Growth and Maturation, Plenum Press, New York, Chapter 4.
Vassetzky, S. G., Sekirina, G. G., Skoblina, M. N., and Bilinkis, A. A., 1983, "The fusion of oocytes of the starfish *Aphelasterias japonica*. II. The capacity of cell hybrids for maturation and cleavage," Cell Differ., *12*, 73-76.
Vassetzky, S. G., Sekirina, G. G., Veisman, B. L., Skoblina, M. N., and Bilinkis, A. A., 1984, "The fusion of oocytes of the

starfish *Aphelasterias japonica*. III. Reconstruction of oocytes from cell and cell fragments (cytoplasts)," Cell Differ., *14*, 47-52.

Vassetzky, S. G., Skoblina, M. N., and Sekirina, G. G., 1986, "Induced fusion of echinoderm oocytes and eggs," in: Methods in Cell Biology, Vol. 27, pp. 359-378.

von der Harr, B., Sperling, K., and Gregor, D., 1981, "Maturing *Xenopus* oocytes induce chromosome condensation in somatic plant nuclei," Exp. Cell Res., *134*, 477-481.

Wang Yu-lan, Zhu Guang, Chai Shi-ping, Tso Jia-ke, and Zhao Jian-xing, 1982a, "Experimental analysis of egg maturation induced by pituitary extracts," Acta Biol. Exp. Sin., *15*, 68-74.

Wang Yu-lan, Hsu Kuo-giang, and Zhu Guang, 1982b, "Morphological changes in toad oocyte envelopes during maturation, and their relationship with ovulation," Acta Biol. Exp. Sin., *15*, 265-283.

Wartenberg, H., and Schmidt, W., 1961, "Elektronenmikroskopische untersuchungen der strukturellen veranderungen in bindenbereich des amphibieneies im ovar und nach befruchtung," Z. Zellforsch., *54*, 118-146.

Wasserman, W. J., 1982, "The role of the maturation-promoting factor in controlling protein synthesis in *Xenopus* oocytes," Dev. Biol., *90*, 445-447.

Wasserman, W. J., and Houle, J. G., 1984, "The regulation of ribosomal protein S6 phosphorylation in *Xenopus* oocytes: a potential role for intracellular pH," Dev. Biol., *101*, 436-445.

Wasserman, W. J., and Masui, Y., 1974, "A study of gonadotropin action in the induction of oocyte maturation in *Xenopus laevis*," Biol. Reprod., *11*, 133-144.

Wasserman, W. J., and Masui, Y., 1975a, "Effects of cycloheximide on a cytoplasmic factor initiating meiotic maturation in *Xenopus laevis* oocytes," Exp. Cell Res., *91*, 381-388.

Wasserman, W. J., and Masui, Y., 1975b, "Initiation of meiotic maturation in *Xenopus laevis* oocytes by the combination of divalent cations and ionophore A23187(1)," J. Exp. Zool., *193*, 369-375.

Wasserman, W. J., and Masui, Y., 1976, "A cytoplasmic factor promoting oocyte maturation: its extraction and preliminary characterization," Science, *191*, 1266-1268.

Wasserman, W. J., and Smith, L. D., 1978a, "Oocyte maturation: nonmammalian vertebrates," in: The Vertebrate Ovary, R. E. Jones (ed.), Plenum Press, New York, pp. 443-468.

Wasserman, W. J., and Smith, L. D., 1978b, "The cyclic behavior of a cytoplasmic factor controlling nuclear membrane breakdown," J. Cell Biol., *78*, 15-22.

Wasserman, W. J., and Smith, L. D., 1981, "Calmodulin triggers the resumption of meiosis in Amphibian oocytes," J. Cell Biol., *89*, 389-394.

Wasserman, P. M., Hollinger, T. G., and Smith, L. D., 1972, "RNA polymerase in the germinal vesicle contents of *Rana pipiens* oocytes," Nature, *240*, 208-210.

Wasserman, W. J., Richter, T. P., and Smith, L. D., 1982, "Protein synthesis during maturation-promoting factor- and progesterone-induced maturation in *Xenopus* oocytes," Dev. Biol., *89*, 152-158.

Weissbach, A., 1979, "The functional roles of mammalian DNA polymerase," Arch. Biochem., Biophys., *198*, 286-396.

Williams, M. A., Trendelenburg, M. F., and Franke, W. W., 1984, "Patterns of transcriptional activity of nuclear genes during progesterone-induced maturation of oocytes of *Xenopus laevis*," Differentiation, *20*, 35-44.

Wilson, E., 1903, "Experiments on cleavage and localization in the nemertine egg," W. Roux's Arch. Entwicklungsmech. Org., *16*, 411-461.

Wilson, E. B., 1925, The Cell in Development and Heredity, Macmillan, New York.

Wischnitzer, S., 1958, "An electron microscope study of the nuclear envelope of the amphibian oocyte," J. Ultrastruct. Res., *1*, 201-202.

Wischnitzer, S., 1966, "The ultrastructure of the cytoplasm of the developing amphibian egg," Adv. Morphogen., *5*, 131-179.

Woodland, H. R., and Adamson, E. D., 1977, "The synthesis and storage of histones during the oogenesis in *Xenopus* oocytes," Dev. Biol., *57*, 118-135.

Wright, P. A., 1945, "Factors affecting in vitro ovulation in the frog," J. Exp. Zool., *100*, 565-575.

Wright, P. A., 1961a, "Induction of ovulation in vitro in *Rana pipiens* with steroids," Gen. Comp. Endocrinol., *1*, 20-23.

Wright, P. A., 1961b, "Influence of estrogens on induction in vitro in *Rana pipiens*," Gen. Comp. Endocrinol., *1*, 381-385.

Wu, M., and Gerhart, J., 1980, "Partial purification and characterization of the maturation-promoting factor from eggs of *Xenopus laevis*," Dev. Biol., *79*, 465-477.

Wylie, A. H., Gurdon, J. B., and Price, J., 1977, "Nuclear localization of an oocyte component required for the stability of injected DNA," Nature, *268*, 150-152.

Yamamoto, K., 1985, "Germinal vesicle contents are required for the cytoplasmic cycle during meiotic division of starfish oocytes," Dev. Biol., *107*, 213-219.

Yamauchi, K., and Yamamoto, K., 1973, "In vitro maturation of the oocytes in the medaka *Oryzias latipes*," Annot. Zool. Jpn., *46*, 144-153.

Yatsu, H., 1904, "Aster formation in enucleated egg fragments of *Cerebratulus*," Science, *20*, 889-890.

Yatsu, H., 1905, "The formation of centrosomes in enucleated egg fragments," J. Exp. Zool., *2*, 287-312.

Yatsu, H., 1907, "An experimental study on the cleavage of the stenophore egg," Proc. 7th Int. Zool. Conf., pp. 1-4.

Yatvin, M. B., and Anderson, J. W., 1967, "Control of ovulation in frog ovaries in vitro," Fed. Proc., Fed. Am. Soc. Exp. Biol., *26*, 535.

Ziegler, D., and Masui, Y., 1973, "Control of chromosome behavior in amphibian oocytes," Dev. Biol., *35*, 283-292.
Ziegler, D., and Masui, Y., 1976, "Control of chromosome behavior in amphibian oocytes. II. The effect of inhibitors of RNA and protein synthesis," J. Cell Biol., *68*, 620-628.
Zotin, A. Y., 1961, Physiology of Water Exchange in the Embryos of Fishes and Cyclostomata [in Russian], Izd. Akad. Nauk SSSR, Moscow.

Chapter 4

MEIOTIC DIVISIONS

S. G. Vassetzky

Most studies of meiosis devote more attention to the meiotic prophase than to the subsequent phases of both meiotic divisions ("maturation divisions") which follow; this is justified, considering the importance of the morphological events and biochemical processes taking place during prophase (see, for example, John and Lewis, 1965; Khvostova and Bogdanov, 1975; Bogdanov and Vassetsky, 1982). Nevertheless, both meiotic divisions, which proceed quite rapidly compared with the meiotic prophase, warrant special consideration, since it is during this period that the meiotic spindle forms and migrates, nuclear transformations take place, morphological organization of gonocytes arises, and their genetic constitution is determined.

The special interest in meiotic divisions is due to the fact that, on the one hand, defects of meiosis arising both spontaneously and in response to a number of external factors lead to genetic diseases and, on the other hand, experimental influences on the germ cells in the process of meiosis make it possible to control development and, in particular, the change in the ploidy of the organism, and pave the way to artificial sex control.

The specificity of meiosis is determined by two basic processes: 1) reduction of the chromosome number by half (this process will be considered in detail below), and 2) recombination of the genetic material. The latter is connected with disjunction of allelic genes to different gametes, independent recombination of the chromosomes from different pairs, and exchange of parts of chromosomes in the crossing-over process.

In oogenesis meiosis proceeds in a similar fashion to spermatogenesis, but there are essential differences between the two processes. The main difference is that of the fate of the cells arising as a result of meiosis. In spermatogenesis four equivalent haploid cells differentiate into spermatozoa; in oogenesis one haploid egg cell and three haploid polar bodies form. Moreover, in oogenesis the prophase of meiosis is characterized by a number of features which are considered in detail by Gruzova in Chapter 2. Lastly, the oogenesis of different animals is characterized by reversible arrests of meiosis (at the end of meiotic prophase and during one of the meiotic divisions) which are completed only upon fertilization or activation of the egg.

4.1. MEIOSIS AND FERTILIZATION

One of the peculiarities of female meiosis among different animals is its arrest at different phases of the meiotic divisions (see Korschelt and Heider, 1903; Lillie, 1923; Wilson, 1925; Dalcq, 1952; Rothschild, 1956; Costello and Henley, 1971; Austin and Walton, 1960; Austin, 1965; John and Lewis, 1965; Ginsburg, 1968; D'Amato, 1977; Masui, 1985).

In accordance with the stage at which the meiosis is arrested, all animals may be divided into four groups (Fig. 36).

I. Meiosis is arrested before germinal vesicle breakdown, i.e., at the diakinesis stage of meiotic prophase. This group comprises the sponges and some representatives of flatworms (*Dicyena*), roundworms (*Ascaris*), annelids (*Nereis*, *Thalassema*, *Urechis*), mollusks (*Ilyanassa*, *Ostrea*, *Spisula*), and chaetognaths (*Sagitta*). It contains three representatives of mammals, i.e., the dog, the fox, and the horse.

II. Meiosis is arrested at metaphase I. This group comprises some sponges (*Ciona*), nemertines (*Cerebratulus*), annelids (*Chaetopterus*), mollusks (*Dentalium*, *Mytilus*, *Pectinaria*), and nearly all the insects. Animals in which meiosis reaches anaphase I, such as *Ophyotrocha*, are sometimes set apart as a separate group (Wilson, 1925; John and Lewis, 1965).

III. Meiosis is arrested at metaphase II. This group comprises nearly all the chordates. It may also be said to contain those animals (*Siredon*, bats) in which meiosis reaches anaphase II (see John and Lewis, 1965).

IV. Meiosis is completed and the development is arrested at the stage of female pronucleus. This group comprises some coelenterates and echinoderms (sea urchins).

In most animals the sperm penetrates the egg soon after the arrest of meiosis. In rare cases (for example, in a nematode *Brachycoelium*, annelids *Dinophilus* and *Histriobdella*, and onychophore *Peripalopsis*), the sperm penetrates into the oocyte at the diplotene stage, i.e., at the meiotic prophase before vitellogenesis has terminated. In these animals the sperm nucleus remains in the cytoplasm unchanged until the oocyte has become fully grown, completed meiotic prophase, and begun its meiotic divisions; only at the end of the II meiotic division does it become transformed into the male pronucleus.

In the overwhelming majority of animals meiosis is completed after activation of the egg by the sperm, which eliminates the arrest of meiosis. This arrest can also be removed by artificial activation with various stimuli (see Ginsburg, 1974).

Of considerable interest are the mechanisms of meiotic arrest and its elimination upon activation. This problem was studied using mainly echinoderms, mollusks, amphibians, and mammals (Heilbrunn, 1920; Runnstrom, 1949, 1952; Sawada, 1952, 1954; Allen, 1953; Runnstrom and Immers, 1956; Raven, 1966; Masui and Markert, 1971; Masui, 1974, 1985; Schuetz, 1975; Tsafriri and Channing, 1975; Tsafriri, 1978; Masui and Clarke, 1979; Channing et al., 1982; Tsafriri et al., 1982). In many animals meiosis is arrested twice. It is arrested in prophase I (and resumed after hormonal stimulation) and at either the I or II meiotic division (Fig. 36), prior to sperm penetration (it is resumed after sperm penetration or egg activation). In some animal species meiosis is blocked before germinal vesicle breakdown and, after resumption, proceeds to the formation of the female pronucleus (Fig. 36, 1), while in the others the prophase arrest of meiosis is relieved by hormonal stimulation and the spermatozoon penetrates into the egg after the completion of meiosis, at the female pronucleus stage. However, the mechanisms of meiotic arrest are as yet unclear. Some aspects of this problem are considered in greater detail by Dettlaff (Chapter 3) and Skoblina (Chapter 5) in this book.

4.2. MORPHOLOGY OF MEIOSIS

There are morphological descriptions of meiotic divisions (see, for example, John and Lewis, 1965; Sokolov, 1966; Kiknadze and Vysotskaya, 1975); however, all of these descriptions are, as a rule, given on the basis of studies of male meiosis. A description of the successive phases of female meiosis (beginning with germinal vesicle breakdown) is therefore given based on the studies of the sturgeon (Dettlaff, 1961; Vassetzky, 1970), the South African clawed frog (Vassetzky and Betina, 1986), and the giant oyster (Vassetzky, 1973). Prometaphase of the I meiotic division

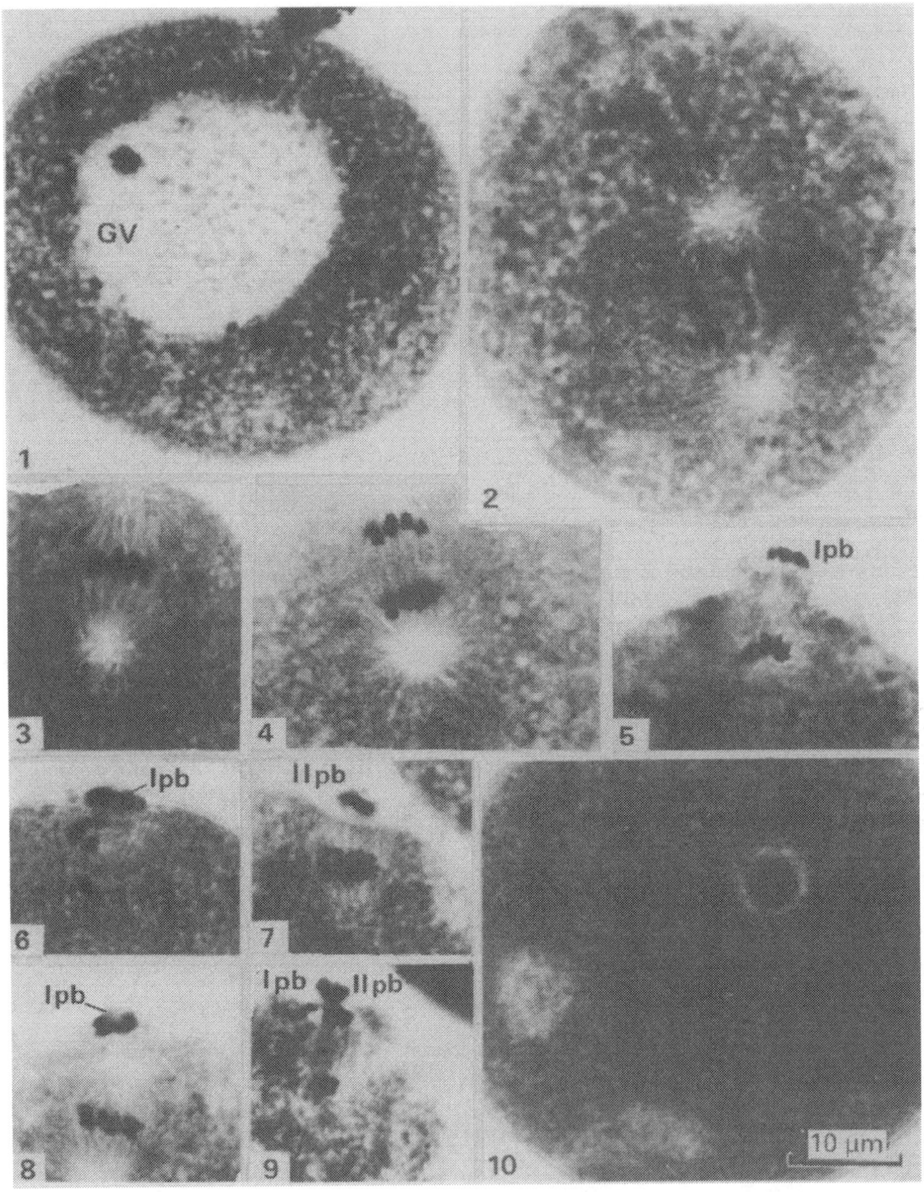

Plate XX. Meiosis in oogenesis of the giant oyster *Ostrea gigas* (Vassetzky, 1973). 1. Prophase I. 2. Prometaphase I. 3. Metaphase I. 4. Anaphase I. 5. Telophase I. 6. Prometaphase II. 7. Metaphase II. 8. Anaphase II. 9. Telophase II. 10) Stage of pronucleus. GV) germinal vesicle; Ipb, IIpb) first and second polar bodies.

is indicated in one of the previous descriptions on the basis of studies of male meiosis in insects. However, it is important, due to the fact that during this period marked changes in the cytoplasm of oocytes are observed, and the meiotic spindle forms and migrates to the oocyte surface together with the chromosomes located on it.

Prometaphase of the I Meiotic Division (Prometaphase I). At the end of prophase the germinal vesicle of the oocyte may be localized either in the center of the oocyte or placed eccentrically, but at a considerable distance from the plasma membrane (in the oyster) or, lastly, under the surface of the oocyte in the region of the animal pole (in sturgeons and amphibians) (Plate XX, 1; XXIa, 1). The germinal vesicle is large and, in some cases, for example in amphibians and sturgeons, can be seen in sections even with the naked eye. It is more often ellipsoid; conjugating chromosomes with distinct chiasmata and numerous nuclei can be seen in the karyoplasm. In its basal part the nuclear envelope invaginations can be seen. A layer of yolk-free cytoplasm adheres to this part of the germinal vesicle. The germinal vesicle breakdown marks the boundary between prophase and prometaphase I. Prometaphase I continues from the destruction of the nuclear envelope and the passage of the chromosomes into the cytoplasm until the formation of the metaphase plate. The germinal vesicle becomes somewhat elongated in its basal parts and its envelope is disrupted. The chromosomes can no longer be seen, and the karyoplasm appears less dense. The zone of yolk-free cytoplasm is enlarged. Later, the germinal vesicle is seen only in fragments. The karyoplasm is not yet fully mixed with the cytoplasm, and the plate where the germinal vesicle has been located is clearly seen. Its basal part is occupied by a dense fibrillar layer ("basket") where the chromosomes later appear. The karyoplasm is then mixed with the cytoplasm and this area is filled with yolk. The fibrillar layer with the chromosomes migrates to the animal pole surface, decreases in size, and transforms into the meiotic spindle located near the plasma membrane. The chromosomes move to the spindle center while the spindle itself migrates to the oocyte cortex and becomes anchored there (see Plate XX, 2; XXIa, 2).

Metaphase I. This stage lasts from the formation of the metaphase (equatorial) plate until the beginning of chromosome disjunction. The spindle is elongated and has a typical form, the spindle fibers are clearly seen, and the chromosomes lie in the middle of the spindle and form the equatorial plate. The spindle is localized perpendicular to the surface of the oocyte or at a small angle to it and with its distal end adjoins the plasma membrane of the oocyte (see Plate XX, 3; XXIa, 3).

Unlike the metaphase of mitosis where the chromosomes have one centromere each and lie in the equatorial plane, in the meta-

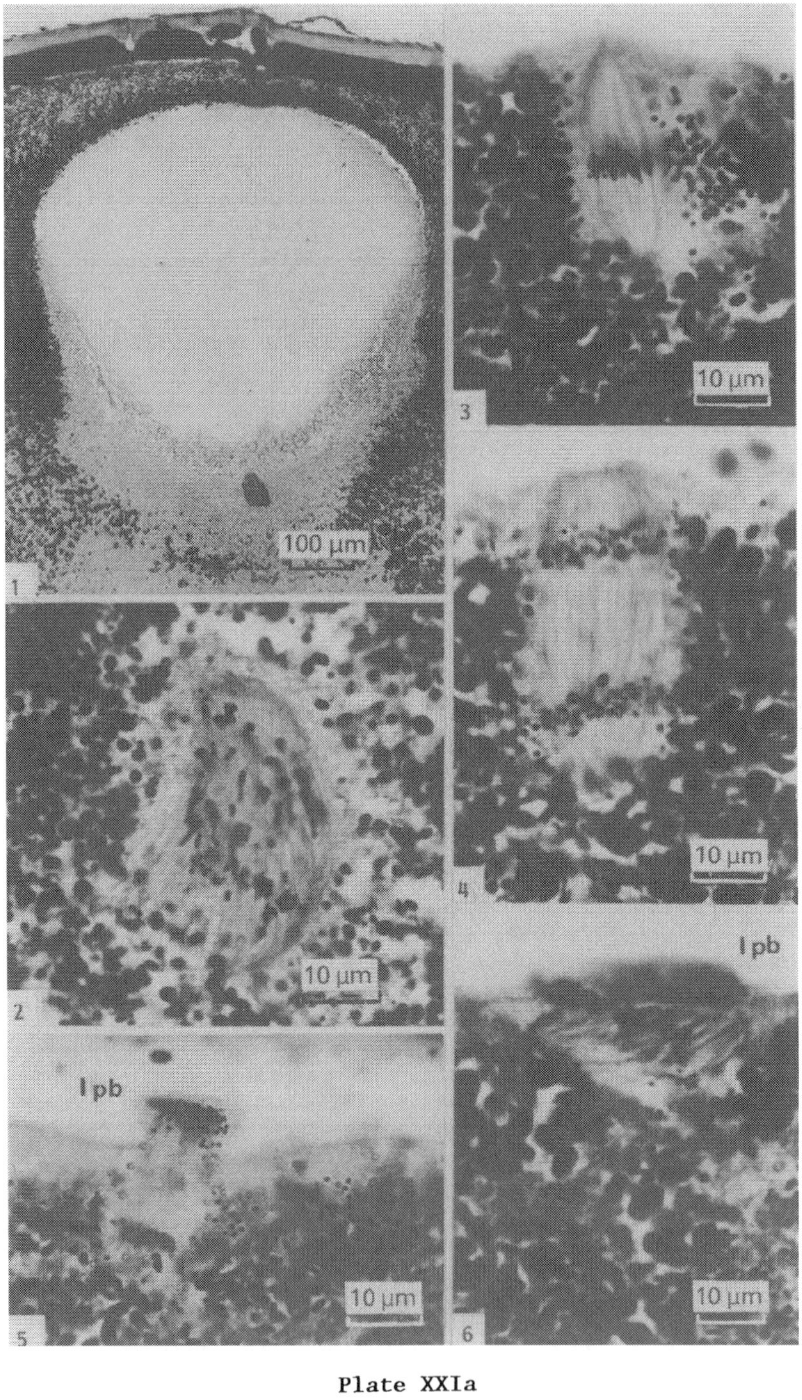

Plate XXIa

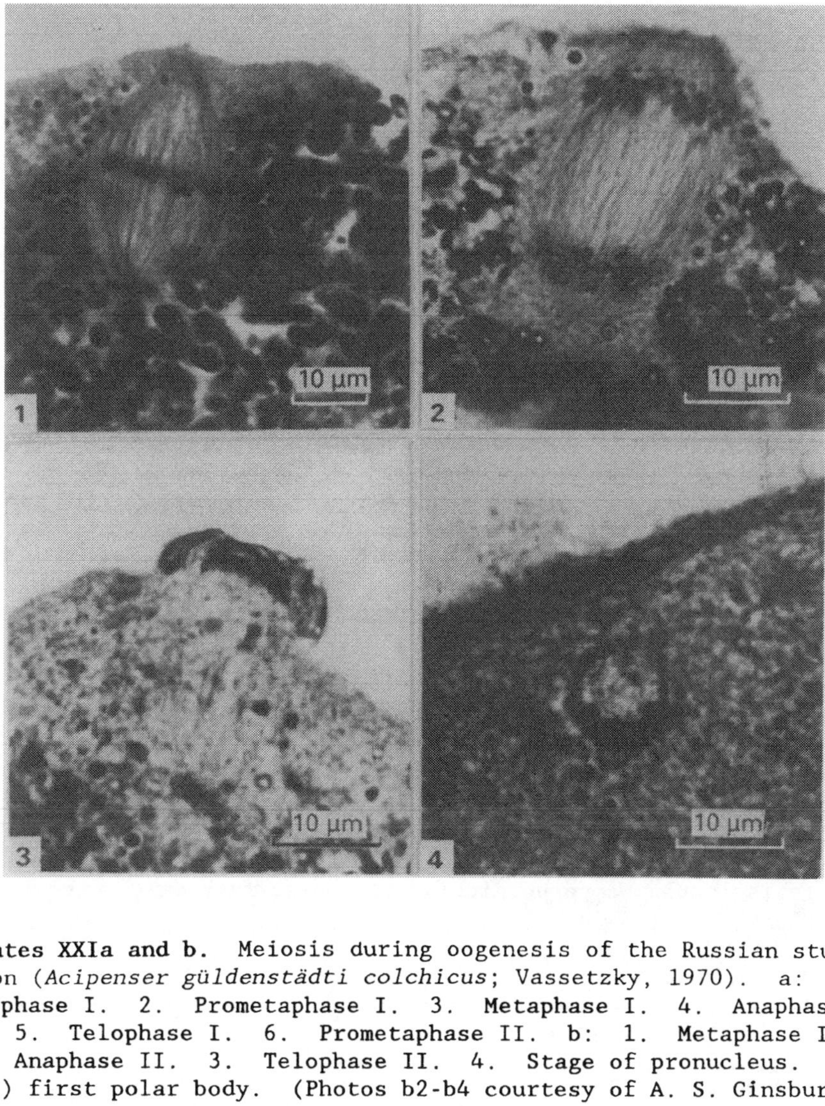

Plates XXIa and b. Meiosis during oogenesis of the Russian sturgeon (*Acipenser güldenstädti colchicus*; Vassetzky, 1970). a: 1. Prophase I. 2. Prometaphase I. 3. Metaphase I. 4. Anaphase I. 5. Telophase I. 6. Prometaphase II. b: 1. Metaphase II. 2. Anaphase II. 3. Telophase II. 4. Stage of pronucleus. Ipb) first polar body. (Photos b2-b4 courtesy of A. S. Ginsburg.)

phase of the I meiotic division the bivalents consisting of two homologous chromosomes and correspondingly possessing two centromeres are localized symmetrically on both sides of the equatorial plane, like mirror reflections, and form tetrads. Thus, the orientation of the centromeres determines the position of the bivalents and ensures the proper disjunction of the chromosomes.

Anaphase I. This phase lasts from the beginning of the disjunction of the chromosomes until they have formed dense groups

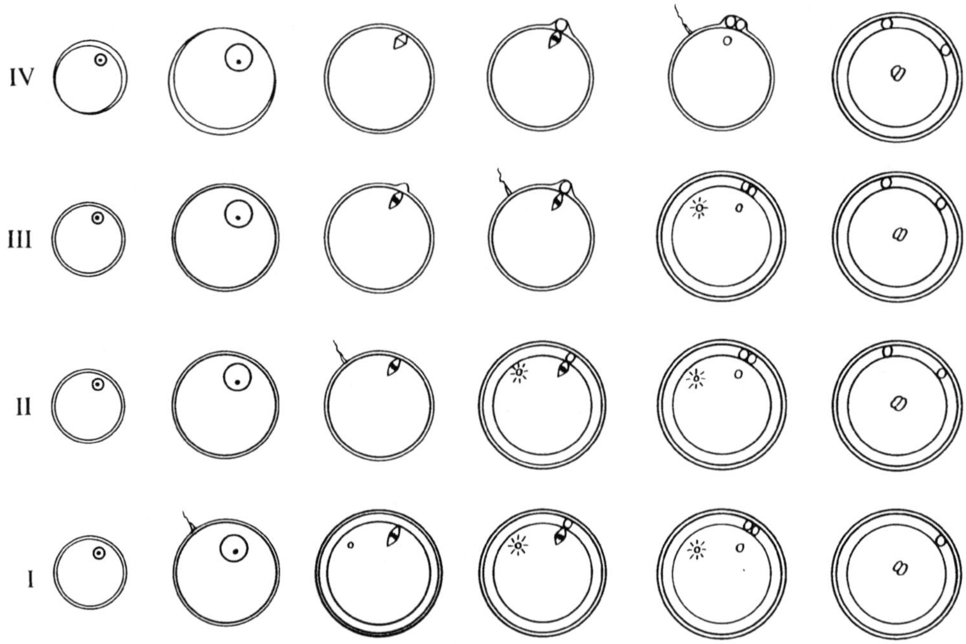

Fig. 36. Stages of oocyte maturation at which meiosis is arrested in different animals. Explanation in text.

at the spindle poles. During anaphase the spindle becomes elongated and the fibers composing it are clearly seen. The chromosomes form two plates which diverge toward the spindle poles. The spindle is localized perpendicular to the surface of the oocyte (see Plate XX, 4; XXIa, 4).

During anaphase I each tetrad divides into two dyads and, as both co-oriented centromeres move toward the opposite poles, the homologous chromosomes diverge.

Telophase I. This phase lasts from the formation of dense groups of the chromosomes at the spindle poles until the extrusion of the I polar body. The spindle narrows down and assumes the form of a strand of parallel fibers. A bulge of the plasma membrane forms over the distal end of the spindle, a furrow is laid perpendicular to the spindle axis, and the distal group of chromosomes with a small amount of cytoplasm separates, forming the I polar body. At the place of its extrusion an invagination, the maturation funnel, remains. The proximal group of chromosomes forms the oocyte nucleus which soon starts the second meiotic division (see Plate XX, 5; XXIa, 5).

During telophase I, the chromosomes still stain intensively with nuclear dyes, for example hematoxylin, and, unlike the mitotic telophase, do not form karyomere structures.

Meiotic division I passes directly into II, which begins with prometaphase II. In the oogenesis of the animals we have studied, the completion of division I is not followed by formation of the nuclear envelope.

Prometaphase II. This stage lasts from the extrusion of the I polar body until the formation of the metaphase plate by the chromosomes. Spindle II forms at the proximal pole of former spindle I. The chromosomes may at first be scattered all over the spindle, but later they concentrate in its central part. The spindle that has just formed is localized at an angle to the surface of the oocyte (see Plate XX, 6; XXIa, 6). Prometaphase II appears to be very short since it is rarely seen in sections of fixed oocytes.

Metaphase II. This stage lasts from the formation of the metaphase plate until the beginning of chromosome disjunction. The spindle is elongated, the spindle fibers are clearly seen, and the chromosomes are arranged in the form of an equatorial plate. The spindle lies perpendicular to the surface of the oocyte. The I polar body is, as a rule, seen on the surface of the oocyte, near the distal pole of the spindle or at some distance from it (see Plate XX, 7; XXIb, 1).

The II spindle is shorter than the I. As already stated, in the overwhelming majority of vertebrates, including sturgeons, meiosis is arrested at this stage.

Anaphase II. This stage lasts from the beginning of chromosome disjunction until the formation of dense groups of the chromosomes at the spindle poles. The spindle fibers are clearly seen; the chromosomes form two plates that separate toward the spindle poles. The spindle is localized perpendicular to the surface of the oocyte (see Plate XX, 8; XXIb, 2).

Telophase II. This phase lasts from the formation of dense groups of chromosomes at the spindle poles until the extrusion of the II polar body (see Plate XX, 9; XXIb, 3). The group of chromosomes at the proximal pole of the spindle forms karyomere vesicles and is then transformed into a typical interphase nucleus characteristic of mitotic divisions, the female pronucleus, which begins to move toward the center of the egg (see Plate XX, 10; XXIb, 4). The formation of the female pronucleus terminates meiosis; this is followed by the juxtaposition of the female and male pronuclei, and in sturgeons by union of chromosomes on the spindle of the first cleavage division (see Ginsburg, 1959).

Meiotic division II is very similar to mitotic divisions in its morphology and other characteristics, especially in its duration (see below). The I polar body usually divides simultaneously with meiotic division II, so that on completion of meiotic division II three polar bodies may be localized on the surface of an egg.

4.3. CHRONOLOGY OF MEIOSIS

The oocyte maturation of amphibians and fishes is a relatively new and rapidly developing area of developmental biology (see Dettlaff, 1977; Masui and Clarke, 1979). After it had been shown that oocyte maturation can be stimulated by gonadotropins (pituitary suspension) not only in vivo but also in vitro in saline, the possibility existed for wide application of experimental embryological and biochemical techniques to the study of oocyte maturation (see Dettlaff et al., 1964, 1965a; Dettlaff and Skoblina, 1969; Dettlaff, 1977). The use of relative characteristics of the development duration allowed for the time parameter to be introduced into these studies and for certain temporal patterns of in vivo and in vitro oocyte maturation in amphibians and fishes to be revealed. The term "oocyte maturation" is defined here as the meiotic maturation of oocytes or the resumption of meiosis after its prophase arrest once vitellogenesis is completed.

In pisciculture (sturgeons were introduced in pisciculture in the USSR long ago), an estimation of the timing of maturation of females after a pituitary injection is of great practical importance, since the untimely obtaining of the eggs can lead to a great loss of eggs due to their under- or overripening. A detailed analysis of many years' data obtained at sturgeon hatcheries has shown that the duration of oocyte maturation and the timing of ovulation markedly differed at different temperatures.

A comparison of the time of oocyte maturation and ovulation in the female body of *Acipenser güldenstädti* and *A. stellatus* at different spawning temperatures with the time of embryogenesis from insemination until hatching has shown that the duration of these periods changes proportionally with temperature (see also Dettlaff and Zuichenko, 1963) and that oocyte maturation at different temperatures takes roughly 1/5 of the duration of embryogenesis (0.19 ± 0.02). By the end of this period most stimulated females mature (Dettlaff and Zubova, 1962; Dettlaff et al., 1965b).

Similar results were obtained when the relative durations of oocyte maturation in the body of the female sturgeon were expressed in numbers of relative units of development time (the duration of one mitotic cycle during the period of synchronous cleavage divisions was taken as such a unit; see Dettlaff and Dettlaff, 1960, 1961) – τ_0. In *A. güldenstädti*, *A. stellatus*, *A. ruthenus*, and *Huso huso*, oocyte maturation in different females takes place within the same interval range, 24 to $37\tau_0$ (Dettlaff and Zubova, 1962; Dettlaff et al., 1968a; Igumnova, 1979; Igumnova and Dettlaff, 1983). On the basis of these results, curves were plotted which allowed the timing of oocyte maturation in injected females to be prognosticated at different temperatures (Dettlaff et al., 1965b, 1968a; Igumnova, 1979, 1985).

In laboratory experimentation with *Misgurnus fossilis*, widely used in developmental biology studies (see Neyfakh, 1962; Kafiani, 1970), mature eggs are obtained from females within roughly $50\tau_0$ after injection with choriogonin (recalculated from the data of Kostomarova, 1975).

In sturgeons, as well as in amphibians (Tchou-Su and Wang Yu-lan, 1958), when oocyte maturation is stimulated by pituitary hormones, the presence of the hormone in the medium is necessary only during the initial, hormone-dependent period of maturation and then, shortly before germinal vesicle breakdown, the oocytes acquire maturation inertia and can complete their maturation after they have been taken from the ovary and placed in a hormone-free Ringer solution (Dettlaff et al., 1964). The duration of the hormone-dependent period also depends on temperature and amounts, in the sturgeons, to 7-$8\tau_0$ after the pituitary injection (Dettlaff, unpublished) at the level of the spawning grounds. In the low river flows or higher (in this case the females are at the incomplete stage IV) this period can last longer. The duration of the hormone-dependent period can itself be used as an index of the state of females (Goncharov, 1976). All these data were a prerequisite for further study of the subsequent changes in oocytes and comparison of their dynamics in vivo and in vitro.

In order to study oocyte maturation, the model of oocyte maturation in vitro under the influence of pituitary suspension developed by Skoblina for *A. güldenstädti* (Dettlaff et al., 1968b; Skoblina, 1970) and *M. fossilis* (Skoblina, 1973) is widely used. The oocyte maturation proceeds in vitro more synchronously and successfully under the influence of progesterone. When using the method of in vitro oocyte maturation, it has been shown that not only is the total duration of oocyte maturation dependent on temperature, but also the duration of the periods from maturation stimulation to germinal vesicle breakdown and from germinal vesicle breakdown to the onset of successive meiotic phases, and that the relative durations of all these periods expressed in τ_0 are characterized by similar values at different temperatures (Vassetzky, 1967a; Goncharov, 1971). The temperature dependence of the duration of the period from germinal vesicle breakdown to meiotic arrest suggests that all changes of the cytoplasm involved in meiotic maturation are also temperature-dependent (see Dettlaff, 1977).

These data are, moreover, of special interest since they indicate that oocyte maturation in the ovary depends on temperature in the same way as does the duration of the cell cycle during the period of synchronous cleavage divisions (τ_0) and those during different periods of embryogenesis. Hence, quantitative evidence was provided for the fact that there are no internal mechanisms of thermoregulation in poikilothermic animals.

These, and some other observations concerning oocyte maturation in fishes and amphibians, confirm the well-known fact that meiosis is a process of longer duration than mitosis in the same species of animals and plant (see, for example, Rhoades, 1961). However, in most studies of the chronology of meiosis only the duration of the meiotic prophase and the total duration of both meiotic divisions have been determined. Since the egg cells of animals may cease their development for several months, even years, in the diplotene stage, the chronology of male meiosis has been mainly studied; a full report on such data is given by Lyapunova and Bogdanov (1975). Data on the duration of meiosis in the oogenesis of animals are limited (see Bennett, 1977); for most of the species studied it has been shown that meiotic division I lasts longer than division II.

The mitotic phase ratio in different animals at different temperatures has been studied in detail using the relative characteristics of the development duration. In these studies it was shown that the relative duration of individual mitotic phases in the course of synchronous cleavage divisions is similar in the most diverse species (sea urchins, amphibians, fishes) and remains unchanged in the same species at different temperatures within optimum limits (Dettlaff, 1962, 1974; Skoblina, 1965; Kostomarova and Ignatieva, 1968; Ignatieva, 1970, 1973, 1979; Rudneva, 1972; Dettlaff and Dettlaff, 1982).

The relative characteristics of developmental duration have also been used for studying meiosis (see Vassetzky, 1982). Special studies of the chronology of meiosis were carried out on sturgeons (Vassetzky, 1970, 1972). The data obtained are shown in Table 4.1, in which the duration of the divisions and their phases are given in τ_0.

The total duration of prometa-, meta-, ana-, and telophase I in sturgeon oocytes is much longer (17 times longer) than that of the corresponding period of synchronous cleavage divisions and meiotic division II. A comparison of the data for the eggs of the newt (*Triturus viridescens*) on the duration of phases of the I (Humphries, 1956) and II meitoic divisions (Fankhauser and Moore,

Table 4.1. Relative Duration (in τ_0) of the Phases of Meiotic and Mitotic (Synchronous Cleavage) Divisions in the Eggs of Sturgeon *Acipenser güldenstädti*

Phase	Divisions		
	I Meiotic (except prophase I) (Vassetzky, 1970)	II Meiotic (Dettlaff, 1961)	Mitotic (Dettlaff, 1962)
Prometaphase	2.2(25)	-	0.1(20)
Metaphase	3.7(43)	0.2	0.2(40)
Anaphase	1.7(20)	0.2	0.14(28)
Telophase	1.0(12)	-	0.06(12)
Total duration of "active" phases	8.6(100)	-	0.5(100)

Note. In parentheses: relative duration of the phases (in % of the total duration of "active" phases).

1941; Fankhauser and Godwin, 1948) shows that meta- and anaphase I are approximately six times as long as metaphase (from the moment of activation) and anaphase II. Similar data are also available for the loach (*Misgurnus fossilis*). A comparison of the data on meiotic division I (Belyaeva and Cherfas, 1965) and mitosis (cleavage divisions) (Ignatieva and Kostomarova, 1966; Kostomarova and Ignatieva, 1968) shows that the duration of the "active" phases of meiotic division I is 8-10 times that of the corresponding period of synchronous cleavage divisions. Recently, the chronology of oocyte maturation in the loach was specially studied by Saat (1982). He showed that prometaphase + metaphase + anaphase I lasted 7.5-7.6τ_0, i.e., their duration exceeded by 18-fold that of the corresponding phases of the second cleavage division — 0.42τ_0 (Ignatieva, 1979). If this duration (from prometaphase to telophase I) is taken as 100%, the relative duration of these phases will be very close both in the loach and the sturgeon (see Table 4.1) for synchronous cleavage divisions (Ignatieva, 1979).

At the same time, the relative duration of the individual phases of meiotic division I and mitosis (synchronous cleavage divisions) in the sturgeon is close (see Table 4.1). This fact indicates that although in the course of oocyte maturation the absolute duration of both the entire division and its phases changes very much, the ratio of the different phases does not undergo any significant change.

Table 4.2 contains data on the duration of meiosis and its phases in the oogenesis of mammals. According to these data, the duration of meiotic division I (except the prophase) is about 9-11 h in the mouse and 4.3-5.4 h in the rat. These values are less

Table 4.2. Duration of Meiotic Divisions and Their Phases in Oogenesis of Mammals, hours

<table>
<tr><th>Species</th><th>Prometaphase I</th><th>Metaphase I</th><th>Anaphase I</th><th>Telophase I</th><th>Prometaphase II</th><th>Telophase II</th><th>Reference</th></tr>
<tr><td>Mouse (<u>Mus musculus</u></td><td>-</td><td>5.5</td><td>1.5</td><td>0.5</td><td>-</td><td>-</td><td>Edwards and Gates (1959)</td></tr>
<tr><td></td><td>1.5</td><td>4.5</td><td>2.0</td><td>-</td><td>-</td><td>-</td><td>Donahue (1969)</td></tr>
<tr><td></td><td>-</td><td>4.0</td><td>-</td><td>-</td><td>-</td><td>-</td><td>McGaughey and Change (1969a)</td></tr>
<tr><td></td><td>6.6</td><td>4.5</td><td>0.2</td><td>0.4</td><td>-</td><td>-</td><td>Sorensen (1973)</td></tr>
<tr><td>Norwegian rat (<u>Rattus norvegicus</u>)</td><td colspan="2">3.9</td><td>0.25</td><td>1.25</td><td colspan="2">3.0</td><td>Odor (1955)</td></tr>
<tr><td></td><td>1.0</td><td>2.3</td><td colspan="2">1.0</td><td>0.5</td><td>1.0</td><td>Mandl (1963a)</td></tr>
<tr><td>Guinea pig (<u>Mesocricetus auratus</u>)</td><td>1.5</td><td>2.0</td><td>2.0</td><td>2.0</td><td>-</td><td>-</td><td>Harvey and Chang (1963)</td></tr>
</table>

than the duration of mitosis during the cleavage period; this is about 13 h in the mouse (Leonov et al., 1972) and about 16 h in the rat (Sembrat, 1955).

The duration of meiotic divisions was also determined in the eggs of some invertebrates by observations of living material, as well as of histological preparations obtained as a result of fractional fixation of developing eggs. In all these animals meiosis is arrested before germinal vesicle breakdown, all the subsequent phases of meiosis taking place under similar conditions in the presence of the sperm. It turned out that in bivalve molluscs *Crassostrea gigas* (Inaba, 1936; Vassetzky, 1973) and *Spisula solidissima* (Sclufer, 1955; Rebhun, 1959), and in the annelid *Urechis caupo* (Gould, 1968) the duration of meiotic division I is 50-100% longer than that of division II. The cytological data for *Crassostrea gigas* (Vassetsky, 1973) show that the relative duration of the prometaphase + metaphase and anaphase + telophase of both meiotic divisions are characterized by close values, viz. 52-62 and 59% for the prometaphase + metaphase and 38-48 and 41% for the anaphase + telophase.

Other ratios are also observed in the meiosis of some invertebrates. Thus, in the eggs of the silkworm *Bombyx mori* metaphase and telophase of the I and II meiotic divisions are of similar duration (Ostryakova-Varshaver, 1958), while in the gastropod *Ilyanassa* the duration of telophase II is three times as long as that of telophase I, so that the total duration of meiotic division II is longer than that of I (Cather, 1963).

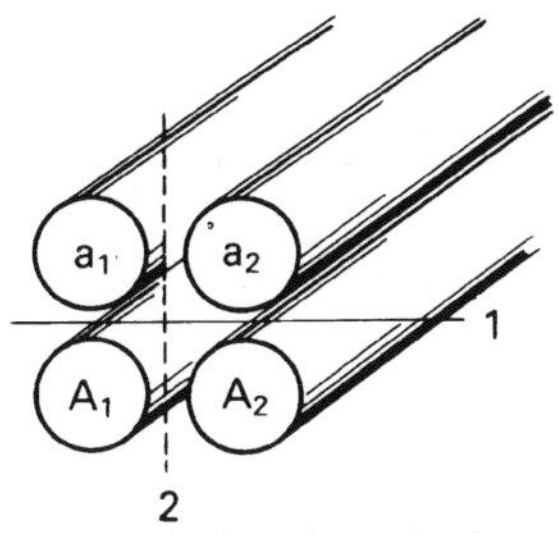

Fig. 37. The bivalent structure. a_1-a_2, A_1-A_2, sister chromatids. 1) Reduction fissure; 2) equation fissure.

Since it is known that in cleavage the rate of nuclear division is determined by the properties of the cytoplasm (Moore, 1933; Aimar et al., 1981), it may be assumed that the differences in the rate of nuclear division in meiotic divisions I and II are also due to the differences in the properties of the cytoplasm of oocytes at these stages. The study of this question is interesting in itself.

The difference in the duration of meiotic divisions I and II may be due to marked differences in the hydration of the cytoplasm of oocytes at these stages in molluscs (Heilbrunn and Wilson, 1955), fish (Zotin, 1961; Oshiro and Hibiya, 1981), and amphibians (Tchou-su and Yen Pai-hu, 1950; Dettlaff, 1966). The much smaller differences in the duration of meiotic divisions I and II in animals in which the sperm penetrates the ooplasm during meiotic division I, as compared with those in which this takes place during division II, may be connected with the fact that both meiotic divisions occur in them under similar conditions (in the activated egg after fertilization), whereas the conditions under which the two meiotic divisions take place in vertebrates differ a good deal, namely, division I occurs in a maturing oocyte and division II is completed in a fertilized egg.

4.4. REDUCTION AND EQUATIONAL CHARACTER OF MEIOTIC DIVISIONS

The cell prepares for the reduction of the chromosome number during the meiotic prophase by the formation of tetrads; the reduction occurs in two successive meiotic divisions. In most animals the disjunction of homologous chromosomes takes place in meiotic division I and that of the sister chromatids in II. Two fissures are distinguished in the tetrads, namely, a reductional fissue along which the disjunction of homologous chromosomes occurs (prereduction), and an equational fissure along which the sister chromatids are separated (postreduction) (see, for example, Kiknadze and Vysotskaya, 1975) (Fig. 37). There are animals in which an inverse succession in meiosis is observed and the disjunction of homologous chromosomes takes place in the second meiotic division, for example in several species of insects of the genus *Aeschna* (Oksala, 1943), and in the mite *Tetranychus urticae* (Feiertag-Koppen, 1980).

However, with the discovery of crossing-over (Darlington, 1937) the ideas of reductional and equational divisions have become almost entirely meaningless since the disjunction of some alleles and parts of homologous chromosomes may occur in I and of the others in meiotic division II. M. J. D. White writes (1973, p. 492): "All that we can say is that the first division will be reductional (separating two paternal strands from two maternal ones) on one side of each point of crossing-over, and equational (separating a paternal and maternal strand from a maternal and paternal one) on the other side; and that regions which separate reductionally in the first division will do so equationally in the second one, and vice versa."

Only in a small number of species listed by John and Lewis (1965) is so-called achiasmatic meiosis observed, i.e., meiosis without crossing-over. It is described only for some representatives of flatworms (*Mesostoma ehrenbergii* – Oakley, 1982), annelids (*Enchytraeidae* – Christensen, 1961), and arthropods – *Copepoda*: *Entocyclops*, *Cyclops*, *Tigriopus* (Beerman, 1954; Beerman, 1959; Ar-Rushdi, 1963; Burton et al., 1981), *Insecta*: *Bombyx mori* (Maeda, 1939), *Cidaria* (Suomalainen, 1953), *Limnophulus decipiens*, *L. borealis* (Suomalainen, 1966). In these animals meiotic division I is truly reductional since the homologous chromosomes disjoin precisely during this division and no crossing-over occurs.

To investigate the absence of chiasma formation, it is necessary to use both cytological and genetic data. Thus, achiasmatic female meiosis was described in *Bombyx mori* (Maeda, 1939). However, White (1954) arrived at the conclusion that the data given by Maeda on the absence of chiasmata are not convincing enough and only the genetic analysis made it possible to arrive at a final conclusion on the absence of crossing-over in the meiosis of *Bombyx mori* (see Astaurov, 1968). The cytological data on the absence of chiasmata in meiosis may be confirmed using biochemical methods as well. Burton et al. (1981) carried out an electrophoretical analysis of the progeny from laboratory crosses in *Tigriopus californicus* and found no recombination in females, thus confirming the earlier observations on the absence of chiasmata in oogenesis of this species (Ar-Rushdi, 1963).

4.5. MEIOSIS IN PARTHENOGENETIC DEVELOPMENT

Quite a few species of animals reproduce parthenogenetically.

There are at least 24 such species among the vertebrates alone (Cuellar, 1971). Meiosis in the oogenesis of such animals is of special interest. The unfertilized egg that gives rise to a new organism has the same chromosome number as the somatic cells, and not half the number as do the eggs of bisexual species. This means that the reduction of the chromosome number is somehow eliminated, or the chromosome number characteristic of the somatic cells (diploid or polyploid chromosome set) is restored. This question is examined in detail in a number of reviews (White, 1954, 1973; Beatty, 1957, 1967, 1970a, b; Peacock and Weidman, 1961a, b; Uzzell, 1970; Cuellar, 1971; Astaurov and Demin, 1972; D'Amato, 1977; Suomalainen et al., 1977; Ivanova, 1978; Mittwoch, 1978; Christensen, 1980; etc.).

In parthenogenesis the chromosome number may be regulated in a number of ways.

1. Reduction may completely drop out, and in this case both meiotic divisions are equational. This phenomenon is observed, for example, in the crustacean *Daphnia pulex* (Schrader, 1926) and mollusc (*Campeloma rufum* (Mattox, 1937). A similar picture is observed in a trematode *Philophthalmus megalurus* in which the oocytes in germinal sacs divide without the extrusion of polar bodies and form diploid embryos of the next generation (Khalil and Cable, 1968). In case of triploid parthogenesis in *Psychoda parthenogenetica* (Diptera), oocyte maturation is realized by simple mitotic divisions, without any sign of meiosis, even abortive (Troiana, 1978). The same "mitotic" parthenogenesis appears to be characteristic of the mite *Histiosoma feroniarum* (Heinemann and Hughes, 1969), several *Pratylenchus* species (Cable, 1971), and *Meloidogyne incognita* (Nematoda) (Abirachned-Darmency, 1982).

2. The chromosome number may double prior to meiosis, most commonly in the process of the last oogonial division or in oocytes directly before the beginning of the meiotic prophase. This process was named "endoduplication" (Muldal, 1952). It is followed by normal meiosis with two divisions and extrusion of two polar bodies with the result that a mature egg is formed with the initial diploid or polyploid chromosome number. Such a mechanism of preserving the initial chromosome set is found in the planarian *Polycelis nigra* (Lepori, 1950), the turbellarian *Bothrioplana semperi* (Reisinger, 1940), earthworms (Lumbricidae) (Muldal, 1952; Omodeo, 1952), some insects: *Ptinus clavipes m. mobilis* (Sanderson, 1960), *Mirabo virgo* (Smith, 1971) and *Carausius morosus* (Pijnacker, 1966), the mite *Brevipalpus obovatus* (Pijnacker et al., 1981), the lizards *Cnemidophorus uniparens* (Cuellar, 1971) and *Heteronotia biholi* (Moritz, 1984), some fishes: *Poeciliopsis* (Schultz, 1967; Cimino, 1972a, and *Poecilia* (Kallman, 1962; Schultz and Kallman, 1968), and amphibians: *Ambystoma jeffersonianum* (Macgregor and Uzzell, 1974), *A. platineum*, and *A. tremblayi* (Asher and Nace, 1971). A similar mechanism is also in-

herent in bisexual *Drosophila melanogaster* giving rise to triploid forms (Kuznetsova and Glotov, 1978; Kuznetsova and Balakireva, 1980; Kuznetsova and Pasyukova, 1980; Kuznetsova, 1983).

3. Meiotic division I may be abortive or may completely drop out. The disjoining, or already disjointed, homologous chromosomes reunite in a single nucleus which starts meiotic division II. In some cases the I anaphase spindle elongates and divides in two halves which later merge, forming the II spindle. The existence of such a mechanism has been demonstrated in the crustacean *Trichoniscus* (Vandel, 1934, 1940), insects: Hymenoptera (Speicher, 1937), Curculionidae (Suomalainen, 1940, 1949), Lepidoptera (Narbel-Hofstetter, 1964), Psocoptera (Mockford, 1971), Coleoptera (Suomalainen, 1965; Lokki et al., 1976; Takenouchi, 1978), mites (Oliver, 1971), and among the Teleostei, in unisexual forms of *Carassius auratus* (Cherfas, 1966, 1969; Kobayasi et al., 1970; Kobayasi, 1971; Cui and Zan, 1982; Zhou et al., 1983). This method of regulating chromosome number is characteristic of all cytologically studied triploid all-female forms (see Schultz, 1980).

4. Meiotic division II may be abortive or drop out altogether. In such cases the disjoining, or already disjointed, sister chromosomes reunite without forming the II polar body, and the initial chromosome number is retained in the nucleus. Such a mechanism was described for crustaceans *Artemia salina* (Barigozzi, 1944) and *Nagara modesta* (Hill, 1948), as well as for some birds (Olsen and Buss, 1972; Olsen, 1974).

5. In a number of cases one more mechanism of restoration of the initial chromosome number has been noted, that of self-fertilization. In these cases the haploid nuclei formed as a result of both meiotic divisions merge. Moreover, the female pronucleus may merge with polar nucleus II located nearby. Such a mechanism of restoration of diploidy has been demonstrated for several species of insects, including *Lecanium hesperidum* (Thomsen, 1927), *Coccus hesperidum*, *Saisettia coffeae* (Nur, 1979), and *Diplolepis rosae* (Stille and Dävring, 1980). The restitution of the initial chromosome number may also be ensured by the fusion of the female pronucleus with one of the progenitors of polar nucleus I. This phenomenon was described in a diploid parthenogenetic form of *Drosophila mangabeirai* (Murdy and Carson, 1959). A special case of restitution of the initial chromosome number in diploid parthenogenesis is described for an insect *Solenobia triquetrella* (Seiler, 1960). In this insect the female pronucleus does not take part in the development of the egg, which is ensured by a so-called "Richtungskopulationskern." This diploid nucleus arises by the fusion of two haploid nuclei forming as a result of meiotic divisions I and II. A so-called restitutive fertilization was described in an insect *Heteropeza pygmae* (Went, 1982). Meiosis

proceeds in the oocytes with the reduction in the number of chromosomes and then the female pronucleus fuses with two somatic nuclei present in the cytoplasm (Zanazzi, 1978; Gandolfi, 1979). Such a mechanism has also been postulated for the lizard *Lacerta saxicola* (Darewski and Kulikova, 1961).

6. Nuclei arising in the process of cleavage may merge. After two normal meiotic divisions the egg begins its development with a nucleus containing a haploid chromosome set. The first and second cleavage divisions take place normally, but after the second cleavage the four nuclei merge in pairs to restore the initial chromosome number. This mechanism is found in some insects, namely Phasmidae (Bergerard, 1962), Coccoidea (Nur, 1971, 1972), Orthoptera (Went, 1982), *Drosophila mercatorum* (Carson et al., 1969), and in mites (Oliver, 1971). A suggestion has been made, based on indirect data, that in parthenogenetic turkeys the restoration of the initial (diploid) number of chromosomes occurs after the first cleavage division (Darcey et al., 1971; DeFord et al, 1979; Harada and Buss, 1981). In a diploid parthenogenetic form of *Carassius auratus* the chromosome number is regulated by chromosome replication without cytokinesis during the first cleavage division (Lieder, 1959).

4.6. GENETIC MECHANISMS OF SEX DETERMINATION

The genetic mechanisms of sex determination are not considered in any detail in this brief summary. Here we consider a special aspect, namely sex determination during meiosis in natural parthenogenesis. There are numerous genetic mechanisms of sex determination in animals (see Myasoedov, 1935; Crew, 1954; Beatty, 1964, 1970a, b; Vorontsov, 1966; Gold, 1979; Bull, 1980; etc.), but in the simplest case they amount to two basic ones (see Rott, 1963; Yurovitsky, 1966; Astaurov and Demin, 1972).

The female sex is homogametic (XX♀) and the male heterogametic (XY♂). In this case, as a result of meiosis, all egg cells carry one X chromosome, and the sperms an X or Y chromosome (in a 1:1 ratio). On fertilization the female pronucleus (X) unites with the male pronucleus carrying either an X or Y chromosome, and either female (XX♀) or male (XY♂) individuals are formed. This type of sex determination occurs in most animals, for example in nematodes, the overwhelming majority of insects, among amphibians in the higher *Anura*, in most reptiles and some fishes, and in mammals, including man.

2. The female sex is heterogametic (ZW♀) and the male homogametic (ZZ♂). In this case, as a result of meiosis, the egg cells carry either a Z or W chromosome, and all sperms one Z chromosome. On fertilization the female pronucleus, carrying

either a Z or W chromosome, unites with a male pronucleus (Z) and either male (ZZ♂) or female (ZW♀) individuals are formed. This mechanism of sex determination is typical of caddis flies and butterflies, among amphibians in *Urodela* and lower *Anura*, and some reptiles and fishes, as well as birds.

In both cases the heterogametic sex may also be determined by one heterochromosome (XO♂, ZO♀), the gametes of heterogametic sex carrying one sex chromosome (X, Z) or no sex chromosomes at all. Mechanisms of sex determination in animals with multiple sex chromosomes (e.g., XXY, XYY) have also been described (Pogosyants, 1975).

Below we shall attempt to present a picture of sex determination in natural parthenogenesis with the restoration of the initial chromosome number since such data may prove useful for approaching the problem of artificial sex control in animals.

In the case of homogametic female sex (XX♀) with all ways of restituting the initial chromosome number, the mature egg at the start of its parthenogenetic development will carry a set of sex chromosomes characteristic of the female sex (XX♀) and such parthenogenesis will be all-female (telytokous).

In the case of heterogametic female sex (ZW♀), the sex determination in parthenogenesis will depend on the way the initial chromosome number is restituted. In parthenogenesis with regulation of the chromosome number by a premeiotic doubling of the chromosome number (endoduplication) the nucleus after meiosis contains the initial chromosome set (ZW♀) and all parthenogenetic individuals will be females, as was shown for reptiles of the genus *Cnemidophorus* (Cuellar, 1971) (Fig. 38A).

In cases where meiosis has completely disappeared (see Fig. 38B), the egg cell beginning its parthenogenetic development has a somatic chromosome set (ZW♀). Hence, in this case, too, all the egg cells produce females as, for example, in the mollusc *Campeloma rufum* (Mattox, 1937).

In cases where meiotic division I is suppressed (see Fig. 38C) and meiotic division II is completed, the nucleus also contains the initial chromosome set (ZW♀) and, as in the two preceding cases, a telytokous (all-female) parthenogenesis is observed. This mechanism is characteristic of the development of a parthenogenetic race of *Carassius auratus* (Cherfas, 1966; Kobayasi, 1971).

On suppression of meiotic division II (see Fig. 38D) egg cells form with a diploid homozygous chromosome set (ZZ♂ and WW-let). In this case, it is an all-male (arrhenotokous) parthenogenesis with a lowered (owing to total homozygosity) viability. Such a mechanism of sex determination is possible in parthenogenetic individuals of some birds (Poole and Olsen, 1957; Sato and Kosin, 1960; Olsen, 1974). However, studies in parthenogenesis of birds have shown that male (homogametic) individuals may be heterozygous by one or several loci. In these cases heterozygosity may be accounted for (Astaurov and Demin, 1972) by the assumption that diploidy is restored by suppression of meiotic division II, the meiotic divisions having been preceded by crossing-over (Fig. 39).

Lastly, in cases where a normal haploid egg is formed in the process of meiosis and its diploidy is restituted, as a result of reduplication of the chromosome with the doubling of their number after the first cleavage division or the union of the cleavage nuclei, the parthenogenetic development proceeds in accordance with the male type (ZZ♂) (see Fig. 38E). This does not exclude that precisely such a method of diploidy restitution and sex determination is characteristic of the insects *Lecanium cerasifex* and *Pulvinaria hydrangeae* (Nur, 1971) and some birds (Astaurov and Demin, 1972; Harada and Buss, 1981).

In cases where both meiotic divisions are complete with the "fertilization" of the egg by the subsequent return of one of the polar bodies (or nuclei) (see Fig. 38F) both female and male individuals may develop (4ZW♀:1ZZ♂:1WW-let). This is bisexual (deuterotokous) parthenogenesis.

An examination of all the possible ways of restituting the initial chromosome number shows that in cases of heterogametic female sex all-female parthenogenesis takes place by endoduplication (premeiotic doubling of the chromosome number) and when meiosis is completely lost, as well as by suppression of meiotic division I. All-male parthenogenesis arises upon suppression of meiotic division II or restoration of the initial chromosome number in the course of cleavage divisions. A combination of these mechanisms, as well as "fertilization" of the egg cell by one of the polar bodies (nuclei), produces a bisexual parthenogenesis.

All ways of restituting the initial chromosome number may occur only in diploid parthenogenetic forms. The triploidy of parthenogenetic forms may be maintained only by premeiotic doubling of the chromosome number, complete loss of meiosis, or suppression of meiotic division I. In all other ways of restituting the initial chromosome number synapsis takes place and this essentially hampers the formation of trivalents (Uzzell, 1970).

An analysis of sex determination in natural parthenogenesis indicates possible ways of artificial sex control in animals. The

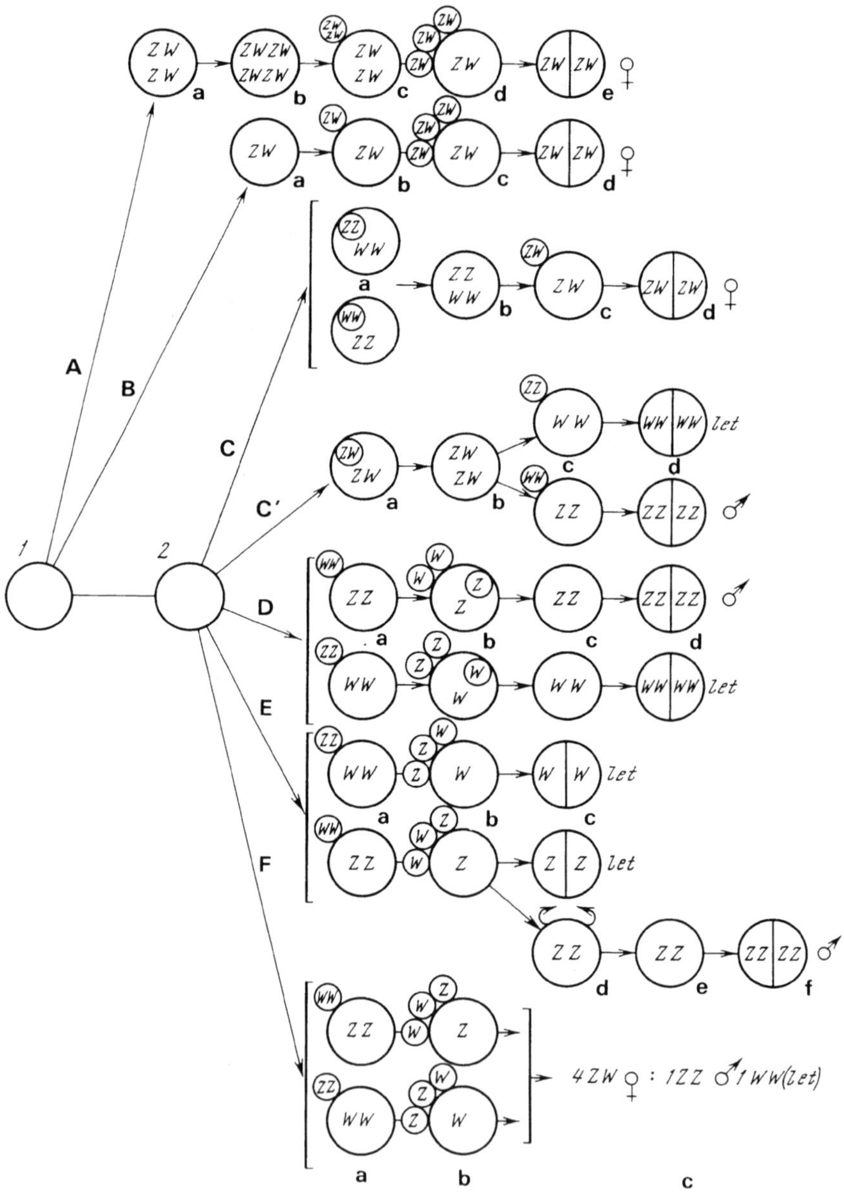
A
B
C
C′
D
E
F
1
2
let
4ZW ♀ : 1ZZ ♂ 1WW(let)

existence of all-female and all-male parthenogenesis in the natural populations of vertebrates and the successful control of sex in the silkworm *Bombyx mori* (Astaurov, 1940, 1958, 1967; Astaurov and Ostryakova-Varshaver, 1957; Strunnikov, 1975, 1983) suggest a promising approach to the problem of artificial sex control in other animals by experimental manipulation of egg cells at various stages of meiosis. This is the reason why data on the influence on meiosis of various factors and on differential sensitivity of the stages of meiosis are of special interest.

4.7. EFFECT OF EXTERNAL FACTORS ON MEIOSIS

The effect of external factors on meiosis has been examined from different points of view (see Rott, 1974; Smirnova et al.,

---

Fig. 38. Diagram of the possible ways of restituting the initial chromosome number and sex determination in parthenogenesis in animals with female heterogamety (ZW) (after Astaurov and Demin, 1972; Strunnikov, 1975, modified). The diagram shows that in animals with heterogametic female sex the appearance of all-female-parthenogenetic progeny is possible via pathways A, B, and C; D and E lead to the appearance of all-male progeny. A combination of pathways C and C′ or F produce bisexual parthenogenesis. A. Diploid meiotic parthenogenesis with premeiotic doubling of the chromosome number (endoduplication): a) oogonium with doubled chromosome number (4n); b) prophase (8n); c) meiotic division I; d) meiotic division II; cleavage division of diploid egg cell (♀). B. Diploid ameiotic parthenogenesis (both divisions equational): a) prophase I (2n); b) division I; c) division II; d) cleavage division of diploid egg cell (♀). C. Diploid ameiotic parthenogenesis with suppression of division I (in cases of prereduction): a) division I incomplete (or absent); b) restoration of tetraploidy; c) division II; d) cleavage division of diploid egg cell (♀). C′. Diploid meiotic parthenogenesis with suppression of division I (in case of postreduction): a) division I incomplete (or absent); b) restoration of tetraploidy; c) division II; d) cleavage division of diploid egg cell (ZZ♂ or WW-let). D. Diploid meiotic parthenogenesis with suppression of division II: a) division I; b) division II incomplete (or absent); c) reduction of diploidy; d) cleavage division of diploid egg cell (ZZ♂ or WW-let). E. Normal manner of appearance of haploid egg cell and meiotic (automyctic) diploid parthenogenesis: a) division I; b) division II; c) cleavage division of haploid egg cell; d, e) restoration of diploidy; f) cleavage division of diploid embryo (♂). F. Restoration of diploidy by "fertilization" with a polar body (nucleus): a) division I; b) division II; c) distribution of sex chromosomes after the union of the female pronucleus with the nucleus of one of the polar bodies; 1) oogonium; 2) prophase I (4n).

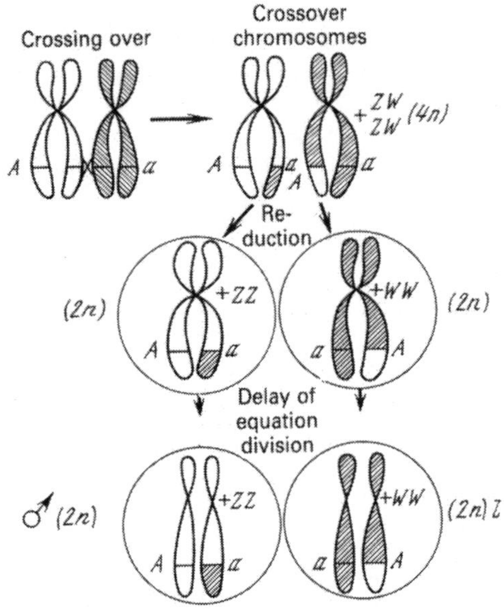

Fig. 39. Appearance of heterozygous parthenogenetic individuals on suppression of meiotic division II (after Astaurov and Demin, 1972). Division I is preceded by crossing-over. All other explanations in the text.

1977). In this section, emphasis is placed on examining the results and mechanisms of the action of external factors on different stages of meiosis.

4.7.1. Temperature

Different doses of heat shock affect the same cells variously, and the same dose variously affects different cell structures (Belehradek, 1935; Nasonov and Aleksandrov, 1940; Heilbrunn, 1952; etc.). The use of heat shock to affect the nuclear apparatus, in particular, is based on the relatively greater thermosensitivity of the division spindle compared with the other cell components. The action of high temperature on dividing cells constitutes an effective means of polyploidization of the embryo (Fankhauser, 1942, 1945; Beatty, 1957; Bungenberg de Jong, 1957; Lieder, 1964a, b; Astaurov, 1965; Rott, 1974; etc.). The biological effect of such a heat shock is determined to a considerable extent by the stage at which it is applied.

The application of a heat shock to mature *Bombyx mori* eggs during metaphase I (Frolova, 1948; Ostryakova-Varshaver, 1958) results in heat damage to the spindle of meiotic division I and prevents the completion of this division and the reduction in the chromosome number. In such eggs, meiotic division II proceeds normally and a diploid female pronucleus with an unreduced chromosome number is formed. Destruction of the I meiotic spindle after heat shock was also noted in the newt (*Triturus viridescens*) oocytes (Humphries, 1956).

Heat treatment of eggs at metaphase and anaphase II in *Triturus viridescens* (Fankhauser and Godwin, 1948; Holmquist, 1952), in the axolotl *Ambystoma mexicanum* (Sladecek and Lanzova, 1959; Rott and Betina, 1964; Betina and Rott, 1966), as well as in the frog *Rana pipiens* (Briggs, 1947) and the mouse (Braden and Austin, 1954), destroys the II meiotic spindle. Irreversible destruction of the spindle gives one or, less frequently, two haploid pronuclei. In cases where the spindle is restored after heat damage, it often sinks deep into the cytoplasm of the egg and, as a result, the polar body is not extruded and two haploid pronuclei are formed.

Thus, in the course of meiotic divisions I and II with certain doses of heat shock the division spindle is destroyed first. Following discontinuation of the heat shock, the spindle may, in some cases, be restored. The rate of the spindle restoration after cessation of the heat shock depends on the temperature used and the duration of the shock.

The data on heat treatment of the egg cells at different stages of meiosis are now examined in more detail. There have been only a few studies in which egg cells were subjected to heat shock during metaphase or other phases of meiotic division I. Heat treatment (46°C for 18 min) of *Bombyx mori* eggs at metaphase I leads to their activation (80-100%) and the hatching of diploid larvae (about 60%) (Astaurov, 1940; Frolova, 1948). Heat treatment of the mature eggs of *Nereis limbata* during metaphase I also suppressed meiotic division and in some cases the polar body did not extrude. As a result, the blastomeres of the developing eggs had a heteroploid chromosome number, and abnormal larvae hatched (Lovelage, 1951).

Heat shock of newt oocytes [37°C during prometaphase to anaphase I (Humphries, 1956)] and sturgeon oocytes (*Acipenser güldenstädti*) [34°C during prometaphase to metaphase I (Vassetzky, 1967a)] led only to a reversible suppression of meiotic division I, apparently due to an insufficient heat dose.

There are many more studies in which eggs were subjected to heat treatment during meiotic division II (Table 4.3). Fankhauser

Table 4.3. Most Effective Doses of Heat Shock to Eggs of Vertebrates during Meiotic Division II Aimed at Altering the Ploidy of Embryos

Species	Temperature, °C	Duration of shock, min	Polyploidy, %	Reference
Triturus	37	5	93.8	Fankhauser and Watson, 1942
	36	8	78.1	
	36	10	50.0	Fankhauser and Godwin, 1948
	38	10	90.5	Blair, 1952
Ambystoma mexicanum	36-37	10	29.0	Sladecek, Lanzova, 1959
	37	10	23.4	Sladecek and Romanovsky, 1960
	36	10	81.0	Rott and Betina, 1964; Rott, 1965
Pleurodeles waltlii	37	5	93.1	Ferrier and Jaylet, 1978
Xenopus laevis	36.1-36.5	5	54.0	Smith, 1958
	36.5-37.0	5-10	90.0	Cassidy and Biackler, 1980
Rana pipiens	37	4	41.0	Briggs, 1947
Bufo vulgaris formosus	33	9	83.7	Muto, 1951, 1957
	34	7	85.0	
Gasterosteus aculeatus	34	5	50.0	Swarup, 1959
Misgurnus fossilis	34	4-5	16.7	Romashov and Belyayeva, 1965
Acipenser güldenstädtl colchicus	34	3	52.3	Vassetzky, 1967a
Oryzias latipes	41	3	6-70	Naruse et al., 1985
Mus musculus	45.0-46.5	4-10	11.0	Fischberg and Beatty, 1952

and his co-workers (Fankhauser and Watson, 1942; Fankhauser, 1945; Fankhauser and Godwin, 1948) applied a heat shock to newt eggs immediately after fertilization at metaphase II and obtained triploid larvae, i.e., meiotic division II was suppressed. These results were confirmed on the same species by others (Erdman and Thompson, 1943; Blair, 1952; Holmquist, 1952), as well as on *Pleurodeles waltlii* (Ferrier and Jaylet, 1978).

Triploid larvae were also obtained in the axolotl by heat treatment of the fertilized eggs at metaphase II (Sladecek and Lanzova, 1959; Sladecek and Romanovsky, 1960). Detailed studies of the effect of heat shocks on fertilized axolotl eggs (Rott and Betina, 1964; Rott, 1965; Betina and Rott, 1966) showed that heating during anaphase II is the most effective; most of the larvae which hatched proved to be triploid.

In experiments on *Rana pipiens* (Briggs, 1947; Humphrey et al., 1950) the maximum number of triploids was obtained by treatment of the eggs 15-20 min after insemination, which apparently corresponds to anaphase II (Porter, 1939). Heating 15-20 min after insemination proved to be the most effective on the toad (*Bufo vulgaris formosus*) eggs (Muto, 1952, 1957); here this stage appears to correspond to anaphase II as well. Heat shock was also successfully applied to the *Xenopus laevis* eggs in metaphase II (Smith, 1958; Cassidy and Blackler, 1980).

As for fish, cases of successful heat treatment of the eggs during meiotic division II are also known. In experiments on the stickleback *Gasterosteus aculeatus* (Swarup, 1959) the highest number of triploids was obtained by heat treatment of the eggs at metaphase to anaphase II. Heating of loach eggs (*Misgurnus fossilis*), inseminated with irradiated sperm, during anaphase I (Romashov and Belyaeva, 1965) yielded diploids (gynogenetic). Heat treatment of fertilized sturgeon eggs at metaphase II also led to the suppression of meiotic division II and development of triploid embryos and larvae (Vassetzky, 1967a). The eggs of the rainbow trout *Salmo gairdneri* subjected to heat shock at metaphase-anaphase II developed also as triploid embryos (Chourrout, 1980; Thorgaard et al., 1981, 1984; Chourrout and Quillet, 1982; Chevassus et al., 1983; Lincoln and Scott, 1983; Refstie, 1983; Lincoln and Bye, 1984). Diploid gynogenesis was also obtained in this species (Thompson and Scott, 1984). Heat shocks were successfully applied to *S. salar* (Refstie, 1983; Benfey and Sutterlin, 1984), *S. trutta*, *Salvelinus fontinalis*, and *Oncorhynchus kisutch* (Chevassus et al., 1983), and *Oryzias latipes* (Naruse et al., 1985).

Heat treatment of mouse eggs at metaphase to anaphase II resulted in the development of triploids due to the suppression of meiotic division II (Fischberg and Beatty, 1952).

Studies of differential thermosensitivity of oocytes at different phases of meiotic division I and II were carried out only on three species of animals; they have shown that the most thermosensitive phases of meiotic divisions (judging by the survival of the embryos or larvae) are anaphase I and metaphase II in the silkworm (Astaurov et al., 1958) and metaphase II in the frog (Briggs, 1947) and the axolotl (Rott, 1965).

A comparison of the thermosensitivity of the same phases of meiotic divisions I and II was made for silkworm eggs (Astaurov et al., 1958) and sturgeon eggs (Vassetzky, 1966). It turned out that (judging by the survival of the embryos) in the silkworm eggs the thermosensitivity of the two divisions was very close, but in the eggs of the sturgeon they differed greatly, the oocytes at metaphase I being much less thermosensitive than the mature eggs at metaphase II (Vassetzky, 1966). The change in thermosensitivity in the course of maturation of sturgeon oocytes correlates with the change in the duration of meiotic divisions and is, possibly, also connected with hydration of the oocyte cytoplasm since it is well known that the sensitivity of the cells to heat shock depends to a very great extent on the viscosity of the cytoplasm and its water content (Belehradek, 1957).

Similar effects are also produced by low temperatures. A cold shock applied to freshly laid fertilized eggs (metaphase II) of the newt (Fankhauser and Griffiths, 1939; Griffiths, 1941) demonstrated that about 50% of the larvae that hatched out after the cold shock (0-3°C for 5 h) turned out to be triploid. Similar results were obtained using other newt species (Kawamura, 1941; Fankhauser et al., 1942; Fischberg, 1944, 1947, 1948; Böök, 1945), the axolotl (Fankhauser and Humphrey, 1943), *Pleurodeles waltlii* (Gallien, 1953, 1959, 1962; Beetschen, 1957, 1960, 1964), *Xenopus laevis* (Kawahara, 1978; Obara et al., 1983; Nakamura et al., 1985), the fish *Gasterosteus aculeatus* (Swarup, 1959), *Pleuronectes platessa* (Purdom, 1972), *Thilapia aurea* (Valenti, 1975), *Salmo gairdneri* (Chourrout, 1980), *Cyprinus carpio* (Makino and Ozima, 1943; Gervai and Csányi, 1978; Nagy and Csányi, 1978; Nagy et al., 1979, 1983; Gervai et al., 1980; Ueno, 1984), *Salvelinus fontinalis* (Lemoine and Smith, 1980), *Ictalurus punctatus* (Wolters et al., 1981), *Rhodeus ocellatus ocellatus* (Ueno and Arimoto, 1982), *Gnathopogon elongatus caerulescens* (Ueno, 1985), and mice (Beatty and Fischberg, 1949; Karp and Smith, 1975). The cold treatment of the fertilized eggs of these species at metaphase-anaphase II yielded triploids. Diploid gynogenesis (all-female progeny) was obtained by cold treatment in *Oncorhynchus kisutch* (Teleostei) (Refstie et al., 1982).

The application of cold to unfertilized silkworm eggs made it possible to obtain all-male diploid parthenogenesis. The genetic and cytological analysis of these individuals has shown that in this case both meiotic divisions take place, and that diploidy is restored in the process of early cleavage by fusion of the haploid nuclei (Terskaya and Strunnikov, 1974, 1975; Terskaya, 1982). In these studies it was established that the same effect (heat shock, cold shock, electric shock, action of CO_2) may, depending on the dose, cause either ameiotic parthenogenesis without the reduction in chromosome number (all-female, in the silkworm) or meiotic parthenogenesis.

Attempts have also been made to apply cold shock to unfertilized eggs of mammals (at metaphase II). There are reasons to assume that, in this case, there is also a suppression of meiotic division II and restitution of diploidy (Pincus and Shapiro, 1940a, b; Thibault, 1949; Chang, 1952, 1954).

A study of the differential sensitivity of the silkworm to cooling (Pokrovskaya, 1958) has shown anaphase I and metaphase II to be the most sensitive, i.e., the same stages that are characterized by increased thermosensitivity (Astaurov et al., 1958).

4.7.2. *Ionizing Radiation*

During meiotic divisions the egg cells are particularly sensitive to irradiation, which may lead to the immediate destruction of the nucleus followed by cytoplasm fragmentation and nuclear pycnosis, cessation of development during early stages of cleavage, or, in cases of continued development, formation of chromosomal aberrations and morphological defects during embryonic development (see Rugh and Grupp, 1959; Delone, 1960; etc.). Radiosensitivity changes markedly in the course of meiosis. Studies of the differential radiosensitivity of insect oocytes at different stages of meiotic division I have shown the metaphase and anaphase to be the most sensitive (*Habrobracon*, Whiting, 1940, 1945; von Borstel and Amand, 1963; *Sciara ocellaris*, Bozeman, 1943; Bozeman and Metz, 1949; *Drosophila melanogaster*, Wurgler et al., 1963; Traut, 1967; Koch et al., 1970; Savontaus, 1980; *Mormoniella*, Ray, 1957, 1958; *Bombyx mori*, Murakami, 1969a, b, 1971; *Carausius morosus*, Pijnancker, 1971).

Studies on vertebrates have shown that in the course of meiosis the radiosensitivity of oocytes increases [*Rana pipiens* (Rollason, 1949); rat (Mandl, 1963b; Mandl and Beaumont, 1964); mouse (Rugh and Grupp, 1959; Edwards and Searle, 1963); Chinese hamster (Sonta, 1981)], reaching its maximum after fertilization at telophase II (mouse, Russell and Saylors, 1963). For the silkworm it was also established that the stages of meiotic division II are more radiosensitive than the corresponding stages of division I (Murakami, 1969b, 1971).

Somewhat different data were obtained for the *Artemia salina* oocytes which proved to be more radiosensitive at prophase I than at metaphase I (Cervini and Giavelli, 1965), and the guinea pig in which the radiosensitivity during meiosis hardly changes from late diplonema of prophase I (Harvey and Chang, 1963).

The increase in radiosensitivity observed in most animals during meiosis may be due to a change in configuration of the genetic material, condensation of chromosomes in particular (Sparrow, 1951), their spiralization (Muller, 1959), chromatid coiling (Koch et al., 1970), and, lastly, the presence or absence of the nuclear envelope (Koch et al., 1970).

It is important to note that when radiosensitivity is determined using different criteria, its maximum levels do not quite coincide. Thus, for the silkworm it was shown that metaphase to anaphase II is the most sensitive as regards mutation frequency, and telophase II as regards mortality (Murakami, 1969a).

4.7.3. *Chemical Substances and Other Factors*

Substances producing an antimeiotic effect, namely colchicine, colcemide, vinblastine, cytochalasin B, etc., are the most commonly used in experiments with oocytes. These substances destroy the microtubules and disintegrate the division spindles blocking the cells at metaphase. The effect of colcemide and vinblastine is completely reversible, that of colchicine to a lesser extent.

The application of these substances during meiosis leads to the suppression of meiotic division I or II, and thus to a doubling of the chromosome set in the egg. Colchicine was successfully used on the eggs of the perch and rainbow trout (Lieder, 1964b), the brook trout (Smith and Lemoine, 1979), the rat (Austin and Braden, 1954; Piko and Bomsel-Helmreich, 1960), the mouse (Edwards, 1954; McGaughey and Chang, 1969b), and the rabbit (Bomsel-Helmreich and Thibault, 1962). Colcemide was also successfully used on the eggs of mice (Jagiello, 1969) and domestic fowl (Wang and Choffer, 1980), vinblastine on the eggs of annelids (Malavista et al., 1968), vincristine on the eggs of mice (Jagiello, 1969), and cytochalasin B on the eggs of starfish (Tsukahara and Ishikawa, 1980), molluscs (Stanley et al., 1981; Allen et al., 1982), fish (Allen and Stanley, 1979, 1981; Allen, 1983), and mice (Balakier and Tarkowski, 1976; Niemerko and Komar, 1976; Opas, 1977; Borsuk, 1982). The sensitivity of oocytes to mitostatics is characterized by certain specific features, the oocytes of the hamster being relatively resistant to the effect of colchicine and vinblastine (Bray and Hancock, 1973).

A mitostatic effect on oocytes during meiosis is also produced by other substances, for example griseofulvin (Malavista et al., 1968), which disturbs the orientation of the spindle microtubules rather than their integrity (Grisham et al., 1973), daunomycin (Jagiello and Lin, 1973), and cytochalasin B (Longo, 1972; Niemerko, 1975).

It was also shown that an antimeiotic effect is produced by lithium chloride (*Limnaea stagnalis*, Groot, 1948) and heavy water (*Asterias forbessi*, Marsland and Hiramoto, 1966).

Some other factors also have an antimeiotic effect. The treatment of the fertilized mouse eggs with distilled water, i.e., osmotic shock, results in the suppression of meiotic division II and development of triploid embryos (Opas, 1977).

Hydrostatic pressure has proved to be one of the most effective and, at the same time, mild antimeiotic agents (Pease, 1946). It was first used by Dasgupta (1962) on fertilized *Rana pipiens* eggs, 5-10 min after insemination (at metaphase to anaphase II). The treatment at metaphase II was more effective; up to 85% of the normally developing embryos were triploid. The cytological analysis has shown that hydrostatic pressure at metaphase II prevents the anaphase chromosome movement and extrusion of the polar body, whereas under the effect of this treatment at anaphase II the spindle submerges into the cytoplasm, but the II polar body still extrudes in most eggs. The antimeiotic effect of hydrostatic pressure also has been demonstrated for some other amphibian species [*Ambystoma mexicanum*, (Gillespie and Armstrong, 1978, 1979; Armstrong, 1984); *Pleurodeles waltlii* (Jaylet and Ferrier, 1978); *Xenopus laevis* (Müller et al., 1978; Tompkins, 1978; Thiébaud et al., 1984)] and fish [*Brachydanio rerio* (Streisinger et al., 1981); *Oncorhynchus masou*, *O. keta* (Onozato, 1981, 1984); *O. tschawytscha*, *O. kisutch*, *Salmo gairdneri* (Allen and Myers, 1983; Chourrout, 1984; Lou and Purdom, 1984a, b; Onozato, 1984; Naruse et al., 1985); *Salmo salar* (Benfey and Sutterlin, 1984); *Misgurnus fossilis* (Vassetzky et al., 1984; Betina et al., 1985; Vassetzky and Betina, 1986)]. The most recent experiments on *M. fossilis* oocytes have shown that meiotic division I can also be suppressed, and after insemination with x-ray-irradiated sperm, gynogenetic diploid larvae can be obtained (Vassetzky, Betina, and Kondratieva, unpublished).

4.8. CONTROL OF MEIOSIS

With respect to the artificial control of meiosis and, hence, sex, of special interest are vertebrates, namely fishes, amphibians, reptiles, birds, and, above all, mammals. Cases of complete parthenogenesis among vertebrates are fairly rare. True parthenogenesis occurs in the first three groups; however, in birds, it is reported only for turkeys; this can hardly be called natural parthenogenesis since it correlated intimately with the susceptibility to fowl pox virus, which can somehow induce parthenogenetic development (Olsen and Buss, 1967). One can hope to obtain parthenogenesis in chickens as well, since the eggs of infected birds tend to be pathenogenetically activated (Kudryavtzev et al., 1981).

No reliable cases of complete parthenogenesis were registered in mammals (see review by Stevens, 1980), although mouse cells of parthenogenetic origin can survive and take part in normal organogenesis when combined with the normal cells in chimeric mice

(Stevens et al., 1977); they can even give rise to germ cells in such chimeric embryos (Stevens, 1978) and to the normal progeny after crossing.

Nature indicates how such barriers may be overcome. There are species among fish, amphibians, and reptiles which reproduce by gynogenesis, i.e., although insemination of sperm occurs, the genetic material does not take part in the subsequent development of the egg (see reviews by Cherfas, 1979; Schultz, 1980; Chourrout, 1982b; Wohlfarth, 1983; Yamazaki, 1983). They include genetically and cytologically well-studied species of the genus *Ambystoma* among amphibians and species of the genera *Poeciliopsis*, *Peocilia*, and *Carassius* among fish. All these species are represented by all-female populations. All-female populations of fish have been discovered by electrophoretic examination of the genus *Menidia* [*M. clarkhubbsi* (Echelle and Mosier, 1981, 1982)]. The genetic structure of *M. clarkhubbsi* is readily accounted for by its origin via hybridogenesis between *M. beryllina* and *M. peninsulae*.

A triploid all-female population of *Cobitis taenia* is known (Vasiliev, 1981, 1985; Vasiliev and Vasilieva, 1982). Polyploid forms of *Cobitis taenia* were earlier reported by Japanese authors (Ueno and Ojima, 1976; Ueno et al., 1980): almost all triploid fish were females. By analogy with the already-studied forms, Vasiliev and Vasilieva (1982) suggested that this triploid all-female population reproduces by gynogenesis.

The discovery of new unisexual forms has shown that unisexual populations (or species) and gynogenesis can occur more widely among the vertebrates than previously accepted. An apparent rarity of unisexual forms can be due to difficulties in finding them (see, e.g., Novikov, 1962).

On the basis of these investigations, a method was elaborated to obtain diploid gynogenetic progenies: the eggs were subjected to thermal shock or hydrostatic pressure before or soon after insemination with the sperm inactivated by chemical substances, UV or ionizing irradiation, or with heterologous sperm (see reviews by Cherfas, 1979; Chourrout, 1982a). Such treatment results in the suppression of the extrusion of the II polar body and diploid chromosome number restoration (see, for example, Golovinskaya et al., 1963; Romashov and Belyaeva, 1964; Golovinskaya and Romashov, 1966; Purdom et al., 1976; Purdom, 1983; Thorgaard, 1983), whereas the fertilization stimulus introduced by the sperm normalizes the subsequent development of the diploid embryo. This method has been successfully applied to a number of fish species: *Acipenser güldenstädti* (Romashov et al., 1963; Vassetzky, 1967a), *Cyprinus carpio* (Cherfas, 1975; Nagy et al., 1978; Cherfas and Ilyasova, 1978; Wu et al., 1981), *Misgurnus fossilis* (Romashov and Belyaeva,

1965; Vassetsky and Betina, unpublished), *Ctenopharyngodon idella* (Stanley, 1976; Nagy et al., 1978), *Salmo gairdneri* (Chourrout, 1980, 1981, 1982a), *Pleuronectes platessa* and *Platichthys flesus* (Purdom, 1969; Purdom and Lincoln, 1974), and *Brachydanio rerio* (Streisinger et al., 1981). Induced diploid gynogenesis has also been achieved in amphibians: *Ambystoma mexicanum* (Trottier and Armstrong, 1976; Gillespie and Armstrong, 1979), *Pleurodeles waltlii* and *P. poireti* (Jaylet and Ferrier, 1978), *Rana pipiens* (Volpe and Dasgupta, 1962; Nace et al., 1970), and *Xenopus laevis* (Kawahara, 1978; Tompkins, 1978).

The preimplantation development of gynogenetic diploid embryos has been observed also in mice (Borsuk, 1982): the eggs were fertilized in vitro and treated with cytochalasin B to suppress the second polar body extrusion, and then the male pronucleus was removed by microsurgery.

Diploid gynogenesis resulting from the suppression of the second polar body extrusion is, however, accompanied by a reduced viability in the gynogenetic embryos and reduced fecundity of gynogenetic females (Cherfas, 1979; Donaldson and Hunter, 1982). This imposes serious limitations on the application of methods affecting the second meiotic division in practical genetics and selection. In any case, the production of all-female populations by induced diploid gynogenesis with suppression of the second meiotic division is hardly advisable (see also Stanley et al., 1975). But there are other ways of utilizing this technique, e.g., the practical selection of strains obtained by chemical mutagenesis (Tsoi, 1981), or its use for inbreeding (Thompson, 1983) and production of clones (Naruse et al., 1985). It is also promising to use triploid forms in practical selection and aquaculture since triploids often show advantages in the rate of growth and maximum weight (Chevassus et al., 1979; Lincoln and Bye, 1984; Lincoln and Scott, 1984).

Of special interest is the application of this method for the mass production of all-female fish populations according to the scheme proposed by Chevassus (Chevassus et al., 1979; Chevassus, 1980). This scheme consists of three steps: 1) All-female progeny are obtained by diploid gynogenesis; 2) the fry are treated with male hormones and, as a result, phenotypical sex inversion occurs and neomales, i.e., phenotypical males with the female genotype, are produced; and 3) neomales are crossed with females and all-female progenies are produced. The first step (induced diploid gynogenesis), as shown above, has already been achieved for a number of species. The second step (hormonal sex inversion) is also quite possible; the introduction of methyl testosterone with food results in a shift in the sex ratio (masculinization) in the rainbow trout (Jalabert et al., 1974, 1975), the carp (Nagy et al., 1981; Wu et al., 1981), and a number of other fish species

(Hunter and Donaldson, 1980; Donaldson and Hunter, 1981; Nagy et al., 1981; Wu et al., 1981; Gomelsky and Cherfas, 1982a, b; Calhoun and Shelton, 1983; Boney et al., 1984; Shelton et al., 1984; Gomelsky, 1985; see also reviews by Gomelsky, 1980, and Stanley, 1981). This means that the scheme for producing all-female viable populations can be realized in fish and probably in other vertebrates.

An interesting phenomenon of experimental sex inversion under the effect of elevated temperatures in amphibians should also be mentioned in this respect (Houillon and Dournon, 1978; Dournon, 1982; Dournon and Houillon, 1983, 1984, 1985; Dournon et al., 1984). A similar type of sex regulation has also been noted in reptiles (Bull, 1980).

We have already mentioned the possibility of diploidization of mammalian eggs by the suppression of the extrusion of the II polar body. But premeiotic endoduplication is also quite plausible. In the Chinese hamster giant oocytes exist (up to 0.5% of the whole population) which proceed through normal meiosis, become fertilized, and develop until at least the second cleavage division (Funaki and Mikamo, 1980a, b; Funaki, 1981), giving rise to triploid digynic embryos. This gives hope for successful application of the method of induced diploid gynogenesis in mammals as well.

4.9. CONCLUDING REMARKS

The factor that causes arrest of meiosis is important but so far the least clear. It would be very interesting to find out why this arrest occurs at different stages of meiotic divisions in different animals.

Of particular importance are further studies in artificial parthenogenesis in animals for the purpose of artificial sex control. A fundamentally important step in this direction, the production of all-female progeny in the silkworm by suppression of meiotic division I in mature eggs by means of heat shock, was made by Astaurov (1940).

An analysis of the published data shows that in vertebrates, as well as in invertebrates, it is possible with temperature shock or hydrostatic pressure to selectively damage the spindle of meiotic division and, as a result, to restitute the diploid chromosome number. The use of temperature shock or hydrostatic pressure to prevent meiotic division I seems most promising. Such influence excludes the reduction in the chromosome number, and leads to all-female parthenogenesis or gynogenesis on fertilization of the eggs with inactivated or foreign sperm in cases of both heterogametic and homogametic female sex.

Application of temperature shock or hydrostatic pressure to the egg cells in the course of meiotic division I involves certain technical difficulties, since in vertebrates meiotic division I takes place in the body of the female. Moreover, effective suppression of meiotic division I requires larger doses of the heat shock than does suppression of division II because of the much lower thermosensitivity of oocytes at these stages. However, methods of surmounting the difficulties under experimental conditions, at least in the lower vertebrates, have now been elaborated; it is now possible to cultivate maturing oocytes in vitro, determine the periods of the different phases of meiotic division I, and obtain gynogenesis (see Vassetzky, 1967b).

The result of suppressing meiotic division II, which is possible even in mammals, will depend, as the published data show, on the genetic mechanism of sex determination. In cases of homogametic female sex, the progeny will be all-male. Successful employment of this method, however, requires one more condition, namely that crossing-over must take place during meiotic division I because suppression of meiotic division II gives rise to complete homozygotes with diminished viability and fertility. This way of affecting meiosis experimentally, nevertheless, seems promising, especially when it is necessary to obtain all-male progeny, the more so since this type of parthenogenesis is in nature apparently encountered in birds (see Olsen, 1974).

Very promising also is production of unisexual progeny by means of experimental androgenesis (Astaurov and Ostryakova-Varshaver, 1957). Studies concerning the problem of conservation of genetic resources of animal species have been progressing, and the application of the methods of artificial parthenogenesis (and gynogenesis) in these studies is of great importance (see Veprintsev and Rott, 1979; Rott and Veprintsev, 1981).

As the research discussed in this chapter shows, studies of meiotic divisions open very promising approaches to the control of the processes of ontogenesis in animals.

REFERENCES

Abirachned-Darmency, M., 1982, "Synaptonemal complex formation and polycomplex occurrence in the ameiotic prophase of *Meloidogyne incognita* (*Nematoda*)," Biol. Cell., *46*, 133-142.

Aimar, C., Delarue, M., and Vilain, C., 1981, "Cytoplasmic regulation of the duration of cleavage in amphibian eggs," J. Embryol. Exp. Morphol., *64*, 259-274.

Allen, R. D., 1953, "Fertilization and artificial activation in the egg of the surf clam *Spisula solidissima*," Biol. Bull., *105*, 213-239.

Allen, S. K., Jr., 1983, "Flow cytometry: Assaying experimental polyploid fish and shellfish," Aquaculture, *33*, 317-328.

Allen, S. K., Jr., and Myers, J. M., 1983, "Chromosome set manipulation in three species of salmonids using hydrostatic pressure," in: Salmonid Reproduction, Ann. Int. Symp., Seattle, Univ. of Washington, pp. 9-10.

Allen, S. K., Jr., and Stanley, J. G., 1979, "Polyploid mosaics induced by cytochalasin B in landlocked Atlantic salmon *Salmo salar*," Trans. Am. Fish. Soc., *108*, 462-466.

Allen, S. K., Jr., and Stanley, J. G., 1981, "Mosaic polyploid Atlantic salmon (*Salmo salar*) induced by cytochalasin B," Rapp. P. V. Réun., Cons. Int. Explor. Mer., *178*, 509-510.

Allen, S. K., Jr., Gagnon, P. S., and Hidu, H., 1982, "Induced triploidy in the soft-shell clam. Cytogenetic and allozymic confirmation," J. Hered., *73*, 421-428.

Armstrong, J. B., 1984, "Genetic mapping in the Mexican axolotl *Ambystoma mexicanum*," Can. J. Genet. Cytol., *26*, 1-6.

Ar-Rushdi, A. N., 1963, "The cytology of achiasmatic meiosis in the female *Tigriopus* (*Copepoda*)," Chromosoma, *13*, 528-539.

Asher, J. H., Jr., and Nace, G.W., 1971, "The genetic structure and evolutionary fate of parthenogenetic amphibian populations as determined by Markovian analysis," Am. Zool., *11*, 381-396.

Astaurov, B. L., 1940, Artificial Parthenogenesis in the Silkworm [in Russian], Izd. Akad. Nauk SSSR, Moscow—Leningrad.

Astaurov, B. L., 1958, "Ways of controlling the development and vital activities of the silkworm by temperature influences," Tr. Inst. Morfol. Zhivotn., *21*, 5-38.

Astaurov, B. L., 1965, "Experimental polyploidy in animals," in: Polyploidy and Selection [in Russian], Nauka, Moscow—Leningrad, pp. 43-56.

Astaurov, B. L., 1967, "Experimental alterations of the developmental cytogenetic mechanisms in mulberry silkworms (artificial parthenogenesis, polyploidy, gynogenesis, and androgenesis)," Adv. Morphogen., *6*, 199-257.

Astaurov, B. L., 1968, Cytogenetics of Development of the Silkworm and Its Experimental Control [in Russian], Nauka, Moscow.

Astaurov, B. L., and Demin, Y. S., 1972, "Parthenogenesis in birds," Ontogenez, *3*, 95-111.

Astaurov, B. L., and Ostryakova-Varshaver, V. P., 1957, "Production of complete heterospermic androgenesis in interspecific hybrids of the silkworm (experimental analysis of the correlative roles of the nucleus and cytoplasm in development and heredity)," Izv. Akad. Nauk SSSR, Biology, *2*, 154-175.

Astaurov, B. L., Ostryakova-Varshaver, V. P., and Strunnikov, V. A., 1958, "Effect of high temperature in the embryonic development of the silkworm (*Bombyx mori* L.). 1. Regular changes in the thermosensitivity of eggs during their maturation and fertilization in connection with elaborating the techniques of experimental androgenesis," Tr. Inst. Morfol. Zhivotn., *21*, 39-80.

Austin, C. R., 1965, Fertilization, Prentice-Hall, Englewood Cliffs, New Jersey.
Austin, C. R., and Braden, A. W. H., 1954, "Induction and inhibition of the 2nd division in the rat egg and subsequent fertilization," Aust. J. Biol. Sci., *7*, 195-210.
Austin, C. R., and Walton, A., 1960, "Fertilization," in: Marshall's Physiology of Reproduction, Vol. 1, Longmans and Green, New York, pp. 310-416.
Bałakier, H., and Tarkowski, A. K., 1976, "Diploid parthenogenetic mouse embryos produced by heat shock and cytochalasin B," J. Embryol. Exp. Morphol., *35*, 25-39.
Barigozzi, C., 1944, "I fenomeni chromosomici delle cellule germinali in *Artemia salina* Leach," Chromosoma, *2*, 549-575.
Beatty, R. A., 1957, Parthenogenesis and Polyploidy in Mammalian Development, Cambridge Univ. Press.
Beatty, R. A., 1964, "Chromosome deviations and sex in vertebrates," in: Intersexuality in Vertebrates Including Man, C. N. Armstrong and A. J. Marshall (eds.), Academic Press, New York, pp. 17-143.
Beatty, R. A., 1967, "Parthenogenesis in vertebrates," in: Fertilization. Comparative Morphology, Biochemistry, and Immunology, Vol. 1, C. Metz and A. Monroy (eds.), Academic Press, New York, pp. 413-440.
Beatty, R. A., 1970a, "Genetic basis for the determination of sex," Philos. Trans. R. Soc. London, Ser. B, *259*, 3-13.
Beatty, R. A., 1970b, "The genetics of the mammalian gamete," Biol. Rev., *45*, 73-120.
Beatty, R. A., and Fischberg, M., 1949, "Spontaneous and induced triploidy in preimplantation mouse eggs," Nature, *163*, 807.
Beermann, S., 1959, "Chromatin-Diminution bei Copepodeni," Chromosoma, *10*, 504-514.
Beermann, W., 1954, "Weibliche Heterogametie bei Copepoden," Chromosoma, *6*, 381-396.
Beetschen, J. C., 1957, "Quelques aspects cytologiques et morphogénétiques de l'hétéroploïdie expérimentale chez le triton *Pleurodeles waltlii* Michah," Bull. Soc. Zool. Fr., *81*, 189.
Beetschen, J. C., 1960, "Récherches sur l'hétéroploïdie expérimentale chez un amphibien urodèle, *Pleurodeles waltlii* Michah," Bull. Biol. Fr. Belg., *94*, 12-127.
Beetschen, J. C., 1964, "Pentaploïdie expérimentale chez l'amphibien urodèle *Pleurodeles waltlii* Michah," C. R. Acad. Sci. Paris, *258*, 1641-1643.
Belehradek, J., 1935, Temperature and Living Matter, Berlin.
Belehradek, J., 1957, "A unified theory of cellular rate processes based upon an analysis of temperature action," Protoplasma, *48*, 53-71.
Belyaeva, V. N., and Cherfas, N. B., 1965, "On the processes of maturation and fertilization in the eggs of the loach *Misgurnus fossilis* L.," Vopr. Ikhtiol., *5*, 82-90.

Benfey, T. J., and Sutterlin, A. M., 1984, "Triploidy induced by heat shock and hydrostatic pressure in landlocked Atlantic salmon (*Salmo salar*)," Aquaculture, *36*, 359-368.
Bennett, M. D., 1977, "The time and duration of meiosis," Philos. Trans. R. Soc. London, Ser. B, *277*, 201-226.
Bergerard, J., 1962, "Parthenogenesis in the *Phasmidae*," Endeavour, *21*, 137-143.
Betina, M. I., and Rott, N. N., 1966, "Cytomorphological analysis of maturation and fertilization of axolotl eggs after heat shock resulting in the appearance of heteroploid embryos," Tsitologiya, *8*, 475-484.
Betina, M. I., Vassetzky, S. G., and Kondratieva, O. T., 1985, "Effects of hydrostatic pressure on fertilized eggs of the loach," Ontogenez, *16*, 365-374.
Blair, C. B., Jr., 1952, "The effect of temperature treatment on *Triturus* eggs as expressed in larval tail-tip epithelium," J. Exp. Zool., *120*, 343-374.
Bogdanov, Yu. F., and Vassetzky, S. G. (eds.), 1982, Genetics, Biochemistry, and Cytology of Meiosis [in Russian], Nauka, Moscow.
Bomsel-Helmreich, O., and Thibault, C., 1962, "Développement d'oeufs triploïdes expérimentaux chez la lapine," Ann. Biol., Anim., Biochim., Biophys., *2*, 265-266.
Boney, S. E., Shelton, W. L., Yang, S.-L., and Wilken, L. O., 1984, "Sex reversal and breeding of grass carp," Trans. Am. Fish. Soc., *113*, 348-353.
Böök, J. A., 1945, "Cytological studies in triton," Hereditas, *31*, 177-220.
Borsuk, E., 1982, "Preimplantation development of gynogenetic diploid mouse embryos," J. Embryol. Exp. Morphol., *69*, 215-222.
Bozeman, M. L., 1943, "Effects of irradiation on oocytes of *Sciara*," Genetics, *28*, 71.
Bozeman, M. L., and Metz, C. W., 1949, "Further studies on sensitivity of chromosomes to irradiation at different meiotic stages in oocytes of *Sciara*," Genetics, *34*, 285-314.
Braden, A. W. H., and Austin, C. R., 1954, "Fertilization of the mouse egg and the effect of delayed coitus and of hot shock treatment," Aust. J. Biol. Sci., *7*, 552-565.
Bray, M. A., and Hancock, J. L., 1973, "Effects of antimitotic agents on polar-body formation and first cleavage of hamster ova," J. Reprod. Fertil., *23*, 255-261.
Briggs, R., 1947, "The experimental production and development of triploid frog embryos," J. Exp. Zool., *106*, 237-266.
Bull, J. J., 1980, "Sex determination in reptiles," Q. Rev. Biol., *55*, 3-21.
Bungenberg de Jong, C. M., 1957, "Polyploidy in animals," Bibl. Genet., *17*, 111-128.
Burton, R. S., Feldman, M. W., and Swisher, S. G., 1981, "Linkage relationships among five enzyme-coding gene loci in the

copepod *Trigriopus californicus*: A genetic confirmation of achiasmatic meiosis," Biochem. Genet., *19*, 1237-1246.
Cable, R. M., 1971, "Parthenogenesis in parasitic helminths," Am. Zool., *11*, 267-272.
Calhoun, W. E., and Shelton, W. L., 1983, "Sex ratios of progeny from mass spawnings of sex-reversed broodstock of *Tilapia nilotica*," Aquaculture, *33*, 365-371.
Carson, H. L., Wei, I. V., and Niederkorn, J. A., 1969, "Isogenicity in parthogenetic strains of *Drosophila mercatorum*," Genetics, *63*, 619-628.
Cassidy, D. M., and Blackler, A. W., 1980, "Induction of triploidy in *Xenopus laevis* by heat shocking of the egg before and after fertilization," J. Exp. Zool., *211*, 379-385.
Cather, J. N., 1963, "A time schedule of the meiotic and early mitotic stages of *Ilyanassa*," Caryologia, *16*, 663-670.
Cervini, A., and Giavelli, S., 1965, "Radiosensitivity of different meiotic stages of oocytes in parthenogenetic diploid *Artemia salina* Leach," Mutat. Res., *2*, 452-456.
Chang, M. C., 1952, "Fertilizability of rabbit ova and the effects of temperature in vitro on their subsequent fertilization and activation in vivo," J. Exp. Zool., *121*, 351-381.
Chang, M. C., 1954, "Development of parthenogenetic rabbit blastomeres induced by low-temperature storage of unfertilized ova," J. Exp. Zool., *125*, 127-150.
Channing, C. P., Anderson, L. D., Hoover, D. J., Kolena, J., Osteen, K. G., Pomerantz, S. H., and Tanabe, K., 1982, "The role of nonsteroidal regulation in control of oocyte and follicular maturation," Recent Prog. Horm. Res., *38*, 331-408.
Cherfas, N. B., 1966, "Analysis of meiosis of unisexual and bisexual crucian carp," Tr. VNIIPRKh, *14*, 68-82.
Cherfas, N. B., 1969, "Main results of the cytogenetic analysis of unisexual and bisexual forms of the crucian carp," in: Genetics, Selection, and Hybridization of Fish [in Russian], Nauka, Moscow, pp. 85-98.
Cherfas, N. B., 1975, "Studies on diploid radiation gynogenesis in the carp. I. Experiments on mass production of diploid gynogenetic progenies," Genetika, *11*, No. 7, 78-86.
Cherfas, N. B., 1979, "Gynogenesis in fishes," in: Genetic Bases of Fish Selection, V. S. Kirpichnikov (ed.), Springer-Verlag, Berlin, pp. 255-273.
Cherfas, N. B., and Ilyasova, V. A., 1978, "Results of studies on diploid radiation gynogenesis in the carp," in: Karyological Variation, Mutagenesis, and Gynogenesis in Fish [in Russian], Leningrad, pp. 61-68.
Chevassus, B., 1980, "Les possibilités d'intervention génétique en pisciculture d'étang," in: La Pisciculture en Étang, INRA Publ., Paris, pp. 253-266.
Chevassus, B., Chourrout, D., and Jalabert, B., 1979, "Le contrôle de la réproduction chez les poissons. I. Les populations 'monosexes'," Bull. Fr. Piscic., *274*, 18-31.

Chevassus, B., Guyomard, R., Chourrout, D., and Quillet, E., 1983, "Production of viable hybrids of salmonids by triploidization," Gén., Sél. Évol., *15*, 519-531.

Chourrout, D., 1980, "Thermal induction of diploid gynogenesis and triploidy in the eggs of the rainbow trout (*Salmo gairdneri* Richardson)," Reprod. Nutr. Dev., *20*, 727-733.

Chourrout, D., 1981, "Induction de la gynogénèse diploïde et de la triploïdie chez la truite arc-en-ciel *Salmo gairdneri*," These, Univ. P. et M. Curie, Paris.

Chourrout, D., 1982a, "Gynogenesis caused by ultraviolet irradiation of salmonid sperm," J. Exp. Zool., *223*, 175-181.

Chourrout, D., 1982b, "La gynogénèse chez les vertébrés," Reprod. Nutr. Dev., *22*, 713-734.

Chourrout, D., 1984, "Pressure-induced retention of second polar body and suppression of first cleavage in rainbow trout: Production of all-triploids, all-tetraploids, and heterozygous and homozygous diploid gynogenetics," Aquaculture, *36*, 111-126.

Chourrout, D., and Quillet, E., 1982, "Induced gynogenesis in the rainbow trout: Sex and survival of progenies, production of all-triploid generations," Theor. Appl. Genet., *63*, 201-205.

Christensen, B., 1961, "Studies on cytotaxonomy and reproduction in the *Enchytraeidae*," Hereditas, *47*, 385-449.

Christensen, B., 1980, Animal Cytogenetics, Vol. 2, *Annelida*, Gebruder Borntraeger, Berlin.

Cimino, M. C., 1972a, "Meiosis in triploid all-female fish (*Poeciliopsis*, *Poeciliidae*)," Science, *175*, 1484-1486.

Cimino, M. C., 1972b, "Egg production, polyploidization, and evolution in a diploid all-female fish of the genus *Poeciliopsis*," Evolution, *26*, 294-306.

Costello, D. P., and Henley, C., 1971, Methods for Obtaining and Handling Marine Eggs and Embryos, 2nd edn., Woods Hole Marine Biol. Lab.

Crew, F., 1954, Sex Determination, Methuen, London.

Cuellar, O., 1971, "Reproduction and the mechanism of meiotic restitution in the parthenogenetic lizard *Cnemidophorus uniparens*," J. Morphol., *133*, 139-166.

Cui, Y., and Zan, R., 1982, "Studies on gynogenesis of the high-backed form of *Carassius auratus* in Kunming Lake," Zool. Res., *3*, Suppl., 237-242.

Dalcq, A., 1952, Initiation à l'Embryologie Générale, Masson, Paris.

D'Amato, F., 1977, Nuclear Cytology in Relation to Development, Cambridge Univ. Press.

Darcey, K. M., Buss, E. G., Bloom, S. E., and Olsen, M. W., 1971, "A cytological study of early cell populations in developing parthenogenetic blastodiscs of the turkey," Genetics, *69*, 479-489.

Darewski, I. S., and Kulikova, V. N., 1961, "Naturliche Parthenogenese in der polymorphen Gruppen der kaukasischen Felseidesche (*Lacerta saxicola* Eversman)," Zool. Jahrb. Abt. Syst. Oekol. Geogr. Tiere, *89*, 119-176.

Darlington, C. D., 1937, Recent Advances in Cytology, Saunders, Philadelphia.

Dasgupta, S., 1962, "Induction of triploidy by hydrostatic pressure in the leopard frog *Rana pipiens*," J. Exp. Zool., *151*, 105-116.

DeFord, L. S., Buss, E. G., Todd, P., and Wood, J. C. S., 1979, "Estimation of haploid cell content of parthenogenetic turkey embryos: A cytofluorometric study," J. Exp. Zool., *210*, 301-306.

Delone, N. L., 1960, "Sensitivity of different phases of mitosis and meiosis to ionizing radiation," in: Ionizing Radiation and Heredity [in Russian], Nauka, Moscow, pp. 155-175.

Dettlaff, T. A., 1961, "Rate of propagation of the fertilization impulse and dynamics of completion of the second maturation division in Acipenserid eggs," Dokl. Akad. Nauk SSSR, *140*, 967-969.

Dettlaff, T. A., 1962, "Dynamics of mitosis of the first cleavage divisions in sturgeon and trout eggs," Zh. Obshch. Biol., *23*, 401-409.

Dettlaff, T. A., 1966, "Experimental data on the mechanisms of oocyte maturation processes," Zh. Obshch. Biol., *27*, 401-410.

Dettlaff, T. A., 1974, "Estimation of the mitotic cycle duration during synchronous cleavage divisions," in: Methods of Developmental Biology [in Russian], Nauka, Moscow, pp. 136-139.

Dettlaff, T. A., 1977, "Development of the mature egg organization in amphibians and fishes at the final stages of oogenesis, during the maturation period," in: Current Problems of Oogenesis [in Russian], Nauka, Moscow, pp. 99-144.

Dettlaff, T. A., and Dettlaff, A. A., 1960, "Dimensionless characteristics of developmental duration in embryology," Dokl. Akad. Nauk SSSR, *134*, 199-202.

Dettlaff, T. A., and Dettlaff, A. A., 1961, "On relative dimensionless characteristics of the development duration in embryology," Arch. Biol., *72*, 1-16.

Dettlaff, T. A., and Dettlaff, A. A., 1982, "Dimensionless criteria as a method of quantitative characterization of animal development," in: Mathematical Developmental Biology [in Russian], Nauka, Moscow, pp. 25-39.

Dettlaff, T. A., and Skoblina, M. N., 1969, "The role of germinal vesicle in the process of oocyte maturation in *Anura* and *Acipenseridae*," Ann. Embryol. Morphogen., Suppl., *1*, 133-151.

Dettlaff, T. A., and Zubova, C. E., 1962, "Ratio of durations of the periods of maturation and embryogenesis in the Russian and stellate sturgeons," Dokl. Akad. Nauk SSSR, *143*, 746-748.

Dettlaff, T. A., and Zuichenko, S. I., 1963, "Metaphase of the first maturation division in sturgeon oocytes," Dokl. Akad. Nauk SSSR, *152*, 246-248.

Dettlaff, T. A., Nikitina, L. A., and Stroeva, O. G., 1964, "On the role of the germinal vesicle in oocyte maturation in anurans as revealed by the removal and transplantation of nuclei," J. Embryol. Exp. Morphol., *12*, 851-873.

Dettlaff, T. A., Nikitina, L. A., and Stroeva, O. G., 1965a, "Analysis of the role and specificity of germinal vesicle in maturation of anuran oocytes using its removal and replacement by somatic nuclei," Dokl. Akad. Nauk SSSR, *160*, 1441-1443.

Dettlaff, T. A., Vassetzky, S. G., and Davydova, S. I., 1965b, Recommendations on Time of Obtaining Eggs in Sturgeons after Hypophyseal Injection [in Russian], Glavrybvod, Moscow.

Dettlaff, T. A., Vassetzky, S. G., and Davydova, S. I., 1968a, "Maturation of sturgeon oocytes under different temperatures and time of obtaining eggs after hypophyseal injection," in: 6th Congr. Int. Reprod. Anim. Insem. Artif., Paris, Vol. 1, pp. 129-131.

Dettlaff, T. A., Skoblina, M. N., and Davydova, S. I., 1968b, "Intercellular influences in oocyte maturation of sturgeon," in: Cell Interactions in Differentiation and Growth, Sympos. Abstr. [in Russian], Tbilisi, pp. 5-7.

Donahue, R. P., 1969, "Maturation of the mouse oocyte in vitro. 1. Sequence and timing of nuclear progression," J. Exp. Zool., 237-250.

Donaldson, E. M., and Hunter, G. A., 1981, "Control of sex ratio in Pacific salmon broodstock," in: Salmonid Broodstock Maturation, Proc. Workshops, Seattle, p. 80.

Donaldson, E. M., and Hunter, G. A., 1982, "Sex control in fish with particular reference to salmonids," Can. J. Fish. Aquat. Sci., *39*, 99-110.

Dournon, C., 1982, "Développement somatique et inversion sexuelle sous l'action de la température chez les amphibiens," in: Proc. 1st Int. Colloq. Pathol. Reptil., Amphib., pp. 227-231.

Dournon, C., and Houillon, C., 1983, "Determinisme genetique du sexe: Demonstration a partir d'animaux à phenotype sexual inverse sous l'action de la température chez l'Amphibien Urodèle, *Pleurodeles waltlii* Michah," C. R. Acad. Sci. Paris, III, *296*, 779-782.

Dournon, C., and Houillon, C., 1984, "Démonstration génétique de l'inversion fonctionelle du phenotype sexual femelle sous l'action de la témperature d'élévage chez l'Amphibien Urodèle: *Pleurodeles waltlii* Michah," Reprod. Nutr. Dév., *24*, 361-378.

Dournon, C., and Houillon, C., 1985, "Thermosensibilite de la differenciation sexuelle chez l'Amphibien Urodele *Pleurodeles waltlii* Michah. Conditions pour obtenir l'inversion du phenotype sexuel de toutes les femelles genetiques sous l'action de la temperature d'elevage," Reprod. Nutr. Dev., *25*, 671-688.

Dournon, C., Guillet, F., Boucher, D., and Lacroix, J. C., 1984, "Cytogenetic and genetic evidence of male sexual inversion by heat treatment in the newt *Pleurodeles waltlii*," Chromosoma, *90*, 261-264.

Echelle, A. A., and Mosier, D. T., 1981, "All-female fish: A cryptic species of *Menidia* (*Atherinidae*)," Science, *212*, 1411-1413.

Echelle, A. A., and Mosier, D. T., 1982, "*Menidia clarkhubbsi* n. sp. (*Pisces*: *Atherinidae*), an all-female species," Copeia, *3*, 533-540.

Edwards, P. I., and Searle, A. G., 1963, "Genetic radiosensitivity of specific post-dictyate stages in mouse oocytes," Genet. Res., *4*, 389-400.

Edwards, R. G., 1954, "Colchicine induction of heteroploidy in early mouse embryos," Nature, *174*, 276-277.

Edwards, R. G., and Gates, A. H., 1959, "Timing of the stages of the maturation divisions, ovulation, fertilization, and the first cleavage of eggs of adult mice treated with gonadotropins," J. Endocrinol., *18*, 292-304.

Erdman, L. H., and Thompson, J. S., 1943, "Production of triploidy on *Triturus viridescens* by heat treatment," Senior Thesis, Princeton Univ. (cit. from Bungenberg de Jong, 1957).

Fankhauser, G., 1942, "Induction of polyploidy in animals by extremes of temperatures," Biol. Sympos., *6*, 21-35.

Fankhauser, G., 1945, "The effects of changes in chromosome number on amphibian development," Q. Rev. Biol., *20*, 20-78.

Fankhauser, G., and Godwin, D., 1948, "The cytological mechanism of the triploidy inducing effect of heat on eggs of *Triturus viridescens*," Proc. Natl. Acad. Sci. USA, *34*, 11, 5-44.

Fankhauser, G., and Griffiths, R. B., 1939, "Induction of triploidy and haploidy in the newt *Triturus viridescens* by cold treatment of unsegmented eggs," Proc. Natl. Acad. Sci. USA, *25*, 4, 233-238.

Fankhauser, G., and Humphrey, R. R., 1943, "The relation between the number of chromosome sets in animal cells," Proc. Natl. Acad. Sci. USA, *29*, 11, 344-350.

Fankhauser, G., and Moore, C., 1941, "Cytological and experimental studies of polyspermy in the newt *Triturus viridescens*. 1. Normal fertilization," J. Morphol., *68*, 347-385.

Fankhauser, G., and Watson, R. C., 1942, "Heat-induced triploidy in the newt *Triturus viridescens*," Proc. Natl. Acad. Sci. USA, *28*, 10, 436-440.

Fankhauser, G., Crotta, R., and Perrot, M., 1942, "Spontaneous and cold-inducted triploidy in the Japanese newt *Triturus pyrrhogaster*," J. Exp. Zool., *89*, 167-181.

Feiertag-Koppen, C. C. M., 1980, "Cytological studies of the two-spotted spider mite *Tetranychus urticae* Koch (*Tetranychi-*

dae, Trombidiformes). II. Meiosis in growing oocytes," Genetica, *54*, 173-180.
Ferrier, V., and Jaylet, A., 1978, "Induction of triploidy in the newt *Pleurodeles waltlii* by heat shock or hydrostatic pressure. Interpretation of the different types of ploidy using a chromosomal marker," Chromosoma, *69*, 47-63.
Fischberg, M., 1944, "Veranderung der Chromosomenzahl bei *Triton alpestris* nach Kältebehandlung der Eier," Rev. Suisse Zool., *51*, 430-436.
Fischberg, M., 1947, "Experimentelle Auslösung von haploider und diploider Parthenogenese bei den Urodelen *Triton palmatus* und *Triton alpestris*," Arch. Kalus-Stift. Vererbungsforsch., *22*, 331-336 (cit. from Bungenberg de Jong, 1957).
Fischberg, M., 1948, "Experimentelle Auslösung von Heteroploidie durch Kältebehandlung der Eier von *Triton alpestris* aus verschiedenen Populationen," Genetica, *24*, 213-329.
Fischberg, M., and Beatty, R. A., 1952, "Heteroploidy in mammals. II. Induction of triploidy in pre-implantation mouse eggs," J. Genet., *50*, 455-470.
Frolova, S. L., 1948, "Cytology of artificial parthenogenesis of the silkworm. II. Cytology of maturation and development upon activation with high temperature," Tr. Inst. Tsitol., Gistol., Embriol., *3*, 1, 162-185.
Funaki, K., 1981, "Active development in preimplantation stages of giant digynic triploids in the Chinese hamster," Proc. Jpn. Acad., *57B*, 18-22.
Funaki, K., and Mikamo, K., 1980a, "Cytogenetic evidence of the giant diploid oocytes capable of developing into digynic triploids in the Chinese hamster," Proc. Jpn. Acad., *56B*, 135-140.
Funaki, K., and Mikamo, K., 1980b, "Giant diploid oocytes as a cause of digynic triploidy in mammals," Cytogenet. Cell Genet., *28*, 158-168.
Gallien, L., 1953, "L'heteroploïdie expérimentale chez les amphibiens," Année Biol., Ser. 3, *29*, 5-22.
Gallien, L., 1959, "Récherches sur quelques aspects de l'hétéroploïdie expérimentale chez le triton *Pleurodeles waltlii* Michah," J. Embryol. Exp. Morphol., *7*, 380-393.
Gallien, L., 1962, "Héteroploïdie et sexualité chez les vertébrés," Année Biol., *1*, 2, 53-70.
Gandolfi, P. G., 1979, "Die Ultrastruktur der Chromosomen-Aufregulation in männlich determinierten Eiern der Gallmucke *Heteropeza pygmae*," Biol. Zentralbl., *98*, 409-428.
Gervai, J., and Csányi, V., 1978, "Induced triploidy in carp (*Cyprinus carpio* L.)," in: Int. Seminar on Increasing the Productivity in Selection and Hybridization, Szarvas, Hungary, pp. 31-36.
Gervai, J., Péter, S., Nagy, A., Horváth, L., and Csányi, V., 1980, "Induced triploidy in carp, *Cyprinus carpio* L.," J. Fish Biol., *17*, 667-671.

Gillespie, L. L., and Armstrong, J. B., 1978, "Pressure induction of triploids in the axolotl," Axolotl Newlett., Vol. 5.
Gillespie, L. L., and Armstrong, J. B., 1979, "Induction of triploid and gynogenetic diploid axolotls (*Ambystoma mexicanum*) by hydrostatic pressure," J. Exp. Zool., *210*, 117-122.
Ginsburg, A. S., 1959, "Fertilization in Acipenserid fishes. I. Fusion of gametes," Tsitologiya, *1*, 510-526.
Ginsburg, A. S., 1968, Fertilization in Fishes and the Problem of Polyspermy, Israeli Program for Sci. Transl., Ltd.
Ginsburg, A. S., 1974, "Methods of stimulation and investigation of the cortical reaction in the egg," in: Methods of Developmental Biology [in Russian], Nauka, Moscow, pp. 57-74.
Gold, J. R., 1979, "Cytogenetics," Fish Physiol., *8*, 353-405.
Golovinskaya, K. A., and Romashov, D. D., 1966, "Splitting by scale cover features upon diploid gynogenesis in the carp," Tr. VNIIPRKh, *14*, 227-235.
Golovinskaya, K. A., Romashov, D. D., and Cherfas, N. B., 1963, "On radiation gynogenesis in the carp," Tr. VNIIPRKh, *12*, 149-168.
Gomelsky, B. I., 1980, "Hormonal sex inversion in fish and possibilities of its application in fish husbandry (A review), Tr. VNIIPRKh, *28*, 117-136.
Gomelsky, B. I., 1985, "Hormonal sex inversion in the carp (*Cyprinus carpio* L.)," Ontogenez, *16*, 398-405.
Gomelsky, B. I., and Cherfas, N. B., 1982a, "Hormonal sex inversion as a method of studying the genetic control of natural gynogenesis in fish," Dokl. Akad. Nauk SSSR, *263*, 467-470.
Gomelsky, B. I., and Cherfas, N. B., 1982b, "Hormonal sex inversion in females of the crucian carp unisexual form," Ontogenez, *13*, 235-242.
Goncharov, B. I., 1971, "Studies of transition of amphibian and sturgeon oocytes from growth to maturation," Author's abstract of Candidate's dissertation, Moscow.
Goncharov, B. F., 1976, "Physiological condition of the follicle of older generation in the stellate sturgeon during spawning migration," in: Ecological Physiology of Fishes [in Russian], Vol. 2, 139-140.
Gould, M. C., 1968, "Completion of the meiotic division after fertilization in *Urechis caupo* eggs despite inhibition of protein synthesis," J. Cell Biol., *38*, 220-224.
Griffiths, R. B., 1941, "Triploidy (and haploidy) in the newt *Triturus viridescens*, induced by refrigeration of fertilized eggs," Genetics, *26*, 69-88.
Grisham, L. M., Wilson, L., and Bensch, K. G., 1973, "Antimeiotic action of griseofulvine does not involve disruption of microtubules," Nature, *244*, 294-296.
Groot, A. P. de, 1948, "The influence of higher concentrations of lithium chloride on maturation and first cleavages of the eggs of *Limnaea stagnalis*. 1.," Proc. Koninkl. Nederl. Akad. Wet., *51*, 588.

Harada, K., and Buss, E. G., 1981, "The chromosomes of turkey embryos during early stages of parthenogenetic development," Genetics, *98*, 335-345.

Harvey, E. G., and Chang, M. C., 1963, "Effects of x-ray-irradiation of ovarian ova on the morphology of fertilized ova and development of embryos," J. Cell. Comp. Physiol., *61*, 133-143.

Heilbrunn, L. V., 1920, "Studies in artificial parthenogenesis. III. Cortical changes and the initiation of maturation in the egg of *Cumingia*," Biol. Bull., *38*, 317-339.

Heilbrunn, L. V., 1952, An Outline of General Physiology, Saunders, Philadelphia.

Heilbrunn, L. V., and Wilson, W. L., 1955, "Changes in the protoplasm during maturation," Biol. Bull., *109*, 271-275.

Heinemann, R. L., and Hughes, R. D., 1969, "The cytological basis for reproductive variability in the *Anoetidae* (*Sarcoptiformes*: *Acari*)," Chromosoma, *28*, 346-356.

Hill, R., 1948, "Parthenogenesis bei *Nagara modesta* Dolff. (*Isopodae*)," Chromosoma, *3*, 232-256.

Holmquist, D. C., 1952, "The cytology of triploid production following heat treatment of the egg of the newt *Triturus viridescens*," Doctoral Dissertation, Univ. North Carolina (cit. from Bungenberg de Jong, 1957).

Houillon, C., and Dournon, C., 1978, "Inversion du phénotype sexuel femelle sous l'action d'une température élévee chez l'Amphibien Urodèle *Pleurodeles waltlii* Michah," C. R. Acad. Sci. Paris D, *286*, 1475-1478.

Humphrey, R. R., Briggs, R., and Fankhauser, G., 1950, "Sex differentiation in triploid *Rana pipiens* larvae and the subsequent reversal of females to males," J. Exp. Zool., *115*, 399-428.

Humphries, A. A., Jr., 1956, "A study of meiosis in coelomic and oviductal oocytes of *Triturus viridescens* with particular emphasis on the origin of spontaneous polyploidy and the effects of heat shock on the first meiotic divisions," J. Morphol., *99*, 97-131.

Hunter, G.A., and Donaldson, E. M., 1980, "Experimental production of all-female and sterile groups of coho salmon (*Oncorhynchus kisutch*)," in: Proc. 31st Annual Northwest Fish Culture Conference, Courtenay, pp. 234-241.

Ignatieva, G. M., 1970, "Regularities of early embryogenesis in salmonid fishes as revealed by the method of dimensionless characterization of development time," Ontogenez, *1*, 20-32.

Ignatieva, G. M., 1973, "Relative duration of karyokinesis and cytokinesis during synchronous cleavage divisions in the carp and pike at different temperatures," Ontogenez, *4*, 15-21.

Ignatieva, G. M., 1979, Early Embryogenesis of Fishes and Amphibians [in Russian], Nauka, Moscow.

Ignatieva, G. M., and Kostomarova, A. A., 1966, "Duration of the mitotic cycle during synchronous cleavage divisions (τ_0) and its dependence on temperature in loach embryos," Dokl. Akad. Nauk SSSR, *168*, 1221-1224.

Igumnova, L. V., 1979, "Temporal patterns of maturation and embryogenesis in *Huso huso* with respect to the development of the biotechnics of its breeding," Author's abstract of Candidate's Dissertation, Moscow.

Igumnova, L. V., 1985, "Temporal regularities of embryogenesis in the sterlet *Acipenser ruthenus* L," Ontogenez, *16*, 67-72.

Igumnova, L. V., and Dettlaff, T. A., 1983, "Temporal regularities of maturation in the stellate sturgeon *Acipenser stellatus* Pallas (*Acipenseridae*) of the Volga spawning population," Vopr. Ikhtiol., *23*, 76-80.

Inaba, F., 1936, "Studies on the artificial parthenogenesis of *Ostrea gigas* Thunberg," J. Sci. Hiroshima Univ., Ser. B, Div. 1, *5*, 29-46.

Ivanova, L. S., 1978, "Parthenogenesis in insects," in: Apomixis in Plants and Animals [in Russian], Nauka, Novosibirsk, pp. 175-187.

Jagiello, G. M., 1969, "Inhibition of reduction in female mouse meiosis," Experientia, *25*, 695-696.

Jagiello, G. M., and Lin, J. S., 1973, "Studies on the process of mouse meiosis and ovulation with daunomycin," Experientia, *29*, 114-117.

Jalabert, B., Billard, R., and Chevassus, B., 1974, "Tentative de contrôle expérimentale du sexe chez la truite arc-en-ciel: Production d'animaux stériles et d'hermaphrodites simultanés autofécondales," in: Actes de Colloques, Vol. 1 (Colloque sur l'aquaculture), pp. 161-174.

Jalabert, B., Billard, R., and Chevassus, B., 1975, "Preliminary experiments on sex control in trout: Production of sterile fishes and simultaneous self-fertilizable hermaphrodites," Ann. Biol., Anim., Biochim., Biophys., *15*, 19-28.

Jaylet, A., and Ferrier, V., 1978, "Experimental gynogenesis in the newt species *Pleurodeles waltlii* and *P. poireti*," Chromosoma, *69*, 65-80.

John, B., and Lewis, K. R., 1965, "The meiotic system," in: Protoplasmotologia, Vol. 6, Springer, Wien.

Kafiani, C., 1970, "Genome transcription in fish development," Adv. Morphogen., *8*, 209-284.

Kallman, K. D., 1962, "Gynogenesis in the teleost *Mollienesia formosa* (Girard), with a discussion of the detection of parthenogenesis in vertebrates by tissue transplantation," J. Genet., *58*, 7-24.

Karp, L. E., and Smith, W. D., 1975, "Experimental production of aneuploidy in mouse oocytes," Gynecol. Invest., *6*, 337-341.

Kawahara, H., 1978, "Production of triploid and gynogenetic diploid *Xenopus* by cold treatment," Dev., Growth, Differ., *20*, 227-236.

Kawamura, T., 1941, "Polyploidy in the Japanese newt *Triturus pyrrhogaster*," Zool. Mag., *53*, 550-552.

Khalil, G. M., and Cable, R. M., 1968, "Germinal development in *Philophthalmus megalurus* (Cort, 1914) (*Trematoda*: *Digenea*)," Z. Parasitenkd., *31*, 211-231.

Khvostova, V. V., and Bogdanov, Yu. F. (eds.), 1975, Cytology and Genetics of Meiosis [in Russian], Nauka, Moscow.

Kiknadze, I. I., and Vysotskaya, L. V., 1975, "Microscopic morphology of meiosis and its modifications," in: Cytology and Genetics of Meiosis [in Russian], Nauka, Moscow, pp. 15-41.

Kobayasi, H., 1971, "A cytological study on gynogenesis of the triploid ginbuna (*Carassius auratus langsdorfii*)," Zool. Mag., *80*, 316-322.

Kobayasi, M., Kawashima, Y., and Takeuchi, N., 1970, "Comparative chromosome studies in the genus *Carassius*, especially with a finding of polyploidy in the ginbuna," Jpn. J. Ichthyol., *17*, 153-160.

Koch, E. A., Smith, P. A., and King, R. C., 1970, "Variations in the radiosensitivity of oocyte chromosomes during meiosis in *Drosophila melanogaster*," Chromosoma, *30*, 98-108.

Korschelt, E., and Heider, K., 1903, Lehrbuch der vergleichenden Entwicklungsgeschichte der wirbellosen Tiere, Verlag von G. Fischer, Jena.

Kostomarova, A. A., 1975, "The loach *Misgurnus fossilis* L., in: Experimental Species in Developmental Biology [in Russian], Nauka, Moscow, pp. 308-323.

Kostomarova, A. A., and Ignatieva, G. M., 1968, "Correlation of the processes of karyo- and cytotomy during synchronous cleavage divisions in the loach *Misgurnus fossilis* L.," Dokl. Akad. Nauk SSSR, *183*, 490-493.

Kudryavtzev, I. V., Oshchepkova, Z. A., Antal, V. D., and Ptashkina, E. A., 1981, "On susceptibility of the RSV-infected and aleukemic fowl to parthenogenesis," Zh. Obshch. Biol., *42*, 467-469.

Kuznetsova, O. V., 1983, "Formation of diploid eggs in compound-3 females of *Drosophila melanogaster*," Genetika, *19*, 775-777.

Kuznetsova, O. V., and Balakireva, M. D., 1980, "Cytogenetic analysis of triploid progeny in compound-2 females of *Drosophila melanogaster*," Genetika, *16*, 1498-1500.

Kuznetsova, O. V., and Glotov, N. V., 1978, "Investigation into mechanisms of formation of diploid ova in *Drosophila melanogaster*," Genetika, *14*, 463-469.

Kuznetsova, O. V., and Pasyukova, E. T., 1980, "Formation of groups of diploid ova in compound-2 females of *Drosophila melanogaster*," Vestn. Leningr. Gos. Univ., *3*, 114-116.

Lemoine, H. L., Jr., and Smith, L. T., 1980, Polyploidy induced in brook trout by cold shock," Trans. Am. Fish. Soc., *109*, 626-631.

Leonov, B. V., Kamakhin, A. P., and Dmitrieva, L. V., 1972, "Microcinematographic investigation of cleavage in mouse embryos in vitro," Ontogenez, *3*, 158-159.

Lepori, N. G., 1950, "Ciclo cromosomico con poliploidia, gino-

genesi ed endomitosi in popolazioni Italiane di *Polycelis nigra* Ehrenberg," Caryologia, *2*, 301-324.
Lieder, U., 1959, "Uber die Entwicklung bei mannchenlosen Stammen der Silberkaraunche *Carassius auratus gibelio* (Bloch) (*Vertebrata, Pisces*)," Biol. Zentralbl., *78*, 284-291.
Lieder, U., 1964a, "Uber den gegenwartigen Stand und die Methodik der Chromosomenuntersuchungen bei Fischen," Z. Fisch., *11*, 673-684.
Lieder, U., 1964b, "Polyploidisierungsversuche bei Fischen mittels Temperaturschock und Colchizinbehandlung," Z. Fisch., *12*, 247-257.
Lillie, F. R., 1923, Problems of Fertilization, Univ. of Chicago Press.
Lincoln, R., and Bye, V., 1984, "Triploid rainbows show commercial potential," Fish Farmer, *9*.
Lincoln, R. F., and Scott, A. P., 1983, "Production of all-female triploid rainbow trout," Aquaculture, *30*, 375-380.
Lincoln, R. F., and Scott, A. P., 1984, "Sexual maturation in triploid rainbow trout *Salmo gairdneri* Richardson," J. Fish Biol., *25*, 385-392.
Lokki, J., Saura, A., Lankinen, P., and Suomalainen, E., 1976, "Genetic polymorphism in parthenogenetic animals. V. Triploid *Adoxus obscurus* (*Coleoptera*: *Chrysomelidae*)," Genet. Res., *28*, 27-36.
Longo, F. J., 1972, "The effect of cytochalasin B on the events of fertilization in the surf clam *Spisula solidissima*. 1. Polar body formation," J. Exp. Zool., *182*, 321-343.
Lou, Y. D., and Purdom, C. E., 1984a, "Diploid gynogenesis induced by hydrostatic pressure in rainbow trout *Salmo gairdneri* Richardson," J. Fish Biol., *24*, 665-670.
Lou, Y. D., and Purdom, C. E., 1984b, "Polyploidy induced by hydrostatic pressure in rainbow trout *Salmo gairdneri* Richardson," J. Fish Biol., *25*, 345-351.
Lovelage, R., 1951, "Some cytological aspects of heat on *Nereis* eggs," Biol. Bull., *101*, 225.
Lyapunova, N. A., and Bogdanov, Yu. F., 1975, "Physiology, cytochemistry, and biochemistry of meiosis," in: Cytology and Genetics of Meiosis [in Russian], Nauka, Moscow, pp. 138-193.
Macgregor, H. C., and Uzzell, T. M., 1964, "Gynogenesis in salamanders related to *Ambystoma jeffersonianum*," Science, *135*, 1043-1045.
Maeda, T., 1939, "Chiasma studies in the silkworm *Bombyx mori* L.," Jpn. J. Genet., *15*, 118-127.
Makino, S., and Ozima, Y., 1943, "Formation of the diploid egg nucleus due to suppression of the second maturation division induced by refrigeration of eggs of the carp *Cyprinus carpio*," Cytologia, *13*, 55-60.
Malavista, E. E., Sato, H., and Besch, K. G., 1968, "Vinblastine and griseofulvine reversibly disrupt the living mitotic spindle," Science, *160*, 770-772.

Mandl, A. M., 1963a, "Pre-ovulatory changes in the oocyte of the adult rat," Proc. R. Soc. London, Ser. B, *158*, 105-118.
Mandl, A. M., 1963b, "The radiosensitivity of oocytes at different stages of maturation," Proc. R. Soc. London, Ser. B, *158*, 119-141.
Mandl, A. M., and Beaumont, H. M., 1964, "The differential radiosensitivity of oogonia and oocytes at different developmental stages," in: Effects of Ionizing Radiation on the Reproductive System, W. D. Cralson and E. X. Cassner (eds.), Pergamon Press, pp. 311-321.
Marsland, D., and Hiramoto, Y., 1966, "Cell division: pressure-induced reversal of the antimeiotic effect of heavy water in the oocytes of the starfish *Asterias forbesii*," J. Cell Physiol., *67*, 13-22.
Masui, Y., 1974, "A cytostatic factor in amphibian oocytes: Its extraction and partial purification (1)," J. Exp. Zool., *187*, 141-147.
Masui, Y., 1985, "Meiotic arrest in animal oocytes," in: Biology of Fertilization, Vol. 1, Academic Press, New York, pp. 189-219.
Masui, Y., and Clarke, H. J., 1979, "Oocyte maturation," Int. Rev. Cytol., *57*, 185-282.
Masui, Y., and Markert, C. L., 1971, "Cytoplasmic control of nuclear behavior during meiotic maturation of frog oocytes," J. Exp. Zool., *177*, 129-146.
Mattox, N. T., 1937, "Oogenesis of *Campeloma rufum*, a parthenogenetic snail," Z. Zellforsch., *27*, 455-464.
McGaughey, R. W., and Chang, M. C., 1969a, "Meiosis of mouse eggs before and after sperm penetration," J. Exp. Zool., *170*, 397-410.
McGaughey, R. W., and Chang, M. C., 1969b, "Inhibition of fertilization and production of heteroploidy in eggs of mice treated with colchicine," J. Exp. Zool., *171*, 405-480.
Mittwoch, U., 1978, "Parthenogenesis," J. Med. Genet., *15*, 165-181.
Mockford, E. L., 1971, "Parthenogenesis in Psocids (*Insecta*: *Psocoptera*)," Am. Zool., *11*, 327-339.
Moore, A. R., 1933, "Is cleavage rate a function of the cytoplasm or of the nucleus?" J. Exp. Biol., *10*, 230-236.
Moritz, C., 1984, "The origin and evolution of pathenogenesis in *Heteronotia binoei* (*Gekkonidae*). I. Chromosome banding studies," Chromosoma, *89*, 151-162.
Muldal, S., 1952, "The chromosomes of the earthworm," Heredity, *6*, 55-76.
Muller, H. J., 1959, Progress in Nuclear Energy, Pergamon Press, London.
Muller, W. P., Thiebaud, C. H., Ricard, L., and Fischberg, M., 1978, "The induction of triploidy by pressure in *Xenopus laevis*," Rev. Suisse Zool., *85*, 20-26.

Murakami, A., 1969a, "Comparison of stage sensitivity to x rays during meiosis in silkworm oocytes," Natl. Inst. Genet. Jpn. Annu. Rep., *20*, 71-72.
Murakami, A., 1969b, "Changes in radiosensitivity during the second meiotic stage of silkworm oocytes," Natl. Inst. Genet. Jpn. Annu. Rep., *20*, 72-73.
Murakami, A., 1971, "Comparison of stage sensitivity to x rays during meiosis in the egg of the silkworm *Bombyx mori*," Int. J. Radiat. Biol., *19*, 167-176.
Murdy, W. H., and Carson, H. L., 1959, "Parthenogenesis in *Drosophila mangabeirai* Malog," Am. Nat., *93*, 355-363.
Muto, Y., 1952, "Production of triploid toads *Bufo vulgaris formosus* (Boulenger) by a temperature shock on fertilized eggs," J. Sci. Hiroshima Univ., Ser. B, Div. 1, *13*, 163-171.
Muto, Y., 1957, "Triploidy in the toad *Bufo vulgaris formosus* (Boulenger), induced by heat treatment of fertilized eggs," J. Sci. Hiroshima Univ., Ser. B, Div. 1, *17*, 143-199.
Myasoedov, S. V., 1935, Phenomena of Reproduction and Sex in the Organic World [in Russian], Sib. Nauchn. Mysl, Tomsk.
Nace, G. W., Richards, C. M., and Asher, J. H., 1970, "Parthenogenesis and genetic variability. I. Linkage and inbreeding estimation in the frog *Rana pipiens*," Genetics, *66*, 349-368.
Nagy, A., and Csányi, V., 1978, "Utilization of gynogenesis in genetic analysis and practical animal breeding," in: Int. Seminar on Increasing Productivity by Selection and Hybridization, Szarvas, Hungary, pp. 16-30.
Nagy, A., Rajki, K., Horvath, L., and Csányi, V., 1978, "Investigation on carp *Cyprinus carpio* L. gynogenesis," J. Fish Biol., *13*, 215-224.
Nagy, A., Rajki, K., Bakos, J., and Csányi, V., 1979, "Genetic analysis in carp (*Cyprinus carpio*) using gynogenesis," Heredity, *43*, 35-40.
Nagy, A., Bercsenyi, M., and Csányi, V., 1981, "Sex reversal in carp (*Cyprinus carpio*) by oral administration of methyltestosterone," Can. J. Fish. Aquat. Sci., *38*, 725-728.
Nagy, A., Monostory, Z., and Csányi, V., 1983, "Rapid development of the clonal state in successive gynogenetic generations of carp (*Cyprinus carpio*)," Copeia, *3*, 745-749.
Nakamura, T., Kawahara, H., and Katagiri, C., 1985, "Rapid production of a histocompatible colony of *Xenopus laevis* by gynogenetic procedure," Zool. Sci., *2*, 71-79.
Narbel-Hofstetter, M., 1964, "Les altérations de la méiose chez les animaux parthénogénétiques," Protoplasmatologia, *6*, 2, 1-160.
Naruse, K., Ijiri, K., Shima, A., and Egami, N., 1985, "The production of cloned fish in the medaka (*Oryzias latipes*)," J. Exp. Zool., *236*, 335-342.
Nasonov, D. N., and Aleksandrov, V. Ya., 1940, Reaction of the

Living Matter to External Influences. Denaturation Theory of Damage and Stimulation [in Russian], Izd. Akad. Nauk SSSR, Moscow—Leningrad.

Neyfakh, A. A., 1962, Problem of Nucleocytoplasmic Interactions in Development [in Russian], Moscow.

Niemerko, A., 1975, "Induction of triploidy in the mouse by cytochalasin B," J. Embryol. Exp. Morphol., *34*, 279-289.

Niemerko, A., and Komar, A., 1976, "Cytochalasin B-induced triploidy in mouse oocytes fertilized in vitro," J. Reprod. Fertil., *48*, 279-284.

Novikov, N. P., 1962, "On possibility of gynogenesis in *Atherestes stomias* Jord. et Gilb in the Bering Sea," Dokl. Akad. Nauk SSSR, *147*, 215-218.

Nur, U., 1971, "Parthenogenesis in Coccids (*Homoptera*)," Am. Zool., *11*, 301-308.

Nur, U., 1972, "Diploid arrhenotoky and automictic thelytoky in soft-scale insects (*Lecaniidae*: *Coccoidea*: *Homoptera*)," Chromosoma, *39*, 381-401.

Nur, U., 1979, "Gonoid thelytoky in soft-scale insects (*Coccidae*: *Homoptera*)," Chromosoma, *72*, 89-104.

Oakley, H. A., 1982, "Meiosis in *Mesostoma ehrenbergii ehrenbergii* (*Turbellaria*, *Rhabdocoela*). II. Synaptonemal complexes, chromosome pairing, and disjunction in achiasmatic oogenesis," Chromosoma, *87*, 133-147.

Obara, N., Kawahara, H., and Katagiri, C., 1983, "Response to skin grafts exchanged among siblings of larval and adult gynogenetic diploids in *Xenopus laevis*," Transplantation, *36*, 91-95.

Odor, D. L., 1955, "The temporal relationship of the first maturation division of rat ova to the onset of heat," Am. J. Anat., *97*, 461-492.

Oksala, T., 1943, "Zytologische Studien an Odonaten. I. Chromosomenverhalt isse bei der Gattung *Aeschna*, mit besonderen Berucksichtigung der postreduktionen Teilung der Bivalente," Ann. Acad. sci. Fenn., A, 4, 1-65.

Oliver, J. H., Jr., 1971, "Sex chromosomes, parthenogenesis, and polyploidy in ticks," in: Evolution and Speciation: Essays in Honor of M. J. D. White, W. R. Atchby and D. S. Woodruff (eds.), Cambridge Univ. Press, pp. 66-77.

Olsen, M. W., 1974, "Frequency and cytological aspects of diploid parthenogenesis in turkey eggs," Theor. Appl. Genet., *44*, 216-221.

Olsen, M. W., and Buss, E. G., 1967, "Role of genetic factors and fowl pox virus in parthenogenesis of turkey eggs," Genetics, *56*, 727-732.

Olsen, M. W., and Buss, E. G., 1972, "Segregation of two alleles for color of down in parthenogenetic and normal turkey embryos and poults," Genetics, *72*, 69-75.

Omodeo, P., 1952, "Cariologia dei *Lumbricidae*," Caryologia, 4, 173-275.

Onozato, H., 1981, "Induced gynogenetic diploidy in salmonids," in: Abstr. Annual Meet. Jpn. Soc. Sci. Fish., p. 142.

Onozato, H., 1984, "Diploidization of gynogenetically activated salmonid eggs using hydrostatic pressure," Aquaculture, *43*, 91-98.

Opas, J., 1977, "Effects of extremely low osmolarity on fertilized mouse eggs," J. Embryol. Exp. Morphol., *37*, 65-77.

Oshiro, T., and Hibiya, T., 1981, "Water absorption of oocytes in the plaice *Limanda yokohamae* during meiotic maturation and its role in rupturing follicles," Bull. Jpn. Soc. Sci. Fish., *47*, 835-841.

Ostryakova-Varshaver, V. P., 1958, "Effects of high temperatures in the embryonic development of the silkworm (*Bombyx mori* L.). II. Cytological changes in the processes of maturation and fertilization of silkworm (*Bombyx mori* L.) eggs under the influence of high temperature," Tr. Inst. Morfol. Zhivotn., *21*, 81-103.

Peacock, A. D., and Weidman, U., 1961a, "Recent work on the cytology of animal parthenogenesis. I," Przegl. Zool., *5*, 5-27.

Peacock, A. D., and Weidman, U., 1961b, "Recent work on the cytology of animal parthenogenesis. II," Przegl. Zool., *5*, 101-122.

Pease, D. C., 1946, "Hydrostatic pressure effects upon the spindle figure and chromosome movement. 11. Experiment on the meiotic divisions of *Tradescantia* pollen mother cells," Biol. Bull., *91*, 145-169.

Pijnacker, L. P., 1966, "The maturation divisions of the parthenogenetic stick insect *Carausius morosus* Br. (*Orthoptera*, *Phasmidae*)," Chromosoma, *19*, 99-112.

Pijnacker, L. P., 1971, "Effects of x rays on different meiotic stages of oocytes in the parthenogenetic stick insect *Carausius morosus* Br.," Mutat. Res., *13*, 251-262.

Pijnacker, L. P., Ferwerda, M. A., and Helle, W., 1981, "Cytological investigations on the female and male reproductive system of the parthenogenetic privet mite *Brevipalpus obovatus* Donnadieu (*Phytoptipalpidae*, *Acari*)," Acarologia, *22*, 157-163.

Piko, L., and Bomsel-Helmreich, O., 1960, "Triploid rat embryos and other chromosomal deviants after colchicine treatment and polyspermy," Nature, *186*, 737-739.

Pincus, G., and Shapiro, H., 1940a, "The comparative behavior of mammalian eggs in vivo and in vitro. VIII. Further studies on the activation of rabbit eggs," Proc. Am. Philos. Soc., *83*, 631-647.

Pincus, G., and Shapiro, H., 1940b, "Further studies on the parthenogenetic activity of rabbit eggs," Proc. Natl. Acad. Sci. USA, *26*, 163-165.

Pogosyants, Y. Y., 1975, "Heteromorphic chromosomes in meiosis (mammals taken as an example)," in: Cytology and Genetics of Meiosis [in Russian], Nauka, Moscow, pp. 42-57.

Pokrovskaya, G. A., 1958, "Control of the development of the silkworm (*Bombyx mori* L.) by low temperatures. I. Age changes in cold resistance of silkworm eggs at early developmental stages," Tr. Inst. Morfol. Zhivotn., *21*, 126-156.

Poole, H. K., and Olsen, M. W., 1957, "The sex of parthenogenetic turkey embryos," J. Hered., *48*, 217-218.

Porter, K. R., 1939, "Androgenetic development of the eggs of *Rana pipiens*," Biol. Bull., *77*, 233-257.

Purdom, C. E., 1969, "Radiation-induced gynogenesis and androgenesis in fish," Heredity, *24*, 431-444.

Purdom, C. E., 1972, "Induced polyploidy in plaice (*Pleuronectes platessa* L.) and its hybrid with the flounder (*Platichthys flesus* L.)," Heredity, *29*, 11-24.

Purdom, C. E., 1983, "Genetic engineering by the manipulation of chromosomes," Aquaculture, *33*, 287-300.

Purdom, C. E., and Lincoln, R. F., 1974, "Gynogenesis in hybrids within the Pleuronectidae," in: The Early Life History of Fish, J.H.S. Blaxter (ed.), Springer-Verlag, Berlin, pp. 537-544.

Purdom, C. E., Thompson, D., and Dando, P. R., 1976, "Genetic analysis of enzyme polymorphisms in plaice (*Pleuronectes platessa*)," Heredity, *37*, 193-206.

Raven, C. P., 1966, Morphogenesis. The Analysis of Molluscan Development, Pergamon Press, New York.

Ray, D. T., 1957, "X-ray sensitivity of meiotic stages of *Mormoniella* eggs," Genetics, *38*, 391.

Ray, D. T., 1958, "Sensitivity of different stages of the eggs of *Mormoniella* to x rays," Proc. 10th Int. Congr. Genet., *2*, 230.

Rebhun, L. I., 1959, "Studies of early cleavage in the surf clam *Spisula solidissima* using methylene blue and toluidine blue as vital stains," Biol. Bull., *117*, 518-545.

Refstie, T., 1983, "Induction of diploid gynogenesis in Atlantic salmon and rainbow trout using irradiated sperm and heat shock," Can. J. Zool., *61*, 2411-2416.

Refstie, T., Stoss, J., and Donaldson, E. M., 1982, "Production of all-female coho salmon (*Oncorhynchus kisutch*) by diploid gynogenesis using irradiated sperm and cold shock," Aquaculture, *29*, 67-82.

Reisinger, E., 1940, "Die cytologische Grundlage der parthenogenetischen Dioogonie," Chromosoma, *1*, 531-553.

Rhoades, M. M., 1961, "Meiosis," Cell, *3*, 1-76.

Rollason, G. S., 1949, "X-radiation of eggs of *Rana pipiens* at various maturation stages," Biol. Bull., *97*, 169-186.

Romashov, D. D., and Belyaeva, V. N., 1964, "On cytology of radiation gynogenesis and androgenesis in the loach (*Misgurnus fossilis* L.)," Dokl. Akad. Nauk SSSR, *157*, 964-967.

Romashov, D. D., and Belyaeva, V. N., 1965, "Increase in production of diploid gynogenetic larvae in the loach (*Misgurnus fossilis*) by using heat shocks," Byull. Mosk. Ova. Ispyt. Prir., Otd. Biol., *70*, No. 5, 93-109.

Romashov, D. D., Nikolyukin, N. I., Belyaeva, V. N., and Timofeeva, N. A., 1963, "On the possibility of diploid radiation gynogenosis in Acipenserid fishes," Radiobiologiya, *3*, 104-110.

Rothschild, N., 1956, Fertilization, Wiley, New York.

Rott, N. N., 1963, "Sex determination in amphibians," Byull. Mosk. Ova. Ispyt. Prir., Otd. Biol., *68*, 4, 118-134.

Rott, N. N., 1965, "Changes in sensitivity of the eggs to heat shock during maturation, fertilization, and early cleavage in the axolotl," Tsitologiya, *7*, 205-211.

Rott, N. N., 1974, "Alterations of ploidy," in: Methods of Developmental Biology [in Russian], Nauka, Moscow, pp. 14-32.

Rott, N. N., and Betina, M. I., 1964, "Production of triploid larvae of the axolotl by heat shock," Tsitologiya, *6*, 95-98.

Rott, N. N., and Veprintsev, B. N., 1981, "Experimental and theoretical foundations for solution of the problem of preservation of animal genetic resources by conservation," Usp. Sovrem. Biol., *91*, 451-466.

Rudneva, T. B., 1972, "Duration of karyomitosis and cell divisions in cleavage divisions II-IV in the clawed frog *Xenopus laevis*," Ontogenez, *3*, 526-530.

Rugh, R., and Grupp, E., 1959, "Response of the very early mouse embryo to low levels of ionizing radiation," J. Exp. Zool., *141*, 571-587.

Runnstrom, J., 1949, "The mechanism of fertilization in *Metazoa*," Adv. Enzymol., *9*, 241-327.

Runnstrom, J., 1952, "The cell surface in relation to fertilization," Sympos. Soc. Exp. Biol., *6*, 39-88.

Runnstrom, J., and Immers, J., 1956, "The role of mucopolysaccharides in the fertilization of the sea-urchin egg," Exp. Cell Res., *10*, 354-363.

Russell, L. B., and Saylors, C. L., 1963, "The relative sensitivity of various germ cell stages of the mouse to radiation-induced nondisjunction chromosome losses and deficiencies," in: Repair from Genetical Radiation Damage and Differential Radiosensitivity of Germ Cells, F. H. Sobels (ed.), Pergamon Press, Oxford, pp. 313-340.

Saat, T. V., 1982, "The chronology of oocyte maturation in the loach and the acquisition of developmental ability," Ontogenez, *13*, 257-264.

Sanderson, A. R., 1960, "The cytology of a diploid bisexual spider beetle *Ptinus clavipes* Panzer and its triploid gynogenetic form *mobilis* Moore," Proc. R. Soc. Edinburgh, *67*, 333-350.

Sato, I., and Kosin, I. L., 1960, "A cytological study of the parthenogenetically developing turkey germ disc and embryos," Cytologia, *25*, 256-266.

Savontaus, M.-L., 1980, "Effect of radiation on the segregation of chromosomes in the oocytes of *Drosophila melanogaster*," Rep. Dep. Biol., Univ. Turku, *2*, 1-27.

Sawada, N., 1952, "Experimental studies on the maturation division

of eggs in *Mactra veneriformis*. I. Inhibitory effect of body fluid," Mem. Ehime Univ., Sec. II (Sci.) B, *1*, 63-68.
Sawada, N., 1954, "Experimental studies on the maturation division of eggs in *Mactra veneriformis*. IV. On the effect of certain polysaccharides," Mem. Ehime Univ., Sec. II (Sci.) B, *2*, 89-92.
Schrader, P., 1926, "The cytology of pseudosexual eggs in a species of *Daphnia*," Z. Indukt. Abstammungs- Vererbungslehre, *40*, 1-27.
Schuetz, A. W., 1975, "Induction of nuclear breakdown and meiosis in *Spisula solidissima* oocytes by calcium ionophore," J. Exp. Zool., *191*, 433-440.
Schultz, R. J., 1967, "Gynogenesis and triploidy in the viviparous fish *Poeciliopsis*," Nature, *157*, 1564-1567.
Schultz, R. J., 1980, "Role of polyploidy in the evolution of fishes," in: Polyploidy: Biological Relevance, W. H. Lewis (ed.), Plenum Press, New York, pp. 313-340.
Schultz, R. J., and Kallman, K. D., 1968, "Triploid hybrids between the all-female teleost *Poecilia formosa* and *Poecilia sphenops*," Nature, *219*, 280-282.
Sclufer, E., 1955, "The respiration of *Spisula* eggs," Biol. Bull., *109*, 113-122.
Seiler, J., 1960, "Untersuchungen uber die Entstehung der Parthenogenese bei *Solenobia triquetrella* T. R. (*Lepidoptera*, *Psychidae*). II. Analyse der diploid parthenogenetischen *S. triquetrella*. Verhalten, Aufzuchresultate und Zytologie," Chromosoma, *11*, 29-102.
Sembrat, K., 1955, "Studies on the cleavage of rat eggs," Zool. Polon., *6*, 139-181.
Shelton, W. L., Yang, S.-L., and Wilken, L. O., 1984, "Sex reversal and breeding of brass carp," Trans. Am. Fish. Soc., *113*, 348-353.
Skoblina, M. N., 1965, "Dimensionless characteristics of the duration of mitosis of the first cleavage divisions in the axolotl," Dokl. Akad. Nauk SSSR, *160*, 700-703.
Skoblina, M. N., 1970, "Experimental studies of the role of nucleus in maturation of amphibian and sturgeon oocytes," Author's Abstract of Candidate's Dissertation, Moscow.
Skoblina, M. N., 1973, "In vitro maturation of loach oocytes under the influence of choriogonin," Ontogenez, *4*, 309-311.
Sladecek, F., and Lanzova, J., 1959, "Cytological mechanism of the production of heteroploids by cold or heat shock in axolotl," Folia Biol., *5*, 379-393.
Sladecek, F., and Romanovsky, A., 1960, "The effects of cold and heat shock on the eggs of axolotl in the early stages," Folia Biol., *6*, 42-49.
Smirnova, I. B., Graevskii, E. Ya., Konstantinova, M. M., and Neyfakh, A. A., 1977, "Effects of ionizing irradiation on ontogenesis," in: Environment and the Developing Organism [in Russian], Nauka, Moscow.

Smith, L. T., and Lemoine, H. L., 1979, "Colchicine-induced polyploidy in brook trout," Prog. Fish Cult., *41*, 86-88.
Smith, S., 1958, "Induction of triploidy in the South African clawed frog *Xenopus laevis* (Daudin)," Nature, *181*, 290-291.
Smith, S. G., 1971, "Parthenogenesis and polyploidy in beetles," Am. Zool., *11*, 341-349.
Sokolov, I. I., 1966, "The cytological bases of sexual reproduction in multicellular animals," in: Handbook of Cytology, Vol. 2 [in Russian], Nauka, Moscow—Leningrad, pp. 390-460.
Sonta, S., 1981, "Notes on cytogenetic effects of irradiation on the oocytes at meiotic stages in the Chinese hamster," Proc. Jpn. Acad., *57B*, 244-247.
Sorensen, R. A., 1973, "Cinemicrography of mouse-oocyte maturation utilizing Nomarski differential-interference microscopy," Am. J. Anat., *136*, 265-276.
Sparrow, A. H., 1951, "Radiation sensitivity of cells during mitotic and meiotic cycles with emphasis on possible cytochemical changes," Ann. N. Y. Acad. Sci., *51*, 1508-1540.
Speicher, B. R., 1937, "Oogenesis in a thelytokous wasp *Nemeritis canescens*," J. Morphol., *61*, 453-471.
Stanley, J. G., 1976, "Production of hybrid, androgenetic, and gynogenetic grass carp and carp," Trans. Am. Fish. Soc., *105*, 10-16.
Stanley, J. G., 1981, "Manipulation of developmental events to produce monosex and sterile fish," Rapp. P-V. Réun., Cons. Int. Explor. Mer, *178*, 485-491.
Stanley, J. G., Martin, J. M., and Jones, J. B., 1975, "Gynogenesis as a possible method for producing monosex grass carp (*Ctenopharyngodon idella*)," Prog. Fish Cult., *37*, 25-26.
Stanley, J. G., Allen, S. K., Jr., and Hidu, H., 1981, "Polyploidy induced in the American oyster *Crassostrea virginica*, with cytochalasin B," Aquaculture, *23*, 1-10.
Stevens, L. C., 1978, "Totipotent cells of parthenogenetic origin in a chimaeric mouse," Nature, *276*, 266-267.
Stevens, L. C., 1980, "Teratocarcinogenesis and spontaneous parthenogenesis in mice," in: Differentiation and Neoplasia (Results and Problems in Cell Differentiation, Vol. 11), pp. 265-274.
Stevens, L. C., Varnum, D. S., and Eicher, E. M., 1977, "Viable chimaeras produced from normal and parthenogenetic mouse embryos," Nature, *269*, 515-517.
Stille, B., and Dävring, L., 1980, "Meiosis and reproductive strategy in the parthenogenetic gall wasp *Diplolepis rosae* (L.) (*Hymenoptera*, *Cynipidae*)," Hereditas, *92*, 353-362.
Streisinger, G., Walker, C., Dower, N., Knauber, D., and Singer, F., 1981, "Production of clones of homozygous diploid zebra fish (*Brachydanio rerio*)," Nature, *291*, 293-296.
Strunnikov, V. A., 1975a, "Artificial meiotic parthenogenesis of the silkworm and its scientific and practical significance," Byull. Mosk. Ova. Ispyt. Prir., Otd. Biol., *80*, 4, 14-30.

Strunnikov, V. A., 1975b, "Sex control in silkworms," Nature, *255*, 111-113.
Strunnikov, V. A., 1983, Control of Silkworm Reproduction, Development, and Sex [in Russian], Mir, Moscow.
Suomalainen, E., 1940, "Polyploidy in parthenogenetic *Curculionidae*," Hereditas, *26*, 51-64.
Suomalainen, E., 1949, "Parthenogenesis and polyploidy in the weevils *Curculionidae*," Ann. Entomol. Fenn. (Suppl.), *14*, 206-212.
Suomalainen, E., 1953, "The kinetochore and bivalent structure in the *Lepidoptera*," Hereditas, *39*, 88-96.
Suomalainen, E., 1965, "Die Polyploidie bei dem parthenogenetischen Blattkafer *Adoxus obscurus* L. (*Coleoptera*, *Chrysomelidae*)," Zool. Jahrb., *92*, 183-192.
Suomalainen, E., 1966, "Achiasmatische Oogenese bei Trichopteren," Chromosoma, *18*, 201-207.
Suomalainen, E., Saura, A., and Lokki, J., 1977, "Genetic polymorphism and evolution of parthenogenetic insects," in: Problems of Experimental Biology [in Russian], Nauka, Moscow, pp. 7-20.
Swarup, H., 1959, "Production of triploidy in *Gasterosteus aculeatus* (L.)," J. Genet., *56*, 129-142.
Takenouchi, Y., 1978, "A chromosome study of the parthenogenetic rice water weevil *Lissorhopterus oryzophilus* Kuschel (*Coleoptera*: Curculionidae) in Japan," Experientia, *34*, 444.
Tchou-Su and Yen Pai-hu, 1950, "Rélation de l'augmentation d'eau dans l'oeuf et la maturation," Sci. Sin., *1*, 165-183.
Tchou-Su and Wang Yu-lan, 1958, "Études comparatives sur l'ovulation et la maturation in vivo et in vitro chez le crapaud asiatique (*Bufo bufo asiaticus*)," Acta Biol. Exp. Sin., *6*, 129-180.
Terskaya, E. R., 1982, "Initial developmental processes in silkworm eggs activated to meiotic parthenogenesis," in: Genetics, Biochemistry, and Cytology of Meiosis [in Russian], Nauka, Moscow, pp. 50-55.
Terskaya, E. R., and Strunnikov, V. A., 1974, "Methods of activating silkworm eggs for meiotic parthenogenesis," Dokl. Akad. Nauk SSSR, *219*, 1238-1241.
Terskaya, E. R., and Strunnikov, V. A., 1975, "Artificial meiotic parthenogenesis of the silkworm," Genetika, *11*, No. 3, 54-67.
Thibault, C., 1949, "L'oeuf des mammifères. Son développement parthénogénétique," Ann. Sci. Nat. (Zool.), *11*, 136-219.
Thiébaud, C. E., Colombelli, B., and Muller, W. P., 1984, "Diploid gynogenesis in *Xenopus laevis* and the localization with respect to the centromere of the gene for periodic albinism a^c," J. Embryol. Exp. Morphol., *83*, 33-42.
Thompson, D., 1983, "The efficiency of induced diploid gynogenesis in inbreeding," Aquaculture, *33*, 237-244.
Thompson, D., and Scott, A. P., 1984, "An analysis of recombina-

tion data in gynogenetic diploid rainbow trout," Heredity, *53*, 441-452.
Thomsen, M., 1927, "Studien über die Parthenogenese bei einiger Cocciden und Alleurodiden," Z. Zellforsch., *5*, 1-116.
Thorgaard, G. H., 1983, "Chromosome set manipulation and sex control in fish," Fish Physiol., *9B*, 405-434.
Thorgaard, G. H., Jaswin, M. E., and Stier, A. R., 1981, "Polyploidy induced by heat shock in rainbow trout," Trans. Am. Fish. Soc., *110*, 546-550.
Thorgaard, G. H., Scheerer, P. D., and Parsons, J. E., 1984, "Residual paternal inheritance in gynogenetic rainbow trout," Genetics, *107*, S106.
Tompkins, R., 1978, "Triploid and gynogenetic diploid *Xenopus laevis*," J. Exp. Zool., *203*, 251-256.
Traut, H., 1967, "X-ray induction of 2,3 translocations in mature and immature oocytes of *Drosophila melanogaster*," Genetics, *56*, 265-272.
Troiano, G., 1978, "Triploidy in the natural population of the psychodine moth fly *Psychoda parthenogenetica* Tonnoir (*Diptera*: *Psychodidae*)," Caryologia, *31*, 225-232.
Trottier, T. M., and Armstrong, J. B., 1976, "Diploid gynogenesis in the Mexican axolotl," Genetics, *83*, 783-792.
Tsafriri, A., 1978, "Oocyte maturation in mammals," in: Vertebrate Ovary, R. E. Jones (ed.), Plenum, New York, pp. 409-422.
Tsafriri, A., and Channing, C. P., 1975, "An inhibitory influence of granulosa cells and follicular fluid upon porcine oocyte meiosis in vitro," Endocrinology, *96*, 922-927.
Tsafriri, A., Dekel, M., and Bar-Ami, S., 1982, "The role of oocyte maturation inhibitor in follicular regulation of oocyte maturation," J. Reprod. Fertil., *64*, 541-555.
Tsoi, R. M., 1981, "Artificial mutagenesis and gynogenesis in practical selection of the carp," Genetika, *17*, 1095-1102.
Tsukahara, J., and Ishikawa, M., 1980, "Suppression of polar body formation by cytochalasin B and fusion of the resulting plural female pronuclei in eggs of the starfish *Asterina pectinifera*," Europ. J. Cell Biol., *21*, 288-295.
Ueno, K., 1984, "Induction of triploid carp and their hematological characteristics," Jpn. J. Genet., *59*, 585-591.
Ueno, K., 1985, "Sterility and secondary sexual character of triploid *Ganthopogon elongathus ceurulescens*," Fish Genet. Breeding Sci., *10*, 37-41.
Ueno, K., and Arimoto, B., 1982, "Induction of triploids in *Rhodeus ocellatus ocellatus* by cold shock treatment of fertilized eggs," Experientia, *38*, 544-546.
Ueno, K., and Ojima, Y., 1976, "Diploid—tetraploid complexes in the genus *Cobitis* (*Cobitidae*, *Cyprinida*)," Proc. Jpn. Acad., *52B*, 446-449.
Ueno, K., Swai, S., and Ojima, Y., 1980, "Karyotypes and geographical distribution in the genus *Cobitis* (*Cobitidae*)," Bull. Jpn. Soc. Sci. Fish., *46*, 9-18.

Uzzell, T., 1970, "Meiotic mechanisms of naturally occurring unisexual vertebrates," Am. Nat., *104*, 439-445.
Valenti, R. J., 1975, "Induced polyploidy in *Tilapia aurea* (Steindacher) by means of temperature shock treatment," J. Fish. Biol., *7*, 519-528.
Vandel, A., 1934, "La parthénogénèse géographique. II. Les males triploïdes d'origine parthénogénétique de *Trichoniscus* (*Spiloniscus*) *elizabethae* Herold," Bull. Biol., *68*, 419-463.
Vandel, A., 1940, "La parthénogénèse géographique. IV. Polyploïdie et distribution géographique," Bull. Biol., *74*, 94-100.
Vasiliev, V. P., 1981, "A possible mode of origin of tetraploid forms in vertebrates," Priroda, 4, 112-113.
Vasiliev, V. P., 1985, Evolutionary Karyology of Fish [in Russian], Nauka, Moscow.
Vasiliev, V. P., and Vasilieva, E. D., 1982, "A new diploid–polyploid complex in fishes," Dokl. Akad. Nauk SSSR, *266*, 250-252.
Vassetzky, S. G., 1966, "The thermosensitivity of sturgeon eggs at the stages of their maturation and cleavage," Zh. Obshch. Biol., *27*, 283-295.
Vassetzky, S. G., 1967a, "Studies of oocyte maturation and early embryogenesis of Acipenserids with respect to possible application of the method of thermal sex control in fish," Author's Abstract of Candidate's Dissertation, Moscow.
Vassetzky, S. G., 1967b, "Changes in ploidy in sturgeon larvae induced by heat treatment of eggs at different stages of their development," Dokl. Akad. Nauk SSSR, *172*, 1234-1237.
Vassetzky, S. G., 1970, "Dynamics of the first maturation division in sturgeon oocytes," Zh. Obshch. Biol., *31*, 84-93.
Vassetzky, S. G., 1972, "Duration and synchrony of the first maturation division in sturgeon oocytes after hypophyseal injection," in: Sturgeons and Problems of Sturgeon Fishery [in Russian], Pischevaya Promyshlennost, Moscow, pp. 251-256.
Vassetzky, S. G., 1973, "Dynamics of the divisions of maturation in eggs of the giant oyster *Ostrea gigas*," Ontogenez, 4, 416-422.
Vassetzky, S. G., 1982, "Duration of meiosis and its phases in animals," in: Genetics, Biochemistry, and Cytology of Meiosis [in Russian], Nauka, Moscow, pp. 81-87.
Vassetzky, S. G., and Betina, M. T., 1986, "Experimental influence on meiosis as a means of control of ontogenesis," in: Developmental Biology and Control of Inheritance [in Russian], Nauka, Moscow.
Vassetzky, S. G., Betina, M. I., and Kondratieva, O. T., 1984, "Production of triploids by hydrostatic pressure on loach (*Misgurnus fossilis*) fertilized eggs," Dokl. Akad. Nauk SSSR, *279*, 212-215.
Veprintsev, B. N., and Rott, N. N., 1979, "Conserving genetic resources of animal species," Nature, *280*, 633-634.

Volpe, E., and Dasgupta, S., 1962, "Gynogenetic diploids of mutant leopard frogs," J. Exp. Zool., *151*, 287-302.
von Borstel, R. C., and Amand, W. S., 1963, "Stage sensitivity to x-radiation during meiosis and mitosis in the egg of the wasp *Habrobracon*," in: Repair from Genetical Radiation Damage and Differential Radiosensitivity of Germ Cells, F. H. Sobels (ed.), Pergamon Press, Oxford, pp. 87-97.
Vorontsov, N. N., 1966, "Evolution of the karyotype," in: Handbook of Cytology, Vol. 2 [in Russian], Nauka, Moscow-Leningrad, pp. 359-383.
Wang, N., and Choffner, R. N., 1980, "Induction of heteroploidy in *Gallus domesticus*," Mutat. Res., *69*, 263-273.
Went, D. F., 1982, "Egg activation and parthenogenetic reproduction in insects," Biol. Rev., *57*, 319-344.
White, M. J. D., 1954, Animal Cytology and Evolution, 2nd edn., Cambridge Univ. Press.
White, M. J. D., 1973, Animal Cytology and Evolution, 3rd edn., Cambridge Univ. Press.
Whiting, A. R., 1940, "Sensitivity to x rays of different meiotic stages in unlaid eggs of *Habrobracon*," J. Exp. Zool., *83*, 249-269.
Whiting, A. R., 1945, "Effects of x rays on hatchability and on chromosomes of *Habrobracon* eggs treated in the 1st meiotic prophase and metaphase," Am. Nat., *79*, 193-227.
Wilson, E., 1925, The Cell in Development and Heredity, Macmillan, New York.
Wohlfarth, G. W., 1983, "Genetics of fish: Applications to warm-water fishes," Aquaculture, *33*, 373-381.
Wolters, W. R., Libey, G. S., and Chrisman, C. L., 1981, "Induction of triploidy in channel catfish," Trans. Am. Fish. Soc., *110*, 310-312.
Wu, C., Chen, R., Ye, Y., and Ke, H., 1981, "Investigation on carp gynogenesis with reference to establishing a pure line," Acta Genet. Sin., *8*, 50-55.
Wurgler, F. E., Ulrich, H., and Scheider-Minder, A., 1963, "Variation of radiosensitivity during meiosis and early cleavage in newly laid eggs of *Drosophila melanogaster*," in: Repair from Genetical Radiation Damage and Differential Radiosensitivity of Germ Cells, F. H. Sobels (ed.), Pergamon Press, pp. 101-104.
Yamazaki, F., 1983, "Sex control and manipulation in fish," Aquaculture, *33*, 329-354.
Yurovitsky, Yu. G., 1966, "Sex determination in fish," Usp. Sovrem. Biol., *62*, 148-160.
Zanazzi, M., 1978, "Lebendbeobachtungen der Chromosomen-Aufregulation bei der Gallmucke *Heteropeza pygmae* (*Cecidomyiidae*, *Diptera*)," Chromosoma, *66*, 309-330.
Zhou, J., Shen, J., and Liu, M., 1983, "A cytological study of the gynogenesis of Fang-Zheng crucian carp of Heilongjiang Province [China]," Acta Zool. Sin., *29*, 11-16.
Zotin, A. I., 1961, Physiology of Water Metabolism in the Embryos of Fish and Cyclostomates [in Russian], Nauka, Moscow.

Chapter 5

MATURATION OF MAMMALIAN OOCYTES IN VITRO

M. N. Skoblina

The oocytes of most mammalian species reach the diffuse diplotene or dictyate stage of the I maturation division at birth or shortly after it and persist at this stage until puberty. In each sexual cycle, gonadotropic hormones induce maturation and ovulation of a certain number of oocytes (depending on the species). The term maturation as used here refers to the period from the beginning of stimulation of preovulatory follicles (or from oocyte isolation into the culture medium, in case of spontaneous maturation) to the metaphase II stage. This stage (M II) is considered as the completion of maturation although, as is well known, meiosis is completed only after fertilization. In this review we shall consider only oocyte maturation. Ovulation of oocytes, which is coordinated with maturation in time but does not seem to be causally related to it, is not discussed [see reviews by Espey (1978) and Strickland and Beers (1979) on ovulation mechanisms].

A large number of reviews have been devoted to various aspects of oocyte maturation in mammals (Donahue, 1972; Dyban and Baranov, 1977; Mangia and Canipari, 1977; Thibault, 1977; McGaughey, 1978; Moor and Warnes, 1978, 1979; Tsafriri, 1978b; Masui and Clarke, 1979; Tsafriri et al., 1982; Channing et al., 1982; Tsafriri, 1985). Here we discuss various aspects of the process, including morphological and biochemical changes of the maturing oocyte and the changes in the preovulatory follicle, including the possible effect of substances synthesized there (cAMP, steroids, etc.) on oocyte maturation. Some attention is given to the acquisition of the capability for fertilization by oocytes matured in vitro, as well as to factors affecting this process.

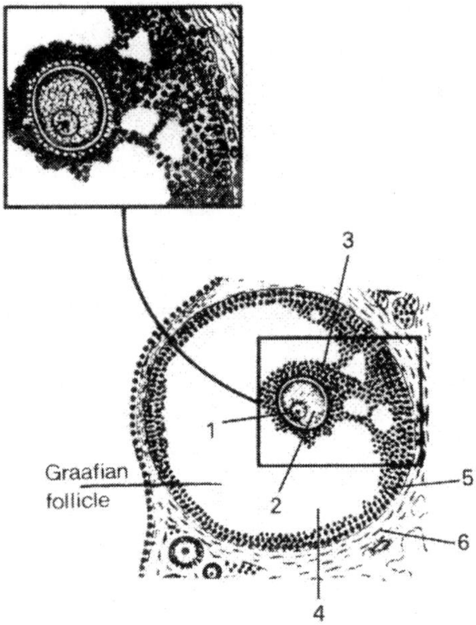

Fig. 40. Diagram of a preovulatory Graafian follicle in mammals (Baker, 1972). 1) Germinal vesicle (oocyte nucleus); 2) oocyte; 3) cumulus; 4) antrum (follicle cavity); 5) granulosa cells; 6) theca.

Extensive studies of mammalian oocyte maturation began after the process was reproduced in vitro. This was first done by Pincus and Enzmann (1935) using rabbit oocytes. It was shown that a cumulus-enclosed oocyte removed from the follicle (Fig. 40) matures without any hormonal stimulation; this was called spontaneous maturation. However, oocyte maturation both in vivo and follicle-enclosed in vitro is induced by gonadotropins. The differences and similarities between spontaneous and gonadotropin-induced maturation are noted below.

5.1. ACQUISITION OF MATURATION COMPETENCE

In mammals there seems to be no correlation between oocyte size and capability to resume meiosis. In a number of species such as the mouse (Sorensen and Wassarman, 1976), rat (Ericksen and Ryan, 1976), and hamster (Iwamatsu and Yanagimachi, 1975) germinal vesicle breakdown (GVBD) during spontaneous maturation occurs in oocytes which have not completed their growth, while in the pig it occurs only in full-grown oocytes (Motlik et al., 1984). Apparently, the oocyte first acquires the ability to resume meiosis and then to complete it (Sorensen and Wassarman, 1976; Sorensen, 1979; Motlik et al., 1984). In most mammal species oocyte growth is arrested in an early antral follicle. Some oocytes from the antral follicles of pubertal animals usually complete meiosis. However, in some animals, e.g., the rabbit (Smith et al., 1978b), sheep (Moor and Trounson, 1977), and pig (Channing and Tsafriri, 1977; McGaughey et al., 1979; Motlik et al., 1984), there is a correlation between the size of antral follicles and the frequency of oocytes completing maturation, while in others, e.g., mouse (Erickson and Sorensen, 1974), hamster (Iwamatsu and Yanagimachi, 1975), cow (Thibault et al., 1975c; Leibfried and First, 1979a), and macaque (Smith et al., 1978a) this correlation seems to be absent. The reason for this difference remains unknown.

Factors affecting oocyte maturation competence are studied in mice and rats. Contradictory results have been obtained. In rats, the competence is dependent upon gonadotropins (Bar-Ami and Tsafriri, 1981, 1986). The effect of gonadotropins is prevented by steroidogenesis inhibitors (Channing and Bar-Ami, 1984); estradiol-17β mediates the effect of gonadotropins only partially (Tsafriri and Bar-Ami, 1981). It is suggested that the acquisition of maturation competence is associated with and dependent upon differentiation of the somatic cells of the follicle (Bar-Ami and Tsafriri, 1986). Mouse oocytes acquired meiotic competence in the absence of gonadotropins and granulosa cells (Canipari et al., 1984; Eppig, 1985, 1986). Mouse oocyte meiotic competence follows a definite time program (Canipari et al., 1984).

5.2. MORPHOLOGY OF MATURATION

One of the first marked changes during maturation is the transformation of the nucleolus in an electron-dense body which gradually diminishes in size and eventually disappears (Zamboni, 1972; Motlik and Fulka, 1981a). In cow oocytes the nucleolus is rarely seen (Motlik and Fulka, 1981a). The GV is displaced to the oocyte periphery. Localization of chromatin in the oocyte nucleus [germinal vesicle (GV)] and the degree of its condensation in full-grown oocytes vary with the species. The nucleolus transformation and chromatin condensation are paralleled by the disappearance of nuclear pores and the fragmentation of the GV membrane (Zamboni et al., 1972; Calarco et al., 1972). In the rat oocytes, the fragmentation is preceded by a perinuclear accumulation of lysosomes; after GV breakdown, lysosomes are once again uniformly distributed in the oocyte (Ezzel and Szego, 1979). The karyoplasm mixes with the cytoplasm, and the I and, later on, II meiotic spindles are formed. In all studied species, no centrioles were found at the spindle poles. Aggregates of compactly arranged vesicles at the spindle poles were described in the mouse (Zamboni, 1970), as well as the tubular form of the endoplasmic reticulum in the rabbit (Zamboni, 1972) and multiple organizing centers in the rat (Albertini, 1984).

The mitochondria behave in very different ways. For example, in the mouse and rat they are uniformly distributed all over the oocyte in the beginning of oocyte maturation and translocated to definite regions after GVBD. In the mouse, the mitochondria are aggregated around the forming spindle (Calarco et al., 1972; Van Blerkom and Runner, 1984), while in the rat they are also aligned under the cell membrane (Albertini, 1984). In the pig and cow, on the contrary, the mitochondria are concentrated in the oocyte periphery in the beginning of maturation, and later become uniformly distributed (Cran, 1985; Kruip et al., 1983; Hyttel et al., 1986). In the pig, the number of mitochondria is markedly reduced and their volume increases by more than 300%, due, apparently, to their fusion (Cran, 1985). In the mouse and cow, the rough endoplasmic reticulum disappeared and free ribosomes appeared in the cytoplasm of maturing oocytes (Merchant and Chang, 1971; Kruip et al., 1983). The number of cortical granules increases during oocyte maturation (Merchant and Chang, 1971; Cran, 1985; Cran and Cheng, 1985) and they are removed to the oocyte periphery.

In the preovulatory follicle, the oocyte and the cumulus cells are intimately connected. The cumulus cell projections penetrate the *zona pellucida* and establish gap junctions with oolemma (Zamboni et al., 1972; Gilula et al., 1978; Szöllösi et al., 1978). During oocyte maturation, the cumulus cell projections are contracted and the contact between the cumulus and the oocyte is disrupted. The contraction of the cumulus cell projections takes place at different times in various species (cf. Zamboni, 1970). The ultrastructure of oocytes matured in vivo is indistinguishable from that of oocytes matured spontaneously in vitro (Merchant and Chang, 1971; Calarco et al., 1972).

In rats, the so-called "cumulus maturation" begins later than oocyte maturation, and proceeds as follows (Hillensjö et al., 1976; Dekel and Phillips, 1979; Dekel et al., 1979): cells located at the cumulus periphery disperse and form a hyaluronidase-sensitive material and during extrusion of the I polar body (2-4 h prior to ovulation) the cumulus turns into a clot of mucus containing suspended cells. The temporal course of these changes in the cumulus is similar in vivo and, when exposed to LH, in vitro.

5.3. DURATION OF MATURATION

Table 5.1 presents data on the duration of maturation in vivo and in vitro for mouse, rat, rabbit, sheep, and pig. Zero time in

Table 5.1. Duration of Maturation (h) in Different Mammal Species in vivo and in vitro

Species	in vivo	Reference	in vitro	Reference
Mouse	9	McGaughey and Chang, 1969	8	Zeilmaker, 1978
	12	Merchant and Chang, 1971	12	Neal and Baker, 1973
	12.5-13	Edwards and Gates, 1959	12-13	Cross and Brinster, 1970
			10-15	Schultz et al., 1978a
			12-17	Bloom and Mukherjee, 1972
	17-18	Cross and Brinster, 1974	17	Donahue, 1968a
	11-18	Neal and Baker, 1973	17	Merchant and Chang, 1971
Rat	7-8	Odor, 1955	7-10	Magnusson et al., 1977
	9	Ahrén et al., 1978		
	9	Tsafriri and Kraicer, 1972		
	13	Niwa and Change, 1975	13	Niwa and Chang, 1975
Rabbit	8	Pincus and Enzmann, 1935		
	9-9.5	Thibault, 1972	9-9.5	Thibault, 1972
	10	Chang, 1955 a,b	12	Van Blerkom and McGaughey, 1978b
Sheep	18-24	Hay and Moor, 1973	17-25	Quirke and Gurdun, 1971
	21-24	Dziuk, 1965		
Pig	36-37	Hunter and Polge, 1966	32-42	Motlík, 1972

vivo corresponds to human chorionic gonadotropin (hCG) injection, and in vitro, to transfer of oocytes into the culture medium. Maturation was considered to be complete after the extrusion of the I polar body or when the oocytes reached metaphase II in vivo and in vitro. The table shows that the duration of maturation, as determined in different studies, varies considerably for the same species, but the range of variation is the same both in vitro and in vivo.

5.4. BIOCHEMISTRY OF MATURATION

When rat oocytes mature in vivo or spontaneously in vitro, oxygen uptake increases. Maximal oxygen uptake is attained at the time of the extrusion of the I polar body. Oxygen consumption in cumulus cells significantly decreases (Magnusson and Hillensjö, 1977; Magnusson et al., 1977; Ahrén et al., 1978). In the absence of oxygen, the oocytes of mice (Zeilmaker et al., 1972), rats (Zeilmaker and Verhamme, 1974), and hamsters (Gwatkin and Haidri, 1974) do not mature.

RNA content of full-grown mouse oocytes and ovulated eggs ranges from 0.35 to 0.6 ng/oocyte (Sternlicht and Schultz, 1981; Kaplan et al., 1982; Bachvarova et al., 1985; Bachvarova, 1985). Nuclear RNA is synthesized throughout the maturation period, as shown for the mouse and pig, both in vivo (Rodman and Bachvarova, 1976) and in vitro (in cases of spontaneous maturation, Bloom and Mukherjee, 1972; Wassarman and Letourneau, 1976a), until GVBD (Rodman and Bachvarova, 1976). Very recently, transcription of mRNA was found during the first 2 h after initiation of ovine oocyte maturation (Moor and Crosby, 1986). However, about 10-20% of the total RNA is degraded during mouse meiotic maturation (Bachvarova, 1985; Bachvarova et al., 1985). Kaplan et al. (1982) found no significant decrease in the amount of rRNA between the oocyte and the ovulated egg. It appears that there is about twice as much mRNA in the full-grown oocytes as in the egg. Close to one-half of the accumulated polyadenylated RNA is totally degraded or deadenylated. mRNA for actin, for example, is present in full-grown oocytes and eggs but is largely deadenylated during oocyte maturation. This change accompanies a large increase in β-actin synthesis (cf. Bachvarova, 1985). Mitochondrial RNA synthesis continues throughout maturation (Moor and Warnes, 1979).

Actinomycin D does not suppress GVBD in the follicle-enclosed oocytes regardless of whether it is injected into the antrum (Thibault, 1972) or added to the culture medium (Tsafriri et al., 1973). The data on the actinomycin effect on spontaneous maturation in the mouse are contradictory although similar concentrations were used. Some workers did not find any inhibition of GVBD (Donahue, 1968b; Jagiello, 1969; Golbus and Stein, 1976; Crozet and Szöllösi, 1980) while others did (Bloom and Mukherjee, 1972). There are also contradictory data on the effect of α-amanitin, although obtained on different animal species: according to Crozet and Szöllösi (1980), the drug has no effect on GVBD in mouse oocytes, while according to Moor and co-workers (Osborn and Moor, 1983; Moor and Crosby, 1986) it inhibits GVBD in ovine oocytes. Even when actinomycin D failed to inhibit GVBD, the oocytes did not reach M II (Donahue, 1968b; Tsafriri et al., 1972). Taking into account these contradictory results, one cannot exclude the possibility that some RNA synthesized during maturation is required for protein synthesis accompanying it. The first evidence has been obtained very recently (Moor and Crosby, 1986). Most of the proteins, however, are synthesized from templates stored beforehand. It has been shown, for instance, that the patterns of proteins synthesized in mouse oocytes blocked by actinomycin D and M I and M II are the same (Golbus and Stein, 1976).

Contradictory results have also been obtained with regard to variations in the rate of protein synthesis during maturation, which seem to be due to the use of different labeled amino acids. Thus, when valine (Stern et al., 1972) or leucine (Cross and Brinster, 1974; Mangia and Canipari, 1977) were used, the rate of synthesis was found to be constant, while with labeled methionine it decreased (the methionine pool remained unchanged) (Schultz et al., 1978a). When tyrosine was used, however, the rate of protein synthesis changed in a complex manner (Mangia and Canipari, 1977). In the latter case, dibutyryl cyclic AMP (db cAMP), which reversibly suppresses maturation (see below), made tyrosine incorporation linear. The kinetics of tyrosine incorporation suggested to the authors that GVBD is preceded by synthesis of short-lived proteins in the oocyte.

Variations in the protein synthesis pattern during maturation have been studied in the oocytes of mouse, rabbit, sheep, and pig. As revealed by polyacrylamide gel electrophoresis, the number of protein classes synthesized in the oocytes of mouse (Golbus and Stein, 1976) and sheep (Warnes et al., 1977) increases during maturation. It remained unclear, however, whether these changes result from synthesis of different proteins or just an altered rate of synthesis. The use of two-dimensional gel electrophoresis allows a more complete picture. The pattern of protein synthesis changes considerably in the oocytes of mice (Schultz and Wassarman, 1977a, b), rabbits (Van Blerkom and McGaughey, 1978a), and pigs (McGaughey and Van Blerkom, 1977): some proteins appear, others cease to be synthesized, and still others are produced at a varying rate. In the mouse (Schultz and Wassarman, 1977a, b) and sheep (Warnes et al., 1977), but not in the rabbit or pig, the major changes occur after GVBD. Attempts have been undertaken to identify proteins synthesized in full-grown and maturing oocytes of various animals. Maturing oocytes of the mouse (Wassarman and Letourneau, 1976b), pig, and rabbit (Motlik et al., 1978) have been shown to produce lysine-rich histones which accumulate in the GV, chiefly above the chromosomes. Inhibition of the pig oocyte maturation by follicular fluid does not affect the density of GV labeling. The mouse (Schultz et al., 1978a), rabbit and pig (Motlik et al., 1978), and sheep (Warnes et al., 1977) oocytes also synthesize nonhistone proteins. Evidence is also available on the synthesis of actin in sheep oocytes (Moor and Osborn, 1983), actin and tubulin in pig oocytes (McGaughey and Van Blerkom, 1977), and actin, tubulin, putative intermediate filament proteins, LDH, calmodulin, and ribosomal proteins in mouse oocytes (LaMarca and Wassarman, 1979; Schultz et al., 1979; Cascio and Wassarman, 1982; Kaplan et al., 1982; Moor and Osborn, 1983).

Full-grown mouse oocytes produce and accumulate in their GV a protein with a molecular weight of 28,000. The synthesis and phosphorylation of this protein ceases with GVBD; no information on its nature is available (Wassarman et al., 1979).

Attention has been recently given to changes in protein phosphorylation during oocyte maturation. In the mouse, the concentration of one low-molecular-weight phosphoprotein increased, while those of two others relatively decreased. But the state of phosphorylation changes, rather than the synthesis, since puromycin did not inhibit the above changes (Schultz et al., 1983a). Qualitative and quantitative changes in protein phosphorylation have also been found in sheep oocytes, e.g., the appearance of a new phosphoprotein with a molecular weight of 33,000 (Crosby et al., 1984).

Mouse oocytes were also used to investigate the effect of karyoplasm on the pattern of protein synthesis during maturation (Schultz et al., 1978b): nuclear fragments underwent the same changes as intact oocytes while anuclear fragments underwent only some of them.

Protein synthesis pattern has also been studied in rabbit and sheep oocytes maturing in vivo and spontaneously in vitro (Van Blerkom and McGaughey, 1978a; Warnes et al., 1977). The latter study also involved the maturation of follicle-enclosed oocytes in vitro. The experiments indicated that spontaneously maturing oocytes differed from those that were follicle-enclosed in vivo and in vitro; particularly, they lacked the characteristic decline in the production of some proteins and increase in others. Rabbit oocytes matured in vivo were likewise somewhat different from those matured spontaneously: some proteins appeared in both oocyte types, but continued to be synthesized only in vivo or, vice versa, only in vitro; four polypeptides were not found in vitro at all. Moor and Osborn (1983) correlate these changes in protein synthesis with the involvement of follicle cells in its control. They consider as an example the post-translational modulation of actin synthesis. This accounts for about 10% of total protein synthesis in dictyate oocytes and is depressed tenfold in eggs in which somatic cell contact is entirely disrupted. The authors suggest that the nature of the signals that are transmitted from the follicle cell to the oocyte is uncertain. However, since inhibition of 17α-hydroxylase with SU 10603 significantly altered protein synthesis, they hypothesize that steroids (especially estrogen) play a significant role as mediators of gonadotropin-induced changes in protein synthesis.

It should, however, be stressed that a very small number of proteins appear to be synthesized nonsequentially during meiotic maturation. Most proteins continue to be produced in the early embryo. These proteins are ubiquitously required by all cells for general maintenance function and structure (Van Blerkom, 1985).

The protein synthesis inhibitors puromycin and cycloheximide suppressed the luteinizing hormone-induced rat oocyte maturation in Graafian follicles (Tsafriri et al., 1972), but failed to prevent GVBD when oocytes matured spontaneously (Donahue, 1968b; Stern et al., 1972; Golbus and Stein, 1976; Wassarman and Letourneau, 1976b; Wassarman et al., 1976; Schultz and Wassarman, 1977a). It was, therefore, believed that GVBD requires no protein synthesis. Ekholm and Magnusson (1979) suggest, however, that the absence of inhibitor effects is due to a low rate of inhibitor penetration. When they employed db cAMP, which reversibly arrests maturation of rat oocytes, and increased the time of exposure to inhibitors, they observed a reversible inhibition of GVBD. The conclusion was that resumption of meiosis depends on the synthesis of short-lived proteins; these must have been produced by the oocyte itself as the experiments were performed on denuded oocytes. According to Ekholm and Magnusson (1979) these may be the proteins that incorporate labeled tyrosine (Mangia and Canipari, 1977). Fulka et al. (1986) have recently shown different inhibitory effects of cycloheximide on oocyte maturation: suppression of GVBD in pig oocytes and absence of suppression in mouse oocytes. The authors believe that these differences are due to different durations of oocyte maturation in these species. Finally, Moor and Crosby (1986) described the protein required for GVBD ("m" protein) in ovine oocytes. This protein has a relative molecular mass of about 47,000: Inhibition of transcription prevents its synthesis and also blocks GVBD. Porcine and bovine oocytes require new protein synthesis for GVBD as well (R. Moor, A. Hunter, and F. Gandolfi, unpublished observation). The treatment of oocytes after GVBD with puromycin prevents M II (Schultz and Wassarman, 1977a; Ekholm and Magnusson, 1979), suggesting that the completion of maturation also requires protein synthesis.

5.5. HORMONAL CONTROL OF MATURATION

5.5.1. Effect of Gonadotropins on the Maturation of Follicle-Enclosed Oocytes In Vivo and In Vitro

As already mentioned, maturation of oocytes in vivo requires gonadotropins. On the day of proestrus a preovulatory gonadotropin (luteinizing hormone, LH, and follicle-stimulating hormone, FSH) surge is observed (Ayalon et al., 1972; Hay and Moor, 1975). In some mammals, follicle-enclosed oocytes treated with hormones can mature in vitro (see below), and both purified LH and purified FSH were effective in this respect (Lindner et al., 1974; Tsafriri et al., 1976c). Purified FSH is also an effective inducer of in vivo maturation of rat oocytes (Tsafriri et al., 1976c). However, suppression of the FSH surge on the day of proestrus in rats (Hoffman et al., 1979), as well as administration of antiserum against FSH to rats or hamsters, does not affect oocyte maturation (ovulated eggs are capable of fertilization and development), while antiserum against LH does (Schwartz et al., 1975; Rani and Moudgal, 1977). The authors conclude that oocyte maturation in vivo is induced by LH. Therefore, our subsequent discussion of hormonal effects on in vitro oocyte maturation will be focused chiefly on LH or a closely related compound, hCG.

First of all, LH binds to a receptor. The use of labeled hCG has shown that the most heavily labeled cells in preovulatory follicles of rat and sheep are those of the *theca interna*, the bulk of the label being localized at the plasma membrane (Zeleznik et al., 1974; Han et al., 1974; Amsterdam et al., 1975; Moor, cit. by Weiss et al., 1976). In the rat and porcine follicles the label was also found over the granulosa cells (Zeleznik et al., 1974; Channing and Kammerman, 1974; Amsterdam et al., 1975). Later studies have shown that treatment of the oocyte—cumulus complex with LH or hCG also results in the hormone binding to cumulus cells (Amsterdam and Tsafriri, 1979; Channing et al., 1979; Lawrence et al., 1980). In none of these reports was the label found over the oocyte. The results of Jagiello and Ducayen (1977), who observed hCG incorporation into the nucleus, cytoplasm, and *zona pellucida* of sheep or monkey oocytes but not into the *theca interna* or granulosa cells, stand apart from others. These data disagree with the evidence of LH induction of follicular steroidogenesis.

A number of processes are induced by LH in the preovulatory follicle, including cAMP production, steroidogenesis, glycolysis, and oocyte maturation and ovulation (see Channing and Tsafriri, 1977). The induction of steroidogenesis in granulosa cells harvested from 3- to 5-mm follicles of pigs (Channing and Tsafriri, 1977) or from preovulatory follicles of rats (Doerr, 1979) requires the binding of LH by only 5-10 or 15-20% of receptors, respectively. It appears that either the follicles have an excess of LH receptors or the receptors are not equivalent (Ryan, cit. by Channing and Tsafriri, 1977).

The time of follicle exposure to LH varies with the process to be induced. For example, treatment of rabbit preovulatory follicles with 5 μg/ml LH for only 1 sec is sufficient to stimulate steroidogenesis for 5-6 h (Younglai, 1975). Other processes need longer contact to be stimulated. Thus, cAMP accumulates in pig granulosa cells induced by 1 μg/ml LH for less than 40 min (Channing and Tsafriri, 1977), whereas the resumption of meiosis in preovulatory follicles of the rat in vivo seems to occur only after 1.5-2 h contact with LH (Lindner et al., 1974). The processes induced by LH are considered in detail in the following section.

5.5.1.1. Production of cAMP

At present it is generally accepted that the induction of steroidogenesis by LH is mediated by cAMP (see Marsh, 1976), although LH concentrations that induce steroidogenesis and cAMP production differ markedly. For example, a half-maximum stimulation of steroidogenic response in rat preovulatory follicles is achieved when only 3% of the available LH receptors are occupied by hCG. At the same time, the earliest detectable cAMP response was seen at a hCG concentration which elicited about half-maximum progesterone response (Doerr, 1979). However, the same study indicates that in the presence of isobutyl methyl xanthine (IBMX) – an inhibitor of phosphodiesterase – the synthesis of progesterone is stimulated by a lower hCG concentration. This finding leads the author to believe that the system already possesses a sufficiently high cAMP level, and hence steroidogenesis in order to be stimulated needs but a slight increase above this level, which escapes detection by the methods used.

The significant rise in cAMP level after LH treatment has been observed in the preovulatory follicles of the rabbit (Marsh et al., 1972), rat (Tsafriri et al., 1971; Tsafriri et al., 1976c; Nilsson et al., 1977), sheep (Moor et al., 1975; Weis et al., 1976), mouse (Schultz et al., 1983a; Hashimoto et al., 1985), and hamster (Hubbard, 1985a, b); it was followed by a decrease (Hashimoto et al., 1985; Hubbard, 1985a, b; Rhodes et al., 1985). Some workers associate this effect mainly with theca cells (Moor et al., 1975; Weis et al., 1976) while others have observed it in granulosa cells as well (Hamberger et al., 1978; Makris and Ryan, 1978; Lindsey and Channing, 1979). In the hamster preovulatory follicles, as the cAMP level rises, that of the cGMP drops (Hubbard, 1983).

5.5.1.2. Steroidogenesis

Before the LH surge, preovulatory follicles of mammals in vivo predominantly secrete estrogens. Following the LH surge in vivo or its addition in vitro, there occurs a rapid and transient increase of estrogen (and sometimes androgen) secretion, which then declines and is replaced with progestin (predominantly progesterone) secretion. The sequence of events has been described for the rat (Hillensjö, 1975; Goff and Henderson, 1979), sheep (Moor et al., 1973; Seamark et al., 1977; Baird et al., 1981), hamster (Saidapur and Greenwald, 1979), and pig (Gérard et al., 1979; Dimino and Berman, 1979).

Prior to ovulation, the incorporation of precursors into all steroids falls to the control level (Mills and Savard, 1973; Suzuki et al., 1977; Goff and Henderson, 1979), and the cells of the theca undergo degenerative changes (Hay and Moor, 1975; Mori et al., 1979).

5.5.1.3. Glycolysis

LH stimulates glycolysis in Graafian follicles both in vivo and in vitro (Hillensjö, 1976; Tsafriri et al., 1976b). This effect can be reproduced in vitro by cAMP (Hillensjö et al., 1978b).

5.5.1.4. In Vitro Maturation of Follicle-Enclosed Oocytes

Without LH, the follicle-enclosed oocytes do not mature (Edwards, 1962; Tsafriri et al., 1971; Hillensjö, 1976; Moor and Trounsen, 1977). After the exposure of the follicles to gonadotropins in vitro, resumption of meiosis was observed in the following animals: the mouse (Tyler and Collins, 1979, 1980; Hashimoto et al., 1985), rat (Tsafriri et al., 1971; Hillensjö, 1976, 1977), hamster (Gwatkin and Andersen, 1976; Plachot and Mandelbaum, 1978), rabbit (Thibault and Gérard, 1973), sheep (Hay and Moor, 1973; Moor and Trounson, 1977), pig (Gérard et al., 1979; Meinecke and Meineke-Tillmann, 1981), cow (Thibault et al., 1975c; Ménézo et al., 1978), and macaque (Ménézo et al., 1978).

5.5.2. Mechanisms of Gonadotropin Effect on Maturation of Mammalian Oocytes

Having discovered spontaneous maturation of rabbit oocytes, Pincus and Enzmann (1935) supposed that the effect of gonadotropins on maturation consisted merely in removing the inhibitory influence of the follicle. Since then, numerous studies have been devoted to the mechanism of gonadotropin action. Quite naturally, attention has been focused on the effects of substances that are induced in the follicle by LH. The maturation criterion used in most of the experimental studies was the GVBD.

5.5.2.1. Effects of cAMP, db cAMP, and Adenylate Cyclase Activations on the Maturation of Follicle-Enclosed Oocytes

The data on these effects are rather contradictory. Initially, no stimulation by either cAMP ($2.9 \cdot 10^{-4}$ to $1.1 \cdot 10^{-3}$ M) or db cAMP ($2 \cdot 10^{-5}$ to $1 \cdot 10^{-3}$ M) added to the incubation medium was observed. A significant increase in the proportion of matured oocytes was found only after the injection of db cAMP in the antrum (Tsafriri et al., 1971, 1972). Later it was found that 1 mM of db cAMP induced GVBD in one-third of the oocytes; polar bodies were not extruded (Hillensjö, 1975-1977). However, in the most recent reports of the same authors (Hillensjö et al., 1978, 1982), there was evidence that neither cAMP nor db cAMP (apparently when constantly present in the medium) could induce maturation, but the treatment of Graafian follicles with db cAMP at 10 mM for 2 h followed by washing resulted in maturation of 68% of the oocytes. Another analog of cAMP, 8Br-cAMP, when added to the incubation medium, induces the maturation of follicle-enclosed rat oocytes (Hillensjö et al., 1982). Adenylate cyclase activators, cholera toxin (activator of regulatory subunit), and forskolin (activator of catalytic subunit), have been widely used recently in order to increase the cAMP level in tissues. It was shown that 1-3 μM forskolin, after continuous application, reinitiates meiosis in the follicle-enclosed rat oocytes (Dekel and Sherizly, 1983; Ekholm et al., 1984). The maturation of follicle-enclosed mouse (Hashimoto et al., 1985) and rabbit oocytes (Hubbard, 1985b) could be induced only by short-term treatments with higher forskolin concentrations (10-60 μM). Cholera toxin (10 μg/ml) induced the maturation of follicle-enclosed mouse oocytes (Freter and Schultz, 1984). On the other hand, db cAMP and forskolin reversibly inhibit the LH-induced maturation of follicle-enclosed oocytes in the mouse (Hashimoto et al., 1985), rat (Hillensjö, 1977; Hillensjö et al., 1978), and hamster (Hubbard, 1985b). The treatment of follicles with two agents that induce cAMP accumulation appears to result in its excessive production and suppression of oocyte maturation (see below). Phosphodiesterase inhibitors affect oocyte maturation differently. Theophylline and IBMX do not stimulate maturation and reversibly suppress LH effects, while ICI 63197 induces GVBD in 66% of the oocytes (Hillensjö, 1976; Hillensjö et al., 1978).

5.5.2.2. Effects of Steroids on Maturation of Follicle-Enclosed Oocytes

The effects of LH upon steroidogenesis in Graafian follicles has been discussed above (see Section 5.5.1.2). Briefly, LH inhibits estrogen secretion and stimulates progestagene secretion. However, when preovulatory follicles of rats were treated with progesterone, 20β-dihydroprogesterone ($1.6 \cdot 10^{-5}$ to $1.6 \cdot 10^{-4}$ M), or estradiol-17β (10^{-5}-10^{-9} M), no maturation was observed (Tsafriri et al., 1971, 1972; Lindner et al., 1974). How steroids affect the gonadotropin-induced oocyte maturation in Graafian follicles is not exactly clear. Estradiol-17β does not alter the percentage of matured oocytes derived from mice (Tyler and Collins, 1979, 1980) or rats (Tsafriri et al., 1972), while progesterone (applied at similar concentrations) has no effect on rat oocytes (Tsafriri et al., 1972) but reduces mouse oocyte maturation (Tyler and Collins, 1980).

The inhibitors of steroidogenesis employed (aminoglutethimide, cyanoketone, SU 10603, and 4-hydroxyandrostenedione) had different sites of attack in the steroidogenic pathway. None of the inhibitors, however, changed the proportion of oocytes resuming meiosis in response to LH. It was concluded that steroids are not required for initiation of maturation (Tsafriri et al., 1971, 1972; Lindner et al., 1974; Lieberman et al., 1976; Moor, 1978; Billig et al., 1983). The data on the ability of oocytes to complete meiosis during inhibition or alterations to the steroid profile are contradictory. Sheep oocytes do not complete meiosis (Moor et al., 1980a), while rat oocytes, according to some authors, complete meiosis (Billig et al., 1983) and, according to others, do not complete it (Tsafriri et al., 1971). As shown by Cran and Moor (1980), exogenous steroids reduce the number of chromosomal aberrations at telophase I and metaphase II.

5.5.2.3. Effects of Pyruvate, Lactate, and Iodoacetic Acid on Oocyte Maturation of Follicle-Enclosed Oocytes

Iodoacetic acid (IA) at a concentration of $2.5 \cdot 10^{-5}$ M inhibits the LH-induced stimulation of lactate production by rat Graafian follicles, but does not affect GVBD. At higher concentration, the inhibitor suppresses oocyte maturation as well. However, addition of pyruvate and lactate does not eliminate the suppressive effect of IA and in its absence fails to induce oocyte maturation (Tsafriri et al., 1976b). The high lactate concentration in the preovulatory follicle prior to the LH surge also suggests that lactate production cannot be a maturation inducer (Zeilmaker and Verhamme, 1974; Zeilmaker, 1978).

5.5.2.4. Effects of Ions on Maturation of Follicle-Enclosed Oocytes

The Ca^{++} ionophore A 23187 induces 61% GVBD in rat follicle-enclosed oocytes while without Ca^{++} and hormone 66% of the oocytes mature. Addition of the ionophore in this case has no effect on maturation, whereas addition of LH increases the rate of maturation up to 86%. Neither EDTA nor EGTA prevent maturation (Tsafriri and Bar-Ami, 1978). Thus, oocyte maturation is induced by both the absence of Ca^{++} and the presence of the ionophore. The authors suppose that the control of meiosis involves sequestrated Ca^{++} (Tsafriri and Bar-Ami, 1978).

5.5.3. Spontaneous Maturation

Maturation is referred to as spontaneous if it is induced by the removal of oocytes from the follicle. The phenomenon has been observed for many species of animals: mouse (Biggers et al., 1967; Cross and Brinster, 1970; Binor and Wolf, 1979), rat (Zeilmaker and Verhamme, 1974; Shalgi et al., 1979), hamster (Gwatkin and Haidri, 1973; Plachot and Mandelbaum, 1978), guinea pig (Jagiello, 1969; Yanagimachi, 1974), rabbit (Pincus and Enzmann, 1935; Chang, 1955a; Van Blerkom and McGaughey, 1978a), sheep (Quirke and Gurdon, 1971; Moor, 1978), pig (McGaughey and Polge, 1971; Leibfried and First, 1980a, b), cow (Sreenan, 1970; Fukui and Sakuma, 1980), horse (Fulka and Okalski, 1981; Desjardins et al., 1985), dog (Mahi and Yanagimachi, 1976), and macaque (Suzuki and Mastroianni, 1966; Smith et al., 1978a).

5.5.3.1. Effect of Cumulus Cells on Spontaneous Maturation

The oocyte removed from a nonatretic follicle is embedded in a compact cumulus. Cumulus removal is reported to have different effects on oocyte maturation. In the mouse and pig it does not affect the percentage of matured oocytes (Cross, 1973; Binor and Wolf, 1979; Leibfried and First, 1980b; Schroeder and Eppig, 1984; Fleming et al., 1985a); in the rat, according to some authors, it does not affect it (Zeilmaker and Verhamm, 1974) and, according to others, decreases it (Fleming et al., 1985b). A decrease in the percentage of maturation was also noted in denuded cow oocytes (Fukui and Sakuma, 1980). The data on the influence of cumulus removal on the rate of oocyte maturation are also contradictory; no influence was found in the mouse (Freter and Schultz, 1984; Eppig et al., 1983), while in the rat some authors did not find any influence (Hillensjö et al., 1982), but others noted the acceleration of GVBD (Dekel and Beers, 1980). Below we will distinguish between the effects on oocyte-cumulus cell complex (OCCC) and on denuded oocytes.

5.5.3.2. Effects of LH, FSH, db cAMP, and Drugs Elevating cAMP on Spontaneous Maturation of OCCC

Unlike its effect on the follicles, FSH, after continuous exposition, inhibits or suppresses the maturation of OCCC in the mouse (Schultz et al., 1983b; Eppig et al., 1983), pig (Racowsky, 1983), and rat (Miller and Behrman, 1985). On the other hand, short-term treatment with LH or FSH accelerates the maturation of mouse OCCC (Freter and Schultz, 1984). Exposure of rat cumulus-enclosed oocytes removed from preovulatory follicles to 10 μg/ml LH accelerates GVBD without affecting the duration of final stages of maturation (Lopata et al., 1977; Kaplan et al., 1978).

The influence of db cAMP on spontaneous oocyte maturation has been repeatedly studied. In some cases we only presume that the OCCC was used since no cumulus removal was described (Stern and Wassarman, 1973; Cho et al., 1974; Magnusson and Hillensjö, 1977), while in the other ones it is known (Nekola and Smith, 1975; Eppig et al., 1983). In any event db cAMP at 100 μg/ml, after continuous application, reversibly blocks GVBD in more than 90% of the mouse oocytes (Stern and Wassarman, 1973; Cho et al., 1974; Nekola and Smith, 1975; Eppig et al., 1983) and rat oocytes (Magnusson and Hillensjö, 1977). The short-term treatment of mouse OCCC with db cAMP accelerates their maturation (Freter and Schultz, 1984). After GVBD, db cAMP does not affect the oocyte maturation in these animals, although it penetrates the oocyte at all stages equally (Cho et al., 1974). The db cAMP effect on maturation of oocytes of other animals is not so straightforward. For example, in the pig OCCC (Rice and McGaughey, 1981; Racowsky, 1983), and cow and sheep (OCCC?) (Jagiello et al., 1981), GVBD is blocked only in a small fraction of oocytes (Jagiello et al., 1981). In addition, the maturation of pig and cow oocytes was unaffected except for higher db cAMP concentrations (500-750 μg/ml), while the response of sheep oocytes is concentration-independent. In the hamster (OCCC?), db cAMP does not suppress maturation (Jagiello et al., 1981), while 8-Br-cAMP does (Hubbard and Terranova, 1982). In the hamster, cow, and sheep, db cAMP also suppresses to some extent the formation of the metaphase II plate (Jagiello et al., 1981). It has recently been shown that hamster OCCC maturation is also suppressed by cGMP, but at a very high concentration (6 mM) (Hubbard and Terranova, 1982).

Cholera toxin inhibits the OCCC maturation in the mouse (Schultz et al., 1983b), but does not affect that in the rat (Dekel and Beers, 1980). However, it potentiates the effect of IBMX on the OCCC maturation in the rat (Dekel and Beers, 1980). Another adenylate cyclase activator, forskolin, reversibly suppresses the OCCC maturation in the rat (Olsiewski and Beers, 1983; Ekholm et al., 1984; Dekel et al., 1984a; Racowsky, 1984), hamster (Racowsky, 1985a), and pig (Racowsky, 1985c).

Phosphodiesterase inhibitors, theophylline and IBMX, reversibly suppress oocyte (OCCC?) maturation in the mouse (Cho et al., 1974) and rat (Magnusson and Hillensjö, 1977; Dekel and Beers, 1978). Evidence on whether gonadotropins can overcome the inhibition by cyclic nucleotides is contradictory. Neither FSH (0.01-10 μg/ml) nor LH (0.05-5 μg/ml) can overcome inhibition in mouse oocyte maturation induced by db cAMP (Nekola and Smith, 1975;

Schultz et al., 1983b), whereas LH potentiated rat oocyte maturation suppressed by cAMP or IBMX. A mixture of LH and FSH was especially effective. Estradiol-17β and progesterone (10^{-6} M) do not mediate the effect of gonadotropins (Dekel and Beers, 1978, 1980). LH (200 ng/ml and 10 μg/ml) overcomes the suppressing effect of 8-Br-cAMP and reduces that of 8-Br-cGMP on oocyte maturation in the hamster (Hubbard and Terranova, 1982).

5.5.3.3. Effects of FSH, db cAMP, and Drugs Elevating cAMP on Maturation of Denuded Oocytes

FSH does not suppress the maturation of denuded oocytes in the mouse (Freter and Schultz, 1983; Schultz et al., 1983b), rat (Olsiewski and Beers, 1983), and pig (Racowsky, 1983), although rat oocyte cAMP increases (Olsiewski and Beers, 1983). Therefore, Olsiewski and Beers suggest the presence of receptors for FSH at the oocyte surface (see also Jagiello and Ducayen, 1977). They account for the failure of FSH to suppress oocyte maturation by the assumption that, in this case, the cAMP level is not sufficient for suppression during the critical period.

db cAMP and 8-Br-cAMP suppress the maturation of denuded oocytes in the mouse (Wassarman and Turner, 1976; Eppig et al., 1983; Barret and Powers, 1983), pig (Rice and McGaughey, 1981), and hamster (Hubbard and Terranova, 1982). In the mouse, higher concentrations of cAMP are required to suppress the maturation of denuded oocytes than that of OCCC (Eppig et al., 1983), while in the pig the dependence is the reverse (Rice and McGaughey, 1981). In the hamster, the maturation of denuded oocytes, unlike that of OCCC, could not be induced by cGMP (Hubbard and Terranova, 1982).

Cholera toxin did not affect mouse oocyte maturation in some experiments (Schultz et al., 1983b) and delayed their maturation in others (Vivarelli et al., 1983). Forskolin reversibly suppresses oocyte maturation in the mouse (Barret and Powers, 1983; Urner et al., 1983; Sato and Koide, 1984a), hamster (Racowsky, 1985a), and pig (Racowsky, 1985c). A higher concentration of forskolin is, however, required to suppress the maturation of denuded oocytes and the sensitivity of the latter dramatically varies in different animals, e.g., 2.5 μM forskolin is required for 50% suppression of oocyte maturation in the mouse, 4.84 μM in the pig, and 65 μM in the hamster (see Racowsky, 1985c). Contradictory data on the influence of forskolin on the maturation of denuded oocytes in the rat (absence — Dekel et al., 1984a; Racowsky, 1984; or presence — Olsiewski and Beers, 1983) appear to be accounted for by a very low sensitivity of the oocytes to forskolin: ID_{50} 250 μM (Olsiewsky and Beers, 1983). Since the suppression of oocyte maturation by forskolin in the mouse, rat, pig, and hamster is accompanied by an elevation of oocyte cAMP, it was concluded that adenylate cyclase is present in the oolemma (Barret and

Powers, 1983; Urner et al., 1983; Racowsky, 1985a, c). This conclusion became quite convincing after Bornslaeger and Schultz (1984) had shown the increase in cAMP level even in a zona-free mouse oocyte. Adenylate cyclase was found cytochemically on the plasma membrane of human oocytes (Suzuki et al., 1983). Cholera toxin-responsible adenylate cyclase was found in sheep oocytes (Crosby et al., 1985).

Phosphodiesterase activity was discovered in mouse (Bornslaeger et al., 1984) and sheep oocytes (Crosby et al., 1985). No increase in its activity during oocyte maturation in the mouse was found. However, degradation of cAMP in the oocytes from FSH- and forskolin-stimulated OCCC was markedly enhanced as compared with that in nonstimulated OCCC (Salustri et al., 1985; Bornslaeger et al., 1984). Changes in the relative percentage of free versus bound intracellular cAMP might be responsible for the different rates of hydrolysis (Salustri et al., 1985).

Evidence on changes in the cAMP content of maturing oocytes is contradictory, e.g., the cAMP in follicle-enclosed sheep oocytes rises and then declines to basal level (Moor and Heslop, 1981), whereas it remains practically unchanged in mouse oocytes under the same conditions (Hashimoto et al., 1985). During spontaneous maturation of OCCC the cAMP content of the mouse oocytes declines (Schultz et al., 1983a), but remains unchanged in the pig oocytes (Racowsky, 1985c).

5.5.3.4. Effect of Steroids on Spontaneous Maturation of OCCC

The effect of steroids on the fertilizability of oocytes maturing in vitro will be considered in Section 5.5.5. Here we shall discuss only their effects upon oocyte maturation. However, the results are contradictory; some examples are cited. Most authors state that progesterone at a concentration of 1-2 μg/ml does not affect the spontaneous oocyte maturation in the mouse (Eppig and Koide, 1978; Barak et al., 1982; Eppig et al., 1983; Eppig and Downs, 1984), but there is some evidence that at the same concentration it exerts an inhibitory effect, and results in necrosis at higher concentrations (Smith and Tenny, 1978, 1980). For rabbit oocytes, progesterone may exert either an inhibitory (32 μg/ml) (Smith et al., 1978b) or a stimulatory (at similar concentrations) effect (Bae and Foote, 1975). An increase in progesterone content of the follicular fluid (FFl) during oocyte maturation can hardly be correlated to its inhibitory effect (Goff and Henderson, 1979; Ainsworth et al., 1980). Another example: Schaerf et al. (1982) found no effect of steroids at a concentration of 1.4 μg/ml on the maturation of OCCC in the pig, and Rice and McGaughey (1981) observed the inhibitory effect of a much lower concentration of testosterone. The latter authors treated pig OCCC with a mixture of db cAMP (1 mM) and testosterone (0.144 μg/ml) and observed a more pronounced inhibition of GVBD. Simultaneous treatments of the pig OCCC with androgens and drugs elevating cAMP were repeated by Racowsky (1983) and Daniel et al. (1986). These latter used, besides testosterone, 19-norandrostenedione, a major steroidal component of porcine FFl (Khalil and Watson, 1985). Androgens also enhanced the inhibitory effect of db cAMP (Racowsky, 1983; Daniel et al., 1986), but weakened that of FSH (Racowsky, 1983). The inhibitory effects of androgens proved to be reversible, dose-dependent, and mediated by the follicle cells (Rice and McGaughey, 1981; Racowsky, 1983; Daniel et al., 1986). Hydroxyglutamide, known to block androgen receptors, abolished the effect of steroids (Daniel et al., 1986).

The mouse OCCC were treated with FSH, db cAMP, or cholera toxin with steroids (progesterone, testosterone, estradiol-17β, etc.) (Eppig et al., 1983; Schultz et al., 1983b; Eppig and Downs, 1984; Eppig, 1985) and the hamster and rat OCCC with forskolin and steroids (Racowsky, 1985b). In the mouse OCCC, unlike the pig OCCC, all studied steroids potentiated the inhibitory effect of FSH (Eppig et al., 1983; Schultz et al., 1983b; Eppig and Downs, 1984). In the mouse OCCC, as in the pig OCCC, steroids (progesterone and testosterone) enhanced the inhibitory effects of db cAMP and cholera toxin (Schultz et al., 1983b; Eppig and Downs, 1984). In the rat OCCC, testosterone did not affect the inhibitory influence of forskolin, while in the hamster OCCC it enhanced it, and an antagonist of estradiol, tamoxifen, reduced it (Racowsky, 1985b).

Suggestions were put forward that the inhibitory effect of the combined influence of an increased cAMP level and testosterone may be mediated by estrogens, an altered progesterone/estradiol ratio (Rice and McGaughey, 1981; Racowsky, 1983), or a disturbance of coupling between the oocyte and the cumulus cells (Racowsky, 1983). However, estradiol did not inhibit spontaneous oocyte maturation in most of the studied species (Eppig and Koide, 1978; McGaughey, 1977; Richter and McGaughey, 1979). When testosterone was substituted by estradiol-17β in combination with db cAMP, no inhibitory effect on the pig OCCC was noted (Racowsky, 1983) and the inhibitory effect on the mouse OCCC was reduced (Eppig et al., 1983). As to the possible influence of steroid profile on maturation, it is, as noted above, either absent (Billig et al., 1983) or expressed at later stages (Tsafriri et al., 1971; Moor, 1978; Moor et al., 1980a). It has also been shown that testosterone affects neither the intraoocyte cAMP nor its transfer from the cumulus cells to the oocyte (Eppig et al., 1983; Schultz et al., 1983b).

The mechanism of the inhibitory effect of steroids and db cAMP on the OCCC maturation is as yet unclear. It is, however, suggested that 19-norandrostenedione may be of physiological importance in regulation of porcine oocyte maturation (Daniel et al., 1986).

5.5.3.5. Effects of Steroids on Maturation of Denuded Oocytes

Steroids do not affect the maturation of denuded oocytes in the mouse (Eppig et al., 1983; Eppig and Downs, 1984) or pig (Rice and McGaughey, 1981). The maturation of denuded hamster oocytes was suppressed by estradiol in a dose-dependent way (Racowsky, 1985b); but the mechanism of inhibitory influence is not clear. In any case, it is not reduced by tamoxifen. Evidence on the combined effect of db cAMP and forskolin with steroids on the maturation of denuded oocytes is also contradictory. In the pig (Rice and McGaughey, 1981; Racowsky, 1983; Daniel et al., 1986) and hamster (Racowsky, 1985b), steroids did not alter the inhibitory influence of db cAMP and forskolin, while in the mouse they enhanced it (Eppig et al., 1983). In the mouse, unlike in the pig, the site of synergistic action for steroids in augmenting the maturation-suppressing action of the cAMP-dependent process is within the oocyte (Eppig et al., 1983; Eppig and Downs, 1984). It has been suggested that testosterone (rather than the other steroids), while interacting with db cAMP in suppression of oocyte maturation in the mouse, acts in an unusual way. It promotes the breakdown of phosphatidylinositol in the oocyte membrane, thus leading to an increase in the C-kinase activity and in the level of phosphorylated proteins maintaining meiotic arrest (see also Ecay and Powers, 1985).

5.5.3.6. Effects of Granulosa Cells and Follicular Fluid (FFl) on Spontaneous Maturation

We have already mentioned that Pincus and Enzmann, back in 1935, suggested inhibition of maturation by the follicle. Numerous studies have been carried out since then to check this hypothesis. At present, most investigators seem to believe that such inhibition indeed occurs, and that the source of it is the granulosa cells (see Thibault, 1977; Moor and Warnes, 1978, 1979; Tsafriri, 1978a; Sato et al., 1982, 1984; Anderson et al., 1985). As to all the rest, the opinions differ. Some workers observed inhibition only after contact of oocytes with granulosa cells (Foote and Thibault, 1969; Sato and Ishibashi, 1977; Leibfried and First, 1980a, b), while others maintain that it occurs in the absence of direct contact, in conditioned medium, or in the FFl (Tsafriri, 1978a, 1985; Downs and Eppig, 1984).

The inhibitory effect of FFl on the maturation of OCCC in the culture medium was demonstrated in the rabbit (Chang, 1955a), hamster (Gwatkin and Andersen, 1976), pig (Tsafriri et al., 1975, 1976a), rat (Tsafriri et al., 1977), and mouse (Channing and Tsafriri, cit. by Tsafriri et al., 1979). Similar results were obtained when OCCC were injected into the cavity of preovulatory porcine follicles (Meineke and Meineke-Tillmann, 1981). Inhibition does not seem to be species specific. Maturation of hamster oocytes can be inhibited by the FFl of cows (Gwatkin and Andersen, 1976) and maturation of the oocytes of the rat and pig by the FFl of man (Tsafriri et al., 1982; Chari et al., 1983). The activity of the oocyte maturation inhibitor (OMI) from FFl depends on the size of follicles used (Gwatkin and Andersen, 1976; Centola and Anderson, 1979); as oocytes mature they become less sensitive to the effect of FFl (Tsafriri et al., 1977; Tsafriri, 1978a; Channing et al., 1983).

The nature of OMI has been studied with FFl of the cow and pig (Gwatkin and Andersen, 1976; Tsafriri et al., 1975), and man (Chari et al., 1983). The active agents of FFl in the pig and cow proved to be similar and were referred to as OMI. Removal of steroids from the FFl did not reduce the inhibitory activity, while proteolytic enzymes did (Tsafriri et al., 1975; Tsafriri et al., 1976a; Gwatkin and Andersen, 1976). The authors came to the conclusion that the inhibitor is a peptide and its molecular weight is about 2000 (Tsafriri et al., 1976a), or ranges from 1000 to 10,000 daltons (Gwatkin and Andersen, 1976).

OMI was chemically identical in porcine FFl, the extract of granulosa cells, and conditioned medium (Tsafriri et al., 1982). The suppression of oocyte maturation by human FFl was not abolished either by ether extraction or by trypsin treatment, heating to 56°C for 1 h or boiling for 5 min, whereas heating to 105°C for 18 h decreased the effect (Chari et al., 1983). Addition of LH partially overcomes the inhibitory effects of porcine and human FFl (Tsafriri, 1978a; Chari et al., 1983). LH completely overcomes the inhibition by the FFl only in the presence of granulosa cells; the partial effect could be due to the presence of cumulus cells (Tsafriri, 1978a; Tsafriri et al., 1979). FSH induces OMI secretion and androgens inhibit it (Anderson et al., 1985). Prolactin appears to bear some influence on OMI secretion or action. In any case, the prolactin treatment potentiates the inhibitory activity. Prolactin itself is not active (Channing and Evans, 1982) but antiserum against prolactin abolishes the OMI effect (Channing et al., 1982). It is interesting that the hamster follicles lose receptors to prolactin after a preovulatory surge of gonadotropins (Oxberry and Greenwald, 1982).

Whereas OMI inhibits the maturation of oocytes cultured within their intact cumuli, it did not interfere with the maturation of fully denuded oocytes of the hamster (Gwatkin and Andersen, 1976), pig (Hillensjö et al., 1979), rat (Lindner et al.,

1983), or mouse (Bar-Ami and Tsafriri, unpubl., cit. by Lindner et al., 1983). Moreover, addition of the low-molecular-weight fraction of porcine FFl (Amicon PM-10 filtration) even tended to facilitate the maturation of denuded rat oocytes. OMI exerts its inhibitory action through the mediation of the cumulus cells (Lindner et al., 1983). But the inhibitory effect of human FFl on meiosis was seen in denuded oocytes (Chari et al., 1983).

In the experiments of Eppig and Downs, a PM-10 fraction of porcine FFl did not suppress, but only inhibited, the maturation of mouse OCCC and denuded oocytes, its effect on the latter being much weaker. The authors studied the synergistic action of PM-10 with the drugs elevating cAMP in OCCC (FSH, db cAMP) and denuded oocytes (db cAMP and forskolin) (Eppig and Downs, 1984; Downs and Eppig, 1984). The inhibitory agent of porcine FFl, active in the presence of cAMP, proved to be a small (with a molecular mass of 1000) hydrophobic molecule (Downs and Eppig, 1984), apparently hypoxanthine and adenosine, which have a synergistic action (Downs et al., 1985; Eppig, 1986). Unfortunately, this agent was also called OMI, thus creating a certain confusion.

However, not all authors succeeded in reproducing the inhibitory effect of FFl on maturation of the pig and cow OCCC cultivated in vitro (Sato and Ishibashi, 1977; Liebfried and First, 1980a, b) and the rat OCCC cultivated in a porcine follicle (Fleming et al., 1985b). It is possible that these difficulties are in part due to the apparently temporal nature of both oocyte sensitivity to OMI and the OMI production, the instability of OMI, the rapid initiation of meiosis, and the need for continued maintenance of cumulus-oocyte coupling to keep meiosis in abeyance (Lindner et al., 1983). Inhibition was, however, obtained following the contact with granulosa cells (Sato and Ishibashi, 1977; Leibfried and First, 1980a, b). Oocyte maturation-preventing factor (OMPF) was extracted from porcine (Sato et al., 1982) and bovine granulosa cells (Sato and Koide, 1984b; Sato et al., 1984, 1986), and was similar to the granulosa cell factor (GCF), judging by the method of extraction. OMPF is a heat-stable polypeptide with a molecular mass of 1450 to 3000 (Sato et al., 1982) or less than 6000 (Sato and Koide, 1984b). OMPF inhibited GVBD in denuded and cumulus-enclosed mouse oocytes. Heparin and heparin sulfate blocked the inhibitory effect of OMPF (Sato et al., 1986).

And, finally, a suggestion has recently been put forward that OMI is an artifact. Fleming et al. (1983) found no OMI activity in PM-10 fraction of porcine FFl when all manipulations concerned with its extraction and purification were carried out at 4°C. The authors believe that OMI arises as a result of degradation of some FFl component at room temperature. This statement agrees poorly with the existence of OMI activity in the conditioned medium of granulosa cells and the chemical identity of OMI from these two sources (Tsafriri et al., 1982).

Adenosine (Miller and Behrman, 1985) and retinoids (Urner and Schorderet-Slatkine, 1984) have recently been proposed as putative maturation inhibitors.

5.5.3.7. Influence of Ions on Spontaneous Maturation

Ca ions are indispensable for survival of full-grown mouse oocytes (Paleos and Powers, 1981; De Felici and Siracusa, 1982). Evidence on the role of Ca^{++} in maturation of porcine OCCC is contradictory. In the absence of Ca^{++} in the medium, or upon blocking of Ca^{++} transfer by verapamil, GVBD does not take place (Bae, 1981; Bae and Channing, 1985). However, the treatment of porcine OCCC with EDTA did not suppress their maturation (Sato et al., 1982). GVBD was not dependent upon Ca^{++} in OCCC of the mouse (Jagiello et al., 1982), rat (Tsafriri and Bar-Ami, 1978), or cow (Leibfried and First, 1979; Jagiello et al., 1982; Maruska et al., 1984).

A decrease in the concentration or absence in the medium of Ca^{++} and Mg^{++} inhibits the maturation of oocytes and OCCC in the mouse and cow (Iwamatsu et al., 1979; Leibfried and First, 1979; Jagiello et al., 1982).

Ca ions are necessary for meiosis to be completed in mouse oocytes (Powers and Paleos, 1982) and cow OCCC (Maruska et al., 1984). Having reached metaphase II, the mouse oocytes do not respond to the absence of Ca^{++} (De Felici and Siracusa, 1982), but in the pig oocytes chromosomes are extremely condensed or degraded (Bae and Channing, 1985).

Data on the effects of A23187 on spontaneous oocyte maturation are highly contradictory. It does not affect the maturation of rat OCCC (Tsafriri and Bar-Ami, 1978), inhibits the maturation of OCCC in the mouse and cow (Jagiello et al., 1982) and pig (Bae and Channing, 1985), and abolishes the inhibitory effect of db cAMP on oocyte maturation in the mouse (Paleos and Powers, 1982). Calmodulin inhibitors reversibly suppress the maturation of oocytes and OCCC in the mouse and cow (Jagiello et al., 1982; Schultz et al., 1983c; Bornslaeger et al., 1984). According to Jagiello et al. (1982), calmodulin per se also suppresses oocyte maturation in the mouse and cow at the stage of spindle assembly.

Lack of sodium ions or presence of the K^+ ionophore valinomycin in the medium gave a 2-h delay of GVBD in mouse oocytes (Powers and Biggers, 1976).

5.5.4. *Possible Mechanisms of Gonadotropin Induction of Oocyte Maturation in Mammals*

Maturation of mammalian oocytes in vivo thus seems to be induced by LH, which stimulates the production of cAMP, steroidogenesis, and glycolysis in the Graafian follicles. We have already seen that neither steroidogenesis nor glycolysis is directly related to the resumption of meiosis in the follicle-enclosed oocytes, whereas spontaneous maturation can proceed without any hormone and, in some cases, without exogenous sources of energy (Zeilmaker and Verhamme, 1974).

Despite a long history of investigations devoted to the hormonal control of mammalian oocyte maturation, there is no generally accepted view concerning its mechanism. There are two main viewpoints: 1) spontaneous oocyte maturation suggests the presence of an inhibitory factor in the follicle, and 2) spontaneous maturation is an artifact of cultivation. Most authors support the former viewpoint. Let us consider it in more detail. If an inhibitor exists, how does it get into the oocyte, how can its transfer be affected by LH, and what is its nature? All supporters of this viewpoint appear to hold that the inhibitor is produced in granulosa or cumulus cells, and most of them believe that it is transferred into the oocyte. How can the inhibitor be transferred into the oocyte? It has already been mentioned that in the preovulatory follicle the cumulus cell projections penetrate the *zona pellucida* and establish gap junctions with the oolemma (Zamboni et al., 1972; Gilula et al., 1978; Szöllösi et al., 1978).

The junctions are permeable to ions and low-molecular-weight compounds (Moor et al., 1980a, b). Before ovulation, the gap junctions are disrupted after the LH surge (Szöllösi et al., 1978; Gérard et al., 1979; Moor et al., 1980a; Larsen et al., 1981); the inhibitor can no longer enter the oocyte, and the latter matures.

However, the data on changes in permeability of gap junctions during oocyte maturation in different species are far from being unambiguous. For example, in the pig (Gérard et al., 1979; Motlik et al., 1986), mouse (Eppig, 1982), and sheep (Moor et al., 1980a), no marked changes in permeability of gap junctions before GVBD were found. Moreover, in the sheep, LH, which induces oocyte maturation, does not affect gap junction permeability, while FSH at a low concentration reduces it without inducing maturation (Moor et al., 1981). The method of estimating the coupling index has recently been modified (Dekel et al., 1984b), since under conditions in which the uptake changes in the cumulus cells may occur, it seems that expressing uptake as a fraction of total cumulus uptake is a more reliable index of oocyte-cumulus metabolic coupling than the comparison of absolute values of label incorporated in treated versus control oocytes, as done before. When using this method, it has been shown that in the rat in vitro within 2 h of culture in the presence of LH, coupling decreases to 45% of its initial level; maturation at this time could be observed in only 25% of the oocytes examined (Salustri and Siracusa, 1983; Dekel et al., 1984b). Unfortunately, the significance of differences was not indicated. When using the same method, it has been shown that, in the hamster, cumulus-to-oocyte coupling progressively decreased with time after release from meiotic arrest both in vitro and in vivo (Racowsky and Satterline, 1985) and, in the mouse, GVBD precedes, in all cases, the termination of metabolic cooperation (Salustri and Siracusa, 1983), and acceleration of the OCCC maturation under the influence of LH is not related to uncoupling (Freter and Schultz, 1984). Varied results may be attributable to species variation, to differences in preparation of the OCCC, or to variation in culture systems (Racowsky and Satterline, 1985). Hence, there is as yet no convincing evidence in favor of the role of gap junctions in the release of meiotic arrest. Recently obtained freeze-fracture data are of interest in this respect. A dramatic loss of cumulus cell gap junctions has been shown in the rat between 2 and 3 h after hCG injection. More than 90% of total gap junction membrane per cell is lost, and even faster concomitant reduction in the mean area of individual gap junction is observed. The projected area of cumulus-oocyte coupling did not change significantly over the first 2.5 h. GVBD is virtually complete within 3 h (Larsen et al., 1986).

What is the nature of the inhibitor? It has been proposed that in the follicle in vivo, a portion of the oocyte cAMP is derived from the cumulus cells (Dekel and Beers, 1978; Dekel et al., 1981). Later, it was shown that cAMP appears to be transferred from the cumulus cells into the oocyte in the sheep (Moor and Heslop, 1981), but not in the mouse (Schultz et al., 1983b). And, finally, after a high phosphodiesterase activity was found in the mouse oocytes (Bornslaeger et al., 1984) and cAMP was determined in the presence of a phosphodiesterase inhibitor, IBMX, or an adenylate cyclase activator, forskolin, cAMP transfer from the cumulus cells to the oocyte was shown for the rat (Racowsky, 1984), hamster (Racowsky, 1985a), pig (Racowsky, 1985c), sheep (Crosby et al., 1985), and mouse (Salustri et al., 1985). But after forskolin is washed off, the hamster (Racowsky, 1985c) and mouse oocytes (Hashimoto et al., 1985) mature at a higher intraoocyte cAMP level than in the normal oocytes. The authors who discovered the possibility of cAMP transfer into the oocyte consider the conditions used to detect the cAMP transfer to be nonphysiological. Under normal conditions, the follicle cell contribution of cAMP to the oocyte may be quite small (Bornslaeger and Schultz, 1985).

Most authors support at present the viewpoint on the existence of at least two factors maintaining meiotic arrest: cAMP and OMI or OMPF (Schultz et al., 1983b; Eppig et al., 1983; Racowsky, 1985a; Tsafriri, 1985; Sato et al., 1985). Unfortunately, as already noted, two factors were called OMI: OMI_1, a peptide (cf. numerous papers of Tsafriri and co-authors), and OMI_2, hypoxanthine with adenosine (Eppig and Downs, 1984; Eppig et al., 1985).

The effect of OMI_1 in the presence of granulosa cells is abolished by LH (Tsafriri, 1978a; Tsafriri et al., 1979). The mechanism of this phenomenon is unknown. It is implied that granulosa or cumulus cells exposed to LH produce substances that are released into the FFl and can overcome the OMI action upon maturation. In addition, these substances can exert a stimulating influence upon the oocyte through cumulus cells (Tsafriri et al., 1976a, 1982). Very interesting results have recently been published by Freter and Schultz (1984). These authors have shown that LH accelerates the maturation of mouse OCCC. This influence is mediated by an elevation of the cAMP level in cumulus cells and is related neither to oocyte-cumulus uncoupling nor to a decrease in the intraoocyte cAMP level. In the follicle, such an "acceleration of maturation" appears to be related to a decreasing influx of inhibitor into the oocyte, the latter being due to a dramatic loss of cumulus cell gap junctions (Larsen et al., 1986). Freter and Schultz (1984) suggest that during spontaneous maturation the oocyte may secrete or catabolize the inhibitor.

OMI_2 consists of hypoxanthine and adenosine (Eppig and Downs, 1984; Downs et al., 1985; Eppig, 1986). Although OMI_2 was originally purified from FFl, it has been shown that the concentrations of hypoxanthine and adenosine in mouse FFl do not increase immediately before the gonadotropin-induced GVBD (Eppig et al., 1985). It is therefore suggested that if hypoxanthine and adenosine and/or their derivatives or precursors are physiologically important, their concentrations within the oocyte-granulosa cell complex are more important than their concentrations in the extracellular FFl (Eppig, 1986).

A very similar conclusion bearing, however, on the factors affecting the OMPF activity was drawn by Sato et al. (1986). Since LH reduces rather than enhances the production of sulfated glycosaminoglycans, it is suggested that the glycosaminoglycan matrix may influence OMPF action on the oocyte (Sato et al., 1986). These results are used to support and develop the hypothesis put forward by the same authors, according to which the resumption of meiosis is controlled by a factor located on the surface of granulosa cells juxtaposed to the cumulus-oocyte complex (Sato et al., 1982). But quite recently the authors appear to hold to a less orthodox viewpoint, since they study the mechanisms sustaining meiosis arrested by two factors of FFl (Sato et al., 1985).

Although direct evidence of a decrease in the cAMP level during oocyte maturation in mammals is scarce, it is of interest that an experimental decrease in the intraoocyte cAMP (by microinjection of phosphodiesterase or a heat-stable inhibitor of protein kinase) initiates meiotic maturation of mouse oocytes in the presence of either db cAMP or IBMX. In addition, specific changes in oocyte phosphoproteins that preceded meiosis resumption were induced by microinjection of the inhibitor of protein kinase (Bornslaeger et al., 1986).

Finally, we have the suggestion that spontaneous maturation is an artifact of cultivation. Biggers and Powers (1979) believe the spontaneous maturation to result from changes in the ionic composition of the medium. Fleming et al. (1983), as was already noted, consider the OMI to be a product of degradation of one of the FF1 components. There is overwhelming evidence in favor of the existence of an inhibitor.

Studies of the mechanisms of oocyte maturation in mammals are so numerous and diverse that this question might be clarified in the near future. It is not excluded that essential species-specific differences will be found, so that what is now considered as controversial may prove to be species-specific peculiarities of oocyte maturation.

5.5.5. *The Capability of Oocytes Matured In Vitro to Undergo Fertilization and Subsequent Development*

The capability of a matured egg to be fertilized and to subsequently develop is undoubtedly the most adequate criterion of maturation.

The question of whether spontaneous oocyte maturation is complete was first raised by Thibault (1970). After insemination of rabbit oocytes spontaneously matured in vitro, the spermatozoa which had penetrated the ovum did not change for several hours; and later, although the sperm heads sometimes swelled, the pronuclei never reached their normal size. Such eggs cleaved but did not develop any further (Thibault et al., 1975a, b). A similar picture was apparently observed by Thibault and co-authors in the oocytes of pigs and cows (Thibault et al., 1975c; Thibault, 1977). Abnormal development of the male pronucleus in some in vitro matured rabbit oocytes has also been reported by Motlik and Fulka (1974a).

Thibault and Gérard (1973) compared the fertilization of spontaneously matured oocytes, oocytes kept in vivo for several hours after the LH surge (and only then transferred to the culture medium), with follicle-enclosed oocytes cultivated in vitro with gonadotropins. Their conclusion was that spontaneously maturing oocytes fail to develop the male pronucleus growth factor (MPGF) in their cytoplasm, and that in vivo MPGF appears approximately 7 h after the LH surge. That spontaneously matured hamster oocytes are defective has been reported by Mandelbaum and co-authors (Mandelbaum et al., 1977; Plachot and Mandelbaum, 1978). In their experiments the sperm was completely unable to penetrate in vitro matured oocytes. If the oocytes were isolated from the follicles 6 h after LH surge, some of them proved to be fertilizable (Mandelbaum et al., 1977; Plachot and Mandelbaum, 1978). The percentage of rat oocytes capable of fertilization increases with the time of their contact with the follicle after the gonadotropin surge (Niwa and Chang, 1975). As a result, the concept was formulated that the oocytes of the hamster, rabbit, and rat cannot be fertilized if transferred in vitro before reaching the tetrad stage at least (see Trounsen et al., 1977). Let us see whether this is really so.

Data on the fertilizability and further development of spontaneously matured oocytes of different animals are summarized in Table 5.2. Pronuclei formation and cleavage is used as a criterion of fertilization. As can be seen from Table 5.2, the percentage of fertilized eggs varies not only with the species but also in the experiments of different authors using the same species. Thus, the frequency of rabbit pronucleus formation varies from 0 to 75%. Since oocyte cultivation always started from the same stage, Motlik and Fulka (1974a) relate the differences to the different composition of the media used. However, we are particularly interested not in the existence of variations, and not even in their possible cause, but in the fact that in almost all species spontaneously matured oocytes turn out to be fertilizable. This is true of the mouse, rat, guinea-pig, rabbit, sheep, cow, dog, and pig. In other words, the MPGF is produced in spontaneously matured oocytes as well. It is also important that in some cases such oocytes developed into normal embryos in the mouse (Cross and Brinster, 1970; Schroder and Eppig, 1984), pig (Leman and Dziuk, 1971), and cow (Newcomb et al., 1978). However, the frequency of normal development of embryos in these experiments is very low, especially when compared with the fraction of normal embryos recovered from follicle-enclosed oocytes matured in vitro. On the basis of a relatively recent examination of more than 1500 oocytes of mouse, rabbit, sheep, and cow, Moor and Warnes (see Moor, 1978) have shown that oocytes matured with and without the follicle yield 40-60% and less than 1% of normal embryos, respectively. Now the mouse oocytes should be omitted from this list, since Schroeder and Eppig (1984) have developed conditions in which spontaneously matured mouse oocytes are fertilized and develop into normal embryos in the same percentage as the in vivo matured oocytes. But the above differences hold true for most animals. What is the cause of such a profound difference?

Table 5.2. Ability of Spontaneously Matured Oocytes to be Fertilized and Undergo Further Development

Species	Fertilization		Pronuclei formed %	Cleavage, %	Blastocysts, %	Reference
	in vivo	in vitro				
Mouse		+	-	42.5	-	Cross, 1973
		+	-	33	-	Cross and Brinster, 1970
		+	-	22.8	7.8	Mukherjee, 1972
Guinea-pig		+	83	-	-	Yanagimachi, 1974
Rabbit		+	0	50	-	Thibault and Gerard, 1973
	+		60	-	-	Motlik and Fulka, 1974a
		+	75	50	13	Van Blerkom and McGaughev, 1978b
Hamster		+	0	-	-	Plachot and Mandelbaum, 1978
Sheep	+		25	15	-	Quike and Gurdon, 1971
Dog		+	-	19.7	-	Mahi, 1975
Cow	+		45	-	-	Hunter et al., 1972
	+		17	-	-	Shea et al., 1976
	+		15	-	-	Sreenan, 1970
	+		23.3	-	-	Newcomb et al., 1978
		+	-	26.7	-	Iritani, 1980
		+	45.5			Fulka et al., 1982
		+	49	47	indiv.	Trounson et al., 1977
Pig	+		43.3	-	-	Motlik and Fulka, 1974b
	+		45	-	-	Leman and Dziuk, 1971
	+		38.9	-	-	Iritani et al., 1978
	+		-	75	76	Schroeder and Eppig, 1984
Rat	+	+	11	-	-	Shalgi et al., 1979
		+	79	-	-	Barak et al., 1982*

*Ova were considered fertilized if they contained the flagellum of spermatozoon. Most oocytes were bipronucleate or at the two-cell stage.

There is considerable evidence that the fertilizability of oocytes and their capacity to develop depends on their exposure to hormones during maturation. For example, Thibault and co-workers (1975b) treated oocytes with prolactin, gonadotropins, and steroids, and arrived at the conclusion (see Thibault, 1977) that the treatment with testosterone or estradiol stimulates the growth of male pronucleus but does not normalize the process completely.

Even better results were obtained on human oocytes by Soupart (1974), who reproduced in vitro the hormonal environment in the follicle after the LH surge (consecutive treatment with estradiol-17β and 17α-hydroxyprogesterone) and observed normal growth of male pronuclei. As known from the reports of Moor and coauthors, estradiol-17β increased the capability of follicle-enclosed sheep oocytes matured in vitro to develop normally. When estrogen secretion is partially suppressed, sheep oocytes mature and are fertilized, but remain unable to form blastocytes (Cran and Moor, 1980). Addition of 1 μg/ml estradiol leads to a sevenfold increase in the frequency of embryos reaching the blastocyst stage. Particularly interesting results were obtained after treating follicles with gonadotropins and steroidogenesis inhibitors, aminogluthetimide, and SU 10603. Fertilization of such oocytes almost always affected the formation of the male pronucleus (Moor, 1978; Moor et al., 1980a). That steroids are indeed involved in the development of capacity for fertilization is also suggested by the correlation between percentage of fertilized oocytes and steroid content in the FFl, which was revealed in vivo for the oocytes of the pig (Hunter et al., 1976) and man (McNatty et al., 1979a, b; Rannevik et al., 1979; Botero-Ruiz et al., 1984). The steroid ratio appears to play an important role. In any case, Liu et al. (1985) state that the combination of high-follicular estradiol-17β concentrations with an accompanying high estradiol-17β/progesterone ratio was associated with the highest quality and fertilizability of oocytes. Thus, oocytes need steroids to complete maturation and to develop the capacity for fertilization. This is not to imply, however, that steroids also act at later stages. Quite the reverse occurs in the case of pig (Hunter et al., 1976) and sheep oocytes (Moor, 1978), where a high estrogen level at the very beginning seems to be essential. The conclusion that steroids are required for the oocytes to acquire the capability for fertilization and subsequent development is inconsistent with the above-mentioned results of Schroeder and Eppig (1984) on the mouse oocytes which, as the authors believe, acquire this capability without hormonal treatment in vitro. But the authors add to the culture medium FCS or blood serum of adult mice which were previously injected with PMSG. Their conclusion would be more convincing if they showed that the medium used contained no hormones.

There undoubtedly are differences in requirements for the composition of culture medium, pattern, and concentration of steroid hormones necessary for the oocytes of various animals to acquire capability for fertilization and development. It has recently been found that the role of follicle cells in this process varies in different animals, e.g., in the mouse the percentage of in vitro mature denuded oocytes which developed to the blastocyst stage was only somewhat lower than that of matured OCCC (Schroeder and Eppig, 1985); in the rat they are fertilized (12% versus 23-32% in the matured OCCC – Fleming et al., 1985b), and in the sheep denuded oocytes and even corona-enclosed oocytes do not acquire capability for fertilization (Staigmiller and Moor, 1984). For sheep oocytes, not only the presence of a certain number of follicle cells, but also direct contact of some of them with the oocytes, proved to be essential (Staigmiller and Moor, 1984).

Unfortunately, evidence on the influence of various medium components, gonadotropins, and profile and concentration of steroids on the development competence of spontaneously matured oocytes is scarce and contradictory. For example, Barak et al. (1982) and Fleming et al. (1985a) could not confirm the stimulatory effect of LH on the fertilizability and developmental capacity of spontaneously matured rat OCCC (cf. Shalgi et al., 1979).

A very interesting system has been employed by Motlik and Fulka (1981b) in their experiments with rabbit oocytes. The cumulus-enclosed oocytes were co-cultured with granulosa cells and FSH. This enabled the authors to observe normal male pronuclei in 88% and normal cleavage in 85% of the oocytes. Similar experiments on spontaneously maturing corona-enclosed sheep oocytes with FSH, LH, and estradiol-17β were carried out by Staigmiller and Moor (1984). The addition of supplementary granulosa cells increased the incidence of cleaving embryos from 13 to 56% and that of blastocyst from 0 to 37%. The method allows "spontaneously" matured oocytes to be subjected to treatment similar to that occurring in the follicle and to explore variations of these treatments. Systems of this kind present new opportunities for investigations into the mechanism of gonadotropin action on oocyte maturation.

REFERENCES

Ahrén, K., Dekel, N., Hamberger, L., Hillensjö, T., Hultborn, R., Magnusson, C., and Tsafriri, A., 1978, "Metabolic and morphological changes produced by gonadotropins and cyclic AMP in the oocyte and cumulus oophorus of preovulatory rat follicles," Ann. Biol. Anim., Biochim., Biophys., *18*, 409-417.

Ainsworth, L., Tsang, B. K., Downey, B. R., Marcus, G. J., and Armstrong, D. T., 1980, "Interrelationships between follicular fluid steroid levels, gonadotropic stimuli, and oocyte maturation during preovulatory development of porcine follicles," Biol. Reprod., *23*, 621-627.

Albertini, D. F., 1984, "Novel morphological approaches for the study of oocyte maturation," Biol. Reprod., *30*, 13-28.
Amsterdam, A., and Tsafriri, A., 1979, "In vitro binding of ^{125}I-human chorionic gonadotropin (hCG) in the preovulatory follicle: Absence of receptor sites on oocyte," J. Cell. Biol., *83*, 225a.
Amsterdam, A., Koch, Y., Lieberman, M. E., and Lindner, H. R., 1975, "Distribution of binding sites for human chorionic gonadotropin in the preovulatory follicle of the rat," J. Cell Biol., *67*, 894-900.
Anderson, L. D., Stone, S. L., and Channing, C. P., 1985, "Influence of hormones on the inhibitory activity of oocyte maturation present in conditioned media of porcine granulosa cells," Gamete Res., *12*, 119-130.
Ayalon, D. Tsafriri, A., Lindner, H. R., Cordova, T., and Harell, A., 1972, "Serum gonadotrophin levels in proestrous rats in relation to the resumption of meiosis by oocytes," J. Reprod. Fertil., *31*, 37-58.
Bachvarova, R., 1985, "Gene expression during oogenesis and oocyte development in mammals," in: Oogenesis, Plenum Press, New York, pp. 453-524.
Bachvarova, R., DeLeon, V., Johnson, A., Kaplan, G., and Paynton, B. V., 1985, "Changes in total RNA, polyadenylated RNA, and actin mRNA during meiotic maturation of mouse oocytes," Dev. Biol., *108*, 325-331.
Bae, In-Ha, 1981, "Role of calcium in resumption of meiosis of cultured porcine cumulus-enclosed oocytes," Biol. Reprod., *24*, Suppl. 1, 92A.
Bae, In-Ha, and Channing, C. P., 1985, "Effect of calcium ion on the maturation of cumulus-enclosed pig follicular oocytes isolated from medium-sized Graafian follicles," Biol. Reprod., *33*, 79-87.
Bae, In-Ha, and Foote, R. H., 1975, "Effects of hormones on the maturation of rabbit oocytes recovered from follicles of various sizes," J. Reprod. Fertil., *42*, 357-360.
Baird, D. T., Swanston, I. A., and McNeilly, A. S., 1981, "Relationship between LH, FSH, and prolactin concentration and the secretion of androgen and estrogen by the preovulatory follicle in the ewe," Biol. Reprod., *24*, 1013-1025.
Barak, Y., Kaplan, R., Kraicer, P. F., and Shalgi, R., 1982, "Influence of luteinizing hormone and progesterone on maturation and fertilizability of rat oocytes in vitro," Gamete Res., *5*, 257-262.
Bar-Ami, S., and Tsafriri, A., 1981, "Acquisition of meiotic competence in the rat: role of gonadotropin and estrogen," Gamete Res., 4, 463-472.
Bar-Ami, S., and Tsafriri, A., 1986, "The development of meiotic competence in the rat: Role of hormones and of the stage of follicular development," Gamete Res., *13*, 39-46.

Barrett, C. B., and Powers, R., 1983, "Inhibition of germinal vesicle breakdown in mouse oocytes by testosterone and agents affecting intracellular cyclic AMP and calcium," J. Cell Biol., *97*, 23a.

Biggers, J. D., and Powers, R. D., 1979, "Comments on control of meiotic maturation in mammals," in: Ovarian Follicular Development and Function, Raven Press, New York, pp. 365-373.

Biggers, J. D., Whittingham, D. G., and Donahue, R. P., 1967, "The pattern of energy metabolism in the mouse oocyte and zygote," Proc. Natl. Acad. Sci. USA, *58*, 560-567.

Billig, H., Hillensjö, T., Tsafriri, A., Magnusson, C., and Brodie, A., 1983, "Nuclear maturation of follicle-enclosed rat oocytes during inhibition of steroidogenesis," Gamete Res., *8*, 79-86.

Binor, Z., and Wolf, D. P., 1979, "In vitro maturation and penetration of mouse primary oocytes after removal of the zona pellucida," J. Reprod. Fertil., *56*, 309-314.

Bloom, A. M., and Mukherjee, B. B., 1972, "RNA synthesis in maturing mouse oocytes," Exp. Cell Res., *74*, 577-582.

Bornslaeger, E. A., and Schultz, R. M., 1984, "Adenylate cyclase activity in zona-free mouse oocytes," Exp. Cell Res., *156*, 277-281.

Bornslaeger, E. A., and Schultz, R., 1985, "Regulation of mouse oocyte maturation: Effect of elevating cumulus cell cAMP on oocyte cAMP levels," Biol. Reprod., *33*, 698-704.

Bornslaeger, E. A., Wilde, M. W., and Schultz, R. M., 1984, "Regulation of mouse oocyte maturation: Involvement of cyclic AMP phosphodiesterase and calmodulin," Dev. Biol., *105*, 488-499.

Bornslaeger, E. A., Mattei, P., and Schultz, R. M., 1986, "Involvement of cAMP-dependent protein kinase and protein phosphorylation in regulation of mouse oocyte maturation," Dev. Biol., *114*, 453-462.

Botero-Ruiz, W., Laufer, N., DeCherney, A. H., Pilan, M. L., Haseltine, F. P., and Behrman, H. K., 1984, "The relationship between follicular fluid steroid concentration and successful fertilization of human oocytes in vitro," Fertil. Steril., *41*, 820-826.

Calarco, P. G., Donahue, R. P., and Szöllösi, D., 1972, "Germinal vesicle breakdown in the mouse oocyte," J. Cell Sci., *10*, 369-385.

Canipari, R., Palombi, F., Riminucci, M., and Mangia, F., 1984, "Early programming of maturation competence in mouse oogenesis," Dev. Biol., *102*, 519-524.

Cascio, S. M., and Wassarman, P. M., 1982, "Program of early development in the mammal: Posttranscriptional control of a class of proteins synthesized by mouse oocytes and early embryos," Dev. Biol., *89*, 397-408.

Centola, G. M., and Anderson, L. D., 1979, "Porcine granulosa cells as a possible source of oocyte maturation inhibitor (OMI)," Biol. Reprod., *20*, Suppl. 1, 89a.

Chang, M. C., 1955a, "The maturation of rabbit oocytes in culture and their maturation, activation, fertilization, and subsequent development in the fallopian tubes," J. Exp. Zool., *128*, 378-405.

Chang, M. C., 1955b, "Fertilization and normal development of follicular oocytes in the rabbit," Science, *121*, 867-869.

Channing, C. P., and Bar-Ami, S., 1984, "Regulation of oocyte meiosis by peptide factors," in: Gonadal Proteins and Peptides and Their Biological Significance, World Sci. Publ. Co., Singapore, pp. 177-190.

Channing, C. P., and Evans, V. W., 1982, "Stimulatory effect of ovine prolactin upon cultured porcine granulosa cell secretion of inhibitory activity of oocyte maturation," Endocrinology, *111*, 1746-1748.

Channing, C. P., and Kammerman, S., 1974, "Binding of gonadotropins to ovarian cells," Biol. Reprod., *10*, 179-198.

Channing, C. P., and Tsafriri, A., 1977, "Mechanism of action of LH and FSH upon the ovary in vitro," Metabolism, *26*, 413-468.

Channing, C.P., Stone, S. L., Fowler, S., Edelson, S., and Schwartz-Kripner, A., 1979, "Comparison of porcine granulosa and cumulus cell properties: LH/hCG receptors, ability to secrete progesterone and ability to respond to LH," Biol. Reprod., *20*, Suppl. 1, 74a.

Channing, C. P., Pomerantz, S. H., Bae, I-H., Evans, V. W., and Atlas, S. J., 1982, "Actions of hormones and other factors upon oocyte maturation," Adv. Exp. Med. Biol., *147*, 189-210.

Channing, C. P., Liu, C-Q., Jones, G. S., and Jones, H., 1983, "Decline of follicular oocyte maturation inhibitor coincident with maturation and achievement of fertilizability of oocytes recovered at midcycle of gonadotropin-treated women," Proc. Natl. Acad. Sci. USA, *80*, 4184-4188.

Chari, S., Hillensjö, T., Magnusson, C., Sturm, G., and Daume, E., 1983, "In vitro inhibition of rat oocyte meiosis by human follicular fluid fractions," Arch. Gynecol., *233*, 155-164.

Cho, V. K., Stern, S., and Biggers, J. D., 1974, "Inhibitory effect of dibutyryl cAMP on mouse oocyte maturation in vitro," J. Exp. Zool., *187*, 383-386.

Cran, D. G., 1985, "Qualitative and quantitative structural changes during pig oocyte maturation," J. Reprod. Fertil., *74*, 237-245.

Cran, D. G., and Cheng, W. T., 1985, "Changes in cortical granules during porcine oocyte maturation," Gamete Res., *11*, 311-319.

Cran, D. G., and Moor, R. M., 1980, "The development of oocytes and ovarian follicles of mammals," Sci. Prog., *66*, 371-383.

Crosby, I. M., Osborn, J. C., and Moor, R. M., 1984, "Changes in protein phosphorylation during the maturation of mammalian oocytes in vitro," J. Exp. Zool., *229*, 459-466.

Crosby, I. M., Moor, R. M., Heslop, J. P., and Osborn, J. C., 1985, "cAMP in bovine oocytes: Localization of synthesis and its action on protein synthesis, phosphorylation, and meiosis," J. Exp. Zool., *234*, 307-318.

Cross, P. C., 1973, "The role of cumulus cells and serum in mouse oocyte maturation in vitro," J. Reprod. Fertil., *34*, 241-245.

Cross, P. C., and Brinster, R. L., 1970, "In vitro development of mouse oocytes," Biol. Reprod., *3*, 298-307.

Cross, P. C., and Brinster, R. L., 1974, "Leucine uptake and incorporation at three stages of mouse oocyte maturation," Exp. Cell Res., *86*, 43-46.

Crozet, N., and Szöllösi, D., 1980, "Effect of actinomycin D and α-amanitin on the nuclear ultrastructure of mouse oocytes," Biol. Cell, *38*, 163-170.

Daniel, S. A., Khalil, M. W., and Armstrong, D. T., 1986, "19-Norandrostenedione (4-estrene-3,17-dione) inhibits porcine oocyte maturation in vitro," Gamete Res., *13*, 173-184.

De Felici, M., and Siracusa, G., 1982, "Survival of isolated, fully grown mouse ovarian oocytes is strictly dependent on external Ca^{2+}," Dev. Biol., *92*, 539-543.

Dekel, N., and Beers, W. H., 1978, "Rat oocyte maturation in vitro: Relief of cyclic AMP inhibition by gonadotropins," Proc. Natl. Acad. Sci. USA, *75*, 4369-4373.

Dekel, N., and Beers, W. H., 1980, "Development of rat oocytes in vitro: Inhibition and induction of maturation in the presence or absence of the cumulus oophorus," Dev. Biol., *75*, 247-254.

Dekel, N., and Phillips, D. M., 1979, "Maturation of rat cumulus oophorus. A scanning electron microscopic study," Biol. Reprod., *21*, 9-18.

Dekel, N., and Sherizly, I., 1983, "Induction of maturation in rat follicle-enclosed oocytes by forskolin," Fed. Eur. Biol. Soc. Lett., *151*, 153-155.

Dekel, N., Hillensjö, T., and Kraicer, P. F., 1979, "Maturational effect of gonadotropins on the cumulus-oocyte complex of the rat," Biol. Reprod., *20*, 191-197.

Dekel, N., Lawrence, T. S., Gilula, N. B., and Beers, W. H., 1981, "Modulation of cell-to-cell communication in the cumulus-oocyte complex and the regulation of oocyte maturation by LH," Dev. Biol., *86*, 356-362.

Dekel, N., Aberdam, E., and Sherizly, I., 1984a, "Spontaneous maturation in vitro of cumulus-enclosed rat oocytes is inhibited by forskolin," Biol. Reprod., *31*, 244-250.

Dekel, N., Sherizly, I., and Galiari, D., 1984b, "Maturation of rat oocytes follows uncoupling in the cumulus-oocyte complex," Biol. Reprod., *30*, Suppl. 1, 144.

Desjardins, M., King, W. A., and Bousquet, D., 1985, "In vitro maturation of horse oocytes," Theriogenology, *23*, 187.

Dimino, M. J., and Berman, S. K., 1979, "Increases in mitochondrial steroidogenesis after short-term incubation of porcine ovarian follicles with luteinizing hormone," Endocrinology, *104*, 780-783.

Doerr, P., 1979, "Relationship between receptor occupancy and stimulation of cAMP and progesterone production in isolated rat granulosa cells," Acta Endocrinol., *91*, Suppl. 225, 289.

Donahue, R. P., 1968a, "Maturation of the mouse oocyte in vitro. I. Sequence and timing of nuclear progression," J. Exp. Zool., *169*, 237-250.

Donahue, R. P., 1968b, "Maturation of the mouse oocyte in vitro. II. Stage-specific protein synthesis," Genetics., *60*, 173-174.

Donahue, R. P., 1972, "The relation of oocyte maturation to ovulation in mammals," in: Oogenesis, Univ. Park Press, Baltimore, pp. 413-438.

Downs, S. M., and Eppig, J. J., 1984, "Cyclic adenosine monophosphate and ovarian follicular fluid act synergistically to inhibit mouse oocyte maturation," Endocrinology, *114*, 418-427.

Downs, S. M., Coleman, D. L., Ward-Bailey, P. F., and Eppig, J. J., 1985, "Hypoxanthine is the principal inhibitor of murine oocyte maturation in a low-molecular-weight fraction of porcine follicular fluid," Proc. Natl. Acad. Sci. USA, *82*, 454-458.

Dyban, A. P., and Baranov, V. S., 1977, "Oogenesis in mammals," in: Modern Problems of Oogenesis [in Russian], Nauka, Moscow, pp. 200-239.

Dziuk, J. F., 1965, "Timing of maturation and fertilization of the sheep egg," Anat. Rec., *153*, 211-224.

Ecay, T. W., and Powers, R. D., 1985, "Changes in phosphatidylinositol metabolism may control germinal vesicle breakdown in the mouse oocyte," J. Cell Biol., *101*, 263a.

Edwards, R. G., 1962, "Meiosis in ovarian oocytes of adult mammals," Nature, *196*, 446-450.

Edwards, R. G., and Gates, A. H., 1959, "Timing of the stages of the maturation divisions, ovulation, fertilization, and the first cleavage of eggs of adult mice treated with gonadotropins," J. Endocrinol., *18*, 292-304.

Ekholm, C., and Magnusson, C., 1979, "Rat oocyte maturation: Effects of protein synthesis inhibitors," Biol. Reprod., *21*, 1287-1293.

Ekholm, C., Hillensjö, T., Magnusson, C., and Rosberg, S., 1984, "Stimulation and inhibition of rat oocyte meiosis by forskolin," Biol. Reprod., *30*, 537-543.

Eppig, J. J., 1982, "The relationship between cumulus cell-oocyte coupling, oocyte meiotic maturation, and cumulus expansion," Dev. Biol., *89*, 268-272.

Eppig, J. J., 1985, "Oocyte-somatic cell interactions during oocyte growth and maturation in the mammal," in: Oogenesis, Plenum Press, New York, pp. 313-347.

Eppig, J. J., 1986, "Mechanisms controlling mammalian oocyte maturation," Res. Reprod., *18*, 1-2.
Eppig, J. J., and Downs, S. M., 1984, "Chemical signals that regulate mammalian oocyte maturation," Biol. Reprod., *30*, 1-11.
Eppig, J. J., and Koide, S. L., 1978, "Effects of progesterone and estradiol-17β on the spontaneous meiotic maturation of mouse oocytes," J. Reprod. Fertil., *53*, 99-101.
Eppig, J. J., Freter, R. R., Ward-Bailey, P. F., and Schultz, R. M., 1983, "Inhibition of oocyte maturation in the mouse: Participation of cAMP, steroid hormones, and a putative maturation-inhibitory factor," Dev. Biol., *100*, 39-49.
Eppig, J. J., Ward-Bailey, P. F., and Coleman, D. L., 1985, "Hypoxanthine and adenosine in murine ovarian follicular fluid: Concentrations and activity in maintaining oocyte meiotic arrest," Biol. Reprod., *33*, 1041-1049.
Erickson, G. F., and Ryan, K. J., 1976, "Spontaneous maturation of oocytes isolated from ovaries of immature hypophysectomized rats," J. Exp. Zool., *195*, 153-158.
Erickson, G. F., and Sorensen, R. A., 1974, "In vitro maturation of mouse oocytes isolated from late, middle, and preantral Graafian follicles," J. Exp. Zool., *190*, 123-127.
Espey, L. L., 1978, "Ovulation," in: The Vertebrate Ovary, Plenum Press, New York, pp. 503-532.
Ezzel, R. M., and Szego, C. M., 1979, "Luteinizing hormone-accelerated redistribution of lysosome-like organelles preceding dissolution of the nuclear envelope in rat oocytes maturing in vitro," J. Cell Biol., *82*, 264-277.
Fleming, A. D., Khalil, W., and Armstrong, D. T., 1983, "Porcine follicular fluid does not inhibit maturation of rat oocytes in vitro," J. Reprod. Fertil., *69*, 665-670.
Fleming, A. D., Evans, G., Walton, E. A., and Armstrong, D. T., 1985a, "Developmental capability of rat oocytes matured in vitro in defined medium," Gamete Res., *12*, 255-263.
Fleming, A. D., Kuehl, T. J., and Armstrong, D. T., 1985b, "Maturation of pig and rat oocytes transplanted into surrogate pig follicles in vitro," Gamete Res., *11*, 107-119.
Foote, W. D., and Thibault, C., 1969, "Récherches expérimentales sur la maturation in vitro des ovocytes de truie et de veau," Ann. Biol. Anim., Biochim., Biophys., *9*, 329-349.
Freter, R. R., and Schultz, R. M., 1983, "Mechanism of gonadotropin regulation of mouse oocyte maturation," J. Cell Biol., *97*, 24a.
Freter, R. R., and Schultz, R. N., 1984, "Regulation of murine oocyte meiosis: Evidence for a gonadotropin-induced cAMP-dependent reduction in a maturation inhibitor," J. Cell Biol., *98*, 1119-1128.
Fukui, Y., and Sakuma, Y., 1980, "Maturation of bovine oocytes cultured in vitro: Relation to ovarian activity, follicular

size, and the presence or absence of cumulus cells," Biol. Reprod., *22*, 669-673.

Fulka, J., and Okalski, A., 1981, "Culture of horse oocytes in vitro," J. Reprod. Fertil., *61*, 213-215.

Fulka, J., Pavlok, A., and Fulka, J., 1982, "In vitro fertilization of zona-free bovine oocytes matured in culture," J. Reprod. Fertil., *64*, 495-499.

Fulka, J., Motlik, J., Fulka, J., and Jélik, F., 1986, "Effect of cycloheximide on nuclear maturation of pig and mouse oocytes," J. Reprod. Fertil., *77*, 281-285.

Gérard, M., Ménézo, Y., Rombatus, P., Szöllösi, D., and Thibault, C., 1979, "In vitro studies of oocyte maturation and follicular metabolism in the pig," Ann. Biol. Anim., Biochim., Biophys., *19*, 1521-1535.

Gilula, N. B., Epstein, M. L., and Beers, W. H., 1978, "Cell-to-cell communication and ovulation. A study of the cumulus-oocyte complex," J. Cell Biol., *78*, 58-75.

Goff, A. K., and Henderson, K. M., 1979, "Changes in follicular fluid and serum concentrations of steroids in PMS-treated immature rats following LH administration," Biol. Reprod., *20*, 1153-1157.

Golbus, M., and Stein, M. P., 1976, "Qualitative patterns of protein synthesis in the mouse oocyte," J. Exp. Zool., *198*, 337-342.

Gwatkin, R. B., and Andersen, O. F., 1976, "Hamster oocyte maturation in vitro: Inhibition by follicular components," Life Sci., *19*, 527-536.

Gwatkin, R. B., and Haidri, A. A., 1973, "Requirements for the maturation of hamster oocytes in vitro," Exp. Cell Res., *76*, 1-7.

Gwatkin, R. B., and Haidri, A. A., 1974, "Oxygen requirements for the maturation of hamster oocytes," J. Reprod. Fertil., *37*, 127-129.

Hamberger, L., Nordeström, K., Rosberg, S., and Sjorgen, A., 1978, "Acute influence of LH and FSH on cyclic AMP formation in isolated granulosa cells of the rat," Acta Endocrinol., *88*, 567-579.

Han, S. S., Rajaniemi, S. J., Cho, M. I., Hirschfield, A. N., and Midgley, A. R., 1974, "Gonadotropin receptors in rat ovarian tissue. II. Subcellular localization of LH binding sites by electron microscopic radioautography," Endocrinology, *95*, 589-598.

Hashimoto, N., Kishimoto, T., and Nagahama, Y., 1985, "Induction and inhibition of meiotic maturation in follicle-enclosed mouse oocytes by forskolin," Dev. Growth, Differ., *27*, 709-716.

Hay, M. P., and Moor, R. M., 1973, "The Graafian follicle of the sheep: Relationships between gonadotropins, steroid production, morphology, and oocyte maturation," Ann. Biol. Anim., Biochim., Biophys., *13*, Hors-serie, 241-248.

Hay, M. F., and Moor, R. M., 1975, "Functional and structural relationshps in the Graafian follicle population in the sheep ovary," J. Reprod. Fertil., *45*, 583-593.
Hillensjö, T., 1975, "Gonadotropic regulation of the preovulatory follicular function. An experimental study on the rat ovary," Doctoral Dissertation, Göteborg.
Hillensjö, T., 1976, "Ooocyte maturation and glycolysis in isolated preovulatory follicles of PMS-injected immature rats," Acta Endocrinol., *82*, 809-830.
Hillensjö, T., 1977, "Dissociation of preovulatory maturational events in rat oocytes and cumuli in the presence of dibutyryl cyclic AMP," Acta Physiol. Scand., *100*, 261-263.
Hillensjö, T., Dekel, N., and Ahren, K., 1976, "Effects of gonadotropins on the cumulus oophorus of isolated rat Graafian follicles," Acta Physiol. Scand., *96*, 558-568.
Hillensjö, T., Batta, S. K., Schwartz-Kripner, A., Wentz, A. C., Salenski, J., and Channing, C. P., 1978, "Inhibitory effect of human follicular fluid upon maturation of porcine oocytes in culture," J. Clin. Endocrinol. Metab., *47*, 1332-1335.
Hillensjö, T., Channing, C. P., and Schwartz-Kripner, A., 1979, "Intrafollicular control of oocyte maturation in the pig," In Vitro, *15*, 32-39.
Hillensjö, T., Magnusson, C., Ekholm, C., Billig, H., and Hedin, L., 1982, "Role of cumulus cells in oocyte maturation," Adv. Exp. Med. Biol., *147*, 175-188.
Hoffmann, J. C., Lorenzen, J. R., Well, T., and Schwartz, N. B., 1979, "Selective suppression of the primary surge of follicle-stimulating hormone in the rat: Further evidence for folliculostatin in porcine follicular fluid," Endocrinology, *105*, 200-203.
Hubbard, C. J., 1983, "The effect of in vivo alterations in gonadotropins and cyclic nucleotides on oocyte maturation in the hamster," Life Sci., *33*, 1695-1702.
Hubbard, C. J., 1985a, "Changes in follicular and oocyte cAMP levels following stimulation by LH," Biol. Reprod., *32*, Suppl. 1, 79.
Hubbard, C. J., 1985b, "The effect of forskolin and LH on cAMP changes and maturation in the follicle-enclosed oocytes of the hamster," Acta Endocrinol., *110*, 413-420.
Hubbard, C. J., and Terranova, P. F., 1982, "Inhibitory action of cyclic guanosine 5'-phosphoric acid (GMP) on oocyte maturation. Dependence on an intact cumulus," Biol. Reprod., *26*, 628-632.
Hunter, R. H., and Polge, C., 1966, "Maturation of follicular oocytes in the pig after injection of human chorionic gonadotropin," J. Reprod. Fertil., *12*, 525-531.
Hunter, R. H., Lawson, R. A., and Rowson, L. E., 1972, "Maturation, transplantation, and fertilization of ovarian oocytes in culture," J. Reprod. Fertil., *30*, 325-328.
Hunter, R. H., Cook, B., and Baker, T. G., 1976, "Dissociation of

response to injected gonadotropin between the Graafian follicle and oocyte in pigs," Nature, *260*, 156-157.

Hyttel, P., Callesen, H., and Greve, T., 1986, "Ultrastructural features of preovulatory oocyte maturation in superovulated cattle," J. Reprod. Fertil., *76*, 645-656.

Iritani, A., 1980, "Fertilization in vitro of follicular oocytes matured in culture in cattle, pig, and human," Arch. Androl., *5*, 77-78.

Iritani, A., Niwa, K., and Imai, H., 1978, "Sperm penetration in vitro of pig follicular oocytes matured in culture," J. Reprod. Fertil., *54*, 379-383.

Iwamatsu, T., and Yanagimachi, R., 1975, "Maturation in vitro of ovarian oocytes of prepubertal and adult hamsters," J. Reprod. Fertil., *45*, 83-90.

Iwamatsu, T., Keino, H., and Murakami, U., 1979, "Divalent cations and mouse oocytes during in vitro maturation," Zool. Mag., *88*, 233-238; Exp. Zool., *197*, 161-172.

Jagiello, G., 1969, "Meiosis and inhibition of ovulation in mouse eggs treated with actinomycin D," J. Cell Biol., *42*, 571-574.

Jagiello, G., and Ducayen, M. B., 1977, "Preliminary evidence for the intracellular localization of luteinizing hormones in mammalian oocytes," J. Exp. Zool., *201*, 345-358.

Jagiello, G., Ducayen, M. B., and Goonan, W. D., 1981, "A note on the inhibition of in vitro meiotic maturation of mammalian oocytes by dibutyryl cyclic AMP," J. Exp. Zool., *218*, 309-311.

Jagiello, G., Ducayen, M. B., Downey, R., and Jonassen, A., 1982, "Alterations of mammalian oocyte meiosis I with divalent cations and calmodulin," Cell Calcium, *3*, 153-162.

Kaplan, G., Dekel, N., and Kraicer, P. F., 1978, "Acceleration of onset of oocyte maturation in vitro by lueteinizing hormone," Gamete Res., *1*, 59-63.

Kaplan, G., Abreu, S. L., and Bachvarova, R., 1982, "rRNA accumulation and protein synthetic patterns in growing mouse oocytes," J. Exp. Zool., *220*, 361-370.

Khalil, M. W., and Watson, J. S., 1985, "Identification and measurement of 4-estren-3,17-dione (19-norandrostenedione) in porcine ovarian follicular fluid using high-performance liquid chromatography and capillary gas chromatography—mass spectrometry," J. Endocrinol., *107*, 375-381.

Kruip, T. A., Cran, D. G., van Beneden, T. H., and Dieleman, S. J., 1983, "Structural changes in bovine oocytes during final maturation in vivo," Gamete Res., *8*, 29-47.

LaMarca, M. J., and Wassarman, P. M., 1979, "Program of early development in the mammal: Changes in absolute rates of synthesis of ribosomal proteins during oogenesis and early embryogenesis in the mouse," Dev. Biol., *73*, 103-119.

Larsen, W. J., Tung, H. N., and Polking, C., 1981, "Response of granulosa cell gap junctions to human chorionic gonadotropin (hCG) at ovulation," Biol. Reprod., *25*, 1119-1134.

Larsen, W. J., Wert, S. E., and Brunner, G. D., 1986, "A dramatic loss of cumulus cell gap junctions is correlated with germinal vesicle breakdown in rat oocytes," Dev. Biol., *113*, 517-521.

Lawrence, T. S., Dekel, N., and Beers, W. H., 1980, "Binding of human chorionic gonadotropin by rat cumuli oophori and granulosa cells: A comparative study," Endocrinology, *106*, 1114-1118.

Leibfried, L., and First, N. L., 1979a, "Effects of divalent cations on in vitro maturation of bovine oocytes," J. Exp. Zool., *210*, 575-580.

Leibfried, L., and First, N. L., 1979b, "Effects of divalent cations on in vitro maturation of bovine oocytes," J. Exp. Zool., *210*, 575-580.

Leibfried, L., and First, N. L., 1980a, "Effect of bovine and porcine fluid and granulosa cells on maturation of oocytes in vitro," Biol. Reprod., *23*, 699-704.

Leibfried, L., and First, N. L., 1980b, "Follicular control of meiosis in the porcine oocytes," Biol. Reprod., *23*, 705-709.

Lieberman, M. E., Tsafriri, A., Bauminger, S., Collins, W. P., Ahrén, and Lindner, H. R., 1976, "Oocyte meiosis in cultured rat follicles during inhibition of steroidogenesis," Acta Endocrinol., *83*, 151-157.

Leman, A. D., and Dziuk, P. J., 1971, "Fertilization and development of pig follicular oocytes," J. Reprod. Fertil., *26*, 387-389.

Lindner, H. R., Tsafriri, A., Lieberman, M. E., Zor, U., Koch, Y., Bauminger, S., and Barnea, A., 1974, "Gonadotropin action on cultured Graafian follicles: Induction of maturation division of the mammalian oocytes and differentiation of the luteal cell," Recent Prog. Horm. Res., *30*, 79-138.

Lindner, H. R., Bar-Ami, S., and Tsafriri, A., 1983, "Control of resumption of meiosis in mammals," in: Ovary, Raven Press, New York, pp. 83-94.

Lindsey, A. M., and Channing, C. P., 1979, "Influence of follicular maturation upon the effect of ovine follicle-stimulating hormone and luteinizing hormone upon cyclic AMP accumulaton by isolated granulosa cells," Biol. Reprod., *20*, 473-482.

Liu, H. C., Bernardus, R., Jones, G. S., Veeck, L., and Rosenwaks, Z., 1985, "Follicular hormonal concentration and oocyte behavior in vitro," Biol. Reprod., *32*, Suppl. 1, 99.

Lopata, A., Fonseca, J. R., and Szego, C. M., 1977, "Cinematographic analysis of cytoplasmic and nuclear events associated with germinal vesicle breakdown in rat oocytes exposed to LH in vitro," J. Reprod. Fertil., *50*, 211-216.

Magnusson, C., and Hillensjö, T., 1977, "Inhibition of maturation and metabolism in rat oocytes by cyclic AMP," J. Exp. Zool., *201*, 139-147.

Magnusson, C., Hillensjö, T., Tsafriri, A., Hultborn, R., and

Ahren, K., 1977, "Oxygen consumption of maturing rat oocytes," Biol. Reprod., *17*, 9-15.

Mahi, C. A., 1975, "Maturation and fertilization in vitro of canine ovarian oocytes," in: 8th Ann. Meet. Soc. Study Reprod., Colorado State Univ., Fort Collins, pp. 74-75.

Mahi, C. A., and Yanagimachi, R., 1976, "Maturation and sperm penetration of canine ovarian oocytes in vitro," J. Exp. Zool., *196*, 189-193.

Mandelbaum, J., Plachot, M., and Thibault, C., 1977, "Role of the follicle in final zona pellucida maturation of hamster oocytes," Ann. Biol. Anim., Biochim., Biophys., *17*, 389-391.

Mangia, F., and Canipari, R., 1977, "Biochemistry of growth and maturation in mammalian oocytes," in: Development in Mammals, North-Holland Publ. Co., Amsterdam—New York—Oxford, Vol. 2, pp. 1-29.

Makris, A., and Ryan, K. J., 1978, "Cyclic AMP and cyclic GMP accumulation by isolated hamster granulosa cells stimulated by LH and FSH," Acta Endocrinol., *89*, 173-181.

Marsh, J. M., 1976, "The role of cyclic AMP in gonadal steroidogenesis," Biol. Reprod., *14*, 30-53.

Marsh, J. M., Mills, T. M., and LeMaire, W. J., 1972, "Cyclic AMP synthesis in rabbit Graafian follicles and the effect of luteinizing hormone," Biochim. Biophys. Acta, *273*, 389-394.

Maruska, D. D., Leibfried, M. L., and First, N. L., 1984, "Role of calcium and the calcium-calmodulin complex in resumption of meiosis, cumulus expansion, viability, and hyaluronidase sensitivity of bovine cumulus—oocyte complexes," Biol. Reprod., *31*, 1-6.

Masui, Y., and Clarke, H. J., 1979, "Oocyte maturation," Int. Rev. Cytol., *57*, 185-282.

McGaughey, R. W., 1977, "The culture of pig oocytes in minimal medium, and the influence of progesterone and estradiol-17β on meiotic maturation," Endocrinology, *100*, 39-45.

McGaughey, R. W., 1978, "In vitro oocyte maturation," in: Methods in Mammalian Reproduction, Academic Press, New York—San Francisco—London, pp. 1-20.

McGaughey, R. W., and Chang, M. C., 1969, "Meiosis of mouse eggs before and after sperm penetration," J. Exp. Zool., *170*, 397-410.

McGaughey, R. W., and Polge, C., 1971, "Cytogenetic analysis of pig oocytes matured in vitro," J. Exp. Zool., *176*, 383-391.

McGaughey, R. W., and Van Blerkom, J., 1977, "Patterns of polypeptide synthesis of porcine oocytes during maturation in vitro," Dev. Biol., *56*, 241-254.

McGaughey, R. W., Montgomery, D. H., and Richter, J. D., 1979, "Germinal vesicle configurations and patterns of polypeptide synthesis of porcine oocytes from antral follicles of different size, as related to their competence for spontaneous maturation," J. Exp. Zool., *209*, 239-254.

McNatty, K. P., Smith, D. M., Osatanondh, R., and Ryan, K. J., 1979a, "The human antral follicle: Functional correlates of

growth and atresia," Ann. Biol. Anim., Biochim., Biophys., *19*, 1547-1558.
McNatty, K. P., Smith, D. M., Makris, A., Osathanondh, R., and Ryan, K. J., 1979b, "Microenvironment of the human antral follicle interrelationships among the steroid levels in antral fluid, the population of granulosa cells and status of the oocyte in vivo and in vitro," J. Clin. Endocrinol. Metab., *49*, 851-860.
Meinecke, B., and Meinecke-Tillmann, S., 1981, "Induction and inhibition of meiotic maturation of follicle-enclosed porcine oocytes in vitro," Theriogenology, *15*, 581-589.
Ménézo, Y., Gérard, M., Szöllösi, D., and Thibault, C., 1978, "In vitro exchange between the follicle and its culture medium," Ann. Biol. Anim., Biochim., Biophys., *18*, 471-476.
Merchant, H., and Chang, M. C., 1971, "An electron microscopic study of mouse eggs matured in vivo and in vitro," Anat. Rec., *171*, 21-25.
Miller, J. G., and Behrman, H. R., 1985, "Adenosine — the oocyte maturation inhibitor of the follicle?" Biol. Reprod., *32*, Suppl. 1, 48.
Mills, T., and Savard, K., 1973, "Steroidogenesis in ovarian follicles isolated from rabbits before and after mating," Endocrinology, *92*, 788-791.
Moor, R. M., 1974, "The ovarian follicle in the sheep: Inhibition of estrogen secretion by luteinizing hormone," J. Endocrinol., *61*, 455-463.
Moor, R. M., 1978, "Role of steroids in the maturation of ovine oocytes," Ann. Biol. Anim., Biochim., Biophys., *18*, 477-482.
Moor, R. M., and Crosby, I. M., 1986, "Protein requirements for germinal vesicle breakdown in ovine oocytes," J. Embryol. Exp. Morphol., *94*, 207-220.
Moor, R. M., and Heslop, J. P., 1981, "Cyclic AMP in mammalian follicle cells and oocytes during maturation," J. Exp. Zool., *216*, 205-209.
Moor, R. M., and Osborn, J. C., 1983, "Somatic control of protein synthesis in mammalian oocytes during maturation," Ciba Found. Sympos., *98*, 178-191.
Moor, R. M., and Trounson, A. O., 1977, "Hormonal and follicular factors affecting maturation of sheep oocytes in vitro and their subsequent developmental capacity," J. Reprod. Fertil., *49*, 101-109.
Moor, R. M., and Warnes, G. M., 1978, "Regulation of oocyte maturation in mammals," in: Control of Ovulation, Butterworths, London—Boston—Sydney, pp. 159-176.
Moor, R. M., and Warnes, G., 1979, "Regulation of meiosis in mammalian oocytes," Brit. Med. Bull., *35*, 99-103.
Moor, R. M., Hay, M. F., McIntosh, J. E., and Caldwell, B. V., 1973, "Effect of gonadotropins on the production of steroids by sheep ovarian follicle cultured in vitro," J. Endocrinol., *58*, 599-611.
Moor, R. M., Hay, M. F., and Seamark, R. F., 1975, "The sheep

ovary: Regulation of steroidogenic, hemodynamic, and structural changes in the largest follicle and adjacent tissue before ovulation," J. Reprod. Fertil., *45*, 595-604.

Moor, R. M., Polge, C., and Willadsen, S. M., 1980a, "Effect of follicular steroids on the maturation and fertilization of mammalian oocytes," J. Embryol. Exp. Morphol., *56*, 316-335.

Moor, R. M., Smith, M. W., and Dawson, R. M., 1980b, "Measurement of intercellular coupling between oocytes and cumulus cells using intracellular markers," Exp. Cell Res., *126*, 15-29.

Moor, R. M., Osborn, J. C., Cran, D. G., and Walters, D. E., 1981, "Selective effect of gonadotropins on cell coupling, nuclear maturation, and protein synthesis in mammalian oocytes," J. Embryol. Exp. Morphol., *61*, 347-365.

Mori, T., Takenaka, A., Yoshida, Y., Suzuki, A., Fujita, Y., and Nishimura, T., 1979, "Preovulatory changes in morphology of rabbit ovarian follicles," Endocrinol. Jpn., *26*, 379-388.

Motlik, J., 1972, "Cultivation of pig oocytes in vitro," Folia Biol., *18*, 345-349.

Motlik, J., and Fulka, J., 1974a, "Fertilization and development in vivo of rabbit oocytes cultivated in vitro," J. Reprod. Fertil., *40*, 183-186.

Motlik, J., and Fulka, J., 1974b, "Fertilization of pig follicular oocytes cultivated in vitro," J. Reprod. Fertil., *36*, 236-237.

Motlik, J., and Fulka, J., 1981a, "Maturation of mammalian oocytes in vitro," Ontogenez, *12*, 435-443.

Motlik, J., and Fulka, J., 1981b, "Fertilization of rabbit oocytes co-cultured with granulosa cells," J. Reprod. Fertil., *63*, 425-429.

Motlik, J., Kopecny, V., and Pivko, J., 1978, "The fate and role of macromolecules synthesized during mammalian oocyte meiotic maturation. I. Autoradiographic topography of newly synthesized RNA and protein in the germinal vesicle of the pig and rabbit," Ann. Biol. Anim., Biochim., Biophys., *18*, 735-747.

Motlik, J., Crozet, N., and Fulka, J., 1984, "Meiotic competence in vitro of pig oocytes isolated from early antral follicles," J. Reprod. Fertil., *72*, 323-328.

Motlik, J., Fulka, J., and Fléchon, J. E., 1986, "Changes in intercellular coupling between pig oocytes and cumulus cell during maturation in vivo and in vitro," J. Reprod. Fertil., *76*, 31-37.

Mukherjee, A. B., 1972, "Normal progeny from fertilization in vitro of mouse oocytes matured in culture and spermatozoa capacitated in vitro," Nature, *237*, 397-398.

Neal, P., and Baker, T. G., 1973, "Response of mouse ovaries in vivo and in organ culture to pregnant mare's serum gonadotropin and human chorionic gonadotropin. I. Examination of critical time intervals," J. Reprod. Fertil., *33*, 399-410.

Nekola, M. V., and Smith, D. M., 1975, "Failure of gonadotropins to induce in vitro maturation of mouse oocytes treated with dibutyryl cyclic AMP," J. Exp. Zool., *194*, 529-533.

Newcomb, R., Christie, L. S., and Rowson, L. E., 1978, "Birth of calves after in vivo fertilization of oocytes removed from follicles and matured in vitro," Vet. Rec., *102*, 461-462.

Nilsson, L., Hillensjö, T., and Ekholm, C., 1977, "Preovulatory changes in rat follicular cyclic AMP and sensitivity to gonadotropins," Acta Endocrinol., *86*, 384-393.

Niwa, K., and Chang, M. C., 1975, "Fertilization of rat eggs in vitro at various times before and after ovulation with special reference to fertilization of ovarian oocytes matured in culture," J. Reprod. Fertil., *43*, 435-451.

Odor, D. L., 1955, "The temporal relationship of the first maturation division of rat ova to the onset of heat," Am. J. Anat., *97*, 461-492.

Olsiewski, P. J., and Beers, W. H., 1983, "cAMP synthesis in the rat oocyte," Dev. Biol., *100*, 287-293.

Osborn, J. C., and Moor, R. M., 1983, "Time-dependent effects of α-amanitin on nuclear maturation and protein synthesis in mammalian oocytes," J. Embryol. Exp. Morphol., *73*, 317-338.

Oxberry, B. A., and Greenwald, G. S., 1982, "Down regulation of prolactin receptors in the hamster ovary by the preovulatory gonadotropin surge," Anat. Rec., *202*, 142a.

Paleos, G. A., and Powers, R. D., 1981, "The effect of calcium on the first meiotic division of the mammalian oocyte," J. Exp. Zool., *217*, 409-416.

Pincus, G., and Enzmann, E. V., 1935, "The comparative behavior of mammalian eggs in vivo and in vitro. I. The activation of ovarian eggs," J. Exp. Med., *62*, 665-675.

Plachot, M., and Mandelbaum, J., 1978, "Comparative study of extra- and intrafollicular hamster oocyte maturation." Ann. Biol. Anim., Biochim., Biophys., *18*, 1237-1246.

Powers, R. D., and Biggers, J. D., 1976, "Inhibition of mouse oocyte maturation by cell membrane potential hyperpolarization," J. Cell Biol., *70*, 352a.

Powers, R. D., and Paleos, G. A., 1982, "Combined effects of calcium and dibutyryl cyclic AMP on germinal vesicle breakdown in the mouse oocyte," J. Reprod. Fertil., *66*, 1-8.

Quirke, J. F., and Gurdon, I., 1971, "Culture and fertilization of sheep ovarian oocytes. III. Evidence on fertilization in the sheep oviduct based on pronucleate and cleavage eggs," J. Agric. Sci., *76*, 375-377.

Racowsky, C., 1983, "Androgenic modulation of cyclic adenosine monophosphate(cAMP)-dependent meiotic arrest," Biol. Reprod., *28*, 774-787.

Racowsky, C., 1984, "Effect of forskolin on the spontaneous maturation and cyclic AMP content of rat oocyte-cumulus complexes," J. Reprod. Fertil., *72*, 107-116.

Racowsky, C., 1985a, "Effect of forskolin on the spontaneous maturation and cyclic AMP content of hamster oocyte-cumulus complexes," J. Exp. Zool., *234*, 87-96.

Racowsky, C., 1985b, "Antagonistic actions of estradiol and tamoxifen upon forskolin-dependent meiotic arrest, intercellular

coupling, and the cyclic AMP content of hamster oocyte-cumulus complexes," J. Exp. Zool., *234*, 251-260.
Racowsky, C., 1985c, "Effect of forskolin on maintenance of meiotic arrest and stimulation of cumulus expansion, progesterone, and cyclic AMP production by pig oocyte-cumulus complexes," J. Reprod. Fertil., *74*, 9-21.
Racowsky, C., and Satterline, R. A., 1985, "Metabolic, fluorescent dye and electrical coupling between hamster oocytes and cumulus cells during meiotic maturation in vivo and in vitro," Dev. Biol., *108*, 191-202.
Rani, C. S., and Moudgal, N. R., 1977, "Examination of the role of FSH in preovulatory events in the hamster," J. Reprod. Fertil., *50*, 37-45.
Rannevik, G., Fernlund, P., Kullande, S., Liedholm, P., Sundström, P., and Wramsbig, H., 1979, "Relation between concentration of steroid hormones in Graafian follicular fluid and in vitro fertilization of the oocyte," Acta Endocrinol., *91*, Suppl. 225, 360.
Rhodes, R. C., III, Fleming, M. W., Murdoch, W. J., and Inskeep, E. K., 1985, "Formation of cyclic adenosine monophosphate (cAMP) in the preovulatory rabbit follicle: Role of prostaglandins and steroids," Prostaglandins, *29*, 217-231.
Rice, C., and McGaughey, R. W., 1981, "Effect of testosterone and dibutyryl cAMP on the spontaneous maturation of pig oocytes," J. Reprod. Fertil., *62*, 245-256.
Richter, J. D., and McGaughey, R. M., 1979, "Specificity of inhibition by steroids of porcine oocyte maturation in vitro," J. Exp. Zool., *209*, 81-90.
Rodman, T. C., and Bachvarova, R., 1976, "RNA synthesis in preovulatory mouse oocytes," J. Cell Biol., *70*, 251-257.
Saidapur, S. K., and Greenwald, G. S., 1979, "Ovarian steroidogenesis in the proestrous hamster," Biol. Reprod., *20*, 226-234.
Salustri, A., and Siracusa, G., 1983, "Metabolic coupling, cumulus expansion, and meiotic resumption in mouse cumuli oophori cultured in vitro in the presence of FSH or db cAMP, or stimulated in vivo by hCG," J. Reprod. Fertil., *68*, 335-341.
Salustri, A., Petrungaro, S., De Felici, M., Conti, M., and Siracusa, G., 1985, "Effect of follicle-stimulating hormone on cyclic adenosine monophosphate level and on meiotic maturation in mouse cumulus cell-enclosed oocytes cultured in vitro," Biol. Reprod., *33*, 797-802.
Sato, E., and Ishibashi, T., 1977, "Meiotic arresting action of the substance obtained from the cell surface of porcine granulosa cells," Jpn. J. Zootech. Sci., *48*, 22-26.
Sato, E., and Koide, S. S., 1984a, "Forskolin and mouse oocyte maturation in vitro," J. Exp. Zool., *230*, 125-129.
Sato, E., and Koide, S. S., 1984b, "A factor from bovine granulosa cells preventing oocyte maturation," Differentiation, *26*, 59-62.

Sato, E., Ishibashi, T., and Iritani, A., 1982, "Meiotic arresting substance separated from porcine ovarian granulosa cell and hypothetical arresting mechanism of meiosis," Adv. Exp. Med. Biol., *147*, 161-173.

Sato, E., Ueno, R., and Koide, S. S., 1984, "Modulation of oocyte maturation-preventing factor of bovine granulosa cells by heparin," J. Steroid Biochem., *20*, 1596.

Sato, E., Wood, H. N., Lynn, D. G., and Koide, S. S., 1985, "Modulation of oocyte maturation by cyclic adenosine 3',5'-pyrophosphate," Cell Differ., *17*, 169-174.

Sato, E., Ueno, H., and Koide, S. S., 1986, "Mouse oocyte maturation modulated by granulosa cell factor and by heparin and heparin sulfate," Gamete Res., *13*, 115-124.

Schaerf, F. W., Anderson, L. D., and Channing, C. P., 1982, "Steroidogenesis by isolated porcine oocyte-cumulus complexes: Lack of an inhibitory effect of low-molecular-weight fraction of porcine follicular fluid on conversion of androgen to estrogen," Gamete Res., *5*, 207-215.

Schroeder, A. C., and Eppig, J. J., 1984, "The development of mouse oocytes that matured spontaneously in vitro is normal," Dev. Biol., *102*, 493-497.

Schroeder, A. C., and Eppig, J. J., 1985, "Developmental capacity of oocytes matured in vitro following isolation from juvenile and adult cycling mice," Biol. Reprod., *32*, Suppl. 1, 50.

Schultz, R. M., and Wassarman, P. M., 1977a, "Biochemical studies of mammalian oogenesis: Protein synthesis during oocyte growth and meiotic maturation in the mouse," J. Cell Sci., *24*, 167-194.

Schultz, R. M., and Wassarman, P. M., 1977b, "Specific changes in the pattern of protein synthesis during meiotic maturation of mammalian oocytes in vitro," Proc. Natl. Acad. Sci. USA, *74*, 538-541.

Schultz, R. M., LaMarca, M. J., and Wassarman, P. W., 1978a, "Absolute rates of protein synthesis during meiotic maturation of mammalian oocyte in vitro," Proc. Natl. Acad. Sci. USA, *75*, 4160-4164.

Schultz, R. M., Letourneau, G. E., and Wassarman, P. M., 1978b, "Meiotic maturation of mouse oocytes in vitro: Protein synthesis in nucleate and anucleate fragments," J. Cell Sci., *30*, 251-264.

Schultz, R. M., Letourneau, G. E., and Wassarman, P. M., 1979, "Program of early development in the mammal: Changes in patterns and absolute rates of tubulin and total protein synthesis during oogenesis and early embryogenesis in the mouse," Dev. Biol., *68*, 341-359.

Schultz, R. M., Montgomery, R. R., and Belanoff, J. R., 1983a, "Regulation of mouse oocyte meiotic maturation: Implication of a decrease in oocyte cAMP and protein dephosphorylation in commitment to resume meiosis," Dev. Biol., *97*, 264-273.

Schultz, R. M., Montgomery, R. R., Ward-Bailey, P. F., and Eppig,

J. J., 1983b, "Regulation of oocyte maturation in the mouse: Possible roles of intercellular communication, cAMP, and testosterone," Dev. Biol., *95*, 294-304.

Schultz, R. M., Heller, D. T., and Bukland, N., 1983c, "Implication of a calmodulin-dependent step in mouse oocyte maturation," Biol. Reprod., *28*, Suppl. 1, 138.

Schwartz, N. B., Cobbs, S. B., Talley, W. J., and Ely, C. A., 1975, "Induction of ovulation by LH and FSH in the presence of antigonadotropic sera," Endocrinology, *96*, 1171-1178.

Seamark, R. F., Phillipou, G., and McIntosh, J. E., 1977, "Ovarian steroidogenesis studied by mass fragmentography," J. Steroid. Biochem., *8*, 885-891.

Shalgi, R., Dekel, N., and Kraicer, P. F., 1979, "The effect of LH on the fertilizability and developmental capacity of rat oocytes matured in vitro," J. Reprod. Fertil., *55*, 429-435.

Shea, B. F., Latour, J. P., Bedirian, K. N., and Baker, R. D., 1976, "Maturation in vitro and subsequent permeability of bovine follicular oocytes," J. Anim. Sci., *43*, 809-815.

Smith, D. M., and Tenney, D. Y., 1978, "Mouse oocyte maturation in vitro: Effects of steroids," Biol. Reprod., *18*, Suppl. 1, 69a.

Smith, D. M., and Tenney, D. Y., 1980, "Effects of steroids on mouse oocyte maturation in vitro," J. Reprod. Fertil., *60*, 331-338.

Smith, D. M., Conawey, C. H., and Kerber, W. T., 1978a, "Influence of season and age on maturation in vitro of rhesus monkey oocytes," J. Reprod. Fertil., *54*, 91-95.

Smith, D. M., Tyler, J. P., and Erickson, G. F., 1978b, "Effects of medium composition and progesterone on maturation in vitro of rabbit oocytes from Graafian follicles of different sizes," J. Reprod. Fertil., *54*, 393-400.

Sorensen, R. A., 1979, "Comments on the control of oogenesis in mammals," in: Ovarian Follicular Development and Function," Raven Press, New York, pp. 375-377.

Sorensen, R. A., and Wassarman, P. M., 1976, "Relationship between growth and meiotic maturation of the mouse oocyte," Dev. Biol., *50*, 531-536.

Soupart, P., 1974, "The need for capacitation of human sperm; functional and ultrastructural observations," in: Biology of Spermatozoa, Karger, Basel, pp. 182-191.

Sreenan, J., 1970, "In vitro maturation and attempted fertilization of cattle follicular oocytes," J. Agric. Sci., *75*, 393-396.

Staigmiller, R. B., and Moor, R. M., 1984, "Effect of follicular cells on the maturation and developmental competence of ovine oocytes matured outside the follicle," Gamete Res., *9*, 221-229.

Stern, S., and Wassarman, P. M., 1973, "Protein synthesis during meiotic maturation of the mammalian oocyte," J. Cell Biol., *59*, 335a.

Stern, S., Rayyis, H., and Kennedy, F., 1972, "Incorporation of amino acids during maturation in vitro by mouse oocyte: effect of puromycin on protein synthesis," Biol. Reprod., *7*, 341-346.

Sternlicht, A. L., and Schultz, R. M., 1981, "Biochemical studies of mammalian oogenesis: Kinetics of accumulation of total and poly(A)-containing RNA during growth of the mouse oocyte," J. Exp. Zool., *215*, 191-200.

Strickland, S., and Beers, W. H., 1979, "Studies of enzymatic basis and hormonal control of ovulation," in: Ovarian Follicular Development and Function, Raven Press, New York, pp. 143-153.

Suzuki, S., and Mastroianni, L., 1966, "Maturation of monkey ovarian follicular oocytes in vitro," Am. J. Obstet. Gynecol., *96*, 723-731.

Suzuki, S., Mori, T., and Nishimura, T., 1977, "Changes in steroidogenesis by isolated rabbit ovarian follicles in vitro during preovulatory interval with emphasis on active androgen production," Endocrinol. Jpn., *24*, 129-137.

Suzuki, S., Endo, Y., Fujiwara, T., Tanaka, S., and Iizika, R., 1983, "Cytochemical study of steroid-producing activities of human oocytes," Fertil. Steril., *39*, 683-689.

Szöllösi, D., Gérard, M., Ménézo, Y., and Thibault, C., 1978, "Permeability of ovarian follicle: corona cell—oocyte relationship in mammals," Ann. Biol. Anim., Biochim., Biophys., *18*, 511-521.

Thibault, C., 1970, "Normal and abnormal fertilization in mammals," in: Advances in Bioscientific Views, Pergamon Press, pp. 63-79.

Thibault, C., 1972, "Final stages of mammalian oocyte maturation," in: Oogenesis, University Park Press, Baltimore; Butterworths, London, pp. 397-411.

Thibault, C., 1977, "Are follicular maturation and oocyte maturation independent processes?" J. Reprod. Fertil., *51*, 1-15.

Thibault, C., and Gérard, M., 1973, "Cytoplasmic and nuclear maturation of rabbit oocytes in vitro," Ann. Biol. Anim., Biochim., Biophys., *13*, Hors-Serie, 145-156.

Thibault, C., Gérard, M., and Ménézo, Y., 1975a, "Preovulatory and ovulatory mechanisms in oocyte maturation," J. Reprod. Fertil., *45*, 605-610.

Thibault, C., Gérard, M., and Ménézo, Y., 1975b, "Acquisition par l'ovocyte de lapine et de veau du facteur de décondensation du noyau duspermatosoïde fécondant (MPGF)," Ann. Biol. Anim., Biochim., Biophys., *15*, 705-714.

Thibault, C., Gérard, M., and Ménézo, Y., 1975c, "Attempts to mature calf oocytes in culture," in: Egg Transfer in Cattle, Cambridge.

Trounson, A. O., Willadsen, S. M., and Rowson, L. E., 1977, "Fertilization and development capability of bovine follicular oocytes matured in vitro and in vivo and transferred to the oviducts of rabbits and cows," J. Reprod. Fertil., *51*, 321-327.

Tsafriri, A., 1978a, "Inhibition of nuclear maturation of isolated rat oocytes by follicular constituents," Ann. Biol. Anim., Biochim., Biophys., *18*, 523-528.

Tsafriri, A., 1978b, "Oocyte maturation in mammals," in: The Vertebrate Ovary, Plenum Press, New York, pp. 409-442.

Tsafriri, A., 1985, "The control of meiotic maturation in mammals," in: Model Systems and Oogenesis, Academic Press, Orlando, pp. 221-252.

Tsafriri, A., and Bar-Ami, S., 1978, "Role of divalent cations in the resumption of meiosis of rat oocytes," J. Exp. Zool., *205*, 293-300.

Tsafriri, A., and Bar-Ami, S., 1981, "The role of gonadotropins and steroids in the acquisition of meiotic competence," Int. Congress Series, *559*, 225-233.

Tsafriri, A., and Kraicer, P. F., 1972, "The time sequence of ovum maturation in the rat," J. Reprod. Fertil., *29*, 387-393.

Tsafriri, A., Lindner, H. R., and Zor, U., 1971, "LH-induced meiotic division in follicle-enclosed ova in culture," Excerpta Med. Int. Congr. Ser., *234*, 471.

Tsafriri, A., Lindner, H. R., Zor, U., and Lamprecht, S. A., 1972, "In vitro induction of meiotic division in follicle-enclosed rat oocytes by LH, cyclic AMP, and prostaglandin E_2," J. Reprod. Fertil., *31*, 39-50.

Tsafriri, A., Lieberman, M. E., Barnea, A., Bauminger, S., and Lindner, H. R., 1973, "Induction by luteinizing hormone of ovum maturation and steroidogenesis in isolated Graafian follicles of the rat: Role of RNA and protein synthesis," Endocrinology, *93*, 1378-1386.

Tsafriri, A., Pomerantz, S. H., and Channing, C. P., 1975, "Inhibition of oocyte maturation by porcine follicular fluid," in: 8th Ann. Meet. Soc. Study Reprod., Colorado State Univ., Fort Collins, pp. 70-71.

Tsafriri, A., Pomerantz, S. H., and Channing, C. P., 1976a, "Inhibition of oocyte maturation by porcine follicular fluid: Partial characterization of the inhibitor," Biol. Reprod., *14*, 511-516.

Tsafriri, A., Lieberman, M. E., Ahren, K., and Lindner, H. R., 1976b, "Dissociation between LH-induced aerobic glycolysis and oocyte maturation in cultured Graafian follicles of the rat," Acta Endocrinol., *81*, 362-366.

Tsafriri, A., Lieberman, M. E., Koch, Y., Bauminger, S., Chobsieng, P., Zor, U., and Lindner, H. R., 1976c, "Capacity of immunologically purified FSH to stimulate cyclic AMP accumulation and steroidogenesis in Graafian follicles and to induce ovum maturation and ovulation in the rat," Endocrinology, *98*, 655-661.

Tsafriri, A., Channing, C. P., Pomerantz, S. H., and Lindner, H. R., 1977, "Inhibition of maturation of isolated rat oocytes by porcine follicular fluid," J. Endocrinol., *75*, 285-291.

Tsafriri, A., Weinstein, Y., Bar-Ami, S., Channing, C. P., Pomerantz, S. H., and Lindner, H. R., 1979, "Control of meiotic maturation of the rat oocyte," Res. Steroids, *8*, 193-198.

Tsafriri, A., Dekel, N., and Bar-Ami, S., 1982, "The role of oocyte maturation inhibitor in follicular regulation of oocyte maturation," J. Reprod. Fertil., *64*, 541-551.

Tyler, J. P., and Collins, W. P., 1979, "Short-term culture of mouse Graafian follicles. Effect of steroids on oocyte maturation and atresia," Res. Steroids, *8*, 251-253.

Tyler, J. P., and Collins, W. P., 1980, "The effect of progesterone and estradiol on gonadotropin-induced oocyte maturation in isolated ovarian follicles from mice," Gamete Res., *3*, 59-66.

Urner, F., and Schorderet-Slatkine, S., 1984, "Inhibition of denuded mouse oocyte meiotic maturation by tumor-promoting phorbol esters and its reversal by retinoids," Exp. Cell Res., *154*, 600-605.

Urner, F., Herrmann, W. L., Baulieu, E-E., and Schorderet-Slatkine, S., 1983, "Inhibition of denuded mouse meiotic maturation by forskolin, an activator of adenylate cyclase," Endocrinology, *113*, 1170-1171.

Van Blerkom, J., 1985, "Protein synthesis during oogenesis and early embryogenesis in the mammal," in: Fertilization Response of the Egg, Academic Press, Orlando, pp. 379-399.

Van Blerkom, J., and McGaughey, R. W., 1978a, "Molecular differentiation of the rabbit ovum. I. During oocyte maturation in vivo and in vitro," Dev. Biol., *63*, 139-150.

Van Blerkom, J., and McGaughey, R. W., 1978b, "Molecular differentiation of the rabbit ovum. II. During the preimplantation development of in vivo and in vitro matured oocytes," Dev. Biol., *63*, 151-164.

Van Blerkom, J., and Runner, M. N., 1984, "Mitochondrial reorganization during resumption of arrested meiosis in the mouse oocyte," Am. J. Anat., *171*, 335-355.

Vivarelli, E., Conti, M., De Felici, M., and Siracusa, G., 1983, "Meiotic resumption and intracellular cAMP levels in mouse oocytes treated with compounds which act on cAMP metabolism," Cell Differ., *12*, 271-276.

Warnes, G. M., Moor, R. M., and Jonson, M. H., 1977, "Changes in protein synthesis during maturation of sheep oocytes in vivo and in vitro," J. Reprod. Fertil., *49*, 331-335.

Wassarman, P. M., and Letourneau, G. E., 1976a, "RNA synthesis in fully-grown mouse oocytes," Nature, *261*, 73-74.

Wassarman, P. M., and Letourneau, G. E., 1976b, "Meiotic maturation of mouse oocytes in vitro. Association of newly synthesized proteins with condensing chromosomes," J. Cell Sci., *20*, 549-568.

Wassarman, P. M., and Turner, P. E., 1976, "Effect of dithiothreitol on meiotic maturation of mouse oocytes in vitro: Dependence of the effect of N^6,O^2-dibutyryl adenosine 5'-cyclic monophosphate," J. Exp. Zool., *196*, 183-188.

Wassarman, P. M., Josefowicz, W. J., and Letourneau, G. E., 1976, "Meiotic maturation of mouse oocytes in vitro: Inhibition of maturation at specific stages of nuclear progression," J. Cell Sci., *22*, 531-545.

Wassarman, P. M., Schultz, R. M., and Letourneau, G. E., 1979, "Protein synthesis during meiotic maturation of mouse oocytes in vitro. Synthesis and phosphorylation of protein localized in the germinal vesicle," Dev. Biol., *69*, 94-107.

Weiss, T. J., Seamark, R. T., McIntosh, J. E., and Moor, R. M., 1976, "Cyclic AMP in sheep ovarian follicles: Site of production and response to gonadotropins," J. Reprod. Fertil., *46*, 347-353.

Yanagimachi, R., 1974, "Maturation and fertilization in vitro of guinea-pig ovarian oocytes," J. Reprod. Fertil., *38*, 485-488.

Younglai, E. V., 1975, "Steroid production by the isolated rabbit ovarian follicle," J. Reprod. Fertil., *45*, 575-582.

Zamboni, L., 1970, "Ultrastructure of mammalian oocytes and ova," Biol. Reprod., Suppl. 2, 44-63.

Zamboni, L., 1972, "Comparative studies on ultrastructure of mammalian oocytes,: in: Oogenesis, University Park Press, Baltimore; Butterworths, London, pp. 5-45.

Zamboni, L., Thompson, R. S., and Smith, D. M., 1972, "Fine morphology of human oocyte maturation in vitro," Biol. Reprod., *7*, 425-427.

Zeilmaker, G. H., 1978, "Observations on follicular lactate concentrations and the influence of granulosa cells on oocyte maturation in the rat (including data on second polar body formation)," Ann. Biol. Anim., Biochim., Biophys., *18*, 529-533.

Zeilmaker, G. H., and Verhamme, C. M., 1974, "Observations on rat oocyte maturation in vitro: Morphology and energy requirements," Biol. Reprod., *11*, 145-152.

Zeilmaker, G. H., Hulsmann, W. C., Wensinek, F., and Verhamme, C., 1972, "Oxygen-triggered mouse oocyte maturation in vitro and lactate utilization by mouse oocytes and zygotes," J. Reprod. Fertil., *29*, 151-152.

Zeleznik, A. J., Midgley, A. R., and Reichert, L. E., 1974, "Granulosa cell maturation in the rat: increased binding of human chorionic gonadotropin following treatment with follicle-stimulating hormone in vitro," Endocrinology, *95*, 818-825.

Chapter 6

SPECIAL FEATURES OF CONTROLLING RESPIRATION DURING OOGENESIS

N. D. Ozernyuk

One of the most important aspects of the metabolism of growing oocytes is the accumulation of energy required for embryonic development. Since the number of mitochondria during embryonic development remains at a constant level (Chase and Dawid, 1972; Abramova and Vasilyeva, 1973), it follows that during oogenesis they must form in quantities sufficient to provide energy for the developing embryo. However, oocytes themselves are characterized by a very high level of respiration; this implies that special mechanisms control this process during oogenesis.

It is well known that oogenesis is associated with changes in metabolism and ultrastructure of the growing oocytes related to the accumulation of reserve nutrients and provision of energy for subsequent embryonic development. In connection with this, the ultrastructural features of the reserve nutrients accumulated by oocytes have been carefully studied (Favard and Carasso, 1958; Ward, 1962; Aizenshtadt, 1965; Srivastava, 1965; Droller and Roth, 1966). The chemical mechanism of glycogen accumulation in oocytes has been investigated in detail (Yurovitsky et al., 1972; Milman and Yurovitsky, 1973). The present chapter deals with the special features of respiration control during oogenesis.

In studying respiration during oogenesis it is necessary to recognize that the mass of the oocyte continuously increases. This is why the increase in oxygen consumption ($\mu l\ O_2/h$ per oocyte) by the growing oocytes appears trivial and offers little information on the peculiarities of energy metabolism in oocytes. Of much greater interest are the data on the respiration intensity ($\mu l\ O_2/h$ per mg of total protein) when oxygen consumption is evaluated per unit of oocyte mass. However, the interpretation of the data on the respiration intensity during oogenesis has a number of limitations mainly because of the fact that the increase in the mass of oocytes at the stage of vitellogenesis takes place not so much on account of the increase in the cytoplasm mass as due to accumulation of a large amount of yolk. This hampers the determination of the true intensity of oocyte respiration. There are methods of removing the yolk from homogenates, but in these cases it is difficult to assess the completeness of yolk removal. Nevertheless, the data on respiration calculated per unit of oocyte mass in the absence of yolk appears to be a measure of the true intensity of respiration during oogenesis.

6.1. OXYGEN CONSUMPTION

6.1.1. Oocyte Growth

Studies devoted to the measurement of oxygen consumption by growing oocytes, employing a variety of animals and different methods, yielded similar results. For example, Platova (1946, 1948), using newt oocytes, showed that oxygen consumption increased many times during the previtellogenic stage of oocyte growth. By measuring the oxygen consumed by the oocytes of the medaka *Oryzias latipes*, Nakano (1953) showed an essential increase in oxygen consumption during vitellogenesis, reaching a maximum in oocytes that had completed growth. Similar data were obtained in the study of the respiration of frog oocytes during the period from vitellogenesis until the completion of growth (Petrucci, 1960a, b). The measurements of oxygen consumption by the oocytes of the loach *Misgurnus fossilis* (during the early stages of oogenesis; from slow to rapid growth) confirmed the general conclusion that during the growth of oocytes the increase in their mass is attended by a multiple increase in oxygen consumption (Ozernyuk, 1970b). An increase in the rate of oxygen consumption is also shown for growing oocytes of *Brachydanio rerio* and *Acipenser stellatus* (Ozernyuk et al., 1980) (Table 6.1).

6.1.2. Oocyte Maturation

Of special interest in studying respiration and its control in oocytes is the terminal period of oogenesis, namely oocyte maturation. As far back as 1941, Lindahl and Holter studied oxygen consumption during oocyte maturation in the sea urchin *Paracentrotus lividus*. It was shown that the oxygen consumption of oocytes is much greater than that of mature unfertilized eggs. Lindahl and Holter concluded that in measuring oxygen consumption during oocyte maturation it is necessary to take into account the

Table 6.1. Change in the Rate of Oocyte Respiration at Different Stages of Oogenesis (μl O_2/h per 100 Oocytes; Brackets Show μl O_2 per 1 mg of the Total Protein)

Oogenesis stages	Loach	Danio	Stellate Sturgeon
Slow growth	0.07 ± 0.1 (1.00 ± 0.18)	0.05 ± 0.01 (0.89 ± 0.06)	0.02 (0.19)
Rapid growth:			
beginning	0.31 ± 0.02 (2.22 ± 0.13)	0.19 ± 0.02 (1.36 ± 0.34)	0.05 (0.40)
middle	0.48 ± 0.02 (1.11 ± 0.05)	0.62 ± 0.10 (0.66 ± 0.11)	0.05 (0.13)
end	4.93 ± 0.40 (0.65 ± 0.02)	1.06 ± 0.13 (0.51 ± 0.07)	4.65 (0.04)
Mature unfertilized oocytes	3.99 ± 0.31 (0.61 ± 0.02)	-	3.63 (0.03)

time between the isolation of the oocytes and eggs from the ovary and the beginning of the experiment. Detailed investigations were continued by Borei (1948). He measured the oxygen consumption of oocytes and mature unfertilized eggs of the sea urchin *Psammechinus*, at different times after they had been taken from the ovary. The oxygen consumption by oocytes, as well as by mature unfertilized eggs, dropped with time. This diminution was of exponential character and reached a plateau in 2 h. However, in the experiment carried out 2 h after the material had been taken from the ovary, the respiration of oocytes was also greater than that of mature unfertilized eggs. Thus, it was demonstrated that during the maturation of oocytes their respiration decreases. A decrease in oxygen consumption was also noted during oocyte maturation in the starfish *Asterias* (Borei, 1948).

A decrease in oxygen consumption during maturation of the oocytes of the medaka (Nakano, 1953), the loach (Ozernyuk, 1970a), and the stellate sturgeon (Table 6.1) also occurs. Unlike in sea urchins (Borei, 1948), the oxygen consumption in fish does not decrease during a few hours of cultivation. Nakano explains this difference by the presence of a large amount of yolk in the fish eggs and oocytes which may maintain oxygen consumption at a constant level, the yolk supplying the respiration substance.

Borei (1948) holds that the decrease in the rate of oxygen consumption during oocyte maturation is of adaptive significance. When mature eggs cannot be fertilized immediately after they are laid, the low level of respiration enables them to retain their viability for a long time.

6.2. INTENSITY OF OOCYTE RESPIRATION

The data on respiration intensity during oogenesis obtained by different authors using various species are largely contradictory (Meshcherskaya, 1935; Platova, 1946, 1948; Gonse, 1953, 1955, 1957a, b). Thus, owing to the difficulty of obtaining large numbers of oocytes at different stages of oogenesis, many authors measured respiration by micromethods, which sometimes reduced the reliability of the data. The development of methods for obtaining large numbers of oocytes from the ovarian stroma eliminates this difficulty, although the presence of follicular epithelium on the surface of the isolated oocytes may affect the results.

Mescherskaya (1935) measured the intensity of frog oocyte respiration at stages of slow growth (the middle of rapid growth) and the end of rapid growth. From the stage of slow growth until the beginning of vitellogenesis the intensity of respiration nearly doubled, but diminished somewhat toward the end of rapid growth. Largely similar data were obtained using newt oocytes (Platova, 1946, 1948) when it was shown that the intensity of oocyte respiration somewhat diminished at the earliest stages of oogenesis. In the sipunculid *Phascolosoma vulgare*, two maxima of intensity of oocyte respiration were discovered, viz. at the stage of slow growth and at the end of rapid growth (Gonse, 1953, 1955, 1975a, b). Moreover, the first peak is higher than the second. Calculations of the intensity of oocyte respiration per volume of yolkless cytoplasm yielded similar results.

The authors of the above studies used micromethods for measuring respiration, which may be one of the reasons for the contradictory results. By obtaining large amounts of oocytes from the ovary, it is possible to employ Warburg's manometric method for recording respiration (Ozernyuk, 1970b; Ozernyuk et al., 1973); this improves the reliability of the data. Using this method, the respiration intensity of loach oocytes was measured during the stage of slow growth, the beginning of rapid growth (vacuolization of the cytoplasm), the beginning of vitellogenesis, and the end of rapid growth. Respiration was calculated per mg of intact oocytes, per mg of total nitrogen of intact oocyes, and per mg of protein of yolkless oocytes. It should be noted that the change in respiration intensity was practically the same in all cases. During the early stages of oogenesis, from the stage of slow growth until the beginning of rapid growth, the respiration intensity more than doubled, and toward the beginning of vitellogenesis, it decreased to the initial level. During the subsequent stages of oogenesis the respiration intensity continued to diminish. Similar results were obtained per mg of protein of yolkless oocytes, but the decrease in respiration intensity during the late stages of oogenesis was not so significant when calculated per intact oocyte.

Overall, the results obtained by different authors using various animals and different methods indicated that the greatest intensity of respiration is characteristic of oocytes at the early stages of oogenesis.

6.3. CONTROL OF OXYGEN CONSUMPTION

The control of oxygen consumption during the growth of oocytes shows a number of peculiarities which distinguish it from that during embryonic development and in adult tissues. The main peculiarity consists in the correlation between oxygen consumption and the number of mitochondria. An increase in the number and sizes of mitochondria during oogenesis was demonstrated by studies of the ultrastructure of oocytes of worms (Tokin, 1961), fish (Ozernyuk and Palmbakh, 1975), amphibians (Karasaki and Komoda, 1958; Ward, 1962; Aizenshtadt, 1965; Srivastava, 1965), and mammals (Anderson and Beams, 1960; Anderson et al., 1970). These morphological data are confirmed by the demonstration of a large increase in the level of mitochondrial protein and the activity of the mitochondrial enzyme, cytochrome oxidase, during oogenesis of fish (Ozernyuk, 1972) and amphibians (Petrucci, 1960a, b). Thus the increase in the number of mitochondria in oocytes is one of the principal reasons for the increased rate of oxygen consumption during oogenesis.

During oocyte maturation the mechanism of the control of oxygen consumption differs from those operating during oocyte growth. Thus, it was already noted that during oocyte maturation the rate of oxygen consumption diminishes (Lindahl and Holter, 1941; Borei, 1948; Nakano, 1953), although the amount of mitochondrial protein hardly changes during these stages (Ozernyuk, 1972). Some authors (Maggio, 1959; Petrucci, 1960a, b) suggest that the decrease in oxygen consumption in maturing oocytes is due to an inhibition of cytochrome oxidase. However, no search for the assumed inhibitor was conducted, and later it was shown that cytochrome oxidase was not a factor that limited respiration during embryonic development (Abramova and Vasiliyeva, 1973). The maturation of sea urchin and loach oocytes is attended simultaneously by a decrease in oxygen consumption and a considerable increase in ATP content (Ozernyuk, 1970a). These data may, apparently, be explained from the standpoint of control of oxygen consumption of the type of "respiratory control" in which an addition of ADP to a suspension of mitochondria stimulates respiration. The ADP/ATP ratio in the mitochondria can therefore control the rate of electron transport in the respiratory chain (see Lehninger, 1964). This hypothesis was, with some modifications, used in explaining respiration control during fertilization (Monroy, 1965), embryonic development (Zotin et al., 1967), as well as some experimental influences on respiration and ATP level, in particular the influence of temperature (Zotin and Ozernyuk, 1966) and x-ray irradiation (Nikolskaya and Grudnitsky, 1970). The "respiratory control" hypothesis may explain the decrease in intensity of respiration during oocyte maturation because the rise in ATP level during this period may hamper the rate of electron transport in the respiratory chain of the mitochondria.

At the same time, however, it is necessary to take into account a number of factors which may preclude the extrapolation of generally accepted ideas to oocytes and eggs. First, control of respiration by the ADP/ATP ratio was shown only on isolated mitochondria, while the studies under discussion concern the intact egg. Such an analogy would be correct if there were a positive correlation between the ATP content of the intact oocyte or egg and the mitochondria from the oocytes or eggs. However, this question was not studied. Moreover, the respiration of mitochondria is stimulated by an addition of exogenous ADP, whereas in intact oocytes or eggs a change in the rate of oxygen consumption occurs during physiological changes in the level of adenine nucleotides controlled, in turn, by numerous factors (Skulachev, 1969).

6.4. CONTROL OF RESPIRATION INTENSITY IN OOCYTES

The suggestion that the rate of oxygen consumption during oogenesis is controlled by the mass of mitochondria was also used in explaining the control of respiration intensity in oocytes. For this purpose the concentration of mitochondrial protein was measured (Ozernyuk, 1972) and the concentration of mitochondria (number of mitochondria per unit of area) was calculated under the electron microscope (Ozernyuk and Palmbakh, 1975; Ozernyuk et al., 1980). The concentration of mitochondrial protein more than doubled from the stage of slow growth to the beginning of rapid growth (stage of vacuolization of the cytoplasm) after which it diminished at the beginning of vitellogenesis and the later stages of oogenesis. The concentration of mitochondria changed in a similar way. As noted above, the respiration intensity similarly reaches a maximum at the beginning of the rapid growth of oocytes. The correlation between respiration and the concentration of mitochondria in oocytes necessitates a search for the cause of the change in intensity of respiration among the factors that control the division of mitochondria. The regulation of mitochondrial division has been ascribed to mitochondrial DNA (Hawley and Wagner, 1967) and the nucleus (Grimes et al., 1974; Kolb-Bachofen and Vogell, 1975; Rinaldi et al., 1977, 1978), as well as to different metabolites, such as riboflavin (Tandler et al., 1969) and ATP (Ozernyuk and Palmbakh, 1975). However, the nature of the factors controlling mitochondrial division remains unclear.

Table 6.2. Concentration of Mitochondria in Different Types of Cells

Type of cell	Number of mitochondria per cell	Number of mitochondria per 100 μm³	Reference
Average rat lymphocyte	6-8	3	
Epithelial cells of the rat duodenum	100-200	-	Tokin, 1972
Cells of the rat liver	1200-1500	10	
Same	1500-3000	24	
Cells of rat hepatoma	1390	10	Allard et al. 1952b
Cells of rat regenerating liver:			
first day	1800	13	Allard et al. 1952a
second day	2500	15	
Amoeba *Chaos chaos*	500,000	3	
Egg cells of sea urchins *Strongylocentrotus purpatus*	14,000	6	Lehninger, 1964
Lytechinus pictus	200,000	32	Piko et al. 1967
Rat oocytes	92,000	-	Pilo and Matsumoto, 1967
Loach oocytes:			
beginning of slow growth	50,000	26	
end of slow growth	600,000	36	Ozernyuk and Palmbakh, 1975
Vacuolization of cytoplasm	4,100,000	50	
beginning of vitellogenesis	12,000,000	39	
Oocytes of *Xenopus laevis*:			
beginning of slow growth	10,000	-	Billett and Adam, 1976
end of slow growth	120,000	-	

The correlation between the intensity of oocyte respiration and the concentration of mitochondria is not universal because, for example, respiration is intensified severalfold during embryonic development (see Zotin, 1972), whereas the concentration of mitochondria during this period does not change (Chase and Dawid, 1972; Abramova and Vasiliyeva, 1973). It is possible that the concentration of mitochondria is not the only factor which controls the respiration intensity of oocytes. A similar point of view on the mechanism of respiration control during oogenesis and during embryonic development was stated by Neyfakh and Abramova (1974).

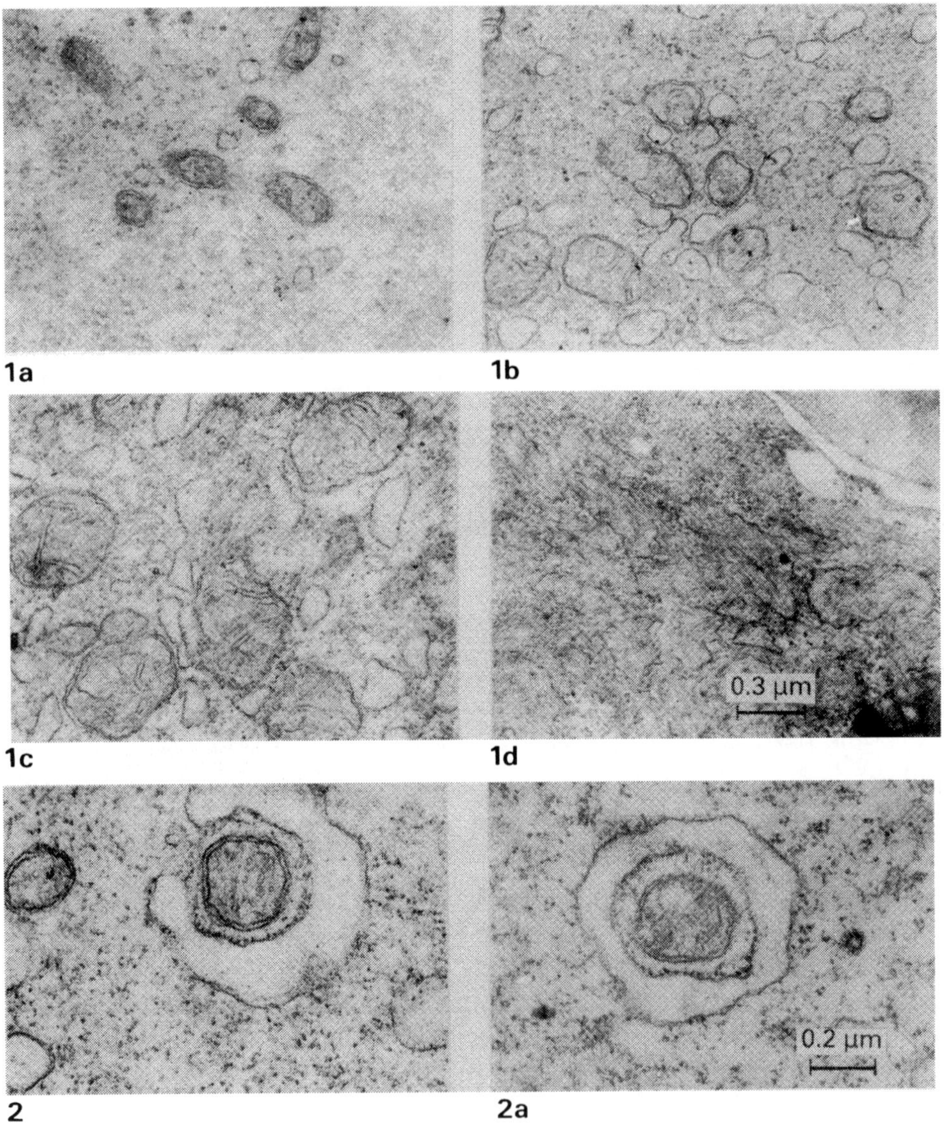

Plate XXII. 1. Changes in the size of mitochondria during oogenesis in the loach: a) beginning of slow growth of the oocytes; b) middle of slow growth; c) beginning of rapid growth; d) end of rapid growth. 2, 2a. Interrelations of mitochondria and granular endoplasmic reticulum in the loach oocytes.

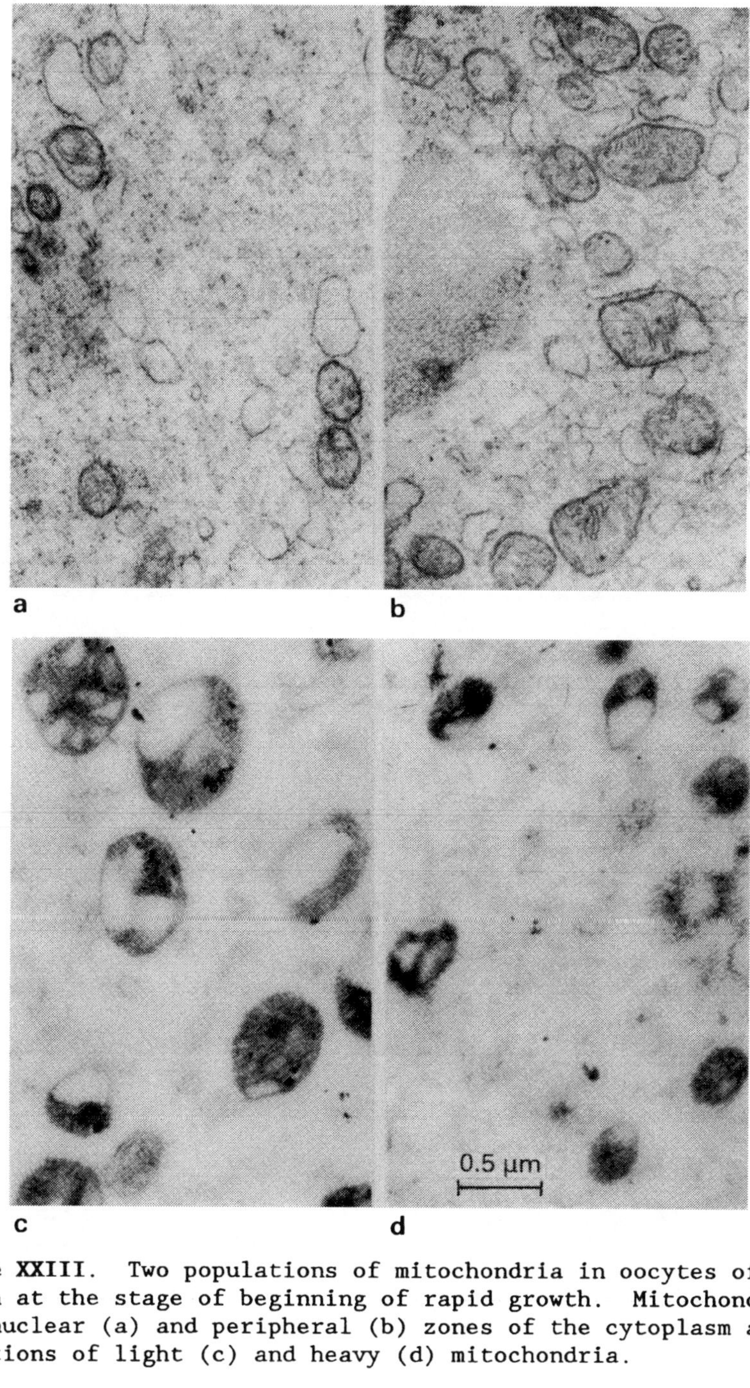

Plate XXIII. Two populations of mitochondria in oocytes of the loach at the stage of beginning of rapid growth. Mitochondria of perinuclear (a) and peripheral (b) zones of the cytoplasm and fractions of light (c) and heavy (d) mitochondria.

Table 6.3. Some Characteristics of Two Mitochondrial Populations from Loach Oocytes

Characteristics	Light mitochondria	Heavy mitochondria
Volume (mm^3)	0.22 ± 0.003	0.45 ± 0.05
Density (g/cm^3)	1.170 ± 0.003	1.180 ± 0.001
Protein/phospholipid ratio	6.67 ± 0.59	8.21 ± 0.79
^{14}C-valine labeling of mitochondrial protein _in vivo_ (imp/min/mg of protein)	7013 ± 209	5659 ± 141
Respiration intensity (ng atoms O_2/mg of protein):		
on endogenous substrates	4.00 ± 0.91	7.07 ± 0.88
succinate	8.05 ± 1.44	33.45 ± 4.99
Cytochrome oxidase activity		
(ΔE/min/mg of mitochondrial protein)	0.0016 ± 0.00006	0.0032 ± 0.00055

A comparison of the concentration of mitochondria in oocytes at the stage of vacuolization of the cytoplasm with their concentration in other types of cells of various organisms shows that it is the highest in oocytes (Table 6.2). This is not unexpected if we consider the high intensity of oocyte respiration.

Such a high concentration of mitochondria in oocytes may lead to quite a high level of energy metabolism, which must affect the rate of metabolic processes, activity of different enzymes, and concentration of various metabolites. Thus, for example, the activity of ATPases is the highest in oocytes at the early stages of oogenesis (Ozernyuk, 1971).

One more characteristic, namely the ATP content in different types of cells and tissues, is shown in Table 6.3.

The increase in the size and number of mitochondria at the early stages of oogenesis is characterized by a number of features (Plate XXII, 1). The increase in the size of mitochondria in the peripheral zone of oocytes occurs faster than in the perinuclear zone (Rinaldi et al., 1977, 1979). Moreover, the growing mitochondria are in contact with the rough endoplasmic reticulum (Kessel, 1966; Sawada et al., 1968; Ozernyuk and Palmbakh, 1974, 1975). Thus, in loach oocytes at the early stages of oogenesis, 15-20% of the mitochondria are surrounded along 25-100% of their perimeter by membranes of rough endoplasmic reticulum (Ozernyuk and Palmbakh, 1974) (Plate XXII, 2).

Table 6.4. ATP Content in Different Types of Cells and Tissues

Object	ATP content, % of dry weight	Reference
Loach oocytes:		
slow growth	1.64	
beginning of vitellogenesis	0.94	Ozernyuk, 1971
Cleaving eggs:		
loach	0.18	
sea urchin	0.27	
Axolotl	0.27	
Fibroblast culture	0.12	
Ehrlich ascites tumor cells	0.12	Zotin et al., 1967
Loach tissues:		
brain	0.034	
liver	0.046	
muscle	0.013	

The different degree of inhibition by cycloheximide of the synthesis of mitochondrial proteins in vivo at different stages of oogenesis warrants the assumption that these contacts apparently play an essential role not only in supplying the synthesis of protein on the ribosomes of cytoplasm with energy, but also in the processes of mitochondrial growth, particularly in the transfer to mitochondria of the proteins synthesized on the cytoplasmic ribosomes (Ozernyuk and Palmbakh, 1974, 1975).

The number and size of the mitochondria in different zones of oocytes increase unequally (Ozernyuk and Palmbakh, 1975). The isolation, ultrastructural, and biochemical characteristics of two mitochondrial populations from loach oocytes were distinguished on the basis of a morphometric analysis. These two categories are: small dividing organelles in the cytoplasmic nuclear zone and large differentiated mitochondria in the periphery of these oocytes (Ozernyuk, 1978a, b, 1980, 1981) (Table 6.4; Plate XXIII).

In connection with these data it is necessary to note the accumulation of mitochondria in the perinuclear zone of *Xenopus laevis* previtellogenic oocytes (Balinsky and Devis, 1963; Billett and Adam, 1976; Callen et al., 1980; Marinos and Billett, 1981). The increase in the mitochondrial mass during the oogenesis of *Xenopus laevis* is mainly due to the increasing number of mitochondria in the mitochondrial cloud (Balbiani body).

Thus it may be concluded that one of the main features of oocyte metabolism is a high level of energy metabolism. The functioning of many metabolic systems, particularly carbohydrate metabolism, which has a glycogen-synthesizing trend in oocytes during oogenesis, is aimed at accumulating nutrients (Milman and Yurovitsky, 1973). The complication of the ultrastructure of mitochondria during oogenesis (Blanchette, 1961; Tokin, 1961; Aizenshtadt, 1965; Srivastava, 1965) and the increase in the concentration of mitochondria also prepare the oocytes for supplying the developing embryo with energy.

REFERENCES

Abramova, N. B., and Vasiliyeva, M. N., 1973, "Some properties of embryonic mitochondria of loach *Misgurnus fossilis*," Ontogenez, *4*, 265-288.

Aizenshtadt, T. B., 1965, "Some features of oocyte ultrastructure in connection with the synthesis of the yolk," Zh. Obshch. Biol., *26*, 230-236.

Allard, C., de Lamirande, G., and Cantero, A., 1952a, "Mitochondria in mammalian cells. II. Variation in the mitochondrial population of the average rat liver cell during regeneration. Use of the mitochondrion as a unit of measurement," Cancer Res., *12*, 580-583.

Allard, C., de Lamirande, G., and Cantero, A., 1952b, "Mitochondrial population in mammalian cells. IV. Preliminary results on the variation in the mitochondrial population of the average rat liver cell during azo dye carcinogenesis," Can. J. Med. Sci., *30*, 543-549.

Anderson, E., and Beams, H. W., 1960, "Cytological observations on the fine structure of the guinea pig ovary with special reference to the oogonium, primary oocyte, and associate follicle cells," J. Ultrastruct. Res., *3*, 423-447.

Anderson, E., Condon, W., and Sharp, D., 1970, "A study of oogenesis in the rabbit *Oryctolagus cuniculus* with special reference to the structural changes of mitochondria," J. Morphol., *130*, 67-91.

Balinsky, B. I., and Devis, R. J., 1963, "Origin and differentiation of cytoplasmic structures in the oocytes of *Xenopus laevis*," Acta Embryol. Morphol. Exp., *6*, 55-108.

Billett, F. S., and Adam, E., 1976, "The structure of the mitochondrial cloud of *Xenopus laevis* oocytes," J. Embryol. Exp. Morphol., *36*, 697-710.

Blanchette, F. J., 1961, "A study of the fine structure of the rabbit primary oocyte," J. Ultrastruct. Res., *5*, 349-363.

Borei, H., 1948, "Respiration of oocytes, unfertilized eggs, and fertilized eggs from *Psammechinus* and *Asterias*," Biol. Bull., *95*, 124-150.

Callen, J. C., Tautre, M., Dennebony, N., and Mounloy, J. C., 1980, "Mitochondrial development in oocytes of *Xenopus laevis*," Biol. Cell., *38*, 13-18.

Chase, J. W., and Dawid, I., 1972, "Biogenesis of mitochondria during *Xenopus laevis* development," Dev. Biol., *27*, 504-518.

Droller, M. J., and Roth, T. F., 1966, "An electron microscope study of yolk formation during oogenesis in *Lebistes reticulatus* guppy," J. Cell. Biol., *28*, 209-232.

Favard, J., and Carasso, A., 1958, "Origine et ultrastructure des plaquettes vitelline de la planorbe," Arch. Anat. Microsc. Morphol. Exp., *47*, 211-219.

Gonse, P. H., 1953, "L'ovogenèse ches *Phascolosoma vulgare*," C. R. Acad. Sci. Paris, *236*, 528-530.

Gonse, P. H., 1955, "Récherches sur la physiologie de l'ovogenèse chez *Phascolosoma vulgare*," Exp. Cell Res., *8*, 550-553.

Gonse, P. H., 1957a, "L'ovogenèse chez *Phascolosoma vulgare*. III. Respiration exogené et endogené l'ovocyte: effet de l'eau de mer," Biochim. Biophys. Acta, *24*, 267-278.

Gonse, P. H., 1957b, "L'ovogenèse chez *Phascolosoma vulgare*. IV. Étude chromatographique des sucres du plasma action de différents substrates et du malonate sur la réspiration de l'ovocyte," Biochim. Biophys. Acta, *24*, 530-531.

Grimes, G. W., Mahler, H. R., and Perlman, P. S., 1974, "Nuclear gene dosage effect on mitochondrial mass and DNA," J. Cell Biol., *61*, 565-574.

Hawley, E. S., and Wagner, R. P., 1967, "Synchronous mitochondrial division in *Neurospora crassa*," J. Cell Biol., *35*, 489-499.

Karasaki, S., and Komoda, 1958, "Electron micrographs of a crystalline lattice structure in yolk platelets of the amphibian embryo," Nature, *181*, 407-408.

Kessel, R. G., 1966, "Some observations on the ultrastructure of the oocyte of *Thyone briareus* with special reference to the relationship of the Golgi complex and endoplasmic reticulum in the formation of yolk," J. Ultrastruct. Res., *16*, 305-319.

Kolb-Bachofen, V., and Vogell, W., 1975, "Mitochondrial proliferation in synchronized cells of *Tetrahymena pyriformis*. A morphometric study of electron microscopy of the biogenesis of mitochondria during the cell cycle," Exp. Cell Res., *94*, 95-105.

Lehninger, A., 1964, The Mitochondrion, W. A. Benjamin, New York–Amsterdam.

Lindahl, P. E., and Holter, H., 1941, "Uber die Atmung der Ovozyten erster Ordnung vor *Paracentrotus lividus* und ihre Veranderung wahrend der Reifung," C. R. Trav. Lab. Carlsberg, Ser. Chim., *24*, 49-58.

Maggio, R., 1959, "Cytochrome oxidase activity in the mitochondria of unfertilized and fertilized sea-urchin eggs," Exp. Cell Res., *16*, 272-278.

Marinos, E., and Billett, F. S., 1981, "Mitochondrial number and cytochrome oxidase and succinic dehydrogenase activity in *Xenopus laevis* oocytes," J. Embryol. Exp. Morphol., *62*, 395-409.

Meshcherskaya, K. A., 1935, "On some physiological properties of oocytes of lower vertebrates at different stages of growth," Archiv. Anat. Gistol. Embriol., *14*, 656-667.
Milman, L. S., and Yurovitsky, Y. G., 1973, Regulation of Glycolysis in Fish Embryos, S. Karger, Basel.
Monroy, A., 1965, Chemistry and Physiology of Fertilization, Holt, Rinehart, and Wilson, New York.
Nakano, E., 1953, "Respiration during maturation and at fertilization of fish eggs," Embryologia, *2*, 21-26.
Neyfakh, A. A., and Abramova, N. B., 1974, "Change in the number of mitochondria and their functions during early ontogenesis," in: Genetic Functions of Cytoplasmic Organelles [in Russian], Nauka, Leningrad, pp. 80-85.
Nikolskaya, I. S., and Grundnitsky, V. A., 1970, "Effect of X-radiation and inhibitors of nucleic acid synthesis on respiration and ATP level in loach embryos," Dokl. Akad. Nauk SSSR, *190*, 478-480.
Ozernyuk, N. D., 1970a, "Role of ATP in controlling respiration during maturation of sea-urchin and loach eggs," Dokl. Akad. Nauk SSSR, *190*, 245-248.
Ozernyuk, N. D., 1970b, "Intensity of respiration and ATP content in loach oogenesis," Dokl. Akad. Nauk SSSR, *192*, 242-245.
Ozernyuk, N. D., 1971, "Change in ATPase activity during loach oogenesis," Ontogenez, *2*, 349-431.
Ozernyuk, N. D., 1972, "Change in the number of mitochondria in the process of loach oogenesis," Dokl. Akad. Nauk SSSR, *207*, 974-977.
Ozernyuk, N. D., 1978a, "The role of mitochondria in regulation of respiration during oogenesis," in: Thermodynamics of Biological Processes, Walter de Gruyter, Berlin—New York, pp. 229-241.
Ozernyuk, N. D., 1978b, The Growth and Reproduction of Mitochondria [in Russian], Nauka, Moscow.
Ozernyuk, N. D., 1980, "A hypothesis of antagonism between the functional maturity of mitochondria and their ability for growth and division," Ontogenez, *11*, 187-191.
Ozernyuk, N. D., 1981, "The activity of cytochrome oxidase in the mitochondria of loach oocytes," Izv. Akad. Nauk SSSR, Ser. Biol., No. 1, 136-138.
Ozernyuk, N. D., and Palmbakh, L. R., 1974, "Connection between mitochondria and cytoplasmic reticulum in loach oocytes," Ontogenez, *5*, 356-404.
Ozernyuk, N. D., and Palmbakh, L. R., 1975, "Growth and reproduction of mitochondria in loach oocytes," Ontogenez, *6*, 442-449.
Ozernyuk, N. D., Zotin, A. I., and Yurovitsky, Y. G., 1973, "Deviation of the living system from the stationary state during oogenesis," W. Roux's Arch. Entwicklungsmech. Org., *172*, 66-74.
Ozernyuk, N. D., Zviadadze, N. G., and Palmbakh, L. R., 1980, "The

intensity of respiration, concentration, and distribution of mitochondria during danio oogenesis," Izv. Akad. Nauk SSSR, Ser. Biol., No. 6, 941-944.

Petrucci, D., 1960a, "La citocrome C-ossidasi nell'oogenesi degli Anfibi," Acta Embryol. Morphol. Exp., *3*, 237-259.

Petrucci, D., 1960b, "Réspiration et cytochrome oxidase au cours de l'oogenèse chez les amphibiens (*Rana esculenta*)," Arch. Int. Physiol. Biochim., *68*, 225-226.

Piko, L., and Matsumoto, L., 1976, "Number of mitochondria and some properties of mitochondrial DNA in the mouse egg," Dev. Biol., *49*, 1-10.

Piko, L., Tyler, A., and Vinograd, J., 1967, "Amount, location, priming capacity, circularity, and other properties of cytoplasmic DNA in sea-urchin eggs," Biol. Bull., *132*, 68-90.

Platova, T. P., 1946, "On the role of the nucleus and cytoplasm in the oxidative metabolism of newt oocytes," Dokl. Akad. Nauk SSSR, *53*, 259-262.

Platova, T. P., 1948, "Oxidative metabolism of newt oocytes in connection with the processes of plasma and yolk synthesis," Dokl. Akad. Nauk SSSR, *60*, 1397-1400.

Rinaldi, A. M., Storace, A., Arzone, A., and Mutolo, V., 1977, "Cell nucleus negativity controls mitochondrial RNA synthesis in early sea-urchin development," Cell Biol. Int. Reports, *1*, 249-254.

Rinaldi, A. M., Salcher-Cillari, J., and Mutolo, V., 1979, "Mitochondrial division in nonnucleated sea-urchin eggs," Cell Biol. Int. Reports, *3*, 179-182.

Sawada, N., Noda, Y., and Ochi, O., 1968, "An electron microscope study on the oogenesis of *Yolfingiya ikedai*," Mem. Ehime Univ. Sci. B, *6*, 25-39.

Skulachev, V. P., 1969, Accumulation of Energy in the Cell [in Russian], Nauka, Moscow.

Srivastava, M. D. L., 1965, "Cytoplasmic inclusions in oogenesis," Int. Rev. Cytol., *18*, 73-98.

Tandler, B., Erlandson, R. A., Smith, A. L., and Wynder, E. L., 1969, "Riboflavin and mouse hepatic cell structure and function. II. Division of mitochondria during recovery from simple deficiency," J. Cell Biol., *41*, 477-493.

Tokin, I. B., 1961, Electron Microscopy of the Ascarid Sexual and Somatic Cells [in Russian], Izd. Leningr. Gos. Univ.

Tokin, I. B., 1972, "Electron microscopic analysis of the process of differentiation and dedifferentiation of cells," Arkh. Anat. Gistol. Embriol., *62*, 46-63.

Ward, B., 1962, "The origin of protein and fatty yolk in *Rana pipiens*. II. Electron microscopical and cytochemical observation of young and matured oocytes," J. Cell Biol., *14*, 309-341.

Yurovitsky, Y. G., Milman, L. S., and Ozernyuk, N. D., 1972, "Some features of controlling glycolysis in loach oocytes," Ontogenez, *3*, 405-414.

Zotin, A. I., 1972, Thermodynamic Aspects of Developmental Biology, S. Karger, Basel.
Zotin, A. I., and Ozernyuk, N. D., 1966, "Effect of temperature on the respiration and ATP level during cleavage of loach eggs," Dokl. Akad. Nauk SSSR, *171*, 1002-1004.
Zotin, A. I., Faustov, V. S., Radzinskaya, L. I., and Ozernyuk, N. D., 1967a, "ATP level and respiration of embryos," J. Embryol. Exp. Morphol., *18*, 1-12.
Zotin, A. I., Milman, L. S., and Faustov, V. S., 1967b, "Change in adenosine triphosphate upon fertilization of sea-urchin and loach eggs," Zh. Evol. Biokhim. Fiziol., *3*, 76-77.

Chapter 7

CONTROL OF CARBOHYDRATE METABOLISM AT DIFFERENT STAGES OF OOGENESIS

L. S. Milman, Yu. G. Yurowitzky, and L. P. Yermolaeva

It is known that in most animals reserve carbohydrates accumulate in the oocyte during oogenesis. These carbohydrates serve as substrates for the metabolism of the embryo and the larva before the onset of active feeding. (For cytological and embryological observations on this topic, see Chapter 1.)

While carbohydrates, mainly glycogen, accumulate during oogenesis, and oocyte metabolism has a "glycogen-deposition" direction, the onset of embryonic development is accompanied by the more or less intensive utilization of glycogen and other reserve compounds (Hutchens et al., 1942; Barth and Barth, 1954; Løvtrup, 1966).

These quite substantial differences in the metabolism of the oocyte and embryo must be conditioned by changes in the complex of enzyme mechanisms relating to the enzymes involved. Constant enzyme activity from the moment of ovulation of the egg right up to the end of early embryogenesis serves to show that the embryo's enzymatic apparatus has already been formed during oogenesis, i.e., before the onset of embryonic development (Milman and Yurowitzky, 1973). It must be assumed that the rearrangement of the control mechanisms determining the direction and character of carbohydrate metabolism takes place at the final stages of oogenesis.

There is little experimental data on oocyte enzyme systems relating to carbohydrate metabolism and the special features of its control in different animals. Here we shall limit ourselves solely to the data obtained on teleost oocytes, namely to those of the loach (*Misgurnus fossilis* L.).

These data are of interest in understanding the control mechanisms, not only in oogenesis, but also at other stages of development at which changes in the metabolism occur. Changes in the correlation between glycolysis and gluconeogenesis and between the accumulation of glycogen and glycogenolysis during oogenesis (including the period of oocyte maturation) differ in principle from the control of similar processes in adult animals. If in the liver or kidney of an adult animal the control of carbohydrate metabolism is mainly determined by the physiological state of the organ (depending on the animal's diet, hormone level, etc.), the rearrangement of the metabolism at certain stages of oogenesis or embryogenesis is genetically programmed. Thus, the study of the rearrangement of metabolism in ontogenesis can shed light on the complex, multistep processes of the metabolism alterations in realizing the "program" of individual development.*

7.1. THE ENZYME SYSTEM OF CARBOHYDRATE METABOLISM IN THE OOCYTES OF THE LOACH

The enzyme system of carbohydrate metabolism in the oocyte includes several functional groups: the enzymes of glycogen synthesis (UDPG-pyrophosphorylase, glycogen synthetase), enzymes of the Embden-Meyerhof pathway (from hexokinase to lactate dehydrogenase, inclusive) (Table 7.1), enzymes of the pentose phosphate cycles or HMPS (glucose-6-phosphate dehydrogenase, 6-phosphogluconate dehydrogenase, transaldolase, and transketolase), the key enzymes in gluconeogenesis (PEP-carboxykinase and fructose diphosphatase), and some others.

Even the first encounter with the ratio of activity of the above enzymes shows that the oocytes of loach are almost devoid of phosphorylase activity even before the end of the period of vitellogenesis, i.e., they are devoid of the enzyme forming the hexosomonophosphates during the degradation of glycogen, but possess relatively high hexokinase activity. This activity is considerably greater than that of the other key enzymes in carbohydrate metabolism, for instance that of phosphofructokinase and glycogen synthetase. It may be suggested that the phosphorylation of glucose transferred to the oocyte from the female's body via hexokinase is the sole reaction supplying the oocyte with hexose phosphates (Fig. 41).

---

*In this chapter, the following abbreviations are used: UDPG, uridinediphosphate glucose; HMPS, hexose monophosphate shunt; HMP, hexose monophosphate; PEP, phosphoenolpyruvate; ATP, ADP, AMP, adenosinetri-, di-, and monophosphate; NAD, nicotinamide adenine dinucleotide; NADP, nicotinamide adenine dinucleotide phosphate; NADH and NADPH, reduced forms of these compounds; PDH, pyruvate dehydrogenase.

Table 7.1. Activity of Carbohydrate Metabolism Enzymes in th
(in U per $1 \cdot 10^4$ cells)

Enzyme	Stages of oogenesis (ooc	
	Onset of vacuolization of cytoplasm (200 µm)	Onset of vit genesis (400
	Glycogen metabolism	
Phosphorylase	traces	traces
α-amylase	-	-
Glycogen synthetase	0.046 ± 0.01	0.42 ± 0.0
UDPG-pyrophosphorylase	-	-
	Glycolysis	
Hexokinase	0.28 ± 0.07	1.8 ± 0.1
Phosphohexose isomerase	1.2 ± 0.1	13 ± 2.0
Phosphofructokinase	0.04 ± 0.005	0.96 ± 0.1
Aldolase	2.4 ± 0.2	16.0 ± 1.3
Triosephosphate dehydrogenase	6.0 ± 0.3	24.0 ± 1.0
Triosephosphate isomerase	3.0 ± 0.1	16.0 ± 0.9
Phosphoglycerate kinase	1.8 ± 0.1	7.0 ± 0.3
Enolase	2.0 ± 0.1	9.6 ± 0.5
Pyruvate kinase	6.6 ± 0.3	38.0 ± 2.0
Lactate dehydrogenase	6.4 ± 0.2	40.0 ± 2.0
	Gluconeogenesis	
Fructose diphosphatase	0.75 ± 0.04	4.0 ± 0.3
PEP-carboxykinase	0.2 ± 0.02	0.8 ± 0.0

Table 7.1 (continued)

Enzyme	Stages of oogenesis (oocyte diameter)			Unfertilized eggs
	Onset of vacuolization of cytoplasm (200 μm)	Onset of vitellogenesis (400 μm)	End of vitellogenesis (800 μm)	
	Pentose phosphate cycle			
Glucose-6-phosphate dehydrogenase	1.2 ± 0.1	3.6 ± 0.2	10.8 ± 0.6	10.8 ± 0.6
6-phosphogluconate dehydrogenase	0.08 ± 0.01	0.4 ± 0.02	3.4 ± 0.2	3.4 ± 0.2
Transketolase	0.04 ± 0.005	0.2 ± 0.02	1.4 ± 0.1	1.4 ± 0.1
Transaldolase	0.2 ± 0.02	1.2 ± 0.1	2.4 ± 0.1	2.4 ± 0.1
Protein (in mg per 10^4 cells)	10	40	1800	1800

Table 7.2. Enzyme Activity (mU per mg Protein) of Glycolysis in Loach Oocytes

Enzyme	Stages of oogenesis			
	Vacuolization of oocyte cytoplasm (200 µm)	Onset of vitellogenesis (400 µm)	End of vitellogenesis (800 µm)	Unfertilized egg
Phosphohexose isomerase	12	45	40	40
Phosphofructokinase	4	24	1.9	1.9
Aldolase	240	400	145	145
Triosephosphate dehydrogenase	600	600	220	220
Triosephosphate isomerase	300	400	134	134
Phosphoglycerate kinase	180	175	44	44
Enolase	200	240	39	39
Pyruvate kinase	660	950	27	60
Lactate dehydrogenase	640	990	22	120

The relatively high activity of the key enzymes of gluconeogenesis, which is comparable in intensity to that of the key enzymes of glycolysis (see Table 7.1), indicates the considerable part played by gluconeogenesis in oocyte metabolism.

The growth of the oocyte is accompanied by a considerable enhancement of the activity of all the enzymes of carbohydrate metabolism if the calculations are made according to the number of oocytes (see Table 7.1). However, if the calculation is made on the protein content (Table 7.2), the increase in enzyme activity is only observed up to the onset of vitellogenesis. Once vitellogenesis begins, the enzyme activity drops markedly. This reduction is explained by the fact that yolk granules which are intensively formed in the oocyte after the onset of vitellogenesis exhibit hardly any enzymatic activity, but constitute a considerable part of the oocyte's protein content. Hence, a calculation based on protein content is misleading. Similar phenomena have been described during investigations of enzyme induction and repression in liver (Weber et al., 1965a, b) and in adipose tissue (Pogson and Denton, 1967; Gumaa et al., 1969). Due to the changes in the protein and glycogen content in the liver, Weber's group based their calculations on the number of nuclei (DNA content),

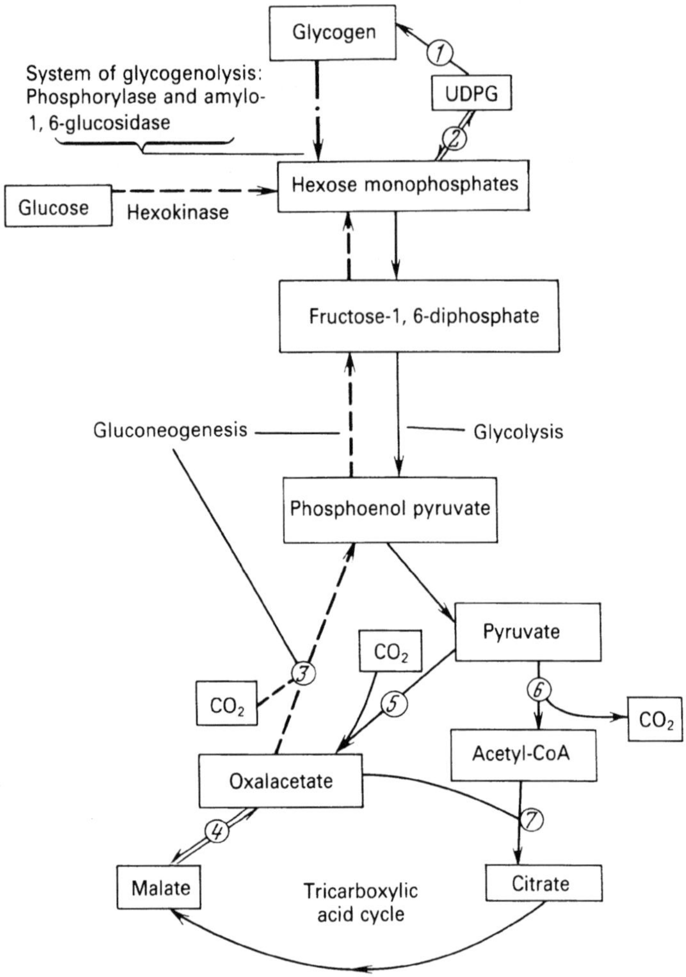

Fig. 41. Carbohydrate metabolism in the loach oocyte and embryo. Solid lines: processes taking place in the oocyte and embryos. Dotted lines: processes occurring in oogenesis, but ceasing with oocyte maturation. Dot-dash lines: processes functioning after the onset of embryonic development. 1) Glycogen synthetase; 2) UDPG-pyrophosphorylase; 3) PEP-carboxykinase; 4) malate dehydrogenase; 5) pyruvate carboxylase; 6) pyruvate dehydrogenase; 7) citrate synthetase. The usual ratio of hexose monophosphates: 70% glucose-6-phosphate, about 30% fructose-6-phosphates, and not more than 3-4% glucose-1-phosphate.

and MacLean's group (Gumaa et al., 1969) used the number of fat pads as the basis of their calculation. Table 7.1 shows that during oogenesis the activity of all the enzymes involved in carbohydrate metabolism is enhanced synchronously.

The principles governing the change in enzyme activity in the central part of the glycolytic chain from aldolase to enolase, inclusive, are extremely interesting. If the activity of triosephosphate dehydrogenase is taken as a unit relative to the activity of the enzymes in this group, it will then be found that during oogenesis an increase in the activity of the above enzymes takes place while a stable relationship between them is maintained. Thus, the activity of aldolase at all the stages of oogenesis studied is 0.4-0.6 of that of triosephosphate dehydrogenase, taken as a unit; the activity of triosephosphate isomerase is 0.5-0.6, the activity of phosphoglycerate kinase is 0.3-0.2, and the activity of enolase is 0.3-0.2 (Table 7.1; Fig. 42).

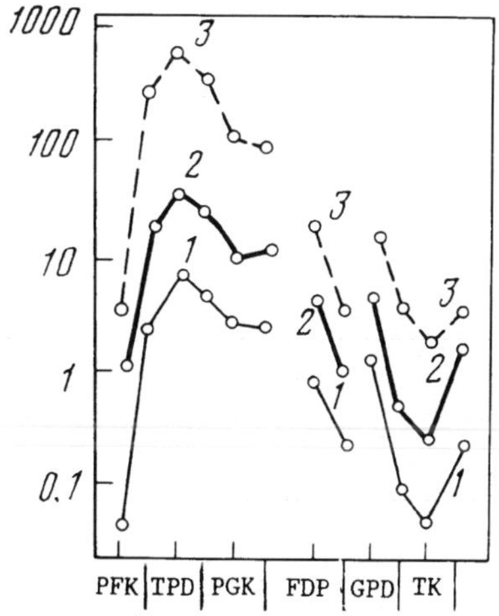

Fig. 42. Coordinate changes in the activity of enzymes of glycolysis, gluconeogenesis, and HMPS during oogenesis in the loach. The enzyme activities have been expressed after Pette (1962) and Brand (1970). Ordinate: activity of the enzymes, U/10^4 cells. Abscissa: enzymes. PFK) Phosphofructokinase; ALD) aldolase; TPD) triosephosphate dehydrogenase; TIM) triosephosphate isomerase; PGK) phosphoglycerate kinase; ENL) enolase; FDP) fructose diphosphatase; PCK) PEP carboxykinase; GPD) glucose-6-phosphate dehydrogenase; PGD) 6-phosphogluconate dehydrogenase; TK) transketolase; TA) transaldolase. 1) Onset of vacuolization of cytoplasm; 2) onset of vitellogenesis; 3) end of vitellogenesis and formation of unfertilized egg.

The retention of a persistent ratio between the activity of enzymes at different stages of oogenesis proves that the enzymes of a given functional group are synthesized in definite molar relationships, i.e., in a coordinated manner. The coordinated enhancement of activity during oogenesis has also been recorded for the key enzymes of gluconeogenesis and for the hexose monophosphate shunt. The relatively constant ratios of the enzyme activity of the above functional groups is depicted in Fig. 42, according to Pette et al. (1962) and Brand et al. (1970). It should be noted that coordinated changes in the activity of the enzymes of triosephosphate-phosphoglycerate metabolism, the hexose monophosphate shunt, and gluconeogenesis are also revealed when calculated per protein content (Yurowitzky and Milman, 1971).

It should moreover be specially stressed that we could not find coordinated changes in the activity of the key glycolytic enzymes. The ratio between phosphofructokinase and pyruvate kinase in oogenesis changes from stage to stage. Thus, at the stage of the oocyte cytoplasm vacuolization, this ratio is 0.006, before the onset of vitellogenesis it is 0.25, and at the end of vitellogenesis it is 0.09. The magnitude of the ratio of hexokinase and phosphofructokinase activity also increases from stage to stage (Table 7.1).

The most likely explanation lies in the fact that the enzymes having isozymes with different kinetic properties (hexokinase, pyruvate kinase, lactate dehydrogenase) do not reveal distinct coordinated changes in the activity due to possible changes in their isozyme pattern. This problem needs further study (Milman and Yurowitzky, 1973).

7.2. GLYCOLYSIS AND GLUCONEOGENESIS IN OOCYTES

As we have already pointed out, a combination of two processes acting in opposite directions, glycolysis and gluconeogenesis, occurs in oocytes. The first process, glycolysis, takes place in every tissue and organ of all animals. The second process, gluconeogenesis (i.e., the formation of carbohydrates from noncarbohydrate precursors), is known to occur in only a limited number of tissues, namely liver, renal cortex, and adipose tissue.

The development of the oocyte and the considerable increase in the enzyme activity of carbohydrate metabolism is accompanied by an enhanced rate of glycolysis, gluconeogenesis, respiration, and apparently by a number of other processes throughout oogenesis. The rate of anaerobic glycolysis, using endogenous substrates in isolated loach oocytes, increased 17 times from the onset of vacuolization of the cytoplasm (oocyte diameter 200-250

Table 7.3. Rate of Anaerobic Glycolysis in Loach Oocytes on Endogenous Substrates

Stage of oogenesis	Oocyte diameter	Formation of lactate, μmoles/10 min		Protein content, μg per oocyte
		per 1000 oocytes	per mg	
Onset of cytoplasm vacuolization	200 μm	0.0014	0.0014	1.0
Onset of vitellogenesis	400 μm	0.024	0.006	4.0
End of vitellogenesis	800 μm	0.15	0.001	180
Unfertilized egg		0.10	0.001	180

μm) up to the stage of vitellogenesis (oocyte diameter 400 μm), and yet another 6 times during the period of vitellogenesis (Yurowitzky et al., 1972). In other words, throughout the period of oogenesis studied, the rate of glycolysis is increased by two orders of magnitude (Table 7.3).

In calculating the rate per mg of oocyte protein, the glycolysis rate prior to the onset of vitellogenesis is enhanced, but it drops during vitellogenesis. As pointed out above, this decrease depends on the accumulation of yolk platelets with a high protein content but hardly any enzymatic activity.

The presence of aerobic glycolysis in isolated oocytes is of some interest. Aerobic glycolysis is only observed in oocytes when they retain a layer of follicular cells, i.e., when isolated in the form of follicles. The rate of aerobic lactate formation on endogenous substrate reaches 0.1 μmoles lactate per 10^3 oocytes per min. However, oocytes themselves are incapable of aerobic glycolysis. Removal of the follicular cells from the oocyte's surface leads to the absence of aerobic glycolysis, while the anaerobic process proceeds at a constant rate (Yurowitzky et al., 1972).

It should be noted in this connection that embryonic tissues are, as a rule, totally devoid of aerobic glycolysis (Warburg, 1956). Aerobic glycolysis of the layer of follicular cells (which are not embryonic tissue) on the surfaces of the isolated oocytes is, in our opinion, caused by the damage that occurs to them during the emergence of the oocytes from the ovary. The layer of follicular cells represents no more than 1-3% of the follicle's mass, and accounts for 10-15% of the total enzymatic activity.

Phosphofructokinase is the pacemaker enzyme of glycolysis. It possesses the lowest activity among the glycolysis enzymes, and the change in the activity of phosphofructokinase correlates very well with the increase in the fructose-1,6-diphosphate level and the glycolysis rate at all stages of oogenesis.

Localization of the control steps in glycolysis may change if the oocytes are incubated with the substrates or the intermediates of glycolysis. When the oocytes are incubated with glucose, their glucose content is markedly enhanced, the glycolysis rate is increased, but there is an insignificant rise in the hexose-monophosphate content. The latter points to the limiting role of hexokinase when there is an excessive glucose influx. The content of fructose-1,6-diphosphate increases no more than 25% at the maximum glucose level, thus showing that phosphofructokinase is the second limiting reaction in the process under these conditions. The level of the 3-C intermediates of glycolysis alters slightly. The ATP level decreases somewhat with the corresponding rise in AMP, while the ADP content remains unchanged. When the oocytes are incubated in a medium to which glucose-1-phosphate has been added, the content of all the intermediates in the glycolytic chain is increased; with the addition of $1\text{-}2 \cdot 10^{-2}$ M glucose-1-phosphate the glycolysis rate is limited by the rate of glycolytic oxidoreduction (Yurowitzky et al., 1972).

The respiration of loach oocytes during the period studied is also considerably enhanced. According to Ozernyuk (1970), oxygen utilization increased tenfold during vitellogenesis, and at least 15 times from the onset of vacuolization of the cytoplasm and until the end of vitellogenesis.

The ratio between the rates of glycolysis and gluconeogenesis in the oocytes depends on the stage of oogenesis. Thus, at the stage of the onset of vitellogenesis the rate of glycogen formation from alanine is two times lower than the rate of glucose transformation into lactate. After the completion of vitellogenesis, the rate of gluconeogenesis is roughly five times less than that of glycolysis (Table 7.4).

The relatively high rate of gluconeogenesis in comparison to glycolysis is also reflected in the other factors controlling the rate of enzymatic processes, namely: the ratio of (ATP)/(ADP + AMP), the value of the phosphorylated state of the adenylic system in the cytoplasm [$ATP/ADP(HPO_4^{-2})$], the ratio $(NAD^+)/(NADH)$ in the cytoplasm and mitochondria, and the ratio of the intermediates in the glycolytic chain. To illustrate this condition, let us compare the content of the basic intermediates of glycolysis, adenine nucleotides, and the nicotinamide adenine dinucleotide couples in loach oocytes at different stages, where gluconeogenesis is negligible (Table 7.5).

In oocytes at all of the stages studied it was noted that the

Table 7.4. Rate of Glycogen Formation (μmoles of glucose/1000 oocytes/10 min) from Different Precursors in the Isolated Loach Oocytes

Process	Onset of vitellogenesis (oocyte diameter 400 μm)	End of vitellogenesis (oocyte diameter 800 μm)
Glycogen formation from alanine (gluconeogenesis without glycogenesis)	0.005 ± 0.001	0.012 ± 0.001
Glycogen formation from glucose (glycogenesis)	0.005 ± 0.001	0.0027 ± 0.001
Anaerobic glycolysis in μmoles of used glucose	0.012	0.07

Note. The ratio between glycolysis and gluconeogenesis has been measured at equal concentrations of the precursors (during incubation of the oocytes with $1\cdot10^{-3}$ M alanine or glucose). These concentrations did not alter the content of the intermediates in the oocytes and were consequently close to the physiological values (Milman and Yurowitzky, 1973).

content of fructose-1,6-diphosphate exceeds that of hexosemonophosphates. This ratio is typical of tissues with a relatively high gluconeogenesis; for instance, it was noted in kidney perfused with gluconeogenic precursors (Weidemann et al., 1969), in liver (Williamson, 1967), and in platelets (Detwiler, 1969).

In tissues without gluconeogenesis (skeletal and cardiac muscles, tumor cells, loach embryos) the level of hexose monophosphate (in particular fructose-6-phosphate) always exceeds that of fructose-1,6-diphosphate. The excess of fructose-6-phosphate over fructose-1,6-diphosphate indicates the presence of the limiting step of glycolysis at the phosphofructokinase level (Hess and Brand, 1965; Williamson, 1965; Wu, 1965; Underwood and Newsholme, 1965).

The relatively intensive gluconeogenesis in the oocytes is ensured by a high ratio of (ATP)/(ADP + AMP). Table 7.5 shows that at all stages of oogenesis studied this ratio is twice as high as in the mature egg with minimum gluconeogenesis. Gluconeogenesis in the oocytes essentially differs from the analogous process in the liver of mammals. In the liver, gluconeogenesis supports the required physiological level of glucose in the blood, but no synthesis of glycogen takes place. In loach oocytes, on the contrary, the free glucose in gluconeogenesis is not formed at all due to the absence of glucose-6-phosphatase, while the hexose monophosphates are transformed into glycogen through uridine-diphosphate glucose (Yurowitzky and Milman, 1973a; see also Fig. 41). The limiting steps of gluconeogenesis are localized: 1)

Table 7.5. Content of Certain Metabolites (in nmoles per 100 oocytes) in Isolated Loach Oocytes

Metabolite	Stages of oogenesis (oocyte diameter)			Unfertilized egg
	Onset of vacuolization of the oocyte (200 μm)	Onset of vitellogenesis (400 μm)	End of vitellogenesis (800 μm)	
Glycogen, mg	-	0.025 ± 0.007	0.95 ± 0.07	1.04 ± 0.07
Glucose	0.8 ± 0.08	3.7 ± 0.3	18.5 ± 3.6	absent
UDP-glucose	-	-	2.2 ± 0.4	4.5 ± 0.6
Hexose monophosphates total	0.07 ± 0.2	0.4 ± 0.07	2.4 ± 0.2	12.0 ± 1.3
Glucose-6-phosphate	0.05	0.25	1.6	8.0
Fructose-6-phosphate	0.02	0.15	0.8	4.0
Fructose-1,6-diphosphate including triosephosphates	0.12 ± 0.01	0.8 ± 0.2	2.9 ± 0.4	2.1 ± 0.3
PEP	-	1.2 ± 0.05	3.0 ± 0.4	5.0 ± 0.4
Pyruvate	-	0.2 ± 0.03	1.5 ± 0.007	3.0 ± 0.3
Isocitrate	0.3 ± 0.02	0.9 ± 0.06	2.2 ± 0.1	2.0 ± 0.1
Citrate	6.0	18.0	44.0	40.0
Malate	-	-	4.0 ± 0.3	30.0 ± 1.5
Oxalacetate	-	-	2	8
Acetyl-CoA	-	-	4.5 ± 0.6	4.5 ± 0.6
ATP	-	10.0 ± 0.35	70.0 ± 1.5	100.0 ± 4.0
ADP	-	1.5 ± 0.1	11.0 ± 0.3	22.0 ± 0.5
AMP	-	1.2 ± 0.1	7.0 ± 0.2	30.2 ± 0.6
NAD^+	-	-	5.5 ± 0.3	9.5 ± 0.5
NAD-H	-	-	1.8 ± 0.1	9.4 ± 0.5
$NADP^+$	-	-	0.9 ± 0.05	1.2 ± 0.06
NADP-H	-	-	0.5 ± 0.05	1.7 ± 0.05
Inorganic phosphate	-	-	605 ± 39	1240 ± 100
(ATP)/(ADP+AMP)	-	3.8	4.0	2.0
(NAD^+/NADH) total	-	-	3.0	1.0

between the pyruvate and PEP; 2) between 3-phosphoglycerate and the phosphotrioses; and 3) between fructose-1,6-diphosphate and fructose-6-phosphate (Yurowitzky and Milman, 1973a).

7.3. GLUCONEOGENESIS AND GLYCOGENESIS IN OOCYTES

Hexose monophosphates, the precursors of glycogen, can be formed not only from noncarbohydrate precursors, but also directly from glucose, through the hexokinase reaction. This pathway of glycogen synthesis is usually called "glycogenesis." What is the relationship between these two pathways of glycogen accumulation in oocytes?

Our observations have shown (Table 7.4) that at the stage of the onset of vitellogenesis the rates of these two processes are approximately the same, and toward the end of vitellogenesis the rate of glycogenesis exceeds the rate of glycogen synthesis from amino acids by more than two times (Yurowitzky and Milman, 1973a).

It is interesting to note that in physiological conditions the rate of glycogen accumulation in oogenesis of loach is very low. This is obvious from the following calculation. The glycogen content in the oocytes prior to vitellogenesis does not exceed 0.25 mg per 10^3 oocytes and, toward the end of vitellogenesis, it is 9.5 mg. However, the accumulation of 9.5 mg glycogen (calculated for 10^3 oocytes) takes no less than 2.5-3 months. In other words, the rate of glycogen accumulation in loach oocytes in vivo is 10 nmoles of glucose equivalents per hour per 10^3 oocytes, or slightly more than 0.16 nmoles per minute.

7.4. SPECIAL FEATURES OF GLYCOGEN SYNTHETASE IN LOACH OOCYTES

As a rule, the rate of glycogen synthesis in animals is directly proportional to the activity of glycogen synthetase (Fig. 41). Control of the enzyme activity is usually exercised through conversion of the inactive form of glycogen synthetase into the active form, and vice versa. The inactive form of the enzyme, called the dependent (D) form, is active catalytically only when there is an excess of glucose 6-phosphate. The dephosphorylation of the enzyme by specific glycogen synthetase phosphatase transforms it into the active, or independent (I), form, glycogen synthetase a (or I form). The catalytic activity of the I form is almost independent of the presence of glucose 6-phosphate and is slightly inhibited by ATP. The reverse conversion of the I form into the D form is ensured by glycogen synthetase kinase.

Table 7.6. Changes in the Enzyme Activity of Glycogen and Hexose Metabolism during Oocyte Maturation (in U/10^4 oocytes)

Enzyme	Time after hormone injection into loach females (h)					Unfertilized egg
	0	16-18	24-25	29-30	34-36	
Hexokinase	4.2±0.3	1.0±0.1	traces	traces	traces	absent
Glycogen synthetase	5.6±0.6	3.2±0.5	3.4±0.4	2.6±0.5	2.6±0.4	2.5±0.5
Phosphorylase with excess AMP	1.0±0.6	2.2±0.2	-	8.2±0.4	10.0±0.5	10.0±0.5
Alpha-amylase	2.4±0.5	-	0.5±0.05	traces	traces	absent
Amylo-1,6-glucosidase	0.026	-	-	-	-	0.050±0.001
"Branching" enzyme (in arbitrary units)	8	-	-	-	-	32

It is interesting to note that the glycogen synthetase of loach only exists in the D form. Neither the oocytes nor the loach embryos or adult fish possess the conversion system of I ⇌ D forms. Hence, the period of most intensive accumulation of glycogen in the oocytes, vitellogenesis, is accompanied by a marked increase in the enzyme activity of more than one order of magnitude (see Table 7.3), but not by its activation. The glycogen content during this period increases by 45 times (Yurowitzky and Milman, 1973a, 1974).

The cessation of glycogen accumulation after the onset of oocyte maturation is correlated with the decrease in enzyme activity, but not with its conversion into the inactive form. During the period of maturation the activity of glycogen synthetase decreases twofold (Table 7.6); this decrease continues in early embryogenesis (Milman and Yurowitzky, 1973).

7.5. OOCYTE MATURATION AND CHANGES IN CARBOHYDRATE METABOLISM

After vitellogenesis ends, oocytes, mature unfertilized eggs, and embryos at the stage of cleavage have an identical enzyme profile of the Embden–Meyerhof pathway, of the pentose phosphate cycle and gluconeogenesis (see Table 7.1), but essentially are different in the activity of the metabolic enzymes of glycogen and glucose.

Oocytes are practically devoid of phosphorylase activity (Tables 7.1 and 7.2) and hence of glycogenolysis. The influx of hexose monophosphates into the carbohydrate metabolism of the oocyte is ensured by the incoming glucose and its phosphorylation during the hexokinase reaction (Fig. 41). Oocytes possess certain α-amylase activity. Gluconeogenesis plays a considerable part in glycogen accumulation. There is no glycogen accumulation in the developing embryo, and glyconeogenesis is the only process ensuring the generation of hexose monophosphates (Fig. 41). The oocyte and the egg differ essentially in the content and ratio of the intermediates of carbohydrate metabolism and adenylic nucleotides (Table 7.7; see Table 7.5). All this points directly to considerable changes in the carbohydrate metabolism during oocyte maturation.

The maturation of oocytes and spawning in fish are stimulated by the luteinizing hormone of the hypophysis. In laboratory conditions, chorionic gonadotropin can serve as an analog of the luteinizing hormones, and its administration to the female loach leads to the oocyte maturation within 34-36 h (at 17-19°C).

Sampling of oocytes during their maturation allows the changes in carbohydrate metabolism to be traced.

Sixteen to eighteen hours after administration of gonadotropin the activity of hexokinase decreases fourfold; at the same time, the influx of glucose into the oocyte decreases, and its content decreases as well (Tables 7.6 and 7.7). After about 29-30 h, only one-fifth of the initial glucose content of the oocyte remains. There is no glucose or hexokinase in the embryo at all. The activity of the glycogen synthetase, mentioned in the previous section, decreases synchronously. During the maturation process the activity of α-amylase also ceases (Table 7.6).

The activity of phosphorylase, on the other hand, doubles 16-18 h after the administration of the hormone: in 30 h it is eight times higher and, at the moment of ovulation, has increased tenfold. The marked increase in phosphorylase activity is not related to the activation of the latent form of this enzyme (Milman and Yurowitzky, 1973). This allows us to suggest that the enhancement of the phosphorylase activity is related to its synthesis de novo.

It is important to note that the basic changes in the activity of the glycogen and glucose enzymes is recorded 29-30 h after administration of the hormone, i.e., after the breakdown of the germinal vesicle (Belyaeva and Cherfas, 1965). Immediately before ovulation, when the oocyte is at metaphase I meiotic division, the activity of these enzymes hardly differs from that in the enzymes of the mature egg.

Table 7.7. Changes in the Metabolite Level (nmoles/100 oocytes) in Loach Oocytes During Maturation

Metabolite	Time after choriogonin injection into female (h)					Unfertilized egg
	0	16-18	24-25	29-30	34-36	
Glucose	18.2±3.6	12.0±1.9	10.0±1.5	4.5±0.3	traces	absent
Hexose monophosphate	2.4±0.2	3.7±0.2	6.4±0.4	8.3±0	10.0±0.6	12.3±1.0
including fructose-6-P	0.8	1.2	2.1	-	-	4.1
Fructose-1,6-DP	2.9±0.4	3.0±0.4	2.5±0.3	1.9±0.3	2.1±0.3	2.1±0.3
PEP	3.0±0.3	4.0±0.3	4.1±0.3	5.0±0.4	5.1±0.4	5.0±0.4
UDPG	2.2±0.3	-	-	-	-	4.5±0.6
Citrate	40.0±1.0	-	-	40.0±1.0	-	40.0±1.0
Isocitrate	2.2±0.2	-	-	2.2±0.3	-	2.2±0.1
Inorganic phosphate	605±39	-	-	-	-	1240±100
Malate	4.0±0.3	-	-	-	-	30±1.5
Oxalacetate	2.0	-	-	-	-	8.0
Acetyl-CoA	4.5±0.6	-	-	-	-	4.5±0.6
ATP	70.0±1.5	78.0±2.6	80.0±3.2	85.0±3.7	100±4.0	100±4.0
ADP	10.0±0.3	12.0±0.3	14.0±0.4	16.0±0.7	18.0±0.7	20.0±0.5
AMP	7.0±0.2	12.3±0.3	14.0±0.3	19.0±0.6	25.5±0.8	30.0±0.6
(ATP)/(ADP + AMP) total	4.0	-	2.8	-	2.3	2.0
(ATP)/(ADP)(HPO_4^{-2}) cytoplasm	1400	-	-	-	-	485
(NAD^+)/(NADH) cytoplasm	1870	-	-	-	-	940
(NAD^+)/(NADH) mitochondria	62	-	-	-	-	395
($NADP^+$)/(NADPH) cytoplasm	1.8	-	-	-	-	0.7

The process of maturation is also accompanied by a marked change in the ratio of (ATP)/(ADP + AMP) and the magnitude of the phosphorylated state of the adenylic system in the cytoplasm. The ratio of fructose 6-phosphate/fructose-1,6-diphosphate increases from 0.2 to 7, which is typical of the tissues with predominating glycolysis. Arrest or a marked decrease in glyconeogenesis is accompanied by a threefold decrease in the index of the phosphorylating state of the adenylic system (Table 7.7). The cytoplasmic ratio (NAD^+)/(NADH) in the mature egg and in the embryo ensures the direction of the flow in the glycolytic chain toward the formation of pyruvate. Hence, it is twofold lower in the embryo than in the oocyte (Table 7.7). Corresponding to this, the mitochondrial ratio (NAD^+)/(NADH) increases sixfold during the maturation of the oocyte; this reflects the decrease in generation of NADH in the cytoplasm as gluconeogenesis stops, and the rate of transport of the reducing equivalents of NADH from the cytoplasm into mitochondria increases (Yermolaeva, 1976).

Thus, oocyte maturation leads to changes in the enzymatic apparatus of glycogen metabolism and, in general, of carbohydrate metabolism in a direction favoring the realization of glycogenolysis and the cessation of gluconeogenesis and the accumulation of glycogen.

7.6. SATELLITE ENZYMES OF GLUCONEOGENESIS DURING OOCYTE MATURATION

Control of the first gluconeogenesis reaction, PEP formation from oxalacetate, is exercised not only by PEP-carboxykinase activity, but by the oxalacetate content in the tissues and its regeneration rate. The level of this metabolite is maintained by a whole group of so-called satellite enzymes of gluconeogenesis (pyruvate carboxylase, malate dehydrogenase, citrate synthetase, etc.), which link gluconeogenesis with the tricarboxylic acid cycle (Fig. 41).

One of the main conditions of gluconeogenesis, according to many authors (Illyin and Usatenko, 1965; Williamson et al., 1968, 1969), is the limitation of the rate of oxalacetate and acetyl-CoA conversion into citrate, i.e., the direction of oxalacetate mainly toward PEP synthesis (Fig. 41). Hence, the enhancement of the rate of gluconeogenesis in animal liver is, as a rule, accompanied by an increase in the oxalacetate and acetyl-CoA content. Acetyl-CoA can be formed in the cell in two main ways: from pyruvate during its decarboxylation through pyruvate dehydrogenase (PDH) and through oxidation of fatty acids. Thus, the increased rate of gluconeogenesis in the liver and in some other tissues does, as a rule, correspond to the enhanced activity of pyruvate carboxylase (the enzyme that transforms pyruvate into oxalacetate) and to inactivation of PDH. With the enhancement of gluconeogenesis the active, dephosphorylated form of PDH is transformed by the corresponding kinase into its inactive phosphorylated form. This decreases the rate of pyruvate transformation into acetyl-CoA, but the level of acetyl-CoA increases at the expense of a more intense splitting of the fatty acids (Linn et al., 1969; Weiland et al., 1971, 1972).

The decrease in gluconeogenesis is naturally accompanied by the reverse phenomena: activation of PDH, decreased activity of pyruvate carboxylase and of the key enzymes involved in gluconeogenesis. Cessation of gluconeogenesis during oocyte maturation is likewise accompanied by some drastic changes in metabolism. The changes in activity of the satellite enzymes are limited to the increase in PDH activity from 60 to 80 mU per 10^4 oocytes. However, as for glycogen synthetase, neither in the oocytes nor in the loach embryos has the reversible conversion of PDH been found; the enzyme has only been found in its active form (Yermolaeva and Milman, 1975). Cessation of gluconeogenesis is accompanied by a marked (more than fourfold) increase in the malate and oxalacetate content (Tables 7.5 and 7.6), but a stable level of acetyl-CoA, citrate, and isocitrate is retained.

The activity of pyruvate carboxylase, malate dehydrogenase, and citrate synthetase remains unchanged; both in the oocytes (at the end of vitellogenesis) and in the embryos it is equal to 1.9, 500, and 4.7 U per 10^4 oocytes (or embryos), respectively.

The oocytes are almost devoid of citrate liase activity, i.e., the enzyme splitting citrate in the cytoplasm into oxalacetate and acetyl-CoA. It is possible that the level of citrate and acetyl-CoA is retained and remains stable during oocyte maturation. Thus, control of this metabolic step is only exercised through the increased activity of PDH, an enzyme possessing the lowest activity among the enzymes of this functional group.

7.7. OOCYTE MATURATION AND CONTROL OF GLYCOGENOLYSIS IN THE MATURE OOCYTE AND EMBRYO

The change in metabolism during oocyte maturation plays a very important part in the embryo energetics. Oocyte maturation stops the utilization of glucose, leads to the disappearance of amylase and hexokinase, and to the onset of glycogenolysis. The latter is the only process of hexose monophosphate formation in the embryo. However, control of the rate of glycogenolysis in embryogenesis, ensuring a gradual increase in the rate of generation of hexose monophosphates, differs in principle from the mechanism of activation of glycogenolysis in adult tissues.

At first sight, the tenfold increase in phosphorylase activity during oocyte maturation (Table 7.6) creates favorable conditions for unlimited glycogenolysis. It is easy to calculate that if the rate of glycogenolysis in vivo is only 5% of the phosphorylase activity, then the glycogen content in the embryo would have been exhausted within a few hours. In actual fact, there is still a store of glycogen in the embryo even after embryonic development is completed right up to the beginning of active nutrition of the larva. According to some data, glycogen consumption throughout embryonic development depletes only a small part of the total glycogen stores (Hutchens et al., 1942; Faustov and Zotin, 1965; Løvtrup, 1966). Thus, some additional mechanism of gradual activation of glycogenolysis (i.e., a gradual increase in the rate of hexose monophosphate generation) must exist in the embryo which, at the same time, uses glycogen very sparingly.

As a rule, in the tissues of all vertebrates the activation of glycogenolysis is conditioned by the activation of phosphorylase, i.e., by the conversion of the "inactive" form of phosphorylase into the "active" one. But at the early stage of development of teleosts (Milman and Yurowitzky, 1969), amphibians (Guessak and Hedrik, 1972), and echinoderms (Bergami et al., 1969), the ratio between the forms of phosphorylase is constant and no changes in phosphorylase activity have been recorded. In this connection, we have suggested that in early embryogenesis a gradual increase in the intracellular activity of phosphorylase takes place.

This increase could have been caused by a gradual decrease in the concentration of metabolite inhibitors (or by an increase in the concentration of metabolite activators). However, as our studies have shown, the changes in the concentration of metabolites in the developing embryo do not favor an increase in the intracellular activity of the phosphorylase (Yurowitzky and Milman, 1973a-c) and this suggestion, therefore, is proved groundless.

We believe that the gradual activation of glycogenolysis in the embryo is ensured in at least two ways. One of them consists in the control of the rate of glycogen phosphorolysis through the "debranching" enzyme amylo-1,6-glucosidase. It is known that glycogen is a "branching" polysaccharide in which the glucose residues are bonded into a polymer by α-1,4-glucoside bonds, and a certain number of lateral branches are bound to the main chain by α-1,6 bonds. Phosphorylase or α-1,4-glucan:orthophosphate glycosyl transferase catalyzes the splitting only of the α-1,4 bonds in the glycogen molecule with the formation of glucose-1-phosphate. At the same time, phosphorylase cannot break the α-1,6-glucoside bonds at the point of glycogen branching and also the α-1,4 bonds

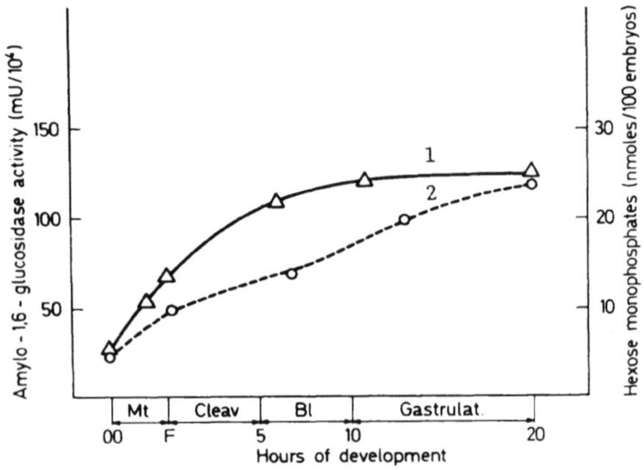

Fig. 43. Relationship between the level of hexose monophosphates (1) and the activity of amylo-1,6-glucosidase (2) after the onset of embryonic development of the loach (at 21°C). Ordinate on the left: activity of amylo-1,6-glucosidase (in mU per 10^4 oocytes or embryos). Ordinate on the right: hexose monophosphate content (in nmoles per 100 oocytes or embryos).

close to the branching points (Cori and Larner, 1951; Illingworth et al., 1956). Hence, interrupted phosphorolysis of glycogen in addition to phosphorylase action also includes the action of enzymes of the so-called "debranching" system. Under physiological conditions the reaction produced by the "debranching" enzyme amylo-1,6-glucosidase generates a certain quantity of free glucose equal to the number of split α-1,6 bonds according to the equation: polysaccharide with an excess of α-1,6 bonds + H_2O → glycogen + glucose.

Our determinations have shown that the activity of amylo-1,6-glucosidase is a limiting factor of glycogenolysis; the activity of this enzyme is gradually enhanced at the early stages of development and the growth of its activity from 50 to 120 mU/10^4 oocytes (or embryos) correlates distinctly with the increase in the hexose monophosphate level (Fig. 43; Table 7.6).

The absence of free glucose in the loach embryo (Table 7.7) also points indirectly to the very low activity of the debranching enzyme, which confirms the viewpoint expressed. It is also important to state that the relatively high activity of the "branching" enzyme, which increases more than twice with the maturation of the oocyte (Table 7.6), apparently ensures the maintenance of the highly branching structure of glycogen in the developing embryo (Yurowitzky and Milman, 1975).

This control mechanism hardly ever functions in the adult tissues as amylo-1,6-glucosidase usually does not limit the rate of glycogenolysis. Only special cases of genetic diseases in man are an exception to this rule (Illingworth et al., 1956).

Another mechanism for controlling the limited rate of glycogenolysis is the transfer of the glycogen–phosphorylase complex from the yolk into the cytoplasm. This process begins at the stage of 32 blastomeres and ends before the beginning of gastrulation (Yurowitzky and Milman, 1973b). Through this transfer the glycogen content in the blastoderm increases almost twofold (without any change in the glycogen content in the intact embryo) and the phosphorylase activity rises from 0.35 to 1.5 U/10^4 embryos. In the activated unfertilized eggs no such process of redistribution of glycogen and phosphorylase takes place. Blastodiscs isolated from unfertilized activated eggs contain the initial amount of glycogen (i.e., 36 mg per 10^4 blastodiscs) and 0.35 U of phosphorylase. These figures remain unchanged until the death of the egg (Yurowitzky and Milman, 1973b).

The biological role of this transfer process, we believe, consists in the fact that the glycogen and phosphorylase are transported to the site of synthesis of amylo-1,6-glucosidase or the debranching enzyme, i.e., from the yolk into the blastoderm.

In other words, the gradual activation of glycogenolysis after the onset of embryogenesis is a complex process which has, on the one hand, been prepared during oocyte maturation and is, on the other hand, initiated by the special trigger mechanisms of embryonic development (Yurowitzky and Milman, 1972, 1973b, 1975).

REFERENCES

Barth, L. D., and Barth, L. J., 1954, The Energetics of Development, Columbia Univ. Press, New York.

Belyaeva, V. P., and Cherfas, N. B., 1965, "On the processes of maturation and fertilization in the eggs of the loach *Misgurnus fossilis* L.," Vopr. Ikhtiol., *5*, 82-90.

Bergami, J., Mansour, T. E., and Scarano, E., 1969, "Properties of glycogen phosphorylase before and after fertilization of sea-urchin eggs," Exp. Cell Res., *49*, 650-655.

Brand, K., Deckner, K., and Musil, J., 1970, "Enzyme pattern of the pentose phosphate pathway in ascites tumor cells," Z. Physiol. Chem., *351*, 213-218.

Cori, G. T., and Larner, J., 1951, "Action of amylo-1,6-glucosidase and phosphorylase on glycogen," J. Biol. Chem., *188*, 17-37.

Detwiler, T. C., 1969, "Levels of intermediates of the glycolytic pathway in rat platelets," Biochim. Biophys. Acta, *177*, 161-164.

Faustov, V. S., and Zotin, A. I., 1965, "Combustion temperatures of the eggs of teleosts and amphibia in embryogenesis," Dokl. Akad. Nauk SSSR, *162*, 965-968.

Guessak, D. J., and Hedrick, J. L., 1972, "Enzymatic characteristics and the control of glycogen phosphorylase during early amphibian development," J. Biol. Chem., *247*, 6603-6609.

Gumaa, K. A., Novello, F., and MacLean, P., 1969, "Pentose phosphate pathway of glucose metabolism in rat adipose tissue," Biochem. J., *114*, 253-264.

Hess, B., and Brand, K., 1965, "Enzyme and metabolic profiles," in: Control of Energy Metabolism, B. Chance and R. W. Estabrook (eds.), Academic Press, New York, pp. 111-123.

Hutchens, J. O., Keltch, A. K., and Krahl, M. E., 1942, "Observations on the glycogen content, carbohydrate consumption, and lactic acid production in sea-urchin embryos," J. Gen. Physiol., *25*, 717-731.

Illingworth, B., Cori, G. T., and Cori, C. F., 1956, "Amylo-1,6-glucosidase of muscle tissue in generalized glycogen shortage disease," J. Biol. Chem., *218*, 123-130.

Illyin, V. S., and Usatenko, M. S., 1965, "The synthesis of phosphoenol-pyruvate, its control and significance in gluconeogenesis," in: Advances in Biological Chemistry, Vol. 7 [in Russian], Nauka, Moscow, pp. 196-205.

Linn, J. C., Petit, F., and Reed, L. J., 1969, "Regulation of the activity of pyruvate dehydrogenase complex from beef kidney mitochondria," Proc. Natl. Acad. Sci. USA, *62*, 234-241.

Løvtrup, S., 1966, "The chemical basis of sea-urchin embryogenesis," Bull. Schweiz. Akad. Med. Wiss., *22*, 201-276.

Milman, L. S., and Yurowitzky, Y. G., 1969, "Phosphorylase of the developing loach embryo," in: Enzymes in Animal Evolution, Vol. 1 [in Russian], Nauka, Leningrad, pp. 126-131.

Milman, L. S., and Yurowitzky, Y. G., 1973, Regulation of Glycolysis in Fish Embryos, S. Karger, Basel.

Ozernyuk, N. D., 1970, "Intensity of respiration and ATP content in loach oogenesis," Dokl. Akad. Nauk SSSR, *192*, 242-245.

Pette, D., Luh, W., and Bucher, T., 1962, "A constant-proportion group in the enzyme activity pattern of the Embden-Meyerhof chain," Biochem. Biophys. Res. Commun., *7*, 419-422.

Pogson, G. I., and Denton, R. M., 1967, "Effect of alloxan diabetes and starvation on glycolytic kinase activities in rat adipose tissue," Nature, *216*, 156-158.

Underwood, A. H., and Newsholme, E. A., 1965, "Some properties of fructose diphosphatase of rat liver and their relation to the control of gluconeogenesis," Biochem. J., *95*, 767-780.

Warburg, O., 1956, "On the origin of cancer cells," Science, *123*, 309-312.

Weber, G., Singhai, R. L., Stamm, N. B., and Srivastava, S. K.,

1965a, "Hormonal induction and suppression of liver enzyme biosynthesis," Fed. Proc., *24*, 746-752.

Weber, G., Singhai, R. L., Stamm, J., and Srivastava, S. K., 1965b, "Insulin-suppressor of biosynthesis of hepatic gluconeogenic enzymes," Proc. Natl. Acad. Sci. USA, *53*, 96-101.

Weidemann, M. S., Hems, D. A., and Krebs, H. A., 1969, "Effects of added adenine nucleotides on renal carbohydrate metabolism," Nephron, *6*, 282-296.

Weiland, O., Siess, E., and Schulze-Wethmar, F. H., 1971, "Active and inactive forms of pyruvate dehydrogenase in rat heart and kidney: effect on diabetes, fasting, and refeeding on pyruvate dehydrogenase interconversion," Arch. Biochem. Biophys., *143*, 593-600.

Weiland, O., Patzelt, Ch., and Loffler, H., 1972, "Active and inactive forms of pyruvate dehydrogenase in rat liver," Eur. J. Biochem., *26*, 426-433.

Williamson, J. R., 1965, "Glycolytic control mechanisms," J. Biol. Chem., *240*, 2308-2316.

Williamson, J. R., 1967, "Effect of fatty acids, glucogen, and anti-insulin serum on the control of gluconeogenesis in rat liver," Adv. Enzyme Regul., *5*, 229-254.

Williamson, J. R., Browning, E. T., and Olson, M. S., 1968, "Inhibition of fatty acid stimulation of gluconeogenesis by decanoylcarnitine in perfused rat liver," Diabetes, *17*, 194-208.

Williamson, J. R., Browning, E. T., and Scholz, R., 1969, "Control mechanisms of gluconeogenesis and ketogenesis," J. Biol. Chem., *244*, 4607-4616.

Wu, R., 1965, "The role of P_i in the control of glycolysis in ascites tumor cells," Biochem. Biophys. Res. Commun., *18*, 402-407.

Yermolaeva, L. P., 1976, "The relation of $(NAD^+)/(NADH)$ in the mitochondria of oocytes and in loach eggs," Ontogenez, *7*, 246-249.

Yermolaeva, L. P., and Milman, L. S., 1975, "The peculiarities of pyruvate dehydrogenase control in the oocytes and embryos of the loach," Ontogenez, *6*, 523-526.

Yurowitzky, Y. G., and Milman, L. S., 1971, " Coordinated changes in the activity of the enzymes of the glycolytic chain in oogenesis of the loach," Biokhimiya, *36*, 1130-1137.

Yurowitzky, Y. G., and Milman, L. S., 1972, "Changes in enzyme activity of glucogen and glucose metabolism during oocyte maturation in the teleost *Misgurnus fossilis* L.," W. Roux's Arch. Entwicklungsmech. Org., *171*, 48-54.

Yurowitzky, Y. G., and Milman, L. S., 1973a, "Interconversion of active and inactive forms of phosphorylase and glycogen synthetase in oocytes and embryos of the loach," W. Roux's Arch. Entwicklungsmech. Org., *173*, 1-8.

Yurowitzky, Y. G., and Milman, L. S., 1973b, "Factors responsible for glycogenolysis acceleration in early embryogenesis of the loach," W. Roux's Arch. Entwicklungsmech. Org., *173*, 9-21.

Yurowitzky, Y. G., and Milman, L. S., 1973c, "Gluconeogenesis in the oocytes of teleosts," Ontogenez, *4*, 182-185.
Yurowitzky, Y. G., and Milman, L. S., 1974, "Glycogen synthetase of loach oocytes and embryos," Biokhimiya, *38*, 86-95.
Yurowitzky, Y. G., and Milman, L. S., 1975, "Enzymes of glycogen metabolism in developing embryos of a teleost," W. Roux's Arch. Entwicklungsmech. Org., *177*, 81-88.
Yurowitzky, Y. G., Milman, L. S., and Ozernyuk, N. D., 1972, "Some peculiarities of glycolytic regulation in loach oocytes," Ontogenez, *3*, 405-414.

SUBJECT INDEX

FSC
www.fsc.org
MIX
Papier | Fördert gute Waldnutzung
FSC® C083411

Kolibri 360 GmbH
Ferdinand-Jühlke-Straße 7
99095 Erfurt, Deutschland
produktsicherheit@kolibri360.de

Safety and Wearable Technology

Impact, Applications, and Implementation in Industry

Gabriel Glynn
Mark Frederick
Tom West

CRC Press
Taylor & Francis Group
Boca Raton London New York

CRC Press is an imprint of the
Taylor & Francis Group, an **informa** business

Designed cover image: image credited to Mākusafe.

First edition published 2025
by CRC Press
2385 NW Executive Center Drive, Suite 320, Boca Raton FL 33431

and by CRC Press
4 Park Square, Milton Park, Abingdon, Oxon, OX14 4RN

CRC Press is an imprint of Taylor & Francis Group, LLC

ISBN: 978-1-032-54988-0 (hbk)
ISBN: 978-1-032-55804-2 (pbk)
ISBN: 978-1-003-43230-2 (ebk)

DOI: 10.1201/9781003432302

Typeset in Times
by SPi Technologies India Pvt Ltd (Straive)

Contents

Part II *The Technology,* Mark Frederick

Part III Leadership, People, and Implementation, Tom West

Part V

Authors

Gabriel Glynn, COSS, CEO and Co-Founder, MākuSafe Corp., I grew up in a small trailer park on the edge of Cedar Rapids, Iowa, a typical Midwest town of about 125,000 people. My father worked as a machinist on 3rd shift at a factory in town, and my mother stayed home with four kids and often watched people with special needs to make extra income. My childhood summers were spent fishing in the nearby creeks and ponds and riding our bikes until the street lights came on. But even as a young child, I found myself fascinated with my father's work at the factory. I loved visiting and seeing the machines, hearing the sounds of grinding, pounding, and the clang of metal on metal. It was hard work, and I remember a couple trips to the hospital to pick him up after an accident at work. I didn't realize how impressionable those visits were on my life trajectory.

I was also fascinated with electronics and computers as a kid. I would take apart various electronics from around the house to repair them or turn them into other creations. By middle school, I had a job working for my neighbor's pawn shop where I would repair broken electronics, getting a cut of anything that was repaired and sold. We also had the first computer in our neighborhood, and I spent a lot of time learning how to build websites for local small businesses. By high school, I had my own tax ID, operating a small company that installed audio equipment and alarm systems in cars. Entrepreneurship was in my blood.

I met my wife, Amanda, while attending college at the University of Northern Iowa. We eventually got married and moved to Des Moines to start our life together. While she found work at a marketing agency, I really wanted to try my hand at building and growing a company. After starting a couple of small businesses and selling them off, I found myself with a great opportunity to join the ownership team at a small web tech company in the suburb of Ankeny. The company, Slash Web Studios, built websites and web software, with a focus on the industrial sector. I spent a great deal of time touring factories, finding ways to improve operations through web technology.

In 2015, we accepted a buyout offer and the search began for my next opportunity. At the time, my father had changed companies and taken on a safety leadership position at a large fulfillment center in Cedar Rapids. It was through spending time with my father that the idea for data-driven proactive safety was born. I took my knowledge of technology and applied it to an industry that was in dire need of a new approach. Thus, MākuSafe was born.

Amanda and I celebrated 18 years of marriage this spring, and we still live in Ankeny with our boys Caleb (13) and Tucker (10). MākuSafe has been an incredible journey that has taken me around the world to places like Australia, Singapore, Malaysia, Germany, France, Belgium, and the United Kingdom. I have met some of my favorite authors, like Dr. Erik Hollnagel, and visited corporate headquarters of some of the world's most iconic companies. It's been a blessing to be a part of such a passionate industry – workplace safety. There's still a lot of work left to do but the journey remains incredibly rewarding.

Mark Frederick, COSS, CTO and Co-Founder, MākuSafe Corp., Growing up in a small town in Minnesota, I had an obsession with taking my toys apart. Armed with screwdrivers, pliers, and sometimes a hammer, I would dismantle even my most cherished possessions. Anything with batteries, switches, or lightbulbs was fair game, and I marveled at the plastic parts, small circuit boards, and the occasional integrated circuit chip. I've always been a curious student of "things."

In 8th grade, the Sears store at our local mall had an Atari 800XL on display. While my parents shopped, I would lose myself in the computer aisle, devouring the programming manuals and writing BASIC code on the display model. Recognizing my passion, my parents gave me an Atari 800XL for Christmas in 1984. My life was never the same – I would stay up late most school nights, programming to my heart's content. I still treasure that computer today.

I pursued computer science in college and found my classes irresistible. I loved my coursework and often went above and beyond what was required. After graduation, my career naturally gravitated toward software development. I started by leading small teams and eventually grew into roles as a software architect, project manager, and thought leader focused on solving complex problems. Along the way, I spent eight years as a "road warrior," consulting for various companies across different geographies and technologies. Eventually, I traded my frequent flier status for more quality time with my family, landing at a Fortune 500 company where I worked on strategic initiatives in cloud computing, IoT, and artificial intelligence. I'm deeply thankful for my family's unwavering support throughout my endeavors.

These experiences laid the foundation for MākuSafe. A few years ago, I decided to push all my chips to the center of the table, embarking on this challenging journey with my co-authors Tom and Gabe to create something truly amazing. I've never looked back. If you ever find yourself with the chance to take a well-calculated risk that could change the course of your life, my advice is simple: do it.

Tom West, SPHR, SHRM-SCP, COSS, Vice President & Global Practice Leader, MākuSafe Corp., After our two co-founders at MākuSafe, I'm the next longest serving member of our team. I was an early investor and advisor to the company in 2016–2017, and then my friend Gabriel called. I'll never forget when he said, "I know it's Saturday, but do you think you can join the company full time on Monday?" I said yes without a hesitation.

Gabe and I had previously been friends for several years. We actually met while he was taking college classes more than 15 years ago and I was his instructor. Gabe was an exceptional student then, always asked great questions, approached things with a unique perspective, and was highly entrepreneurial. I had been an executive leader with many startup ventures, and Gabe later enlisted my help with several companies he was growing from time to time. I had no idea that one day I'd be working so closely at MākuSafe with the guy I mentored long ago.

I also felt fortunate that I finally got to meet and work with Mark, who possesses a number of remarkable traits. But his outstanding ability to conceptualize a solution we would take to the market and execute it so thoughtfully is one of the reasons we've achieved what we have. Together, Mark, Gabe, and I formed what Mark often referred to as the "needed three legs of our stool." Mark built our solution, Gabe provided vision and enlisted the required support, and I was tasked with developing our messaging, talking with industry, and building strategic and professional allies. We were developing a new market, educating company leaders about the potential for MākuSafe to make the world a safer place.

While Mark describes himself as a student of things, I have been a student of management and leadership development throughout my career. My experience is as a business owner and leader, senior human resources professional, trainer, consultant, and college teacher of 29 years. Throughout my career, I've always been focused on helping organizations operate more effectively and efficiently, enabling leaders to create more engaged, productive employees and strong safety cultures. I continue to be excited that this is such a big part of what we're doing here at MākuSafe. I am proud that over this time I have worked closely with and delivered educational presentations to the National Safety Council, ASSP, AIHA, and VPPPA, and been asked to be an Avetta Fellow.

From the outset we planned to co-author this book, with each of us intentionally focusing on our respective strengths. Gabe paints a vision for the future of wearable safety tech. Mark makes the complexities of technology easy to understand, eloquently shedding light on what's possible and why we developed our tech the way we did. My contribution is relaying what we've learned about the essentials for successful implementation, its impact on workers, and what this requires from leaders. We hope that it makes a difference in your operations so you can send more workers home safely every day.

Acknowledgments

To our peers, mentors, and colleagues in the safety industry: many of you laid the groundwork for our technology to fit directly into the hands of safety practitioners. Your advice, feedback, and guidance along the way has helped move our vision for a data-driven approach to safety into a reality.

To our clients from around the world: each of you faces unique challenges based on your type of work, geography, leadership style, and company culture. You helped us mold our vision into something that can work across all types of geographies and industries. Many of you were early adopters and without you we could not have gotten this technology into the field.

To our investors: without your financial support and belief in our team and mission, we would not have gotten a product built and into the marketplace. Your patience as we navigated setbacks, including a global pandemic, has driven our perseverance and success.

To our families: we're so grateful for your love and support. Late nights, long weeks, and months of travel around the world – you kept our spirits high when the road was most challenging. We could not have walked this journey without you.

To the MākuSafe team: your commitment to the mission, to supporting our clients, and to bringing new, innovative, ideas to the table has been what helps set us apart. We're not just a vendor to our clients, we're a partner, and it's because of your creativity and dedication.

To Jennie Morton: for your help in collecting our thoughts, stories, and ideas into this text. You are the glue that stuck it all together.

And finally to our readers: you picked up this book hoping to learn something about a new approach to safety. Thank you for taking this all in. We hope that it provides some practical and applicable insights into how you might approach safety differently.

Foreword

The origin story of MākuSafe has a poignant beginning – with an industrial disaster in 1919.

My great grandfather, John Griffin, was a factory worker in Cedar Rapids, IA. He was employed at Douglas Starch Works, which was a large-scale operation that produced cornstarch. However, the manufacturing process resulted in a fine and highly combustible dust that permeated the site. It was only a matter of time before an accident occurred. An explosion devastated the factory, and over 40 people tragically perished.

My great grandfather only lived to tell the tale because he left work over his lunch break to have a date with the woman who would become my great grandmother. It was by pure happenstance that his life was spared. My entire family trajectory could have been altered if he had kept his usual routine. It now seems like a divine calling that several of his descendants are part of the workplace safety industry.

My father, Brian Glynn, also had a long career in manufacturing, including as a machinist in a printing press factory, and later moved into management roles. In the early 2000's he accepted the role of safety director for a retail fulfillment center in Cedar Rapids, Iowa. One day over lunch, I asked him about the specifics of his job, and he made a joke about the amount of paperwork. Intrigued, I asked what he was documenting. His answer stunned me:

"I pretty much wait for something bad to happen, figure out why it happened, and then put a plan in place so it doesn't happen again. Then I go back to waiting for something else bad to happen."

In a nutshell, my dad had articulated the fundamental problem with workplace safety – it's reactive.

His comment left me wondering – can we use technology to help safety leaders like my father see risk before an accident happens, not after? So many companies rely on lagging indicators, yet an injury or death is the worst type of evidence that there was present. If safety leaders could identify hazardous conditions ahead of time, then there's a strong chance of intervening and mitigating them.

It was this simple premise that launched MākuSafe. Our idea is that when companies gather real-time data from wearable devices, they gain advanced knowledge of when and where risk is present. These red flags call attention to an area or a behavior that requires investigation. These alerts are what ultimately enables management to implement corrective actions that prevent a situation from creating an incident.

Up until recently, uncovering unknowns was the Holy Grail of workplace safety. But wearable sensors are a major turning point in our industry – we finally have the power to bring hidden risks to light. We're stepping into a new era where connected workers are the next evolution in Industry 4.0.

By Gabriel Glynn, founder and CEO of MākuSafe

About MākuSafe Technology

MākuSafe is an award-winning connected worker wearable technology and safety data analytics solution deployed across the globe. Our mission is to improve worker health, safety, and productivity by reducing injuries and mitigating workplace hazards and risk exposures.

The company was founded in 2016 based on a simple premise – because every individual worker's experience is unique to them, data gathered about environmental exposures and ergonomic motions from each individual is critical to understanding risk in the workplace. This data includes industrial hygiene and environmental concerns, slips/trips/falls, strain and exertion, ergonomic risk, and location/proximity.

This approach provides a more complete picture of workplace risk, trends and changes in those risks over time. It also integrates the worker into connected advanced manufacturing environments to accelerate industry 4.0 initiatives. In addition, the push-to-talk MyVoice capability improves communication by enabling real-time reporting from the frontline worker. Enhancing communication contributes positively to organizational culture, participation, and engagement.

MākuSafe offers a robust suite of comprehensive capabilities designed to provide real-time, individualized monitoring and notifications that enhance workplace safety and productivity. Below is an overview of the key features and benefits MākuSafe provides.

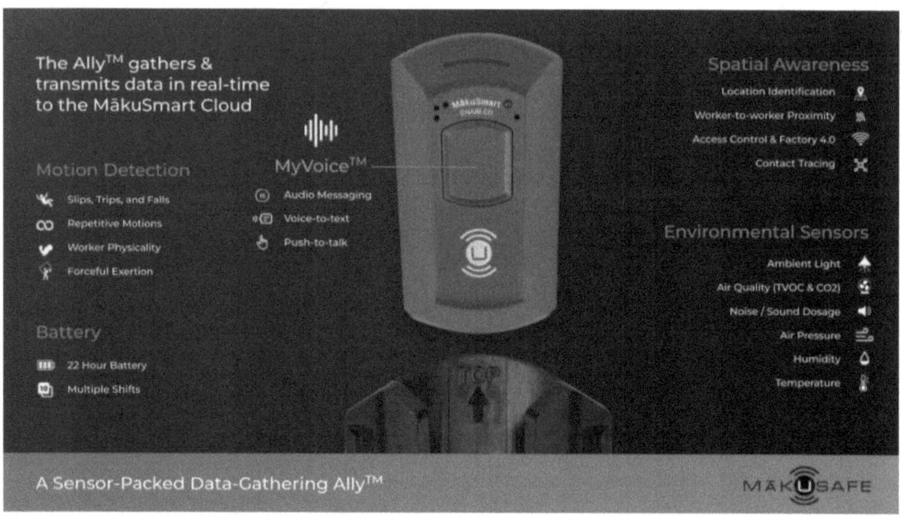

FIGURE 0.1 Strapped to the upper arm, the Ally wearable safety device carries a wide array of sensors and capabilities.

ERGONOMICS MONITORING AND PREVENTION

MākuSafe identifies and mitigates ergonomic concerns that can lead to injuries and reduce productivity. The system provides real-time alerts for potential issues such as slips, trips, and falls as well as identifying instances of strain and forceful exertion. By monitoring repetitive movements and overall physicality, MākuSafe helps prevent musculoskeletal disorders (MSDs) so workers remain healthy.

INDUSTRIAL HYGIENE MONITORING

Maintaining a safe and healthy work environment is crucial, and MākuSafe plays a vital role by offering comprehensive industrial hygiene monitoring. Whether workers are indoors or outdoors, the system tracks temperature, humidity, and heat index on an individual basis to prevent heat-related illnesses. It also includes a full noise dosimeter with seven-octave readings and DBA sound dosage (TWA) for hearing conservation programs. Additionally, MākuSafe monitors ambient light levels, which can impact worker fatigue. It also measures air quality to detect harmful gasses and fumes, including TVOC, CO2, and air pressure.

MYVOICE PUSH-TO-TALK COMMUNICATION

To enhance worker engagement and communication, the MyVoice push-to-talk feature allows workers to send voice messages by holding down a button and speaking into the device. Workers can report near-misses, share good catches, make observations, and conduct inspections with ease. The system also supports voice-to-text transcription, elevating worker participation by ensuring their perspectives are captured and communicated effectively. Audio messages can be delivered to dashboard users with optional notifications.

SPATIAL AWARENESS AND CONTACT TRACING

MākuSafe's spatial awareness capabilities provide detailed location identification by area, room, or workstation, which is overlaid on the facility floor plan. This feature enables the visualization of trends and patterns in real time, helping to identify areas that may require attention. The system also monitors worker-to-worker proximity, ensuring compliance with social distancing guidelines. In the event of a health concern, MākuSafe offers on-demand contact tracing to enable employer notifications, which is particularly relevant for COVID-19 infection prevention requirements.

COMMITMENT TO WORKER PRIVACY

Designed with a high degree of concern for worker privacy, the system does not collect any personal identifiable information (PII), ensuring compliance with HIPAA regulations. No biometric data is gathered, and workers are not subjected to haptic or other feedback. Furthermore, MākuSafe does not continuously track workers, maintaining their privacy and peace of mind.

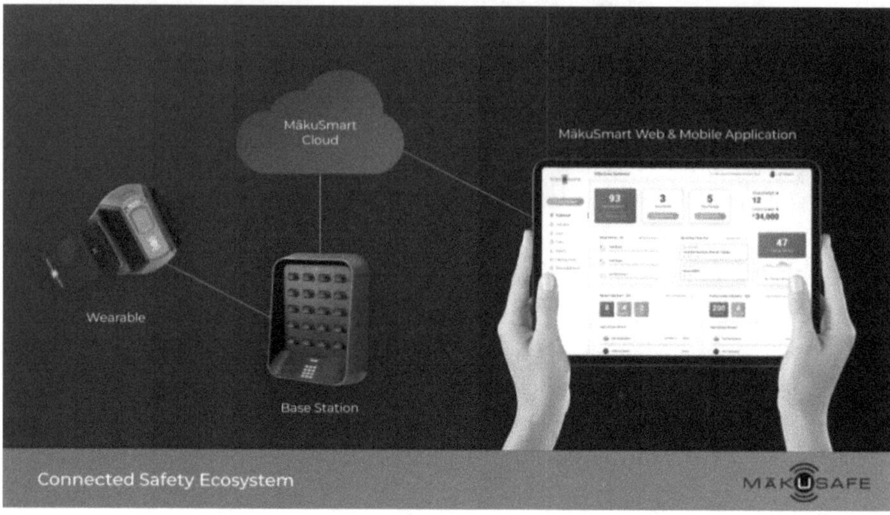

FIGURE 0.2 The Ally wearable device sends data in real time through the MākuSafe ecosystem to our cloud analytics platform. Within seconds, the latest risk and hazard intelligence and custom reporting are available to leaders in the MākuSmart dashboard. Each base station kiosk houses 20 wearable devices, and each device can be used by two different shifts per day.

MĀKUSAFE INTERACT AND INDUSTRY 4.0 INTEGRATION

As part of the Industry 4.0 revolution, MākuSafe Interact offers optional integration with other connected systems and Industrial Internet of Things (IIoT) equipment. This capability enhances productivity and efficiency by enabling worker presence and wearable proximity detection, further embedding the connected worker into the automated workplace.

SUPPORT FOR REMOTE AND FIELD WORKERS

MākuSafe extends its monitoring capabilities beyond the traditional workplace to remote and field workers. Wearables can be paired with mobile devices, providing GPS-enriched data via Android and iOS apps. This real-time feature allows organizations to gain insights into the experiences and exposures of workers outside the facility, ensuring comprehensive safety coverage.

MĀKUSMART: CLOUD SAFETY INTELLIGENCE AND ANALYTICS

The MākuSmart platform provides a powerful cloud-based safety intelligence and analytics solution. It delivers immediately actionable data insights on risks and hazards, allowing for at-a-glance understanding of the latest indicators, their locations, and who is affected. The platform's Location Explorer, SiteMap, and Trends Engine enable the study of data sets across timeframes and workgroups, providing a comprehensive view of safety performance. The system's open architecture supports integration with over 100 APIs and webhooks, ensuring seamless information exchange with other platforms and systems.

MākuSafe offers a comprehensive and integrated approach to workplace safety and productivity, leveraging real-time monitoring, advanced analytics, and robust communication tools to create a safer, more efficient work environment.

1 The Evolution of Safety Management – The Future Is Now

In the dimly lit control room of a manufacturing plant, the quiet hum of machinery was punctuated by the steady beeping of monitors. Claire Jackson, the plant's safety manager, sat at her desk reviewing the latest safety report. The numbers were stark. Despite decades of progress and a steady decline in injury rates, the past few years had seen little to no improvement. More alarmingly, fatalities were beginning to rise.

Claire knew these statistics all too well. The Bureau of Labor Statistics (BLS) data highlighted a troubling trend: While non-fatal Total Recordable Incident Rates (TRIRs) had plateaued since 2013, fatalities were climbing ("Fatal work injuries"). It was a pattern not confined to the United States; across the globe, "work-related deaths have jumped more than 12% since 2000, according to estimates from the International Labour Organization" ("Workplace deaths"). The question that haunted her and countless other safety professionals was simple yet profound: How do we break through this floor? How do we work toward zero?

The traditional approach to safety management has focused on lagging metrics such as incident rates and lost time. These are useful for tracking what had already happened but offer little insight into preventing future incidents. It is becoming clear that a paradigm shift is needed – one that emphasizes prediction and prevention over reaction. This is the dawn of a new era in safety thinking, often referred to as "Safety Differently" or "Safety 2.0."

FROM REACTIVE TO PROACTIVE: A NEW VIEW OF SAFETY

Lagging indicators have been the bedrock of safety management for decades. These metrics, while essential for understanding past performance, provide a rear-view mirror perspective. They told us where we had been but not where we are going. The stagnation of injury rates and the rise in fatalities are stark reminders of the limitations of this approach.

The leading causes of workplace injuries – overexertion, falls, and contact with objects or equipment – remain stubbornly consistent. These incidents are not just statistics; they are stories of human pain and loss. Claire reflected on the irony that most workplace incidents are preventable. The challenge is not in identifying what went wrong but in creating systems that prevent these incidents in the first place.

Leading organizations have long recognized the need for a proactive approach. The journey toward what we now call "Safety Differently" can be traced back to

DOI: 10.1201/9781003432302-1

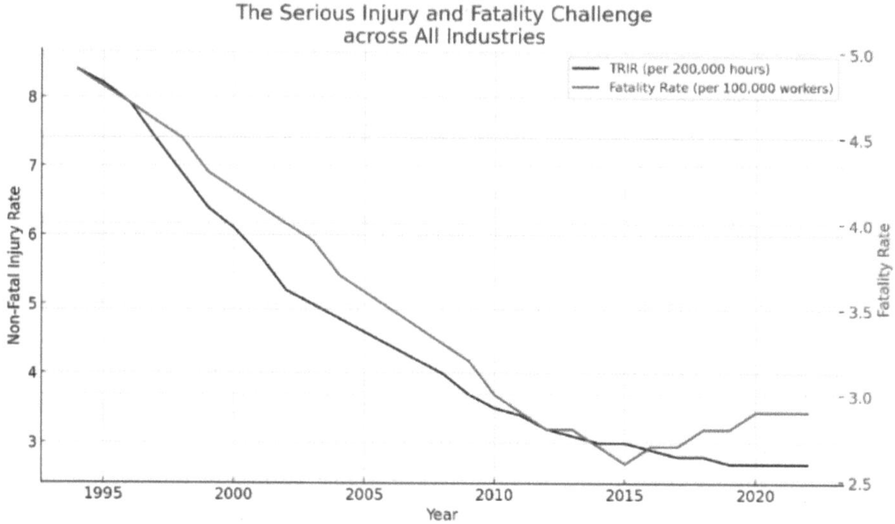

FIGURE 1.1 SIF rates have risen in the past decade, according to the Bureau of Labor Statistics ("Injuries, Illnesses, and Fatalities Databases").

pioneers like Paul O'Neill during his tenure at Alcoa. O'Neill's relentless focus on safety as a core value, rather than a priority that could shift, had set a powerful example. He understood that safety was not just about compliance but about culture and leadership.

Thought leaders, such as Todd Conklin, Thomas Krause, Sidney Dekker, Erik Hollnagel, Bob Edwards, Rob Fisher, and others, have since expanded on these ideas, advocating for a human and organizational performance (HOP) perspective. This approach recognizes that human error is inevitable and that systems must be designed to mitigate the consequences of these errors. It emphasizes the importance of learning and understanding work as it is actually performed, rather than as it is imagined in procedures and policies.

THE ROLE OF WEARABLE SAFETY TECHNOLOGY

As Claire continued to contemplate the future of safety management, her thoughts turned to the potential of emerging wearable safety technologies. The last decade has seen remarkable advancements in this area, with wearables offering transformative potential in preventing Serious Injuries and Fatalities (SIFs) as well as reducing overall incident and injury rates. Leveraging these technologies provides the predictive and preventative capabilities that traditional methods lack.

Predictive analytics, powered by data collected from wearable devices, offers a revolutionary approach to identifying potential hazards before they result in incidents. By analyzing real-time data on worker movements, environmental conditions, and physiological metrics, safety managers can identify patterns and trends that indicate increased risk. This data-driven approach enables more informed decision-making and proactive interventions.

Claire envisioned a future where every worker was equipped with wearables that monitors their vital signs and physical movements. Any anomalies, such as elevated heart rates or unsafe postures, would trigger alerts, allowing for immediate corrective action. This real-time data collection and analysis would be pivotal in preventing incidents before they occurred.

Artificial intelligence (AI) and machine learning (ML) would further enhance the capabilities of predictive analytics. These technologies shift through enormous datasets from wearables, identifying subtle correlations and predicting outcomes with remarkable accuracy. AI-powered safety systems could continuously learn from this new data set, improving their predictive capabilities over time. For instance, an AI system might analyze data from wearables to trend the likelihood of fatigue-related incidents. By understanding the precursors to SIFs, the system could recommend specific preventive measures, such as adjusting work schedules or providing additional breaks.

THE HUMAN ELEMENT: INTEGRATING TECHNOLOGY WITH CULTURE

While technology offers powerful tools for enhancing safety, Claire knew that its successful implementation depends on integrating these tools with a strong safety culture. Technology provides data and predictive insights, but it is the human element – frontline workers, safety managers, and organizational leadership – that ultimately determines how effectively these insights are acted upon.

Creating a culture where workers feel comfortable using new technologies and where their input is valued was crucial. Workers need to be trained not just in how to use new tools, but also in understanding their importance and potential impact on safety. Open communication and collaboration are essential to ensuring that technology complements human efforts rather than replacing them.

As Claire finished her review and prepared for the upcoming safety meeting, she felt a renewed sense of optimism. The integration of advanced wearable technologies with proactive safety management practices represents a powerful strategy for breaking through the current plateau in safety performance. By harnessing the predictive power of data, the practical applications of wearable devices and other new technologies, organizations can move closer to the elusive goal of zero incidents.

In the chapters that follow, we will delve deeper into the specific wearable technologies that are transforming safety management and examine case studies of organizations that have successfully implemented these innovations. By understanding both the potential and the challenges of these technologies, we can chart a path toward a safer future for all workers. The journey to zero is within reach, and technology is a key ally in achieving this vision.

Welcome to "Safety and Wearable Technology: Impact, Applications, and Implementation in Industry." The path to zero incidents starts here, with a commitment to see safety differently and leverage the full potential of wearable technology.

BIBLIOGRAPHY

"Fatal work injuries up in 2022." *Bureau of Labor Statistics*, 29 December 2023, https://www.bls.gov/opub/ted/2023/fatal-work-injuries-up-in-2022.htm

"Injuries, Illnesses, and Fatalities (IIF) Databases." *Bureau of Labor Statistics*, https://www.bls.gov/iif/data.htm

"Workplace deaths up 12% worldwide in past two decades: report." *Safety+Health*, 5 December 2023, https://www.safetyandhealthmagazine.com/articles/24808-workplace-deaths-up-12-worldwide-in-past-two-decades-report

Part I

Evolving Safety, Gabriel Glynn

2 Challenging the Safety Narrative

How pervasive are serious injuries in the workplace? The number of accidents and fatalities is eye-opening. Looking at the United States alone:

- There were 5,190 fatal work injuries recorded in 2021.
- This is an 8.9% increase over 2020, demonstrating a rising problem.
- For every 100,000 full-time workers, the fatal work injury rate was 3.6 fatalities, also up several points from the previous two years.
- These numbers mean that a worker dies every 101 minutes from a work-related injury on average, which is roughly 100 people per week.
- Progress on lowering these rates has been at a standstill for over a decade. Since 2011, the number of fatalities hovers around 5,000 annually ("National Census").

In Iowa, the headquarters of MākuSafe, the top sector is advanced manufacturing ("Advanced Manufacturing"). Yet it's shocking that the state averages about one worker per week lost to a fatality ("Fatal Work"). Nearly half are from preventable accidents at a jobsite, with leading causes from falls and exposure to harmful substances or environments ("Fatal Work"). Every death is a painful reminder that we can do better for our workers.

It's not only the loss of life that is incalculable. From a business perspective, there's also an undeniable financial toll on the economy. A fatal incident creates a ripple effect that impacts individual employers through downtime, lost production, decreased morale, and reputation damage. The likelihood of OSHA fines as well as litigation is also high. Not to mention every family who bears unexpected funeral expenses and outstanding medical bills while facing a future without a loved one's earnings.

It makes one wonder how far we've actually come in modern times. The Industrial Revolution was a major economic and societal shift, yet the advent of machinery also created treacherous working conditions. For example, if you tour a wool mill from the 1900s, you'll shake your head over the lack of safety precautions. Today's industrial facilities are a world apart in terms of protecting workers, yet risk management remains an ever-present concern. All businesses, from a family-owned factory to a Fortune 500 player, share the same challenge – how to minimize the danger and difficulty of the manufacturing process.

The usual advice to improve safety culture includes actions like clarifying expectations, evaluating processes, adding incentives, and following regulations. All of these are worthwhile efforts and will lead to positive impacts. But there's one surprising step that is often overlooked – an attitude check.

DOI: 10.1201/9781003432302-3

Many organizations have limiting mindsets about the origin of an accident. These entrenched beliefs act like blinders, obscuring the nuances of workplace risks. When these assumptions go unchallenged, they create a kind of tunnel vision that hampers safety efforts.

Let's dispel two of the most pervasive safety myths: 1) Work as planned translates perfectly into practice and 2) human error is the first place to lay blame after an accident.

WORK AS DESIGNED vs. DONE

"Work as planned vs. work in practice" is a phrase by Todd Conklin, senior advisor at the Los Alamos National Laboratory, in his book *The 5 Principles of Human Performance*. It acknowledges there's a gap between 1) how a company designed work to be performed and 2) how workers actually complete those tasks. Some variations are completely normal and expected. However, the space between expectations versus reality is where accident risk begins to sharply increase and move toward the failure line (see Figure 2.1). How does this happen?

1. "Work as planned" is disconnected from everyday conditions
   Just because something is designed a certain way doesn't mean that's the safest way.

   This is because work is often created in a vacuum. For example, the process was mapped by people who don't have first-hand experience on a factory floor. Or it was simulated in a virtual environment that didn't fully factor in a wide range of safety scenarios. Whether a human or computer

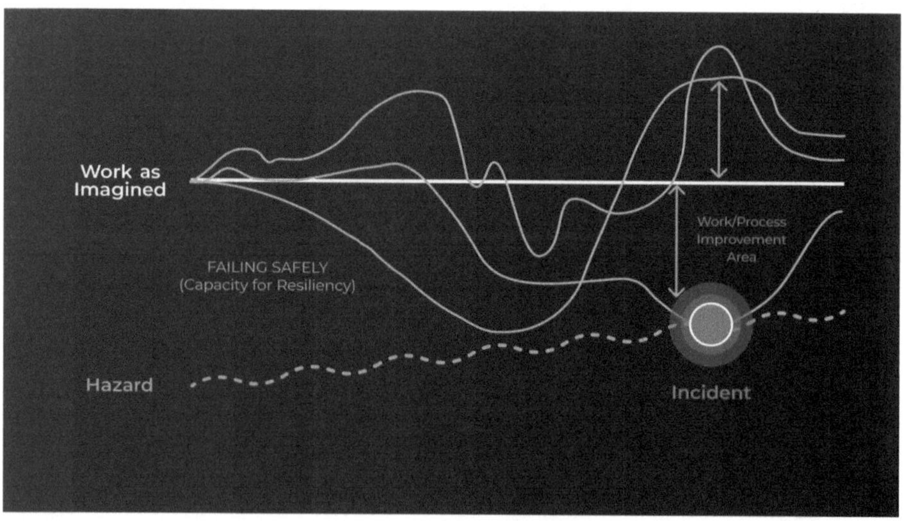

FIGURE 2.1 Work as performed, on good days and bad, does not follow a straight path as it was imagined or designed. Workers are forced to adapt to the circumstances of the day. By truly understanding the realities of how work is performed, we can improve work processes and create capacity to work safely, even when systems fail or people inevitably make errors.

envisions the work, it's extremely easy for the model to fail to predict minute changes that have a massive bearing on risk.

For example, the plan may not recognize that the manufacturing area is 95 degrees. This heat routinely causes workers' hands to be sweaty. In turn, that moisture affects their ability to grasp a lever. If their grip is compromised, their hand might slip. There are 11 levers where this creates no issues. But on lever 12, the wrong movement at the wrong moment will cause a ripple effect that escalates into a safety situation.

2. There's human behavior at play
   Even well-designed work doesn't perfectly match the reality of worker behavior.

   Work can – and often does – fluctuate from the design. This happens for any number of reasons, such as environmental stressors (unexpected glare), a lack of training (employees don't know any better or have forgotten), rushing to meet quotas (taking shortcuts), or variations in individual work styles (people have different body mechanics). Larger organizational issues such as being short staffed or increasing production expectations can also be at play.
3. The planned work doesn't anticipate unexpected changes
   The simplest alteration can have a domino effect.

   All it takes is for a single change to force a modification to the regular workflow. It might be an unexpected event like a machine breaking down or a larger situation like construction that reroutes the forklift path. While the change may look harmless on the surface, it could produce unintended consequences that heighten risk.

WHEN THE PLAN GOES AWRY

No matter their origin, all deviations create a dip away from the design's baseline. To be fair, not all dips are a blunder. Many don't even have the potential to cause an accident, much less represent an immediate safety threat. In fact, mistakes are so common that they should not only be anticipated but built into the process. Consider this argument by Conklin on his podcast *Pre-Accident Investigations*.

> "A worker making a mistake is super normal. Error is so normal, it's not even interesting. And it's never causal. Error is just an operational phenomenon that happens all the time. The reason it can't be causal is because error doesn't always create failure conditions. Workers screw up all the time, but not every screwup has high consequences of failure. Most of them, you can catch or correct, or they just don't matter. That sets a weird precedent … that error is normal. The battle isn't taking error out of a system – it's creating a system that can tolerate the normal error that's always in that system"
>
> (Conklin, "Safety Moment").

It's when error hasn't been factored into the process that variations start to compromise safety. Once these mistakes gain momentum, they predictably move closer to the failure line – the point of no return when an error results in an accident, injury, or fatality.

This is where many organizations have a blind spot – *they have no idea how many times they've brushed the edge of that failure line.*

No accidents, no problems, right? The assumption is that a lack of incidents means that work is being done as described and designed. But what the company doesn't realize is that they had skirted the failure line a hundred times before a SIF finally happened. They were always on track for this accident because no one was looking for dips, much less measuring their impact on safety.

> "If a facility has vulnerabilities for SIFs and is not paying attention to those precursors in a systematic way, focusing on small injuries is counterproductive. It allows SIF rates to stagnate and employees to get hurt. It leaves the most senior leaders flummoxed when a serious injury occurs seemingly out of the blue – while the worker who sees the vulnerabilities day in and day out counts the event as inevitable,"

Argue Krause and Bell, two organizational safety experts, in their book *7 Insights into Safety Leadership* (27).

How do we address this mindset? Companies need to routinely evaluate their safety culture to pinpoint where risk has been overlooked. That means asking questions like:

- Where are things going well?
- Where are there departures from work as designed?
- Where are these variations creating near misses?

Each of these should be followed up with the simple question of "why?" If an area has a stellar safety record, what is driving that? If workers are deviating from the planned work, what's the reason? Knowing is better than assuming. When an organization isolates good safety practices, they can iterate those precautions throughout a site or even across multiple locations.

Once organizations start analyzing precursors, they often stumble across a surprising truth – few accidents are caused by a single failure. Tragic incidents are frequently the result of a chain of events that was influenced by business decisions and organizational behavior. Remember the phrase "when you point one finger, there are three fingers pointing back at you?" Preventing a SIF means taking a hard look in the proverbial mirror and reflecting on how the entire company contributed to safety risk.

THE WORKER ISN'T ALWAYS THE ROOT CAUSE

The root cause mentality assumes there was a single point of failure that occurred at a single point in time and was triggered by a single individual. This default reaction places the blame entirely on the worker. If they did something different than how work was designed, it is simply evidence of their neglect.

Instead, companies need to flip the script and work backwards from an incident. What were all of the decision points that led to this one moment in time? There were likely dozens – if not hundreds – of factors that increased risk until the SIF's tipping point. They often originate from operational changes that trickled down from the corporate level, an unexpected repair, or external complications like supply chain delays.

Let's run a hypothetical SIF that resulted in the death of Jim. A facility has a large robotic arm that picks up steel plates and places them on the CNC machine to be laser cut. The equipment is protected by a locked cage. Furthermore, employees are trained to never enter this area while the machine is running. Jim is an experienced worker and has been at this station for over a year. But for unknown reasons, Jim bypassed the safety gate on the machine and went inside the cage only to be crushed to death.

Root Cause Reaction

The conclusion – "Jim was an idiot."

It's unfortunate that a common refrain in the industry is "you can't fix stupid." The problem is that when one person is declared the root cause, the company feels absolved from examining any circumstances that preceded that event. All of the responsibility is placed on the employee who was hurt or died, and it's back to business as usual. The accident is categorized as an outlier rather than the tip of an iceberg.

End result: The company sees the incident as the result of individual behavior and takes no further action.

Domino Effect Reaction

The conclusion – "We accepted a certain amount of risk, and Jim's the victim of our poor planning and decision making."

The reality is a SIF is rarely the result of a single point of failure. A company that prioritizes world-class safety will use a SIF to peel back the layers. They will start asking questions like:

- Why did Jim do that? What was he trying to accomplish?
- Why did he feel like he had to take this risk? Was he under pressure from other areas to meet deadlines or goals? Was this action perceived as an acceptable shortcut?
- Was this an isolated incident or not? Or were there 500 other times that Jim was able to get in and out through the gate without harm? Did this lead to a false sense of security?
- Was there a machine malfunction? Is there always a part that gets stuck or needs adjusting that would cause someone like Jim to enter the gate?
- Have we created a lax environment that makes people comfortable with bypassing safety protocols?

The company might find that a series of interconnected influences created the perfect storm for this failure. For example, the machine guard broke the previous month. Even though Jim put in a work order, he was told to continue working without it.

This was because corporate made a decision the year earlier that the only way they could bonus people was based on productivity. So they cut back on expenses, which meant they didn't have a secondary machine or spare parts on standby. Meanwhile, maintenance acted on Jim's work order, but it was going to take two months for the new part to ship from overseas. That left at least a month where Jim was working without the proper protection. It was in that window that Jim made a decision that cost his life.

End result: The company sees the incident as the result of risky organizational behavior and forms a plan to correct course.

KEY TAKEAWAYS

- Don't assume work as designed is how it's being done
- Don't assume a lack of accidents means a lack of problems
- Don't assume the worker is the root cause of an accident

BIBLIOGRAPHY

"Advanced Manufacturing: Where Manufacturing Keeps Advancing." *Iowa Economic Development*, 2024, https://www.iowaeda.com/advanced-manufacturing/

Conklin, Todd. *The 5 Principles of Human Performance: A Contemporary Update of the Building Blocks of Human Performance for the New View of Safety*. 2019.

Conklin, Todd. "Safety Moment – Principle Number 1 – Mistakes are normal." *Pre-Accident Investigations*, 2 May, 2018, https://preaccidentpodcast.podbean.com/e/safety-moment-principle-number-1-mistakes-are-normal/

"Fatal Work Injuries in Iowa – 2022." *U.S. Bureau of Labor Statistics*, 7 March 2024, https://www.bls.gov/regions/midwest/news-release/fatalworkinjuries_iowa.htm. Press release

Krause, Thomas R. and Kristen J. Bell. *7 Insights into Safety Leadership*. The Safety Leadership Institute, 2015.

"National Census of Fatal Occupational Injuries in 2022." *U.S. Bureau of Labor Statistics*, 19 December 2023, https://www.bls.gov/news.release/pdf/cfoi.pdf. Press release.

3 Data-Driven Businesses Include Safety

When your organization talks about data-driven decisions, is safety included?

Businesses are using data to inform every area of their operations in ways that would have been unimaginable even a decade ago. Because of real-time insights, companies can calculate everything from the timing of layoffs to investing in capital improvements. They are digging into the nuances of who their consumers are and their buying patterns, then translating those factors into product positioning. They are doing the same profiling on purchasers so they have more negotiation power. They can even monitor daily price fluctuations across the supply chain. Up and down operations, leaders are getting more data savvy.

But safety is left out of the discussion in many cases. In fact, it's often one of the last business units to undergo digital transformation. There are three culprits that keep workplaces from fully embracing the capabilities of safety data.

#1 – DECISIONS ARE DRIVEN BY INTUITION

There are few instances where a business would lean entirely on gut instinct. No leader is going to use wishful thinking to determine if it's going to be a banner year or if they need to order more components to fulfill orders. They're going to approach those considerations armed with verifiable evidence and honed forecasting. Safety decisions should also be founded in statistical data – otherwise blind optimism is in the driver's seat and could lead you astray.

For example, a business wants to increase productivity and revenue, so they add robotic welders to boost output. Their first oversight is assuming the robots won't create any complications, which means they didn't consider the downstream impact on shipping and receiving. Second, they are not gathering any critical data that is useful in understanding the change, like environmental and ergonomic sensors that could alert them to trouble. The reality is that production is outpacing the warehouse, and this has caused an unintended bottleneck that is escalating into a safety issue.

First, product starts piling up in different areas. Pallets are stacked so high that they block out the lighting and reduce visibility. The overflow also causes the team to pack pallets in spaces not intended for this activity, which is problematic because they now overlap with where the forklifts drive. The consequence is the forklift routes are detouring unexpectedly. The combination of decreased light and modified forklift paths is a perfect storm for someone to be seriously injured or killed.

What happened is that new risk was introduced because an operational change was made in a data vacuum. This is a failure twice over because the delays in shipping and receiving cancel out the welding department's productivity as well as significantly increases the likelihood someone will get seriously injured or killed.

DOI: 10.1201/9781003432302-4

What could have halted this drift toward the failure line? First, physical exertion data from workers would make it noticeable when there's a point of departure. Second, mobile sensors that detect ambient light could flag when it has fallen below acceptable levels. If both of these had been treated as leading indicators, the company could have called for a safety investigation.

#2 – ACCIDENT DATA IS THE SOLE INDICATOR OF RISK

Traditionally, many bad things must happen in order to have enough data to identify a safety trend. The paradox is that the more accidents you have, the more data there is to trend. Except that's an incredibly poor way to get data. Businesses shouldn't pass a certain threshold of injuries before taking action. It would be like waiting to plan for workplace violence until a fight broke out – it's a passive approach that doesn't make any sense.

Accident rates also shouldn't be used as the sole predictor of future behavior. Keep in mind SIFs and DARTs exclusively look at past events. SIF reports can also be unreliable in trying to understand what truly happened because other critical factors are often omitted, such as environmental factors, fatigue data, and production pressure. Organizations would have a much better understanding of where risk is if they used near miss (good catch) data and leading indicators rather than relying on multiple SIF reports.

#3 – ONLY STATIC SENSORS MONITOR SAFETY

Smart factories frequently deploy sensors to detect hazardous conditions. However, these devices have limitations because they are fixed to one location and only detect within a set radius. A more advanced and precise method to gather data is to deploy wearable sensors on workers that gather data in real time, giving a much clearer picture of what the individual worker is experiencing.

Consider a wall-mounted temperature sensor. While it generates a good set of data, it's specific to what the wall is experiencing. It can't tell that a person who is 20 feet away is standing next to a machine that's producing a dangerous level of heat. The worker's experience is totally different than the wall reading, even though they are in the same room. This leads to an awareness gap because the stationary sensor doesn't have adequate proximity to a heat source to monitor risk.

Now imagine the worker is wearing a temperature sensor that generates alerts when heat exceeds safe levels. The picture changes completely. Employees could protect themselves by taking breaks and hydrating. They might even stop to inspect the machine for performance concerns.

SAFETY DATA ELIMINATES TRIAL AND ERROR

Risk data takes the adage of "you can't manage what you don't measure" and puts some real teeth behind it. By eliminating guesswork, safety data empowers businesses to make decisions based on evidence. They can:

- Eliminate assumptions about working conditions
- Weigh options that will lower risk

FIGURE 3.1 As the hot equipment example illustrates, risk can be subtle. It can also be cumulative, building over time. Just because accidents aren't happening doesn't mean that work is being done safely or environmental hazards aren't present. It's why it's so transformative to have employees generating safety data regardless of their location. They help an organization identify where work is deviating from the design, prompting a timeout to investigate. It's this ability to stop, assess, and correct that prevents dips from toeing the failure line.

- Compare safety gains against investment costs
- Prove that a work redesign achieved safety goals

Quantifying the impact of a change is where safety data is particularly powerful. Implement too many modifications at once and it is nearly impossible to isolate which one moved the safety needle. It's like making a diet switch by adding and subtracting multiple foods at the same time – there's no way to determine an improvement was the result of one particular change.

But with a baseline of safety data, a business can create control groups for evaluation purposes. They can implement one option and measure the impact against the baseline, do the same for a second option, then compare both outcomes to each other. To determine which course of action is the best, they can then proceed to a cost-benefit analysis and use real data to support it.

CONVEYOR LOADER vs. NEW HIRE

Let's revisit the case of adding robotic welders to increase productivity. In this version, all employees have a wearable safety device that captures motion data. The plant has an established record of how much physical effort is typical for each department. They also monitor for fluctuations that indicate workers are expending more energy than usual.

The ripple effect from the robotic welders is that the two-person dock crew is struggling to keep pace. This is documented as a sharp rise in physical exertion over a two-week period. The change in physical exertion is classified as a leading indicator because it shows increased potential for overexertion, strain, repetitive motion, and injuries. The conclusion is that ergonomics is being compromised in the shipping and receiving department. Since that is not an acceptable risk, corrective action is warranted.

Rather than wait for the loaders to complain or get hurt, leadership proactively asks the affected team for their viewpoint and starts an informed discussion about how to resolve overexertion. Through their discussions, two solutions are proposed: 1) add a conveyor for loading and unloading trucks or 2) hire a third person.

To verify the safety gains associated with either, the factory tests each solution individually and updates the motion data for the new conditions. Now they have a before-and-after picture of exertion for both choices. This data allows the company to balance safety with other factors like cost and long-term planning.

Conveyor	**Third Loader**
Reduces physical expenditure by 50%	Reduces physical expenditure by 70%
One-time investment cost + associated upkeep	Ongoing compensation + benefits but no upfront costs
Less expensive but less effective	More expensive but more effective

Leadership can then embark on sophisticated decision-making that is free from conjecture. If lower cost is a priority, then the conveyor belt is a better option because it will decrease risk by an acceptable margin while satisfying shipment demands. If cash flow is a concern, then hiring a temporary worker will be prudent until the company can invest in equipment down the road. If the labor market is tight and staffing is difficult, then the upfront costs of equipment are more justifiable. It's also plausible the factory will adopt both measures. This might create an exciting opening to add another robotic welder because there's proof that higher output won't sacrifice safety.

All of these considerations are possible because they are rooted in one thing – an IoT connected worker generated data associated with a leading indicator. The best part? The entire decision process was done before there was ever an accident or injury. This is just one of the many ways connected workers and safety data are part of continuous improvement.

KEY TAKEAWAYS

- Operational changes made without data often increase risk
- Accident data alone doesn't give a full picture of risk
- Safety data eliminates guesswork

4 Green Flags in Safety Culture

How likely are your workers to come forward with a safety concern?

Across virtually every organization we talk to, there are few companies that are good at gathering safety information from workers. Communication barriers – perceived or founded – are a significant impairment to collecting critical feedback from the factory floor. Any employee hesitation often boils down to three uncertainties – is it convenient, safe, and worthwhile to report an issue?

First, recognize if there are real-time pressures on employees that take priority over reporting an issue. Stepping away from their workstation for even a few minutes to flag a hazard can have immediate and negative consequences, such as impact their ability to reach quotas, affect other people down the line, or disrupt the entire assembly process. These impacts are a disincentive for them to alert a supervisor about lagging or leading indicators, much less near misses.

Second, the reporting process is often manual and cumbersome. Whether paper or electronic, it takes time to fill out documentation. If the reporting tool is confusing, too complicated, time-consuming, or overly generic, employees won't bother with it.

But the biggest challenge often lies in how empowered employees are to say something in the first place. Leadership's receptiveness to feedback, validation of concerns, and willingness to implement a solution can make or break this aspect of safety culture.

EMPOWERING WORKERS

Consider that the number one fear that prevents someone from speaking up is retaliation. It's like company surveys that claim to be anonymous but clearly are not – no one is fooled. Employees shouldn't be worried that there will be repercussions if they shine a light on a problem. If everybody's afraid to say something, no amount of PPE and advanced reporting tools will reverse the damage. Staff need to see proof that they can freely go to management if they see something unsafe or have a suggestion to reduce risk. They need to know that concerns will not only be protected but celebrated for their willingness to come forward.

There's also nothing more damaging to safety culture than somebody taking the initiative to speak up and then seeing nothing is being done. All it takes is for one person to be ignored and it breeds a perception that the company doesn't care. This is especially toxic if a number of people have raised a concern and management never acts on the information. This also happens when a report is submitted but sits

DOI: 10.1201/9781003432302-5

there for hours with no review or action, to the point where the information is now completely useless due to the delay. You can't blame employees for beginning to think, "I don't know why I say anything because nothing ever gets resolves. No one takes us seriously so I might as well keep quiet and just do my job."

A lack of communication can also unintentionally compound this problem. It's possible there's concrete progress at the leadership level that employees can't see. Management could be having meetings about how to solve an issue or working on the budget to purchase a solution. Maybe they even went ahead and ordered equipment that will reduce risk but are waiting on a two-month shipping delay. But if no one is communicating updates to the floor staff? It still has every appearance that safety is only given lip service.

Trust and transparency are paramount to any safety culture. Without accountability and follow-through, safety data can only be a benchwarmer rather than a most valuable player. One of the most simple yet effective tools we have seen our clients use for transparency is a "You Said, We Did" board. One side has comments and suggestions from frontline employees (You Said), the other side is what has been or is being done (We Did). This is done on a rolling basis, even if resolution is still in progress. For example, the "We Did" update might be as simple as "We took this to our weekly management meeting and discussed it. It's on the agenda for next month to see if this is still a concern." That kind of transparency shows that not only is everyone in the organization being heard, but leadership is making a concrete effort to rectify issues as well.

HOLISTIC SAFETY LEADERSHIP FROM L'ORÉAL

Our team had the privilege of attending a presentation by Malcolm Staves, Global VP of Health & Safety for L'Oréal. With 88,000 employees and over 40 manufacturing facilities, safety is mission critical to this beauty leader ("About L'Oréal"). We were thrilled to learn how L'Oréal prioritizes leading indicators as well as holistic safety approaches. They even go as far to ask a startling but revealing question at every location:

"How are we going to kill somebody with what we're doing?"

This is how they evaluate every square foot of a facility. No location is exempt from this process – even zones where no one has ever been hurt doing a particular job or locations with low incident rates are scrutinized with this question in mind. It's how they force themselves to uncover and understand leading indicators, finding those essential lessons learned that can be applied companywide.

They also include their workforce as "active agents" in their safety culture, "using their acuity and skills to be part of the protection process" ("L'Oréal, A World-Class"). Rather than mandates to passively follow, "L'Oréal is keen to instil [*sic*] in employees a sense that their individual contributions to creating and maintaining a safe workplace, however small, are vital in creating a self-sustaining bottom-up (rather than top-down) safety ethos" ("L'Oréal, A World-Class").

What's even more impressive is how L'Oréal has framed safety as a necessity that extends beyond working hours, encouraging staff to be health and safety ambassadors in their personal lives. As Staves explained in an interview with *Safety & Health Practitioner*:

> "We do not expect anybody to work safely only for L'Oréal – then they are potentially working safely to avoid being a statistic. We have changed the conversation dynamics to the why and the sense of purpose and the need to work safely for themselves and for their families so they can spend time outside of work being with the people they love doing what they love to do" ("'It's now all").

LOOK FOR THE GREEN FLAGS

How are you documenting leading indicators?

Leading indicators are one of the most beneficial metrics in a safety program. They come in many forms, from safety meetings and refresher training to proactive maintenance and PPE compliance. It's an amazing shift in safety culture when an organization moves from responding to injuries (lagging indicators) to proactively mitigating potential hazards (leading indicators).

But we'd like to suggest another type of leading indicator to investigate – areas of no accidents.

Safety professionals are so attuned to red flags that we've forgotten there are green flags too. The absence of data is a compelling story that demands attention. When a zone or a team has a low incident record, it's a strong sign of a healthy safety culture. You want to investigate what a department did or is doing that is positively impacting safety.

This is where data can help you conduct a positive form of root cause analysis – uncovering a safety solution that's already a success in your manufacturing environment. There are two data sets that will reveal best practices that have been flying under the radar: motion data and accident-free employees.

Motion Data

Measuring physicality offers an incredible wealth of insights. Take a factory with a hundred workers doing the same job requiring the same output. Due to differences in height, strength, and body mechanics, it's normal for physicality data to show a range in exertion. Companies tend to focus on workers who are at the top of the bell curve. They often assume – and rightly so – that individuals exerting more energy are either not working efficiently, at risk of an injury, or both.

But they also need to pay attention to the lowest risk workers. There is a reason they are at the bottom of the distribution – it's not pure luck or statistical likelihood. Chances are they've done something intentionally or even subconsciously to improve safety. These employees are learning opportunities.

For example, an employee welded a custom mechanical bar onto their workstation so that they can flip a heavy piece of material over with greater ease and speed. Because they are not having to physically maneuver the part every single time, they've accomplished two safety improvements: lowered injury risk from repetition motion and strengthened ergonomics.

But it would be easy to overlook this type of self-directed safety improvement. No one asked them to make this modification – they simply innovated on their own. There's a small chance the solution will appear on a safety inspection or be mentioned during a briefing. However, the worker or their supervisor isn't thinking about the bigger safety picture – it's not their job to consider how this single improvement could be scalable throughout the plant and at what cost.

So what would bring leadership to this individual in the first place? Motion data. This worker's low physicality ranking would stand out among colleagues. That's a green flag begging to be investigated. Safety managers can connect with this individual to learn what's contributing to their score, keeping an eye out for replicable solutions. Even if the safety solution or habit is one-off and unique to this person, there's benefit in recognizing their efforts and highlighting them as a role model.

Injury-Free Workers

Far too often, safety investigations are only focused on people who got hurt. If your only measurement of safety success is the rate that people are still getting injured, that's a short-sighted way to assess risk.

What if we looked at people who have no history of accidents?

This turns the safety model on its head. Because if you have 300 employees who are free of incidents, their track record is another type of green flag. The data is telling us this worker pool is actively supporting the safety culture – that's worthy of investigation. Their actions or habits are an untapped source of constructive insights.

REAL-TIME REPORTING

What if workers were wearing a device that allowed them to create a quick recording that was immediately sent to the attention of safety leaders?

Our voice reporting function does exactly that – bypassing paper forms, phone apps, and electronic logs. It takes mere seconds to make a recording with the same ease as a voicemail. It can be done in real time, no waiting to remember an issue at the end of the shift or hunt down a supervisor before heading home. The information goes directly and quickly to designated leadership, allowing them to address immediate hazards or start conversations about areas of concern.

This feature is one method to overcome the lack of near-miss (good catch) data, when the failure line was within sight but narrowly avoided through good fortune. It's also a way for workers to note when work is drifting away from the design before there's a situation heading for trouble.

MākuSafe's platform logs the voice recording with a speech-to-text transcription. Additional documentation such as photos, detailed notes, updates after an investigation, and mitigation actions can all be included. This creates a robust and referenceable entry stemming from the initial communication. There is even an option to create a word cloud with commonly used terms from the messages to visually depict what employees are reporting.

One of our global clients uses the voice function to gather information about near misses. Using behavior-based safety observations, employees are asked to spend a few minutes of their day anonymously monitoring a peer while working. Rather than a punitive "gotcha" exercise, the purpose of the observations is to have an objective set of eyes spotting hidden issues. For example, it might take the viewpoint of a coworker to notice that someone is improperly rotating their body while moving materials. The motion is likely subconscious rather than a flagrant safety violation, but it still needs to be treated as a leading indicator because of the likelihood of strain or a repetitive motion injury.

Voice reporting is also a meaningful way to highlight where someone is contributing to safety culture. These are the "good catches" that need to be rewarded. They are anything from proactive behaviors and adherence to policies, to corrective action that reduces workplace hazards. For example, Sam is observed consistently using multiple points of contact while lifting parts – this is conduct that should be recognized and positively reinforced. Voice reporting is an efficient and easy way to report good catches, ultimately contributing to the strength and efficacy of your safety culture.

KEY TAKEAWAYS

- There are many barriers that prevent employees from reporting safety issues
- Trust and transparency are the foundation to any safety culture
- Documenting leading indicators ("green flags") is essential

BIBLIOGRAPHY

"About L'Oréal." *L'Oréal Groupe*, n.d. https://www.loreal.com/en/group/about-loreal/

"'It's now all about putting people back on the right track': SHP speaks to Malcom Staves, Global VP of HS at L'Oréal." *Informa Markets*, 2024, https://www.shponline.co.uk/culture-and-behaviours/its-now-all-about-putting-people-back-on-the-right-track-shp-speaks-to-malcom-staves-global-vp-of-hs-at-loreal/

"L'Oréal, A World-Class Safety System." *L'Oréal Groupe*, n.d. https://s32891.pcdn.co/wp-content/uploads/2023/07/Loreal-A-World-Class-System-digital.pdf

Part II

The Technology, Mark Frederick

5 Why Wearables Aren't a Fad

Let's clear up one of the biggest misconceptions about wearables – they are not a craze.

Considering that the earliest versions arrived in the 12th century, we can't even call them a trend. They predate the arrival of computers, cars, electricity, and even steam power by hundreds of years. Wearables are a well-founded part of the human experience, fundamentally changing societies with every iteration. Wearables are an unstoppable force.

Supported by advancements in consumer electronics, wearables are steadily moving into the workplace. It's not only because the incoming generation is already comfortable with high tech – our "always connected" mindset has normalized wearables across most age groups. In 2022, wearable users were estimated at 1 billion based on sales figures. We're only going to continue down this path – the market share was $27.91 billion in 2019, more than doubled to $61.30 billion in 2022, and is projected to hit $74 billion by 2025 (Beckman).

This widespread adoption is paving the way for wearable safety technology to become a regular, everyday tool. From exoskeletons to adhesive patches, there's a wide aperture on how wearables already are and will be standard on a job site. It's only a matter of time before they become as commonplace and essential as phones and computers. It's a confident projection that more workers will be equipped with wearable technology that helps them do their job not only more efficiently but more safely.

DEFINING A WEARABLE

Wearable technology has two defining qualities. The first is that it is worn on a human. While it may seem like an obvious statement, we can't overlook the historical context of how we got to where we are today. Many mechanical engineering and computing capabilities had to be invented to arrive at this point. Along the way, we have been conditioned to accept assistance from technology on or in our bodies for centuries. In fact, you could argue that the human body has always been a target platform for technology in one way or another.

The second characteristic is that the wearable is participating in an information exchange. It's either a consumer of or contributor to information; sometimes even both in the same device. No matter how it's generated, this data is used to enhance decision-making.

In every version of wearable technology, there's an undercurrent that remains constant – our desire to be autonomous. Wearables provide a certain level of independence because we have essential information directly on our person. Humans have

DOI: 10.1201/9781003432302-7

long benefited from these insights, from the simple ability to check the time or read words more clearly to taking measurements of our body's most vital functions. We're so comfortable with wearables that we've even allowed medical devices to be implanted inside our bodies. We are living more healthy, productive, and satisfying lives because of wearables.

Safety wearables have a long lineage of distinguished forerunners. Each of these categories – cultural disruptors, life savers, personal entertainment, and mobile communication – has a direct influence on our technological ability and societal willingness to wear a smart device at work.

Wearables as Cultural Disruptors

One of the earliest examples of wearable technology is optometry, which debuted around the start of the 1300s. People take it for granted these days, but it required a significant leap in glass technology to adapt and refine handheld lenses for the human body. It was revolutionary when people could wear magnification to see their world more clearly.

Another wearable milestone was the pocket watch in the early 1500s ("A Guide to History of Pocket Watches"). Prior to that point, telling time was a public service in the form of clock towers. Technology that once loomed over a town square was eventually miniaturized so it could hang from a small chain. The first known wrist-watch then landed on the scene in 1810, designed for the Queen of Naples ("1810, The First Wrist Watch"). Once wearable timepieces graduated from a symbol of wealth and were available to the everyday consumer, tracking hours and minutes was democratized.

It's not hard to imagine naysayers in those early days who claimed watches were "just a trend" and "people would soon get tired of having them on their wrists." Yet look at where we are today – you'd be hard-pressed to find someone who isn't carrying a clock on them in some form or another during waking hours. Once the time went on our bodies, we never took it off.

Wearables as Life Savers

Another revolutionary moment in wearable development was the introduction of medical devices. These breakthroughs came with the invention of the transistor in 1947 by Bell Laboratories ("1952: Transistorized"). Technologies that once relied on vacuum tubes could be scaled down into lightweight devices. For example, the first commercial hearing aids worn completely in the ear were released in 1952 from Sonotone ("Inventing the Transistor").

The introduction of the battery-operated pacemaker was a major turning point in healthcare. Medtronic is credited for the first wearable external device in 1957, which adapted pacemakers that had to be plugged in and housed on a cart during open-heart surgeries; they produced their first transplantable model in 1958 ("75th anniversary"). Wilson Greatbatch, an engineer working under the VA hospital system, holds the patent for the first self-contained, fully implantable pacemaker in 1962 ("Wilson Greatbatch"; "The Invention of the Cardiac Pacemaker"). It is no small thing to imagine how many people's lives have been extended thanks to a wearable on one of our most important organs.

This wearable springboards into implantable cardioverter defibrillators (ICDs), which were successfully demonstrated during clinical trials throughout the 1980s and refined to terminate two types of abnormal heart rhythms by the early 1990s (Van Welsenes et al.). It's amazing to think that people at risk of cardiac arrest now have the reassurance that a heart problem is being mitigated by a piece of technology that's doing a better job than their own body.

Even though pacemakers and defibrillators are a closed loop system, they still demonstrate the power of a responsive wearable. They collect information about how our bodies are functioning and can automatically generate a corrective action to alleviate a problem that could lead to death otherwise.

The 2010s and 2020s have also been incredible decades for fitness and health wearables. Today we have multiple brands of smart rings that measure blood pressure and heart rate as proxies for mood and focus. The *Inspire* implant relaxes airway muscles to address sleep apnea as an alternative to CPAP machines and even has an accompanying app that tracks data (*Inspire*). Diabetics can opt for any number of real-time glucose monitors with a sensor on or even under the skin. There have even been inspiring advances in bionic prosthetics equipped with artificial intelligence, allowing people with amputations to feel sensations and control movement that better mimic a natural limb. We are living in a moment of increasingly personalized wellness data – all generated from wearables.

Wearables as Personal Entertainment

Another pivotal moment came in the form of portable electronics in the 1920s. The radio company Zenith introduced the first portable radio in 1924 ("Heritage"). However, models of this era were bogged down by heavy batteries and fragile glass vacuum tubes. This made them bulky enough to require a carrying case, such as the Zenith Trans-Oceanic model (Litwinovich). They were portable but not quite wearable.

It wasn't until the late 1940s and early 1950s that transistor circuitry allowed radios to shrink down into a handheld device. We can thank Sony's TR-63 in 1957 as the world's first pocket-sized radio ("Product & Technology Milestones"). Information that used to be confined to a wood cabinet in your living room or inside your car – news, weather, shows, and music – was now available on the go. It created an entire category of wearables exclusively for personal entertainment, spawning descendants in the form of the Walkman and later the Discman, MP3 players, and every music and podcast app.

Wearables as Always Connected

We can mark the rise of cell phones as the tipping point into constant contact. While the first cell phone in 1973 was the size of a brick, it was the beginning of truly mobile communication (McNamee and Yu). Once we cut the cord, there was no stopping our ability, much less desire, to always be in touch. Cell phones jumped into our cars, evolved into petit digital phones and flip phones for our pockets and purses, and had cousins in the form of pagers and BlackBerries. And then the smartphone arrived in 2007.

With honorable mention to the IBM Simon, Apple was the one that cemented the smartphone as a must-have device. They created a user experience that we hadn't seen before, paired with an ecosystem of devices seamlessly working together. This level of orchestration means that an iPhone is synced with the Apple Watch. It's possible to take a call from my wrist while my phone is in another room, all the while my AirPods are delivering the sound of someone's voice directly into my ears. We see this harmonization continue with the Apple Vision Pro, which accepts FaceTime calls before your eyes ("Apple Vision Pro").

The evolution of wearables is a majestic testament to human ingenuity. All of this progress is now making its way into our work lives. Particularly where safety is concerned, wearable technology has the power to shine a light on hidden risk.

WEARABLES AS PPE

Thanks to wearables, the work of safety is no longer the sole domain of the safety professional – it is now a crowdsourced effort. It's smart PPE, with the power to protect the individual user, everyone in their immediate area, and everyone in the larger workplace.

When we look at traditional PPE, they are exclusively about the individual. Without a doubt, we need static wearables like hard hats, hearing protection, steel-toed boots, and the likes. We want Jerry's head to be safe from falling debris and Jane's feet to be insulated against electricity. But no matter how many of these wearables are issued throughout a work site, they protect one person and only one person at a time.

The same is true of digital wearables, such as those that look for medical symptoms related to heat stress. They use metrics like heart rate, external body temperature, and sweat activity as signs someone is becoming overexerted. But as incredibly useful as these devices are, they continue to focus on the individual.

That's where our safety wearable takes a fundamentally different approach to risk – we look for hazards in the environment.

First, we collect environmental data that has a direct bearing on safety but is otherwise hard to detect. After all, you can't see with your own eyes when decibels are too high or air quality is poor. Our wearable intentionally doesn't have biometric readings or continuous location tracking because we know an accident is rarely the fault of one person or an isolated mistake. We're looking outward from the employee, not inward.

Second, our wearable forms a distributed network of environmental sensors. Each individual becomes a safety contributor, serving as the new "eyes and ears on the ground." Since readings are completely mobile, we can glean environmental insights far better than stationary devices. Every employee is continuously reporting on their environment as they move throughout the workplace performing their job. Companies can use this collective data to gain insights into ever-changing working conditions, allowing them to proactively avoid hazards and eliminate risk.

Wearable safety technology means that every worker is participating in an invaluable group project – one that could very well save a life.

KEY TAKEAWAYS

- Our bodies have worn technology for centuries, starting with the first glasses to today's smart watches
- All wearables create an exchange of information
- Safety wearables are a logical iteration of this technology

BIBLIOGRAPHY

"A Guide to History of Pocket Watches." *Watch Museum*, https://watchmuseum.org/a-guide-to-history-of-pocket-watches/

"Apple Vision Pro." *Apple*, 2024, https://www.apple.com/apple-vision-pro/

Beckman, Jeff. "15 Wearable Technology Statistics [2023 Edition]. *TechReport*, 30 May 2024, https://techreport.com/statistics/wearable-technology-statistics/#:~:text=Wearable%20Technology%20Sales,-5.&text=2022%20showed%20the%20number%20of,billion%20people%2C%20and%20sales%20increased

"1810, The First Wrist Watch." *Breguet*, n.d. https://www.breguet.com/en/history/inventions/first-wristwatch

"Heritage." *Zenith*, https://zenith.com/heritage/

Inspire, 2024, https://www.inspiresleep.com/

"Inventing the Transistor." *Computer History Museum*, 2024, https://www.computerhistory.org/revolution/digital-logic/12/273

"The Invention of the Cardiac Pacemaker." *U.S. Department of Veterans Affairs*, 2 August 2018, https://www.research.va.gov/research_in_action/The-invention-of-the-cardiac-pacemaker.cfm

Litwinovich, Paul. "Zenith Trans-Oceanic, The 'Royalty of Radios.'" *WHSU Public Radio*, 27 March 2015, https://www.wshu.org/vintage-radio/2015-03-27/zenith-trans-oceanic-the-royalty-of-radios

McNamee, Kai and Mallory Yu. "50 years ago, Martin Cooper made the first cellphone call." *NPR*, 3 April 2023, https://www.npr.org/2023/04/03/1167815751/50-years-ago-martin-cooper-made-the-first-cell-phone-call

"1952: Transistorized Consumer Products Appear." *Computer History Museum*, 2024, https://www.computerhistory.org/siliconengine/transistorized-consumer-products-appear/

"Product & Technology Milestones." *Sony*, 2024, https://www.sony.com/en/SonyInfo/CorporateInfo/History/sonyhistory-b.html

"75th anniversary." *Medtronic*, April 2024, https://www.medtronic.com/us-en/our-company/history.html

Van Welsenes, G. H. et al. "Improvements in 25 Years of Implantable Cardioverter Defibrillator Therapy." *Netherlands Heart Journal*, vol. 19, no. 1, 22 December 2010. *National Library of Medicine*, https://www.ncbi.nlm.nih.gov/pmc/articles/PMC3021193/

"Wilson Greatbatch: The Implantable Cardiac Pacemaker." *Massachusetts Institute of Technology*, https://lemelson.mit.edu/resources/wilson-greatbatch

6 Sophisticated Sensors

In our daily lives, we encounter technological marvels that not long ago seemed like mere fiction. Computing has become entirely personal as we now carry in our hands the same processing power that used to require rooms of equipment. This could never have happened prior to this point in history without the emergence of several key technologies.

Up until the past 20 years, collecting and analyzing sensor data at scale and in real time was difficult. We had to overcome a number of limitations with communication infrastructure, data streaming, device size, and power consumption. Now sensors are so small, accurate, and energy efficient that they can reside on the human body. Today's wearable safety sensors process large amounts of data for hours on end. Computing speeds and analytics then allow us to process data about what an employee is experiencing on the job and draw conclusions about working conditions as they're happening.

All of this wouldn't be possible without a number of inventions. It starts with the debut of the semiconductor, which led to integrated circuits. We also have binary code and data processing speeds to thank for the ability to draw insights. These milestones have all led to a powerful network of sensors backed by advanced computing power that is saving lives in the workplace.

SWITCHING TO TRANSISTORS

Today's wearables could never be achieved without the transistor. In 1947, a team at Bell Laboratories led by physicist William Shockley set out to unravel the mysteries of solid-state physics. At the heart of their exploration lay the dual nature of the semiconductor, which is part conductor, part insulator. Among the scientists, John Bardeen and Walter Brattain envisioned a world where these materials could act as tiny switches or amplifiers.

When they created the first functional transistor, it was a humble contraption: a few wires delicately attached to layers of gold foil and semiconductor material. Yet it could control the flow of electricity with unprecedented precision ("The Transistor"). This was a major evolution compared to the prevailing technology, which relied on bulky, power-hungry vacuum tubes (Zorpette). The transistor's ability to regulate currents directly resulted in the creation of electronics that are smaller, more efficient, and more versatile.

BIRTH OF BINARY

Transistors also gave rise to binary code, a language of ones and zeros that computers can process and store. Computers use transistors as switches to control these binary signals so they can be read. A single wire can signify one when it is carrying a current or zero without. Early computers, with their large heat-generating vacuum tubes,

DOI: 10.1201/9781003432302-8

were far less proficient at this two-symbol system. But with the advent of transistors, computers became more sophisticated while shrinking in size.

The rapid miniaturization eventually began to target everyday consumers, such as the first personal computers. Reducing the transistor's footprint was a crucial first step toward the proliferation of internet-connected devices all around us today. It's the reason why an industrial worker can don a safety wearable and seamlessly report data about their environment.

SAND TO CIRCUITS

When sand is processed, the refined silica creates the foundation for modern transistors, often referred to as silicon. Transistors can be laid down onto large sections of silicon to form dense groups, called integrated circuits or chips, working together to accomplish a task. Innovators have continued to find approaches for increasing the density of transistors on silicon for decades.

This rate of miniaturization was predicted by Gordon Moore in 1965 ("1965: 'Moore's Law'"). As the co-founder of Intel, he suggested miniaturization would happen at a rate of doubling transistor density every two years. This has largely held true. For example, the first Intel processor was released in 1971 and contained 2,300 transistors ("Intel's First"). In 1982, Intel released the 80286 processor, which touted 134,000 transistors ("Intel 45"). By comparison, the most recent release of Apple's M2 Ultra processor contains a staggering 134 billion transistors on a single silicon substrate ("Apple Introduces").

Semiconductor manufacturers have continued to develop innovative techniques to stack semiconductor materials to keep the miniaturization dream alive. A byproduct of Moore's Law is that the device housing can shrink as well. We no longer have computers that occupy a closet – they can be worn on our bodies as smart PPE instead.

PROCESSING AT LIGHTNING SPEED

In the early days of computing, machines were tasked with automating processes that were once painstakingly done by hand. As technology progressed, computers could perform increasingly complex and demanding tasks that would have been insurmountable for humans to perform manually in the same amount of time. Thanks to speeds like petaFLOPs, we have the computing bandwidth to collect and process vast amounts of data.

A petaFLOP is a unit of measure describing a quantity of computer calculations. An individual FLOP is a single floating point calculation. In simpler terms, each time you add two prices with dollars and cents together in your head, you perform a FLOP. The prefix "peta" is used to note a multiplier of a quadrillion (10^{15}). To put this into perspective, one petaFLOP is equivalent to doing a single math operation every second for over 31 million years ("Understand measures"). Currently, the fastest computer in the world is at Oak Ridge National Laboratory and can execute over 1,000 petaFLOPs per second; the second fastest comes in 585 petaFLOPS and is owned by the Argonne National Laboratory (Afifi-Sabet).

For safety professionals, these speeds are critical to accomplish the goal of understanding human work in real time. Along with miniaturization, they are what empowers wearable sensors to uncover indicators that are often hidden, like detecting indoor air pollution or risky movements that might lead to a repetitive motion injury or an unexpected fall.

VOC SENSORS FOR AIR QUALITY

Indoor air quality (IAQ) is one of the hardest safety concerns to detect as it is rarely visible to the eye or noticeable to the nose. Volatile organic compounds (VOCs) are chemicals that compromise IAQ when they evaporate. Often synthetic in origin, they are emitted by paints, cleaning agents, adhesives, sealants, and fuels. Even at low levels of exposure, health effects may include headache, dizziness, nausea, and fatigue as well as nasal, skin, eye, and throat irritation. VOCs are particularly insidious because they are often a cumulative hazard through repeat inhalation rather than a single exposure ("Volatile").

Thanks to the evolution of semiconductors, the latest generation of IAQ sensors is the size of a grain of rice. Our wearable uses a metal oxide sensor, which contains a semiconducting material that binds with oxygen. The presence of VOCs removes some of those oxygen molecules, decreasing resistance and increasing current flow. By measuring the resistance of the semiconductor, the wearable sensor can estimate the total VOC (TVOC) concentration in the environment.

Because these mobile IAQ sensors take readings wherever workers go, they are particularly adept at uncovering VOCs in surprising places. Case in point – our wearable helped a food production company find a hidden VOC source from a machine applying glue to cardboard lids. This was a shock as no one had reported fumes or health concerns. Furthermore, the glue's safety data sheet didn't have a VOC warning. However, when the company talked to the adhesive manufacturer, they learned that the glue doesn't off-gas once it cools. Without the IAQ sensor, no one would have realized that VOCs were present while the glue was warm.

ACCELEROMETERS FOR SAFE BODY MECHANICS

Accelerometers are based on Newton's first law of motion that an object will remain at rest unless acted upon by an external force. A type of accelerometer called microelectromechanical systems (MEMS) measures static forces such as gravity as well as dynamic forces resulting from motion or vibration.

MEMS accelerometers detect changes with a high level of accuracy. For instance, when an object experiences acceleration, it exerts a force on the accelerometer's sensing elements, causing them to shift with motion. This movement is measured and converted into an electrical signal, providing information about the object's acceleration. The applications are vast, from features such as screen rotation and stability control in helicopters to health applications such as tracking respiratory rate or fitness activity.

For workplace safety, accelerometers are how MākuSafe flags repetitive motion and slips. We use a three-axis accelerometer that correlates to typical motion: side to side, up and down, and back and forth. These movements are recorded over the course of a day. Using a calculated result called a signal magnitude vector, we can understand the cumulative impact of these three types of accelerations. That data is charted for individual workers in our Motion Explorer, which creates a profile that tracks and ranks motion exposure by hour, day, week, and month. This analysis is what allows a company to identify workers who have high potential for injuries.

For example, the data shows there's a difference in motion exposure for two people who are working in the same area and performing the same job. It's a clue that prompts safety leaders to investigate why there is a disparity. They can then implement meaningful changes, such as ergonomic training or adaptive equipment.

This data is a game changer. It isn't effective for a safety professional to conduct a walkthrough and be in the right place at the right time to catch a glimpse of someone moving in a risky manner. Instead, sensor data can highlight workers who have the greatest amount of risk. After making improvements, data can be collected again to show what impact was made. All thanks to a tiny MEMS accelerometer.

KEY TAKEAWAYS

- Key inventions like transistors and integrated circuits are the foundation for modern sensors
- Device miniaturization allows them to be worn on the body
- Computing speeds enable real-time data processing to uncover safety indicators

BIBLIOGRAPHY

Afifi-Sabet, Keumars. "The 7 most powerful supercomputers in the world right now." *Live Science*, 28 March 2024, https://www.livescience.com/technology/computing/top-7-most-powerful-supercomputers-in-the-world-right-now

"Apple Introduces M2 Ultra." *Apple*, 5 June 2023, https://www.apple.com/newsroom/2023/06/apple-introduces-m2-ultra/

"Intel's First Microprocessor." *Intel*, n.d. https://www.intel.com/content/www/us/en/history/museum-story-of-intel-4004.html#:~:text=Initial%20transistor%20count,technology%20held%20560%20million%20transistors

"Intel 45 Nanometer Technology Gives the Transistor Dramatic Make-Over." *Intel*, 29 January 2007, https://www.intel.com/pressroom/kits/45nm/IntelTransistor_Bkgndr_FINAL.pdf

"1965: 'Moore's Law' Predicts the Future of Integrated Circuits." *Computer History Museum*, 2024, https://www.computerhistory.org/siliconengine/moores-law-predicts-the-future-of-integrated-circuits/

"The Transistor: 75 Years Since the Famed Nokia Bell Labs Invention Changed the World." *Nokia Bell Labs*, 15 December 2022, https://www.bell-labs.com/institute/blog/the-transistor-75-years-since-the-famed-nokia-bell-labs-invention-changed-the-world/#gref

"Understand measures of supercomputer performance and storage system capacity." *Indiana University*, 2024, https://kb.iu.edu/d/apeq#:~:text=A%201%20petaFLOPS%20(PFLOPS)%20computer,every%20second%20for%2031%2C688%2C765%20years

"Volatile Organic Compounds' Impact on Indoor Air Quality." *EPA*, 13 August 2024, https://www.epa.gov/indoor-air-quality-iaq/volatile-organic-compounds-impact-indoor-air-quality

Zorpette, Glenn. "How the First Transistor Worked." *Spectrum IEEE*, 20 November 2022, https://spectrum.ieee.org/transistor-history

7 IoT for Safety

Currently, there are an estimated 17 billion internet-connected devices, which is over twice the global population (Vailshery). If we could visualize this distributed network on a map, it would appear that everything and everyone is connected. The information they generate is transforming our understanding of the world, including workplace safety in industrial settings. When safety sensors are worn by workers, the data can be extremely valuable in understanding how work is actually happening.

This result provides an unparalleled view into atmospheric conditions and ergonomic motion – leading indicators that contribute to the majority of accidents. By analyzing this pool of IoT data, safety professionals can uncover risks and hazards before they become a problem.

IoT RADICALLY CHANGES PROBLEM-SOLVING

One of the ways IoT is a game changer is that it opens up new ways of problem-solving. Thanks to our constant state of connectivity, we are increasingly asking, "why can't I know this continuously for myself and never miss an important piece of information?" Real-time data from IoT empowers us to gain those answers, using insights to tackle issues in innovative ways.

A prime example is weather data. There was an age when only scientists were in charge of collecting readings. They had to physically step outside, measure the relative humidity with a hygrometer, and then submit recordings to be aggregated into a forecast. Today, we have volunteer networks like Weather Underground that are based on IoT. Participants install their own weather station that includes atmospheric sensors such as wind speed, rainfall, pressure, and temperature. A WiFi device then transmits the data for analysis. With over 250,000 contributors, this crowdsourced IoT network continues to change the nature of weather forecasting ("Personal Weather").

We can also look at how people use health and fitness trackers to proactively make decisions based on health data. They don't have to wait for a yearly doctor's appointment to see that their wellness needs attention. They can voluntarily make adjustments to their diet, activity levels, and sleep hygiene to improve their numbers – all because a wearable smart device is providing actionable information.

Commercial and industrial buildings have pioneered IoT networks as well. Major systems such as lighting, HVAC, and energy mechanicals have embedded sensors. Because these devices are reporting on efficiency, maintenance, and performance, they're changing the way facilities are operated and even automated. Even the simplest solutions can be revolutionary. Imagine the labor savings of an IoT paper towel dispenser that can send an alert when it's low rather than a worker manually checking every bathroom on a schedule.

How does IoT support safety professionals? By leveraging workers to create a dynamic network that is always on the move. The downside of conventional sensors

DOI: 10.1201/9781003432302-9

is that they are typically mounted on a wall with a limited radius for detection. Even with handheld or tripod devices, a user may find it difficult to multitask while taking the reading. However, wearable safety devices will monitor for environmental and ergonomic hazards, specific to them, with no impact to workers' jobs. Their only involvement is to wear the device just like any other PPE. As a result, these wearable IoT sensors generate data about the immediate surroundings of your most precious assets – your people.

HOW IoT AUGMENTS SAFETY OBSERVATIONS

Having IoT sensors moving throughout your facility is an incredible advantage over walkthrough observations. Findings from site inspections are often limited because they only occur at specific intervals. Despite their vast training and knowledge, it's rare for a safety professional to be at the right place and the right time to witness a hazard.

Let's meet Josh, an industrial hygienist who is working hard to protect the front-line workers at his facility. There are several reasons why site observations aren't yielding the safety data he needs.

- **People change their behavior when they are being watched**. Workers know who Josh is and what his job entails. The minute he appears, they'll start being more careful. Except that Josh doesn't need people to be on their best behavior – he needs them to act normally. Just by his mere presence, Josh isn't getting an accurate view of everyday working conditions.
- **Josh has a preconceived idea about what risks could be present**. He's been to this facility many times and understands which hazards are inherently unique. He also knows its safety track record through dealing with past incidents. However, this familiarity can start to color his perception of what to expect, and he'll start having more difficulty looking for risk with fresh eyes.
- **Josh is just one person – he can't be everywhere at once**. What are the odds that he will catch something that is a safety issue? Even if Josh systematically performs his observation once a week for 30 minutes, that's just a fraction of time. The law of averages isn't working in his favor.
- **Workers can have a subjective view of safety**. For example, Josh could issue a survey asking about which day is the hardest for physical exertion. The majority of people might cite Monday simply because of its reputation, yet Josh sees that most injuries occur on Friday. Now he's confused because his workforce's perception about risk doesn't match his safety numbers.
- **Because the work environment is continuously changing, risk can be introduced at any moment**. But Josh can't be on the floor all the time – observations are only one part of his job. How can he track indicators when he's elsewhere?
- **Risk can be invisible to humans without instruments**. Josh can only rely on his five senses so much. He will inevitably need specialty tools to check for hidden hazards like decibel levels, VOCs, ambient light, and humidity. These readings all take time and training to perform, all of which are in short supply for Josh.

It leaves Josh wondering why he can't use technology to do his job in the same way as other professionals. He thinks about the warehouse operations manager who has vibration sensors on the conveyor systems. Or the day trader who works from her home office yet knows stock prices on a second-by-second basis.

Josh is thinking to himself, "How come I don't know at any given moment what's happening with the safety position of my company? Shouldn't I be able to know what kind of air quality my workers are experiencing in real time? Isn't it possible to collect data on fatigue contributors and then trend them for patterns? Where are the tools that allow me to see the bigger safety picture?"

Josh is asking the right questions, and he can gain these insights by equipping his workers with safety sensors. The resulting data supercharges his job, allowing him to oversee risk without physically having to be present.

With safety data generated at all hours and throughout the entire facility's work zones, Josh can now drill into the true nature of risk at this location. The details provided by environmental and motion sensors help him uncover several surprising issues.

Heat Stress

The Scenario

Josh's company has a challenge with indoor temperatures because it uses equipment that releases heat as a byproduct. The outside temperature is also impacting the indoor temperature despite their efforts to move air throughout the facility. This has resulted in a heat index that is regularly over 100-degrees Fahrenheit. It is such a pervasive problem that the company tries to stay ahead of dehydration by lining the hallways with stacks of sports drinks.

The Problem

Though the facility had three wall-mounted temperature sensors, they weren't IoT connected. Each thermometer has a clipboard hanging below, and it is somebody's job to walk over every hour and record the temperature. Occasionally, somebody takes those numbers to Josh, who tries to manually draw some conclusions. But ultimately, these readings aren't useful because they aren't automated or aggregated effectively. They also aren't representative of what each worker is experiencing.

The Response

Using wearable sensors like MākuSafe to gather real-time data directly from their employees, Josh has immediate insight into heat exposure and can anticipate when workers are going to need hydration breaks. Protecting workers moved from a manual, brittle process to one that is proactive, ultimately leading to a more productive and safer work environment.

Physical Exertion

The Scenario

Josh's company understands that the line between work as designed versus work as done is always moving. But they need a more intelligent way to measure these fluctuations. For example, they want to know if their risk profile changes throughout the day, such as the difference between eight in the morning versus two in the afternoon. This is driven by a rise in motion injuries among their loading teams.

The Problem

After a period of data collection, a trend emerges – every Thursday at 2pm, there's a significant spike in motion across the whole team. Because this window of time is a complete outlier, it's a concrete clue for Josh to investigate. In talking to the affected team, he learns that this concentration of movement matches the weekly delivery of a large shipment. Because it's a large load to move and the dock isn't air conditioned, everyone rushes to complete the work. Despite their efforts to be efficient, however, the team reports feeling hot and exhausted each time.

The Response

Josh not only knows that this one particular afternoon has higher risk, but more importantly, what is the root cause. The data pointed Josh to a trifecta of lagging indicators: heat exhaustion, dehydration, and physical motion injuries. He can now start to problem solve, such as working with the shipper on its schedule, adding more team members to this shift, or ordering automated conveyors for unloading.

Inadequate Lighting

The Scenario

Part of Josh's site includes a warehouse with forklift drivers. Even with regular safety training and OSHA certifications, there's still a considerable amount of accidents in this zone. Josh is stumped why this is an ongoing issue given that standard precautions are in place.

The Problem

After outfitting workers with wearable safety sensors, the likely root cause is revealed – this area is registering inadequate lighting. Upon visual inspection, Josh finds that due to recent volumes in orders, pallets of product are being staged in this area. The high stacks are significantly reducing visibility. It's also highly likely that workers are struggling to see fork trucks due to the lower lighting and tighter aisles.

The Response

Mitigating this risk started with creating a new staging area and relocating pallets. Josh also met with the forklift drivers to ensure they have a way to report issues like this in the future.

KEY TAKEAWAYS

- Workers with wearable sensors can be added to your IoT network
- Technology and data support the important work of safety professionals
- Data can uncover the true nature of risk

BIBLIOGRAPHY

"Personal Weather Station Network." *Weather Underground*, 2024, https://www.wunderground.com/pws/overview

Vailshery, Lionel Sujay. "Internet of Things (IoT) in the U.S." *Statista*, 4 June 2024, https://www.statista.com/topics/5236/internet-of-things-iot-in-the-us/#topicOverview

8 Machine Learning for Intelligent Insights

How do you handle the volume of data generated by safety sensors? Because IoT devices produce a prolific amount of information, traditional analytics aren't capable of analyzing its volume or complexity. We need artificial intelligence to shift through it and discover actionable information.

Machine learning has the ability to consolidate, organize, and analyze fast-moving safety data. It's what allows safety professionals to uncover important trends connected to risks and hazards. Using these intelligent insights, they can improve working conditions and have a profound impact on safety.

UNDERSTANDING INTELLIGENT INSIGHTS

The defining quality of an intelligent insight is context. Though many smart devices generate tremendous amounts of raw data, we need a framework to give it meaning. Otherwise, all you have is noise. An intelligent insight puts data to work.

- From data, we create information
- From information, we can draw insights
- From insights, we can make decisions
- From decisions, we can measure change

For example, a temperature reading is a type of raw data. By itself, that number doesn't mean anything. But an insight is created when we determine that the temperature is too high and poses a safety risk to workers. This intelligent insight empowers leadership to make adjustments that will reduce the likelihood of heat stress.

Intelligent insights are also a workforce multiplier because they help safety professionals find indicators where they may not have realized they existed in the first place. It's all about uncovering the true nature of risk, which is revealed by deviations in the data. What we've found many times is that what is typical or "normal" in a facility is just the status quo – no one has evaluated whether it's safe or unsafe.

Let's use noise as a safety concern. According to OSHA thresholds, 85 decibels or lower is permissible without additional safeguards ("Occupational Noise"). If your facility's average is above 85 decibels, a hearing conservation program is required. But sometimes companies haven't taken the time to establish their decibel average. They have no idea what their everyday levels are, much less where they fall within OSHA's criteria. Even if they start taking regular readings, they're awash in information that needs context and interpretation. For example, how would they know that

DOI: 10.1201/9781003432302-10

noise reaches harmful levels on Wednesday and Thursday afternoons? They need a way to corral all that data so they can derive intelligent insights from it. This is where they can look to artificial intelligence to help interpret data quickly and efficiently.

MACHINE LEARNING FOR SAFETY DATA

Machine learning, which is a subset of artificial intelligence, uses algorithms to train a model to recognize patterns and classify that data accordingly. It's how large amounts of data are processed so intelligent insights can be drawn in an automated way.

In *unsupervised* learning, we provide a plethora of examples from different data sets and allow the machine learning model to discover the nuances between them. This self discovery is often used for unexplored data sets to find previously unobserved groupings of data.

In *supervised* learning, we show the model examples of known data sets and then have it look for new examples that match. The MākuSafe team illustrates this at conferences with a simple potato and a touch of humor. We show a picture of three bins of potatoes labeled Group A, B, and C to match their colors. Then we show a picture of an unlabeled potato and ask the audience which group it belongs to. When they classify it correctly, we joke that they just performed machine learning. This is because each audience member "trained" themselves by looking at the picture of potatoes in bins. In fact, this is supervised learning because similar data was already grouped into the bins. When presented with an uncategorized potato, they could immediately classify the potato based on the similar ones from the picture.

Machine learning models are trained using thousands of examples that are grouped into known classes. For example, models have been populated with millions of handwritten letters so they have the ability to decipher individual letters. When the model is presented with a new handwritten sample, it knows an "a" from "o" and so on. More importantly, this can be done at scale across millions of examples with amazing speed – no human intervention is required.

This same technique is changing safety. It's essentially how MākuSafe has trained our motion model to recognize slips, trips, and falls. When somebody has a high force motion, our machine learning model can say "I recognize this motion signature as a slip." It is even able to tell us how confident it is with the classification: "I'm 73% confident that is a slip, 17% confident it's a trip, and 10% confident it's a fall." Not only is this useful for uncovering individual instances, but safety leaders might also discover a cluster of slip incidents that signal there is an underlying problem at play.

TRENDS UNCOVER ISSUES

Part of the beauty of machine learning is that it can discover trends that may not be perceived easily by safety professionals or uncovered by traditional data analytics. This is because it's difficult to interpret a single point on an IoT network, much less all of them combined. So, we use a consensus engine to shift through sensor information, tabulating their signals into a voting system that creates meaning.

Air quality is a common example of a hidden issue. How can you truly capture if an area has high VOCs? It can be difficult to discern if there's a genuine problem or an outlier, like a worker who put on too much body spray. By using data to establish a trend, it's easier to show when a pattern has grown into a concern worth exploring.

Let's say there's a trend that shows air quality has decreased by 26% on your loading docks. We can dig into that, noting that CO_2 went up 14.2% over the last three weeks while VOCs went up 25.9%. We can next look at the time frame to see how long this issue has been occurring. We can see that in the last three weeks, it's risen to 600 ppb. Knowing that our baseline is 510 parts per billion, that's a 15% rise in VOCs.

All of this context from the data is what leads us to an intelligent insight – there is proof there is an air quality issue and we need to investigate. A logical step would be to look into the ventilation system or see if something unexpected is offgassing. For example, the root cause is that the dock floors had just been painted. By adding some temporary fans to help dispel fumes, VOCs will soon return to normal.

GETTING AHEAD OF TRIPPING HAZARDS

Slips and trips are notoriously difficult to document, mainly because most people brush it off as inconsequential. They might even assume they are at fault – who hasn't sheepishly tripped because they were shuffling their feet? But what that one individual doesn't realize is that eight of their coworkers have also tripped at this particular spot. It's possible the culprit is something in the environment, like an uneven floor, that needs attention. How do you gain optics into this risk and have the ability to minimize it before it escalates into a fall and someone gets a broken wrist or worse?

Here's how it might look in practice. The data from our wearable device notes an increase in slips in a particular concrete section. An analysis suggests that it may be the result of increased condensation from humidity at a certain time of day. This intelligent insight allows the health and safety manager to collaborate with the facilities team to run the air handler more frequently in that area.

One client experienced this capability firsthand and mitigated a hazard that could have caused a slip or trip. MākuSafe data captured several slipping motions at one entrance. When the safety team went to investigate, they found that snow had blown onto the sidewalk leading up to the building. They worked with their site logistics team to clear off the path and apply ice melt.

This is the power of using advanced analytics and machine learning to handle safety data. Far from replacing safety professionals, a machine learning model allows them to rapidly digest and process data. Armed with intelligent insights, leaders are empowered to make safety decisions in new and profound ways.

KEY TAKEAWAYS

- Data needs context in order to be actionable
- Machine learning models help organize and process data
- Trends in the data can reveal safety issues that are otherwise hard to detect

BIBLIOGRAPHY

"Occupational Noise Exposure." *Occupational Safety and Health Administration*, https://www.osha.gov/noise#:~:text=OSHA%20requires%20employers%20to%20implement,%2Dweighted%20average%20(TWA)

Part III

Leadership, People, and Implementation, Tom West

9 Leading Indicators from Wearables Improve Safety Outcomes

In November of 2019, we had the opportunity to participate in a Public Stakeholder Meeting held by the Occupational Safety and Health Administration (OSHA) on "Using Leading Indicators to Improve Safety and Health Outcomes." The meeting was a follow up to a June 2019 OSHA publication that provides guidance and information to employers on incorporating leading indicators into their safety and health programs. Comments made throughout the session were supportive of the use of leading indicators in safety and health programs. Many trade organizations and larger companies had begun implementing or assessing the use of leading indicators, but there was a high degree of variability. Although there was no real standard for use at that time, and none still now, there is great support for OSHA and other entities globally to further promote their use to improve workplace safety outcomes.

A takeaway from this meeting was that there was agreement in spirit on the importance of leading indicators, but an acknowledgment that industries are different and businesses are unique. Therefore, interpretation may vary, and deciding which leading indicators were most meaningful to each would be challenging. Further, there seemed to be an absence of clear ways to use leading indicators, even when well defined.

Comments included what general attributes "good" leading indicators should possess. Frankly, it was disappointing that the suggested indicators were limited to things like training time, toolbox talks, daily safety conversations, meeting attendance, time spent in plant by senior managers, unsafe acts, scorecards, and the like. While tracking these is undoubtedly a step in the right direction, wearables and sensor monitoring are accessible now and offer a tremendous opportunity for real-time insights on experiences and exposures associated with leading indicators.

DEFINING LEADING AND LAGGING INDICATORS

One particularly insightful written contribution to that OSHA meeting came from the partners of ORC HSE Strategies, global thought leaders in health and safety. Their comments, on behalf of their significant roster of member companies, underscored the need for practical, actionable leading indicators that truly enhance workplace safety. Their perspective was spot on, highlighting the necessity for leading indicators to not be just metrics, but also "tools that drive meaningful safety improvements." They suggested tracking observations, safety suggestions and implementation rate, safety action items completed on time, building transparent cultures of finding and

DOI: 10.1201/9781003432302-12

fixing issues, focus on where the employee meets the work, risk-based approach, taking action and proactivity.

MākuSafe also voiced during that OSHA meeting that technology, specifically wearables, have the potential to generate tremendous amounts of new data that will offer employers easy access to meaningful leading indicators of risk and hazards. So much more is possible now, so easily, and provides significantly more potential to make an immediate difference.

Keep in mind that leading indicators are proactive, preventive, and predictive measures that provide information about the performance of safety and health activities. They focus on what could potentially cause future incidents or injuries. Examples include safety training sessions and safety audits, but more importantly the reporting of near-misses and anything that we can do to identify potential hazards before incidents occur.

Lagging indicators, on the other hand, are reactive measures that provide data on incidents that have already occurred. These include statistics on injury rates, lost workdays, and workers' compensation claims. While lagging indicators are crucial for understanding past performance, they do not predict future incidents. The idea of a balanced scorecard of both types is often suggested as it makes sense that lagging indicators should provide a useful gauge of how well our leading indicators have been selected and are working.

Leading indicators are essential for creating a safer workplace as they help organizations identify and mitigate risks *before* incidents occur. They contribute to a proactive safety culture and can lead to significant improvements in safety performance, such as:

1. **Early Detection of Hazards:** Leading indicators can reveal potential hazards early, allowing for timely interventions.
2. **Improved Safety Culture:** Focusing on leading indicators can promote a culture of continuous improvement and proactive safety management.
3. **Enhanced Employee Engagement:** Employees are more likely to engage in safety programs when they see their efforts in reporting and mitigating hazards are valued and acted upon.
4. **Reduction in Incidents and Injuries:** By addressing potential risks early, organizations can reduce the number and severity of workplace incidents and injuries.

Despite the clear benefits, implementing leading indicators is not without challenges. For starters, defining universal indicators is difficult since different industries and operations have unique risks. Next, management's buy-in to support and act on leading indicator data is crucial for success. The ability to collect and analyze leading indicator data can also be resource intensive and may require new technology or processes. Finally, successful implementation requires a positive safety culture where employees feel comfortable reporting potential hazards without fear of retribution.

OSHA and others have developed guides and materials with recommendations on using leading indicators. However, they acknowledge that, given the diversity of industries and operations, one-size-fits-all practices would be challenging to establish. Further, the vast majority of examples of leading indicators still being discussed in most organizations today seem to revolve around things that may be easy to measure but are not the most beneficial. Nearly every discussion of leading indicators seems to include things like measuring attendance at safety meetings, documenting training completion, and number of toolbox talks delivered. While these measures are important, they often fail to provide the most actionable and important insights needed to prevent incidents before they occur.

Given the work we were doing at MākuSafe and how we could contribute to meaningful leading indicator data, we developed our own acronym for characteristics of useful leading indicators – T.R.U.E^2. To be as useful as possible, leading indicators can and should have these characteristics:

Timely: The data should be available in real-time or near real-time.
Relevant: Indicators should relate directly to worker health and safety, including the most common causes of incidents, injuries, and claims.
Useful: The data should not only provide unique insights not easily obtained from other sources but also be immediately actionable.
Easy and Economical: It should be straightforward to gather and interpret the data without being prohibitively expensive.

The data gathered by the MākuSafe wearable devices includes an array of potential environmental hazards, human movement, and ergonomic risk and facilitates communication of frontline worker observations. Other wearables may only offer a portion of these, have different strengths, or specific applications. But in our case, the data represents leading indicators for a wide array of the most common and frequent workplace incidents and injuries. Additionally, by gathering both ergonomic and industrial hygiene data simultaneously, you can begin to understand if factors in the environment may be directly contributing to ergonomic risks. For example, poor air quality may be accelerating fatigue in workers, leading to a higher likelihood of MSD injuries from those exposed to the poor air quality.

T.R.U.E.^{2} IN PRACTICE

Consider this:

> "In the U.S. alone workplace injuries cost businesses more than $1 billion per week. That's more than $58 billion per year, with the vast majority caused by just 10 common injuries. Liberty Mutual Insurance regularly publishes their Workplace Safety Index (WSI) where they examine the top 10 causes of the most serious disabling workplace injuries – those leading to more than five missed workdays – ranked by their direct cost to employers based on medical expenses and lost wages. The top 10 are responsible for 82% of that $58 billion"
>
> (2023 Workplace).

FIGURE 9.1 Wearable technology can gather leading indicator data on a wide array of common but costly EHS concerns. This can include environmental hazards, such as temperature, heat index, noise exposure, and air quality. In addition, potentially harmful human movement and motion data can be gathered, including forceful strain and exertion, push/pull, and level of physical effort expended to perform work, all of which can contribute to cumulative physical impact and potential for MSDs.

Based on BLS data, the WSI identifies a number of interesting findings. In nearly every instance, the risk of these accidents could very likely have been mitigated by a leading indicator:

- Three of the top ten causes are related to slips, trips, and falls, which constitute 35% of the cost of the top 10 or about a third of the $58 billion total. How many of those were preceded by others who slipped, tripped, or fell in the same place without being hurt and didn't tell anyone? In the case of falls from height, how many of those had movement or motion indicators detectable by sensors from when others caught themselves on a wobbly ladder, or similar, but didn't fall?
- Two more causes out of the ten involve exertion, human movement, or ergonomics and are responsible for approximately another 35% of the cost of the top 10. Sensors like MākuSafe can identify forceful exertion and generate notifications for those workers who rank high in work effort being expended compared to their peers in similar work roles. Some wearables require a multitude of sensors and may be less practical for day-to-day wear, but can help identify sub-optimal postures or bends. In either case, these precursors can provide direction on where an ergonomic assessment is needed or even where video-driven AI can be deployed to further understand and prevent ergonomic injuries.

- Two causes – struck by or caught in – account for 7%. How many of those were near-misses where forceful or accelerated human movement, such as someone jumping out of the way or pulling their hand out, could have been detected before these injuries occurred?
- Pedestrian vehicular incidents also accounted for $1.61 billion. Since these are workplace injuries, it's fair to assume a good portion of those involve fork trucks or similar powered industrial vehicles. How many of those had traffic patterns, near-misses, or accelerated movement indicators that could have been identified by sensors before injuries result?
- One cause responsible for exposure to harmful substances costs $3.35B per year. How many of those could have been detected by VOC or gas monitors in advance of exposure?

A wearable device such as Mākusafe has been developed, which is able to contribute leading indicator data automatically when near-misses occur. This data, made useful, has predictive value that when acted upon can contribute significantly to all of these accident causes.

KEY TAKEAWAYS

- The 2019 OSHA meeting was important for MākuSafe because it reinforced the importance of leading indicators in improving workplace health and safety outcomes.
- It also validated our early thinking on leveraging real-time data to predict and prevent incidents.
- A focus on the characteristics of meaningful leading indicators enabled through sensors and wearables will be crucial in fostering safer and more proactive workplace environments.

BIBLIOGRAPHY

"2023 Workplace Safety Index." *Liberty Mutual*, 10 Aug 2023, https://business.libertymutual.com/insights/2023-workplace-safety-index/

"Comment from Newell, Stephen, et al; ORCHSE Strategies, LLC (ORC HSE)." *Regulations.gov*, 6 Feb 2020, https://www.regulations.gov/comment/OSHA-2019-0005-0023

Occupational Safety and Health Administration. "Meeting Summary: Using Leading Indicators to Improve Safety and Health Outcomes Public Stakeholder Meeting." *Regulations.gov*, 13 Jan 2020, https://www.regulations.gov/document/OSHA-2019-0005-0020

"Using Leading Indicators to Improve Safety and Health Outcomes." *OSHA*, https://www.osha.gov/leading-indicators/research

"Using Leading Indicators to Improve Safety and Health Outcomes – Additional Resources." *OSHA*, https://www.osha.gov/leading-indicators/research

10 Successfully Implementing Innovative Technology in Organizations

Introducing innovative technologies like wearable safety devices into an organization can be a transformative journey. Any digital transformation initiative, including safety technology, is broad and complex, requiring strategic planning and effective change management from leaders.

Organizations will also need to recognize that they may encounter employee resistance when implementing new technology. It's common for concerns like privacy or general apprehension to arise, which is echoed in the Safety Technology 2024 survey by The National Safety Council's Work to Zero initiative ("Safety Technology 2024"). Workers may not even see the value: "46% reported they did not want to change the safety equipment they were currently using and 41% reported current safety equipment and technology meets their needs so there is no need for new adoption" ("Safety Technology 2024"). And yet the report calls out that the biggest motivator for exploring new safety technology is a recent serious injury or fatality.

The foundation of implementing any new technology begins with a willingness to embrace change and a piloting attitude. Organizations must be open to exploring new ideas and experimenting with novel solutions so innovation can thrive. This chapter will explore the essential steps and considerations for successfully implementing such cutting-edge solutions like wearables. We'll delve into the importance of a piloting attitude, ensuring organizational readiness and navigating change management.

HOW TO ESTABLISH A STRATEGIC PILOT

Pilots serve as valuable testing grounds to assess the feasibility and effectiveness of new technologies before committing to full-scale implementation. In the early stages, it can be tempting to say yes to any pilot opportunity. However, it is crucial to recognize that managing successful pilots demands significant resources. Therefore, focus on well-defined pilots where both the vendor and the client can succeed. The pilot should have clear success criteria and the potential for scalability. As Stephen Covey taught us, "begin with the end in mind," or at least have partial clarity around problems to be addressed, and then define what a new solution can and should solve. Also, embrace the realities of the solution being considered, in its current format,

DOI: 10.1201/9781003432302-13

as well as the potential future capabilities that will be useful once developed and released by the vendor.

Both the vendor and the client need to prove the value of the new technology swiftly and efficiently. There is a phrase often used in startup circles called pilot purgatory. If a piloting client and the vendor haven't clearly identified how to proceed with agreed upon milestones, some pilots can go on indefinitely. It is in the best interests of both parties to understand the plan's initial stages, create measurable goals, and set a defined timeline. They should also agree on a long-term vision of what a larger relationship could look like. It's crucial to answer questions like:

- How will we measure success during the pilot?
- What is our time frame?
- Who needs to be involved *DURING* the pilot that would be responsible for scaling the technology if the pilot is deemed successful?

Choosing the right pilot site or department is crucial for a successful implementation. It is essential to identify areas with a high likelihood of success and minimal compliance hurdles. For example, we started one pilot with a food packaging operation in the warehouse with material handling workers. This allowed them to evaluate a trial run first, then scale up to cover more of their workforce in other parts of the operation.

Having an internal champion will also significantly impact the success of a pilot. Often holding titles like VP of Safety or Director of EHS, these advocates promote the new technology within the organization, staking their reputation on its potential benefits. They play a crucial role in driving change, rallying support, and ensuring that the technology is embraced by all stakeholders. Their commitment and dedication can make a substantial difference in overcoming resistance and achieving successful implementation. This support needs to trickle-down to frontline leadership, who will play a crucial role in making sure that the technology is being utilized and that the workers understand their role in the pilot's success. We'll expand on this in a moment, but it bears repeating that the role of the frontline leadership and frontline workers is extremely critical in properly evaluating the new technology.

Lastly, effective implementation requires cross-functional communication and early involvement of all relevant departments. IT, legal, HR, and other stakeholders may need to be engaged from the beginning to address concerns and ensure smooth integration. There are many people within an organization who need the opportunity to provide input or ask questions. A lack of stakeholder involvement can lead to significant delays and potential roadblocks. Open communication and collaboration will allow organizations to navigate these challenges effectively.

ENGAGING FRONTLINE WORKERS

Frontline workers are often the most affected by new technology implementations. Ensuring their inclusion and early communication, especially around why the change is important, is vital to avoid resistance and negative reactions. Listening to their concerns, addressing their questions, and involving them in the process develop a sense of ownership and acceptance. Focus groups and learning sessions can be valuable tools to gather feedback and demonstrate the organization's commitment to a smooth transition.

This is essential to ensuring both initial and ongoing usage or participation. As the saying goes, even the best tool in the world does no good if it sits in the toolbox unused. For technologies like exoskeletons, it may be feasible to obtain only a couple of units and ask for volunteers who want to experiment with them on their own for a few weeks, and have them report back.

In our case, to expedite time to value, we recommend deploying 100 wearables to a group of 200 workers on two different shifts, with client leadership committing to at least 80% participation. In about 100 days, we will have monitored approximately 100,000 to 150,000 hours of work, gathered millions of data points showing what normal work looks like as well as indicators of potential hazards and risk, and typically find loss avoidance evidence.

With good involvement and communication, this can be done positively. When pilots are enforced through mandatory compliance, they can be met with resistance. Clearly communicating the functionalities, limitations, and direct benefits to frontline workers can significantly enhance participation, ultimately ensuring the pilot's success. Leadership's involvement in this process is crucial for building trust and commitment. Early communication campaigns can also play a significant role in a successful launch. When these requirements are addressed, use of a new technology solution can be expected by leadership which leads to faster time-to-value than making use optional.

FIGURE 10.1 Installation of wearable kiosks in a large facility with hundreds of workers. To overcome the challenge of nearly 20 languages being spoken on site, card readers were installed on each kiosk to provide a swipe-and-go experience for workers.

Our solution requires collaboration with the client to preconfigure hardware, set up software instances, and then go onsite to do an installation, which usually takes about a half a day even for a large facility with millions of square feet of coverage area. We have learned that deploying our technology should coincide with an in-person overview directly. We underscore what the technology does, what it does not do, what data is gathered, and how it will be used. This is most successful when leadership is shoulder to shoulder with us, committing to the same tenants. In some instances, management even delivers this orientation themselves, preceding it with lots of early communication that new tech is coming soon, why it is important, and how it will be used. Some clients have built early communication messaging campaigns and were extraordinarily successful at launch as a result.

As the pilot progresses, ongoing communication between leadership and frontline workers is essential to maintain momentum and celebrate progress. Sharing data, demonstrating its use, and reinforcing the importance of the technology help in sustaining enthusiasm and commitment. Celebrating small successes and milestones creates a positive feedback loop that encourages continued adoption and utilization.

NAVIGATING UNIONIZED ENVIRONMENTS

Implementing new technology in unionized environments requires careful navigation. Engaging with both labor leaders and management early in the process can help address concerns and foster collaboration. By highlighting the mutual benefits of the technology, organizations can gain support from both sides and create a conducive environment for successful implementation. One illustrative story from our experience stands out.

We were in talks with a multinational heavy industrial manufacturer with a nearby unionized site about a potential pilot location. They requested an onsite visit and a joint meeting with labor leaders and members of management, which was a great idea as it swiftly led to bridging any gap between the two perspectives. After just a few minutes of discussion, something remarkable happened. Both sides of the table – union representatives and management – began volleying back and forth about who cared more for the workers on the front lines and who believed more strongly in the benefits of our technology.

This experience underscored the importance of clear, transparent communication and collaboration. By addressing the concerns and interests of both labor and management, we were able to help both sides see the value of the new technology from their unique perspectives.

KEY TAKEAWAYS

- Intentional and strategic change management is necessary to successfully implementing innovative technology like Mākusafe's wearable safety devices.
- Pilots should be designed with specific objectives, timelines, and success metrics.

- Gaining the support of workers, including union representatives, is critical when introducing any new technology.

BIBLIOGRAPHY

"Safety Technology 2024: Examining Trends in Technology Solutions Used to Reduce Serious Injuries and Fatalities in the Workplace." *National Safety Council*. n.d. https://www.nsc.org/faforms/work-to-zero-safety-technology-2024

11 Moving Beyond Behavior-Based Safety Approaches

In our conversations with various organizations, it's clear that many remain committed to Behavior-Based Safety (BBS) programs. That's understandable, as the journey of implementing successful programs is lengthy, involving deep organizational changes, and the intent and objectives behind BBS are undeniably worthwhile. However, while the theory of BBS may seem sound, its practical application often falls short of expectations. This chapter delves into the origins and principles of BBS, its potential benefits, and the common pitfalls in its implementation. It also explores why gathering input from frontline workers as a form of a leading safety indicator is invaluable, why that may be the most desirable behavior change, and how wearable tech supports this.

THE RISE OF BEHAVIOR-BASED SAFETY

The origins of behavior-based safety (BBS) can be traced back to the 1930s with the research of Herbert William Heinrich, an assistant superintendent at Travelers Insurance Company. Heinrich's research suggested that around 90% of accidents,

FIGURE 11.1 This illustration of Heinrich's Safety Triangle represents the concept that for every one fatality, there are typically 29–30 minor injuries, 300 recordables, 30,000 no-injury incidents, and 300,000 unsafe acts or at-risk behaviors. While this hierarchy emphasized the importance of addressing even minor incidents to prevent more serious injuries, it also unfortunately led to targeting worker behavior.

DOI: 10.1201/9781003432302-14

injuries, and illnesses were the result of "worker errors." This approach to health and safety, which assumes that most injuries and illnesses are caused by unsafe worker actions, became popular in the United States. Many safety professionals consider Heinrich a pioneer of modern BBS programs.

The term "BBS" was coined in the 1970s by Gene Earnest and Jim Palmer, two Procter & Gamble safety professionals working on a project to reduce injuries at their company. BBS is a proactive safety approach that builds on American engineer William Edwards Deming's principles of continuous improvement. It's also based on scientific research and writings by B.F. Skinner, the father of behavior analysis.

The premise of BBS is simple: By reinforcing positive behaviors and addressing unsafe actions, organizations can reduce workplace incidents. Early adopters saw significant improvements, leading to the widespread adoption of BBS in various industries. The ideal principles of BBS have good intentions:

BBS Actions	BBS Outcomes
Observation and Feedback: Regular observation of employee behavior, followed by constructive feedback.	**Reduction in Accidents:** By encouraging positive behaviors, many organizations have seen a decrease in workplace accidents.
Positive Reinforcement: Encourage safe behaviors through rewards and recognition.	**Improved Safety Culture:** Engaging employees in safety processes.
Employee Involvement: Engage employees in safety processes to foster ownership and accountability.	**Enhanced Communication:** Routine observations and feedback improve communication between employees and management.
Data-Driven Decisions: Use data from observations to identify trends and areas for improvement.	**Continuous Improvement:** Ongoing efforts to enhance safety protocols based on feedback and observations.

THE PROS AND CONS OF BBS

Despite the potential benefits, the reality of BBS implementation is often less than ideal. Instead of fostering a proactive safety culture, BBS can devolve into a bureaucratic exercise. One common issue is the reduction of BBS to a mere checklist. Employees are required to document a certain number of behavioral observations weekly, which can lead to superficial compliance rather than meaningful engagement.

Consider this scenario – you and your colleague Bob are required to document five observations each week. Under pressure and due to a friendly relationship, you both might end up filling out an index card or a form in a smartphone app about each other without diligently performing genuine observations. This act is sometimes called "pencil whipping" because it gives the illusion of a sign-off but there's been no real verification. The result is that Bob documents nothing but exemplary behavior from you, and of course, you return the favor for Bob. This undermines the purpose of BBS as the data collected becomes meaningless.

Another major pitfall is the presence of a blame culture. If employees fear discipline or blame for reporting unsafe behaviors, the entire BBS process becomes ineffective. The goal of BBS is to celebrate positive behaviors and identify improvement opportunities, not to single out individuals for punishment. When fear is present, employees are less likely to report honestly, rendering the system useless.

MOVING BEYOND BBS

While many organizations struggle with BBS, some achieve tremendous results by implementing it correctly. These organizations often engage all levels of employees in the feedback process without the threat of punishment. They'll also promote positive behavior by focusing on catching employees doing the right thing, or even approximately right, akin to the principles outlined long ago by Ken Blanchard in the book *One Minute Manager*.

Perhaps the most crucial behavior that we want and need from our people is open and transparent communication. That starts with creating an environment where employees feel safe to report issues and suggest improvements. Just imagine how your organization would benefit from truly unfiltered knowledge. Imagine what you could routinely learn from the frontlines by asking genuine questions like:

- What's on people's minds?
- What's going well and what needs to be changed?
- When reporting potential problems, what do they worry about?
- What tools or equipment do they need?
- If you had unlimited resources what improvements would you make in your job or your department, and why?
- How should the process be changed?
- What will go wrong next?
- Where do they think the next incident will occur?
- Where might someone die on the job?
- Where do we say "things are done this way" but actually do them another way?
- Where are we forced to do things outside of our plan because of constraints or other unplanned changes?
- Do you feel safe doing this job?
- If you could make this task easier, how would you do it?

Employees should feel comfortable sharing their thoughts, reporting potential problems, and suggesting changes without fear of reprisal. This kind of communication is essential for identifying and addressing risks proactively. Yet leaders continue to struggle to value and embrace this behavior.

We've asked live audiences for years now – how many believe that communication from the workers perspective is the "holy grail" of keeping people safe? Most or all hands in the room go up. Next, with hands remaining up, we ask how many have implemented a simple and structured way to collect this input – most hands go down. The next question is how many of you get as many reports as you would like or learn about as many things as you think actually happen? Now we're down to one or two hands at best. When asked how many reports they collect, the answer is usually around 8–10 while the highest is 20 a month or so. This demonstration leaves the rest of the safety professionals in the room in awe of those who are yielding that many.

FIGURE 11.2 Example of a construction worker using wearable tech to report observations and safety concerns via voice from the field.

This is how we met Mark, who is a remarkably experienced and excellent safety director in a large packaging company. Mark was the sole person with the last hand raised. He said his company received nearly 1,000 frontline reports a month from over 60 sites. They were distributed to senior management who actively responded, creating a culture of open communication and continuous improvement. He even credited this practice as being instrumental in helping them operate their business through the pandemic when no one was traveling.

Mark explained that he had a high degree of interest in wearable technology, especially if it could help facilitate communication from the frontline worker. Making reporting as easy as a push of a button could have the potential to increase the feedback they were already getting. Furthermore, if the software technology associated with the device could also help facilitate where those messages were directed and even automate actions for people in his organization, they could exponentially leverage the communication from the frontline workers. Mark knew that communication was crucial to building their safety culture yet he also knew he could always use more.

THE ROLE OF TECH IN FRONTLINE COMMUNICATION

MākuSafe's human-centric approach to wearable tech allows workers to push a button and report issues verbally through a device on their arm. Even though there's a wide array of sensors monitoring a plethora of experiences and exposures in real time, the MyVoice feature may be one of the most beneficial capabilities the solution offers. This innovation not only makes it easier for employees to report near-misses and observations but also ensures that these reports are captured accurately and managed efficiently.

MyVoice allows frontline workers to send real-time voice reports directly to leadership that is immediately actionable. This push-to-talk capability leverages the natural human ability to sense and report work conditions, making it a critical tool for workplace safety. Workers can report anything they see, smell, or feel that could be a hazard, providing valuable insights that wearable sensors alone cannot capture. This includes everything from near-misses, good catches, and physical or behavioral observations to maintenance requests and tool or workstation needs.

For instance, one of the first MyVoice reports received was a report about a missing guard on a machine. Rather than needing to stop what they were doing, hunt down a supervisor, or walk to another building to fill out a form, they only needed to take a few seconds to communicate their observation. And because leadership had this frontline observation, they could take action and likely prevented a serious injury. Another early message was the voice of a worker that was clearly shaken after she reported that she was nearly "struck by one of the machines" in their facility.

Our technology ecosystem prioritizes these messages above all other data collected by the wearable device. This approach underscores the importance of human observation in maintaining workplace safety. Empowering frontline workers to share their observations reinforces their role as the first line of defense in identifying and mitigating hazards.

Every company will say that they want to hear more and know more from the front lines. They try note cards, open office policies, suggestion boxes, and so many other ways to get clear and immediate information. A sophisticated organization knows that their frontline workers know more than anyone else in the organization about how work is being done. The burden often falls on the worker to find a supervisor, fill out a form, or stop work and report some sort of situation. Company productivity goals are often at odds with these solutions because they require a lot of that

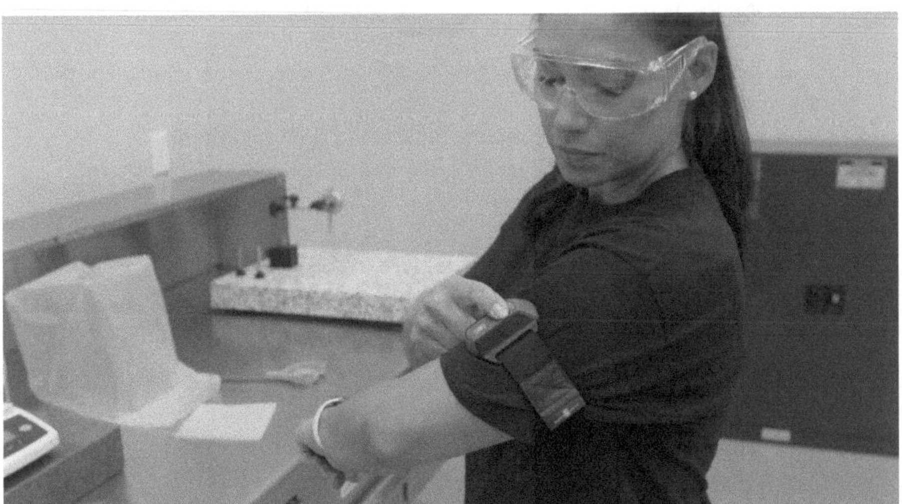

FIGURE 11.3 A worker can communicate observations to leadership by speaking into a wearable device.

worker, including sacrificing productivity. What it boils down to is leadership taking swift action on the observations of a frontline worker. It benefits the worker, the productivity, and ultimately the company. And every worker appreciates when their voice is heard.

It's worth considering also that if used well, these messages inherently possess psychological safety. The worker can volunteer to share what they see without worrying about any repercussions. This will be reinforced when leadership responds to their concern by redesigning work or mitigating a hazard rather than focusing on a person's behavior.

Finally, voice communication provides an incredible opportunity for a leader to respond by saying, "thank you, I heard you, and here's what we're doing about it." Companies should always take the opportunity to positively reinforce the fact that someone took time to communicate something they thought was valuable. We need more of that.

BUILDING A CULTURE OF OPEN COMMUNICATION

To evolve workplace safety, we need to move beyond BBS. The most important behavior to encourage in workers is a willingness to share their perspective. By fostering open communication, leveraging human-centric technology, and focusing on positive reinforcement, organizations can create a proactive safety culture that truly values employee input and continuously strives for improvement. The use of voice communication will further empower frontline workers, prioritizing risk identification and reporting as foundational elements of a successful safety program.

KEY TAKEAWAYS

- Avoid common pitfalls such as the checklist mentality and a culture of blame
- Create an open environment where employees are encouraged to report safety observations
- Focusing on work-related problems rather than individual faults
- Collect frontline observations by implementing real-time voice reporting

12 Haptic Feedback Isn't Always the Answer

The world is full of haptic technology. We're accustomed to devices that vibrate for alerts and touchscreens that click like a button. These sensations are found in everything from consumer goods like phones and watches to the entertainment world with video games and virtual reality. They're helping doctors learn to perform robotic-assisted surgeries and pilots to simulate flights. So wouldn't a mild pulse as a safety warning be just as helpful?

The surprising answer is "it depends." Haptic feedback can be useful for short-term compliance, such as in a life-threatening situation like an alert when dangerous levels of a toxic gas are present, or to evacuate immediately. It can also be helpful when a powered industrial vehicle or forklift is around the corner, but the view is obstructed. But if you want to drive lasting, meaningful behavior change, then haptics could be working against those goals.

This is because most safety devices with haptic feedback produce adverse stimuli. It only activates when there's a perceived problem, making haptics a type of punishment. Often incorrectly referred to as negative reinforcement, which would be the removal of adverse stimuli, instead the delivery of adverse stimuli is appropriately referred to as punishment because of the psychological effect it has. The "buzz" is delivered as a negative consequence in an attempt to condition the worker and shape their behavior as they are automatically targeted as the problem needing to be solved. Furthermore, this approach does nothing to reward or praise users for good behavior. Although leaders often like the idea of an easy and automatic way to improve mindfulness, neither of these aspects contribute to long-term change, which is the ultimate goal of improving any safety culture.

HAPTICS – GREAT FOR CONSUMERS, BUT NOT NECESSARILY SAFETY

To understand appropriate applications for haptic technology, it's helpful to recognize the purpose of this technology. The goal of vibrotactile motion is real-time feedback. A device typically delivers gentle force to the wearer through vibrations, jolts, or buzzes. The logic behind this technology is that since humans aren't always in a position to see or hear a cue, a harmless physical force is a suitable way to get their attention.

A popular example is the Apple Watch, which included its Taptic Engine when first released in 2014 ("Apple Unveils Apple Watch"). For the past decade, users have felt a slight buzz whenever they receive a notification. The feature then evolved from alerts and keyboard impressions into gaming experiences and meditation apps ("Apple Watch").

In the health category, wearers of fitness trackers like FitBit, Garmin, and Amazfit experience nudges to get moving. Similar tactile feedback can be found in sit-stand

DOI: 10.1201/9781003432302-15

FIGURE 12.1 Chapter 12 sidebar QR operant-conditioning. SIDEBAR: Video – Check out these fun clips from TV's The Big Bang Theory, where negative reinforcement vs. punishment is debated, and effective behavior shaping through positive reinforcement is explored. https://makusafe.com/blog/operant-conditioning/

desks, ergonomic office chairs, and back braces. Users experience a pulse whenever it's time to change positions or poor posture is detected. Tesla even has a full-body suit that "provides touch and force feedback to train and develop an athlete's muscle memory," including the ability to reduce "the risk of injury during intense training sessions ("New Level of Sports Training").

These consumer examples have one thing in common – the haptic feedback isn't punitive. It delivers neutral communication (you have a message) or a positive celebration (you hit your step count), with no intention to produce guilt or a "gotcha" moment. Even when the buzzing is a reminder, like it's time to drink more water, it's a low-stakes situation. Missing a movement goal for the day won't lead to serious consequences or an immediate health crisis.

However, haptic feedback in industrial environments has an entirely different context. First, a life-or-death situation is a reality on a manufacturing floor or assembly line. There are threats that can have a dire effect on workers: being caught in, caught between, or struck by machinery, and the likes. Ignoring or not understanding a safety cue is a major risk for an injury or a fatality.

Second, haptic feedback as a safety alert is most often a negative message – it's not usually neutral or positive. There's a vibration or buzz only when something is flagged as a risky action, movement, or threat. The haptic feedback's sole function is to deliver a warning.

Why is haptic feedback a problem when a factory is already filled with alarm systems? For the simple reason that those notifications are a form of mass communication. When they deploy, they are directed at everyone – no individual is called out or targeted. Furthermore, not a single version of those alarms is attached to a human body.

DIFFERENCES BETWEEN ALARM SYSTEMS AND HAPTIC FEEDBACK

Alarm Systems	Haptic Feedback
Facility-wide communication	Individual notification
Mounted on a wall, ceiling, or equipment	Worn directly on a worker's body
Uses sound or light	Delivers buzzes or vibrations

THE DOWNSIDE OF HAPTIC FEEDBACK

Another reason why haptic technology may be counterproductive as a safety warning is the human factor. What goes missing from the discussion is what happens to people when they are constantly subjected to negative consequences or punishment. They begin to equate the sensations with uncertainty, a nuisance, an interruption, or a reprimand.

1. ***Poor Context*** – or *Why on Earth Am I Getting Buzzed Again?* Not all haptic communication provides a clear context. Some devices, such as the wearable we developed at MākuSafe, monitor a wide array of potential hazards. But if we delivered haptic feedback, the same or similar sensation might mean numerous things and is open to interpretation. An often used example of a simple signal being interpreted differently might be a flashing yellow light at an intersection. Does that mean stop, slow down, proceed cautiously, or what exactly? Wearers wouldn't have an immediate way to distinguish which reading was triggered. They'll wonder, "Is it because this area is too hot, fumes were detected, I lifted incorrectly, or I nearly tripped?" If they don't know what the issue is, they can't effectively and quickly determine what exactly to do about it. Any hesitation alone could be a safety issue in a situation where time is of the essence. There are some wearable technologies that buzz multiple times depending on what is detected, possibly adding to the confusion.

   A lack of training can also create a knowledge gap. Let's say a company implements a back harness that vibrates when there is risky motion. They even provide an orientation to show how to wear it, emphasizing the tool is for protection. But they don't talk about the big picture of how many workers are suffering from strains or why it's important for everyone to avoid injuries. Nor do they teach employees about how to respond to a haptic warning, like reviewing proper lifting techniques. Employees may not understand why a behavior change is necessary in the first place, much less how to make corrections in the moment or retrain their bodies to move safely. Further, the worker's behavior may not be the issue at all. They likely know this. An adjustment to a process or elevated workstation may eliminate the need for a particular motion altogether.
2. ***Easy to Ignore*** – or *This is Just As Annoying As My Snooze Button.* Humans are incredibly good at tuning out unwanted interference. Think of how easy it is to disregard pings from chat platforms. Notification desensitization, or alert fatigue, is a real concern. In fact, studies have shown that even nurses in intensive care units can become overwhelmed by and desensitized to alerts, ultimately "interfering with caring for patients and causing reduced trust in alarm systems" (Lewandowska et al.). Especially if workers don't see the value or purpose of haptic sensations, they are likely to stop paying attention to them.

   Imagine wearing a gas monitor in a confined space where fumes are known to be present but not usually at a dangerous concentration. While the device is a sensible precaution, what happens if the tactile sensors are set to vibrate regardless of the reading level? The worker is repeatedly buzzed with what amounts to false alarms. At best, that's mildly annoying and possibly distracting. At worst, the employee disregards a valid reading and puts themselves and possibly others in harm's way.

3. ***Distraction*** – *Now My Concentration Is Broken and I Messed Up.* Haptic feedback is intrusive and disruptive by design. Workers have no idea when they are going to receive a pulse. That inherent randomness doesn't make sense in settings where people need their focus to stay safe.

   The risk of a startle reflex, no matter how minor, can have serious ramifications. Like a welder who has to restart an advanced weave bead, setting back a project's timeline. A food packaging worker who accidentally drops foreign material into a batch, escalating into a costly recall and a PR disaster. A forklift driver who is buzzed when pedestrians are nearby slams on their brakes, and the sudden movement causes the load to be dropped. In some environments, or in the case of a load of hazardous material load, would it be better to alert the pedestrian? Not to mention all the situations where someone unexpectedly jerking could lead to a lost finger or worse.
4. ***Punishment*** – or *I'm In Trouble Again.* There's also a reality that haptic feedback will make users feel like they are receiving a "wag of the finger." If workers feel like they are wearing a version of Big Brother, they'll form an antagonistic relationship with these tools. They'll even associate the delivery of negative consequences with the company rather than the technology, only adding to the negative perception.

   Let's return to the haptic back harness. What happens if the device is issued with the assumption that workers are to blame for improper lifting? For starters, management won't pause to review if there are environmental issues like an obstructed shelf or bulky materials that are causing a team to struggle with ergonomics. Second, they won't ask their workforce for suggestions on how to make this activity easier, like vacuum lifters or lift tables. The haptic feedback is only going to breed resentment because it's not addressing the actual safety issue.

Here's the end result with any four of these scenarios – workers will find a way to tune out, avoid wearing, or disable the haptic feedback. They might even learn how to move or work in a way that avoids the buzzer, ultimately putting themselves in a riskier position. None of this is going to result in long-term safety gains.

REINFORCEMENT THEORY AND SAFETY MANAGEMENT

Let's step away from haptic technology for a moment and into the world of change management and behavior theory. Behaviorism examines how people learn through conditioning. It is especially interested in how individuals are affected by positive or negative elements in their surroundings. One of the classic expressions of this concept is Pavlov and his conditioning with salivating dogs.

As this school of thought progressed, B.F. Skinner added his theory – that people are primarily controlled by their environment and thus their behavior can be controlled to produce a specific outcome ("B. F. Skinner"). Whereas Pavlov used unconscious conditioning, associating the stimulus of a metronome with feeding times, Skinner advocated for "the rewarding of a partial behavior or a random act that approaches the desired behavior" ("B.F. Skinner: 1904–1990"). He further believed that "everything we do and are is shaped by our experience of punishment and reward" ("B.F. Skinner: 1904–1990").

What he found, and many others have explored since, is that there is a fundamental difference between positive reinforcement and the delivery of negative consequences, aka punishment. If the consequences to an action are interpreted as a reward rather than punishment, the desired behavior is more likely to begin to take root as a habit. Penalties only have limited success at producing lasting change. Parents often find this with children – a rebuke can stop someone in their tracks and achieve immediate compliance. But even with adults, if you want someone to permanently adopt a different behavior, a series of rewards is more motivating.

This is the nexus where haptic biofeedback has been misapplied. In consumer devices, the gentle pressure is usually neutral or positive and used to reinforce a beneficial habit. But a safety device with haptics does neither – it delivers negative stimuli as a way to tell a worker to stop something.

Furthermore, haptic feedback does nothing to address the root of many safety issues – an environmental hazard. It once again assumes that the worker is the root problem, not something in their surroundings. It also assumes that the worker knows what to do to mitigate the hazard or reduce the risk. Haptics alone can't replace the insights and impacts employed by EHS professionals, ergonomists, and industrial hygienists. While it can call attention to a risk, that's only the first step. It is not a substitute for the benefits of conducting a safety analysis, implementing corrective measures, increasing training, or reengineering work to minimize hazards.

Haptic feedback may seem like an "easy button" for promoting better safety behavior, but at best it's a short-term fix to a more important underlying problem. To sustain an effective safety culture, companies need to use positive reinforcement instead. This can only be supported by a robust ecosystem of incentives, praise, recognition, communication, technology, and reengineered work.

KEY TAKEAWAYS

- Can be distracting, confusing, or even frustrating to workers
- Does not educate, encourage, or empower
- Ultimately not effective at creating permanent, meaningful change

BIBLIOGRAPHY

"Apple Unveils Apple Watch—Apple's Most Personal Device Ever." *Apple*, 9 September 2014, https://www.apple.com/newsroom/2014/09/09Apple-Unveils-Apple-Watch-Apples-Most-Personal-Device-Ever/

"Apple Watch User Guide." *Apple*,2024, https://support.apple.com/en-au/guide/watch/apd371dfe3d7/watchos

"B. F. Skinner." *Harvard University – Department of Psychology*, 2024, https://psychology.fas.harvard.edu/people/b-f-skinner

"B.F. Skinner: 1904–1990." https://www.pbs.org/wgbh/aso/databank/entries/bhskin.html

Lewandowska, Katarzyna, et al. "Impact of Alarm Fatigue on the Work of Nurses in an Intensive Care Environment—A Systematic Review." *International Journal of Environmental Research and Public Health*, vol. 17, No 22, 13 November 2020. *National Library of Medicine*, https://www.ncbi.nlm.nih.gov/pmc/articles/PMC7697990/

"New Level of Sports Training with Haptic Technology." *TESLASUIT*, 9 October 2023, https://teslasuit.io/blog/sports-training-with-haptic-technology/

13 How Wearables Contribute to the Prevention of Serious Injuries and Fatalities

In any industry, the most valuable asset is its people. The well-being of employees is not only a moral imperative but also a critical factor in the sustained success and reputation of an organization. It is wise for organizations and safety professionals to focus on preventing life-altering serious injuries and fatalities (SIFs) by concentrating on high-hazard types of work. These companies may consider moving to ASTM E2920 as a standard for recording injuries based on severity of incidents and meaningful connection to work. Often we hear that because they and their organizations are committed to an established strategy for focusing on SIFs and potential serious injuries and fatalities (P-SIFs), they don't see how wearable sensor tech will help them with their priorities, major incidents, or saving lives.

Yet research by the Campbell Institute of the National Safety Council pointed to the need for gathering and analyzing data as crucial. They recommended observations, walkarounds, and conversations as important to revealing precursors to SIF ("Designing Strategy for Serious Injury and Fatality Prevention"). We believe using wearable sensor technology to understand the context of work can have a profound impact on improving safety outcomes. Especially when paired with advanced push-to-talk communication capabilities, companies have an opportunity to understand precursors and gain the perspective of the frontline workers, which will contribute to SIF prevention.

THE VALUE OF WORKER INPUT FOR SIFs AND PSIFs

Certain types of work are inherently more hazardous and thus are primary focus areas for SIF prevention efforts. Along with vehicles and roadway work, these types of work are responsible for the most severe incidents and loss of life across industries.

Hazard	Description
Confined Space Entry	Working in spaces with limited entry and exit, such as tanks, silos, and storage bins, especially where hazardous atmospheres or other dangers may be present.
Vehicle Safety	Operation and maintenance of vehicles, including company cars, trucks, and heavy machinery, emphasizing safe driving practices and vehicle condition.

DOI: 10.1201/9781003432302-16

Hazard	Description
Electrical Work	Tasks involving electrical systems and equipment, where the risk of electric shock, arc flash, and fire are significant concerns.
Forklifts and Mobile Powered Equipment	Safe operation of forklifts, cranes, and other mobile machinery to mitigate collisions, tip-overs, and other accidents.
Lockout/Tagout (LOTO) Energy Isolation	Procedures ensuring machinery is properly shut off, de-energized, and unable to start up again prior to the completion of maintenance or repair work.
Hazardous Materials	Handling, storing, and disposing of chemicals and other hazardous substances, where exposure can lead to serious health risks.
Working at Height	Tasks performed at elevated levels, such as on ladders, scaffolds, or rooftops, where falls are a high risk.
Welding and Hot Work	Activities that produce heat, sparks, or flame, such as welding, cutting, and brazing, posing risks of fire and burns.
Excavation and Trench Work	Digging and working in trenches or other excavated areas, where cave-ins and other soil-related hazards are a concern.
Bulk Material Handling and Storage	Movement and storage of large quantities of materials, where the risks include spills, equipment failures, and manual handling injuries.

These workers are the first to encounter potential hazards in their daily tasks. Their direct interaction with machinery, materials, and processes means they are acutely aware of the practical challenges and dangers inherent in their work. This firsthand experience enables them to identify risks that may not be evident to managers or safety officers who are not present in the field.

For instance, a worker operating heavy machinery might notice that a particular safety guard is prone to malfunction under certain conditions. Reporting this observation can prompt a timely intervention that prevents a potential SIF. Without the worker's input, such an issue might only be discovered after an incident occurs, potentially resulting in injury or fatality.

One of the most memorable examples we've heard about was a worker charged with identifying hazards on a ship. He happened to notice boot prints on top of a pipe right next to the railing. It was uncovered that other team members were routinely climbing on top of this elevated pipe so they could reach equipment up high, forcing them to hang over the rail while under way at sea. The potential for someone to go overboard and be lost was extraordinary, yet this risk was only identified because of a boot print.

This perfectly illustrates why preventing SIFs involves not only reacting to incidents but also learning from near-misses and seeking out where serious injuries or fatalities could occur. Workers' insights are crucial in these scenarios. By listening, learning, empowering, and analyzing P-SIF events, organizations can identify underlying hazards and systemic issues that need to be addressed.

For example, a near-miss involving a falling object might reveal weaknesses in a company's material handling procedures. Workers who witness or are involved in such incidents can provide detailed accounts of what occurred, why it happened, and how similar events can be avoided. This feedback loop is essential for continuous improvement in safety practices.

THE ADVANTAGES AND IMPACT OF PUSH-TO-TALK VOICE REPORTING

In recent years, the advent of wearable technology with push-to-talk (PTT) voice communication capabilities has revolutionized the way workers can contribute to safety efforts. This capability, which we chose to include in our MākuSafe device, offers several advantages that enhance the quantity and quality of worker input, leading to better prevention of SIFs and a deeper understanding of P-SIFs.

PTT functionality allows workers to report hazards and incidents in real time, especially in dynamic work environments where conditions can change rapidly. This immediacy ensures that safety issues are documented and addressed promptly, reducing the lag time between hazard identification and mitigation. This elevates the voice of the frontline worker, breaks down barriers, and aids inclusion of needed perspective.

For example, a construction worker on a large multi-contractor job site finds an alarming structural issue. Using the PTT voice communication, they record the following message: "You better come down and see this I-beam – you are not going to like what I found!" The safety manager receives a real-time notification about the report, takes 30 seconds to listen, and immediately arrives to investigate. After confirming the worker's observation, they issue a stop work order until the problem is remedied.

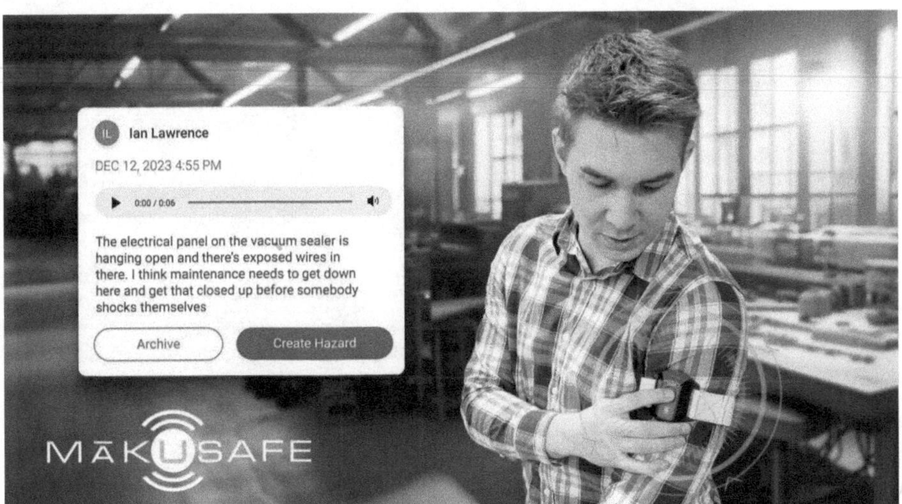

FIGURE 13.1 A worker reporting an electrical hazard using push-to-talk voice communication. Only a few seconds of their time is needed, but their observations can make a critical difference.

The simplicity of PTT devices also encourages more frequent and spontaneous reporting. Workers can quickly relay information without needing to stop their work to fill out forms or access a computer. This ease of use lowers the barrier to reporting, leading to higher engagement and more comprehensive safety data collection.

CONTEXTUAL DATA ON WORKING EXPERIENCES AND EXPOSURES

Wearable sensor devices can provide data about the context in which work is being performed. This data includes variables such as heat index, sound exposure, air quality, physical exertion levels, repetitive movement, location, time on shift, task being performed, and equipment being used. Coupled with input and observations from PTT communication, leaders can begin identifying patterns and developing targeted interventions. Data can be integrated with safety management systems to automatically log reports and track trends over time. This integration provides a rich dataset for analyzing safety issues. This centralized intelligence can enhance predictive analytics, helping organizations to anticipate SIFs more effectively.

Incorporating this contextual data offers numerous benefits:

Improved Learning from Near-Misses

Contextual data provides valuable insights analyzing and learning from near-misses. Understanding the exact conditions under which an incident occurred helps identify

FIGURE 13.2 Wearable sensor data can provide insights into working conditions and how they might impact safety, such as heat, noise, air quality, and low light.

root causes and contributing factors, leading to more effective corrective actions. For instance, knowing that a worker was performing a high-risk task with inadequate equipment and was affected by environmental conditions at the time of a near-miss can prompt a review and improvement of task-specific protocols and equipment standards.

Comprehensive Risk Assessment

Contextual data enables a more comprehensive risk assessment by providing a detailed understanding of the environmental and operational factors contributing to safety hazards. For example, high heat index readings combined with prolonged physical exertion can indicate an increased risk of heat-related illnesses. Addressing these conditions proactively can prevent incidents.

Personalized Safety Measures

Data on individual worker exposures and conditions can help tailor safety measures to specific needs. If certain workers are consistently exposed to high noise levels, targeted interventions such as providing enhanced hearing protection or rotating tasks can be implemented to reduce their risk of hearing loss.

Enhanced Predictive Analytics

Collecting and analyzing contextual data enhances predictive analytics capabilities. By identifying patterns and correlations between various factors and safety incidents, organizations can predict and mitigate risks more effectively. For example, tracking the correlation between air quality and respiratory issues can lead to improved ventilation systems in specific areas.

Better Workforce Management

Data on factors such as time on shift and physical exertion levels can inform workforce management decisions. Monitoring these parameters helps prevent worker fatigue and overexertion, reducing the likelihood of errors and accidents. Ensuring adequate rest periods and task rotation can enhance overall safety and productivity.

THE IMPACT ON SAFETY OUTCOMES

When worker input is systematically incorporated into safety management systems through advanced technologies like wearable PTT devices and contextual data collection, the impact on safety outcomes can be profound. Organizations that prioritize worker feedback tend to have lower rates of SIFs and a stronger safety culture. Workers feel empowered to take ownership of their safety and that of their colleagues, leading to more vigilant and conscientious behavior on the job.

Moreover, the data collected from worker reports and contextual monitoring can be used to track safety trends, identify recurring issues, and develop targeted interventions. This data-driven approach ensures that safety initiatives are based on real-world conditions and experiences, making them more effective and relevant.

KEY TAKEAWAYS

- Workers provide critical insights that are essential for identifying precursors and contributing factors to SIF events.
- Push-to-talk voice reporting via wearable devices, along with the collection of contextual data about working experiences and exposures via sensors, aids learning and analysis to uncover where potential for SIFs is present.
- By establishing robust feedback mechanisms and valuing the contributions of workers, organizations can significantly enhance their safety outcomes.

BIBLIOGRAPHY

"Designing Strategy for Serious Injury and Fatality Prevention." *Campbell Institute*, 2020, https://www.thecampbellinstitute.org/wp-content/uploads/2020/02/Campbell-Institute_Designing-Strategy-for-Serious-Injury-and-Fatality-Prevention.pdf

14 The New View of Safety

The roots of new approaches to safety can be traced back to Paul O'Neill's tenure at Alcoa in the late 1980s and early 1990s. O'Neill's revolutionary focus on worker safety not only transformed Alcoa but also set a new standard for safety management in the corporate world. Many of the professionals who were part of this transformation at Alcoa went on to influence other organizations, contributing to the evolution of safety management that brings us to where we are today.

Several concepts have emerged in recent years that embody this new approach to safety: Human and Organizational Performance (HOP), Safety 2.0 by Erik Hollnagel, Safety Differently from Sydney Dekker, and the New View of Safety. While these concepts have some differences, they share a common foundation: the application of modern management principles to safety practices.

HUMAN AND ORGANIZATIONAL PERFORMANCE (HOP)

HOP is attributed to thought leaders such as Todd Conklin, Bob Edwards, Rob Fisher, and others. It represents a significant shift from traditional safety management approaches. The willingness of these thought leaders to share these concepts and their work openly in the hope of widespread adoption and improved results is appreciated and noble.

The National Safety Council appropriately points out that HOP is science based and risk focused. They define HOP not as a safety and health discipline but as "an operating philosophy combining human factors engineering with organizational psychology and leads to better, more efficient, more resilient and safer workplaces where employees and leaders are both engaged and valued by each other and the organization" (Human & Organizational Performance). Its core principles include the following.

Error Is Normal

People make mistakes, and even the best people do. Error isn't a choice. Ever have your keys in one hand and trash in the other but have to remind yourself which one gets tossed in the bin in front of you? The vast majority of the time, work is successful and mistakes don't have negative outcomes.

Workers Are Adaptive Problem Solvers

They routinely find ways to solve problems and adapt to get work done successfully. This is because job instructions and SOPs never completely address every possible scenario and nuance.

DOI: 10.1201/9781003432302-17

Context Influences Behavior while Systems Drive Outcomes

Organizations put people in the position of having to make adjustments and decisions about the right thing or an acceptable way, every day. You should think of your frontline workers as real-time risk managers.

Our Response to Failure Matters

Blame and punishment fix nothing. The traditional response of being punitive hampers any learning and further destroys any chance for open and transparent communication. Blame is the enemy of improvement.

Learning and Improvement Are Vital

Learning is deliberate and doesn't happen by accident. You can punish or choose to learn, but it's not possible to do both effectively. Consider the difference between an accident investigation and an event analysis. Organizations that embrace learning benefit greatly. In the context of HOP and safety management, many will implement well-designed organizational learning teams, events, and methodologies.

Imagine for a moment that a company fired an employee if they made a mistake or strayed from a SOP. That company would learn nothing about why their employee was compelled to do the work differently than designed. It's also likely the next person they put in that role will find themselves unemployed when they encounter the same conditions. The cost of recruiting and training for that role would be extremely high.

PERSPECTIVE FROM THE INSURANCE INDUSTRY

The evolution of safety management is not occurring in isolation. The insurance industry, particularly in workers' compensation, is undergoing a transformational shift. Forward-thinking leaders in insurance are no longer content to just issue payments to repair or replace losses. Instead, they are seeking to engage with policyholders proactively to help predict and prevent losses before claims occur.

This shift in the insurance industry aligns perfectly with the new approaches to safety management. Companies like MākuSafe, which is based in one of the top insurance hubs in the United States, are well positioned to bridge the gap between these new safety philosophies and the evolving needs of the insurance industry.

LEARNING FROM NORMAL WORK

One common question from organizations exploring wearable safety technology is: "We don't have many incidents or we've already reduced our incident rate significantly – why would we use your wearable solution? What else is there for us to learn?"

This question highlights a crucial aspect of the new view of safety: the importance of learning from normal work. Erik Hollnagel suggests that instead of aiming for "zero incidents" as a goal, we should establish "work to 100%" as an objective. This approach allows organizations to relentlessly develop resilience potential in all aspects of work, even when incident rates are low. Recalling the Blue Line Model,

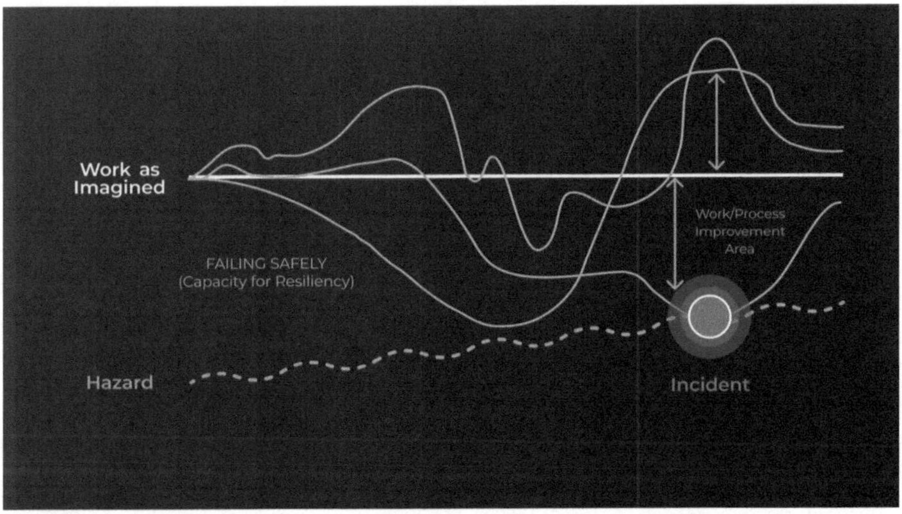

FIGURE 14.1 Work as performed is represented here by the blue lines in the model. On good and bad days alike, the realities of how work gets performed do not follow a straight path as it was imagined or designed. Workers are forced to adapt to circumstances as they arise. By truly understanding the realities of how work is performed, we can improve work processes and create capacity to work safely, even when systems fail or people inevitably make errors.

consider the value of implementing practices like Blue Line Reviews or Learning Team Meetings.

Learning from normal work is essential because:

1. It helps understand how work is actually done, not just how it's imagined or prescribed.
2. It reveals successful adaptations and best practices that can be shared across the organization.
3. It identifies potential risks before they lead to incidents.
4. It builds a culture of continuous improvement and learning.

This is vital because today's work environments are complex systems with numerous interacting components. As Todd Conklin aptly put it, the goal should be to "Fix the work, not the people." This shift in focus from individual behavior to systemic factors is a key principle of the new safety paradigm.

Implementing these new approaches to safety requires a high degree of emotional intelligence, as described by Daniel Goleman. Leaders must be able to understand and manage their own emotions and those of others to create a culture of openness and learning.

Moreover, the concept of humble inquiry, introduced by Edgar Schein, is crucial for organizational learning. The ability to ask good questions of those who do the work every day and gain meaningful insights is essential for leaders. This approach leads to a culture of trust and continuous improvement.

THE ROLE OF TECHNOLOGY IN THE NEW SAFETY PARADIGM

Wearable sensor technology, like that developed by MākuSafe, is particularly well suited to support and accelerate these new safety initiatives. By providing real-time data on environmental conditions, worker movements, and potential hazards, wearable technology enables organizations to:

1. Normalize and anticipate mistakes and provide opportunities for learning without blame.
2. Shift focus from blaming individuals to understanding systemic factors.
3. Enable continuous learning and improvement to build resilience and the capacity to fail safely.
4. Provide context about the work environment, helping understand how systems drive behaviors and outcomes.
5. Allow for immediate and informed responses to potential safety issues.

For example, a production line worker learns her coworker is regularly getting a bruise on her arm from a certain part of a machine. She can use the voice communication feature on her wearable to recommend adding padding to the machine. The safety manager can then investigate the situation further to learn how widespread this issue is among workers. Listening to this team also uncovers a process improvement suggestion, which supports quality standards and eliminates an awkward lifting motion.

As technology facilitates communication and breaks down barriers, it creates opportunities for positive conversations focused on risk in the work. This directly leads to a more positive culture by creating engagement and participation.

The new view of safety, embodied in approaches like HOP and supported by advanced technologies, represents a significant shift in how organizations manage safety. By focusing on understanding human factors, system influences, and the importance of learning from normal work, these approaches promise to create safer workplaces and more resilient organizations. Technology with the right approach can be very human centric and is in alignment with contemporary safety management philosophy.

KEY TAKEAWAYS

- The integration of new safety philosophies with emerging technologies and evolving insurance practices offers exciting possibilities for the future of workplace safety.
- Organizations that embrace these approaches and leverage advanced technologies will be well positioned not only to manage safety but also to manage safely in an increasingly complex world.

BIBLIOGRAPHY

"Human & Organizational Performance: A path to improvement for all organizations." *National Safety Council*, 2021, https://www.nsc.org/getmedia/6dc7b76d-09a2-4ad5-9bc2-0b0365c34b46/nsm-wk4-mi

15 How Organizational Culture Shapes Innovation Adoption

In the ever-evolving landscape of workplace safety, the introduction of innovative technologies holds the promise of safeguarding workers, reducing incidents, and potentially saving lives. However, the successful deployment of such technologies, particularly wearable safety devices, is not solely dependent on their potential benefits. Instead, an organization's existing culture and leadership play a pivotal role in either enabling or limiting the adoption of these innovations, particularly when it involves asking employees to don a wearable device every day. The thoughtful implementation of wearable technology can also contribute to building stronger safety cultures.

THE CULTURE ICEBERG: VISIBLE AND HIDDEN INFLUENCES

To understand the impact of organizational culture on innovation adoption, we must first explore the concept of culture itself. The iceberg model provides an apt analogy for organizational culture, with observable elements above the surface and hidden yet influential factors beneath.

Above the waterline are the tangible aspects and behaviors of an organization:

- How people treat each other
- Physical workspace layout, neatness, and organization
- Dress codes
- Company logos and slogans
- Visible safety measures and signage
- People speaking up and actively contributing
- The willingness of workers to lead or follow

However, it's what lies beneath the surface that truly shapes what you see when you walk around:

- Deeply held attitudes, beliefs, and perceptions
- Shared values and assumptions
- Unwritten rules and norms
- Historical experiences and collective memory
- Power dynamics and relationships

DOI: 10.1201/9781003432302-18

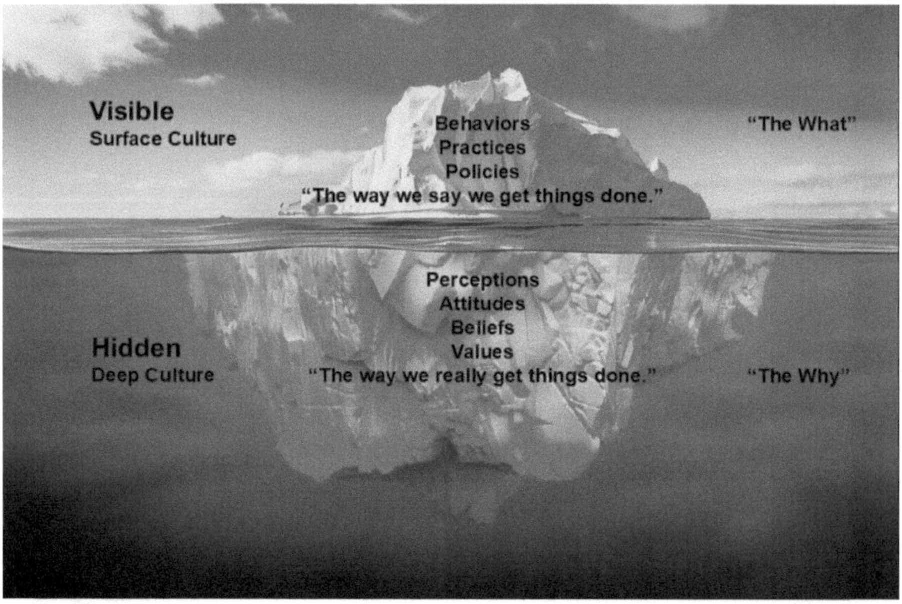

FIGURE 15.1 Workplace culture is like an iceberg – only the most obvious elements are visible. The majority, however, are more subtle and often unwritten. These "silent" cultural marks are what predominantly drives behavior.

These hidden elements form the bedrock of an organization's culture, influencing how employees interpret and respond to new initiatives, including the introduction of wearable safety technology.

LEADERSHIP IS THE CATALYST FOR CULTURAL CHANGE

When discussing organizational culture in the context of safety, the term "safety culture" often emerges. However, this concept has been a subject of debate among scholars and practitioners. The reality is that safety culture is not a separate entity – it is an integral part of the overall organizational culture. It reflects how safety is prioritized, communicated, and integrated into daily operations and decision-making processes.

The debate surrounding safety culture stems from the difficulty in isolating it as a distinct element. Instead, it's more accurate to view safety as a core value that permeates all aspects of organizational culture, shaping behaviors and decisions at every level.

In navigating the cultural landscape, leadership plays a crucial role. Effective leadership is critical in inspiring and shaping how things are done, setting the tone for how employees generally think, and creating alignment between personal views and the organization's values. One compelling perspective on leadership asserts that it involves "inspiring people to see a vision of what can be in such a powerful way that they choose to follow you there." This approach is particularly relevant when introducing innovative safety technologies like wearable devices.

However, leaders face significant challenges in this endeavor. More often than not, they have to start by overcoming the "Big Brother" myth. Employees often have immediate concerns about wearable technology, including:

- Will my employer now know my medical and personal information?
- Is this going to keep buzzing me and interrupting my work?
- Will the company know where I'm at all the time?
- Is the company just trying to find performance reasons to get me in trouble or get rid of me?

To address these concerns, leaders must clearly communicate the limitations and actual capabilities of the technology. They need to be willing and able to persuade workers to consider an alternate viewpoint. For instance, we at MākuSafe have gone to great lengths to develop a wearable safety technology so that we would not:

- Monitor biometrics, health data, or PII
- Constantly track employee location
- Know when someone is on break or how many times they visit the restroom
- Listen to worker conversations

We often point out that the data we gather about heat levels, sound octaves, or slip/trip motion should be viewed as a way to hold leaders accountable for fixing hazards – not anything that could be used against a worker. In organizations with adversarial relationships between management and labor, inspiring a shared vision becomes

FIGURE 15.2 Without a strong culture of trust and respect or leadership commitment and influence, the reaction of workers to wearable tech may be met with distrust and cynicism.

exponentially more difficult. Leaders must work to build trust and open communication channels before introducing potentially controversial technologies. Leaders will need to articulate how the technology aligns with the organization's values and how it benefits both the company and individual employees.

While organizational culture influences technology adoption, the relationship is reciprocal. Technologies like MākuSafe can, in turn, shape and enhance organizational culture when implemented thoughtfully. MākuSafe's MyVoice communication feature can elevate frontline worker perspectives if leaders genuinely value their input. This creates opportunities for meaningful conversations, allowing leaders to seek in-depth employee comments and suggestions, explore needs and concerns, and express gratitude for insights shared. These interactions can foster new behavioral habits, resulting in higher engagement and participation in safety programs. Over time, this can lead to a more robust safety culture and a general improvement in organizational communication and trust.

For example, a safety leader notices readings for high TVOCs and CO_2 detected. They initiate a discussion between the shift supervisor and workers, which reveals batches of raw materials from a supplier that smell different. Together, they created a protocol to leave a voice report whenever materials arrive that smell unusual. This back-and-forth communication allows the company to better track quality concerns through production as well as address any supplier issues.

THE DECISION-MAKING NEXUS

Ultimately, an organization's ability to keep people free from harm, hazards, and risks stems from decision-making throughout the entire organization and all of its interrelated parts. These decisions are fundamentally based on the system of shared values within the organization. When safety is deeply ingrained in the organizational culture, it influences decisions at all levels, from strategic planning to daily operations. This holistic approach ensures that safety considerations are not an afterthought but an integral part of every decision, reducing the likelihood of unintended risks or consequences.

Imagine a worker who can use voice reporting because they note the new flooring throughout their entire department is slick. Triggered by this timely observation, the safety manager can evaluate the situation. They might find a hidden issue, like the wrong paint was used. The maintenance team mistakenly used a gloss paint with no added grit, which now poses a slip or fall risk. Safety and maintenance can then work together to remediate this issue – all because technology and a strong safety culture empowered the worker to record an invaluable 30-second message.

As we look ahead, it's clear that the use of wearables and other innovative safety technologies will become increasingly ubiquitous, expected, and integrated into how work is performed. However, because we are still in the early stages of adoption, there are challenges to successful deployment that companies will need to consider and address. Organizations with positive cultures and engaged workers are better positioned to overcome signs of resistance or hesitation. They have the foundation necessary to address concerns, build trust, and effectively implement new technologies.

KEY TAKEAWAYS

- Only a portion of workplace culture is easy to observe. When strengthening safety attitudes, organizations will need to leverage those that are hidden undercurrents.
- While innovative safety technologies like MākuSafe offer tremendous potential to improve workplace safety, their successful adoption is inextricably linked to an organization's existing culture and leadership.
- By recognizing this relationship and actively working to create a culture that values safety, innovation, and employee input, organizations can pave the way for successful technology adoption.

16 Management Effectiveness, Systems Orientation, and Certifications

In the realm of organizational management, a systemic approach is crucial for achieving both effectiveness (goal attainment) and efficiency (wise resource use). This systems orientation views management as a continuous process, ensuring that organizations can adapt, improve, and thrive in dynamic environments. We believe strongly that wearable tech can strengthen the safety management process, improve outcomes, and provide focus for leaders.

THE MANAGEMENT PROCESS

While various models may differ slightly in their steps, the four core functions of the management process consistently include:

Planning – Where Do We Want to Go?

Planning is the foundational step in any management process. It involves setting objectives and determining the best course of action to achieve them. Despite its critical role, planning is often neglected, resulting in unclear goals and directions. Without proper planning, it is very difficult to understand if the plan's objectives have been met. It can also be challenging to assess where things went wrong. If there's no clear plan in place, then how will you know where and why an action has deviated from an objective?

Organizing – How Will We Align Our Resources to Get There?

Organizing involves arranging resources – human, financial, and material – in a way that effectively supports the achievement of the goals set during the planning stage. This includes creating structures, roles, and responsibilities.

Leading – Who Will Carry Out the Work?

Leading is about inspiring, motivating, and guiding people to work toward the organization's goals. Effective leadership is built on communication, fosters a positive work environment, and encourages collaboration and commitment.

DOI: 10.1201/9781003432302-19

Controlling – How Will We Monitor and Measure to Ensure Desired Outcomes?

Controlling involves monitoring progress and making adjustments as necessary. It ensures that the organization remains on track to achieve its goals by evaluating performance and implementing corrective actions when needed. Much like planning, evaluating is often overlooked, leading to missed opportunities for improvement. Even when things have seemingly gone according to plan, evaluation helps uncover where members of the team may have taken alternative actions that still resulted in the desired outcome. Knowing that can help with planning for future projects or uncovering spots where you may have just gotten lucky with a decision that wasn't according to plan.

The military uses a process called after-action review (AAR). This process can be easily adapted as a project evaluation in the corporate setting. You can conduct your own internal AAR by asking some of the following questions:

1. What was supposed to happen?
2. What actually happened?
3. What went well? Why?
4. What went poorly? Why?
5. What were the consequences?
6. What are our key learnings?
7. What can we apply to future projects?

It's essential to conceptualize this process as a continuous loop, where feedback fuels ongoing improvement.

SAFETY MANAGEMENT: A PARALLEL PROCESS

The principles of effective management apply equally to safety management. A continuous improvement approach in safety ensures both effectiveness in reducing incidents and efficiency in resource allocation. This systemic view of safety and health management systems (SHMS) has been widely advocated by major associations and thought leadership outlets. The American Society of Safety Professionals (ASSP), the National Safety Council (NSC), and the Occupational Safety and Health Administration (OSHA) all promote the implementation of SHMS. OSHA even provides comprehensive guidelines on safety management systems ("Recommended Practices").

Regardless of the source, the safety management process typically involves common elements including:

1. **Hazard identification:** Identifying potential hazards in the workplace is the first step in preventing accidents and injuries.
2. **Risk analysis:** Assessing the identified hazards to understand the level of risk they pose and prioritizing them accordingly.

3. **Implementing a hierarchy of controls:** Applying controls based on the hierarchy of controls, such as elimination, substitution, engineering controls, administrative controls, and personal protective equipment (PPE).
4. **Evaluating results:** Continuously monitoring and evaluating the effectiveness of the implemented controls to ensure they are working as intended.

Crucially, this process cannot be effective without active participation from frontline workers, a pervasive safety mindset, and strong leadership. These elements are essential for identifying hazards, implementing controls, and continuously improving safety practices.

How this looks in reality might be a shift supervisor who receives a notification of a trip motion detected by his MākuSafe wearable device. Looping in the safety director, they discuss the occurrence with the worker and learn he had been tripping on the small ergonomic mat that's new in his work area. They determine larger mats will pose less of a tripping risk, and a replacement order is made.

SAFETY CERTIFICATIONS: PURSUIT OF EXCELLENCE

Many organizations seek safety certifications as a means of validating their commitment to safety and demonstrating the effectiveness of their safety management systems. This often yields quantifiable positive results. A prime example is Cintas, which became the first company to achieve safety certification at over 100 sites and is now well beyond that (Shaw).

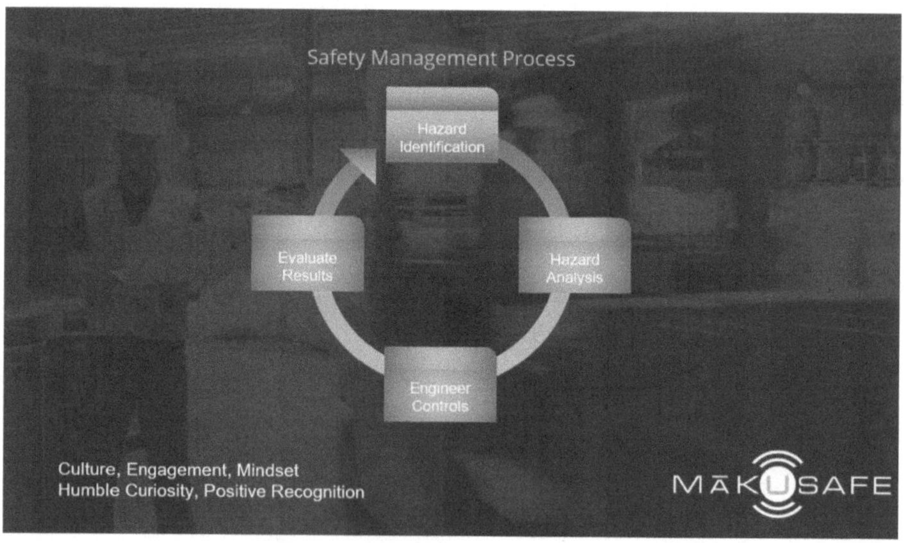

FIGURE 16.1 Improving safety is a continuous loop that cycles between identifying and minimizing risk. Any traditional tendency to view safety management as performing only the first three job functions in a linear fashion and waiting to see what happens is outdated and should be forgotten.

There are various well-recognized alternatives for safety certification, including ANSI Z10, ISO 45001, and VPP Star. While each has its specific requirements, they share common elements that align with general management process steps and basic safety management principles. These include:

1. **Management Leadership:** Commitment from top management to drive the safety program.
2. **Worker Participation:** Involving workers in the safety process to leverage their insights and encourage a culture of safety.
3. **Hazard Identification and Assessment:** Regularly identifying and assessing workplace hazards.
4. **Hazard Prevention and Control:** Implementing measures to prevent and control hazards.
5. **Education and Training:** Providing continuous safety education and training for employees.
6. **Program Evaluation and Improvement:** Regularly evaluating the safety program and making necessary improvements.
7. **Communication and Coordination:** Ensuring effective communication and coordination among all stakeholders, including host employers, contractors, and staffing agencies.

THE ROLE OF TECHNOLOGY IN SAFETY MANAGEMENT

Emerging technologies, such as wearable safety devices, are revolutionizing the field of safety management. Platforms like MākuSafe, when implemented with the right approach, can significantly enhance the effectiveness and efficiency of safety management processes. These technologies can aid in achieving and maintaining certifications by providing:

Automated Hazard Identification: Wearable devices can detect potential hazards in real time, providing immediate alerts to workers and supervisors.

Data-Based Decision-Making: Collecting evidence on workplace conditions helps in making informed decisions to improve safety.

Documentation: Time and date-stamped readings provide a clear record of workplace conditions and incidents. It includes tracking corrective actions and control measures as well.

Human-Centric Reporting: Voice reporting features enable workers to easily report hazards, encouraging greater participation and engagement.

Measurement and Monitoring: Continuous data collection allows for effective monitoring of corrective actions and controls.

Program Improvements: Data insights help in identifying areas for improvement, leading to a more effective safety program.

For example, a worker is responsible for running an industrial power washer and wears ear plugs as hearing protection. However, based on decibel recordings by MākuSafe, it's found that they are routinely experiencing more than the allowable sound dosage within the first couple of hours of his shift. Safety leadership is able to reevaluate hearing protection for this role to ensure protection is commensurate with the exposure level.

By leveraging such technologies, organizations can create a more responsive, proactive, and participative approach to safety management. This not only aids in meeting certification requirements but also supports a culture of continuous improvement in safety practices.

KEY TAKEAWAYS

- Viewing management – and by extension, safety management – as a systemic, continuous process is key to organizational effectiveness and efficiency.
- By embracing this approach, pursuing relevant certifications, and leveraging innovative technologies, organizations can create safer, more productive environments while meeting their strategic objectives.

BIBLIOGRAPHY

"Recommended Practices for Safety and Health Programs." *OSHA*, https://www.osha.gov/safety-management

Shaw, Seyfarth. "Cintas Becomes First Employer to Reach 100 Certified OSHA VPP-Star Worksites." *Seyfarth*, 29 March 2019, https://www.environmentalsafetyupdate.com/2019/03/cintas-becomes-first-employer-to-reach-100-certified-osha-vpp-star-worksites/#:~:text=and%20cost%20savings.%E2%80%9D-,Cintas%20Corporation%20has%20recently%20become%20the%20first%20company%20to%20reach,VPP%20location%20in%20early%202019

17 The Evolution of Management Thinking

Over the past century, our understanding of effective leadership and management has evolved significantly. Early theories on best practices for managing and leading people have been refined and replaced by newer models, each contributing valuable insights into human nature, motivation, and organizational dynamics. Despite this evolution, many of the fundamental challenges in leadership remain. Organizational change efforts often fail, and a significant portion of the workforce remains disengaged, as evidenced by surveys like Gallup's State of the Workforce ("Employee Engagement"). We can all benefit from having a more solid understanding of how to build strong cultures. After all, even world-class athletes practice the fundamentals and have a coach.

This chapter explores the evolution of management thinking, summarizing key concepts and their contributions to today's best practices in leadership. It also highlights contemporary ideas essential for achieving great results with and through people. To successfully introduce new tools and technologies, share visions and values, and expect lasting change, leaders must live and breathe time-tested basics.

"There is no better teacher than history in determining the future. There are answers worth billions of dollars in a $30 history book" – Charlie Munger

"I don't think much of a man who is not wiser today than he was yesterday" – Abraham Lincoln

CLASSICAL APPROACHES

Classical thinking on management still provides valuable lessons. Organizations today cannot operate with the simple carrot-and-stick mentality as they did 100 years ago. Modern organizations must demonstrate concern for people, needs, and satisfaction. However, classical thinkers were some of the originators of those ideas, providing a strong foundation that has evolved over time. Understanding is helpful to today's leaders.

Frederick Winslow Taylor: The Father of Scientific Management

Frederick Winslow Taylor, often called the father of scientific management, was an engineer who introduced a systematic approach to improving efficiency and productivity. In 1909, Taylor published *The Principles of Scientific Management.* He proposed that by optimizing and simplifying jobs, productivity would increase. His key elements included:

- **Harmony and Cooperation:** Emphasized the importance of collaboration between workers and management.

DOI: 10.1201/9781003432302-20

- **Training and Development:** Advocated for the training and development of workers to enhance efficiency.
- **Standardized Work Methods:** Developed scientific methods for performing tasks to ensure consistency and efficiency.

Taylor's focus on systematic training and development laid the groundwork for modern human resource practices and continuous improvement methodologies. However, too much reliance on a scientific approach to productivity without concern for people is outdated and will not work today.

Mary Parker Follett: Transformational Leadership

Mary Parker Follett, rightly recognized as the mother of modern management, was way ahead of her time. Instead of emphasizing a scientific industrial approach, she advocated for what she saw as the far more important human element – regarding people as the most valuable commodity present within any business. In her writings as early as 1927, she introduced the concept of transformational leadership. Her contributions include:

- **Interrelationship of Leadership and Followership:** Emphasized the dynamic relationship between leaders and followers and the power of collective goals.
- **Human Element:** Advocated for the importance of human relations in management, considering people as the most valuable asset in any organization.
- **Positive Human Relations:** Highlighted the need to promote positive human relations to achieve effective management.

Follett's emphasis on the human element and collective goals resonates with contemporary leadership approaches that value employee engagement and empowerment. While this was a shift from what Talyor, Max Weber, and others taught, Follett, Lillian Galbraith, and their contemporaries laid the groundwork for the movement that followed.

Elton Mayo: The Hawthorne Effect

Elton Mayo's experiments at the Hawthorne Works, a Western Electric plant in Cicero, Illinois in the 1920s and 1930s, led to an understanding of what we now refer to as the Hawthorne Effect. Mayo was attempting to discover correlations between lighting levels and productivity, but instead what he found was that in spite of poor environmental conditions, productivity was directly related to how people were treated. Mayo's work underscored the importance of considering human factors in management, paving the way for the human relations movement and modern organizational behavior studies.

The Human Relations Movement and Behaviorism

The era and thinkers that followed paved the way for modern organizational behavior studies and human resource management techniques, emphasizing employee engagement, team building, leadership development, and organizational culture as critical factors for success.

Douglas McGregor: Theory X and Theory Y

Douglas McGregor introduced Theory X and Theory Y in the 1950s and 1960s, exploring assumptions about human motivation and management styles. These beliefs or conclusions not only manifest themselves in a leader's behavior, but have an impact on followers. McGregor depicted a contrast between two opposing ends of a spectrum:

- **Theory X Manager:** Assumes people are inherently lazy and require strict supervision, relies on rules and formal authority, has an "us vs. them" attitude, and has a singular focus on production or outcomes. May achieve short-term compliance out of fear, but likely creates unhappy workers in the long run.
- **Theory Y Manager:** Assumes people are self-motivated and seek responsibility, utilizes encouragement, offers rewards, and strives for satisfaction. Achieves far greater and lasting results with and through people.

McGregor's theories highlighted the impact of management assumptions and leadership style on employee behavior.

Blake and Mouton's Leadership Grid

Blake and Mouton's Leadership Grid evolved from McGregor's work, emphasizing a balanced approach to leadership. Rather than X or Y styles as an either-or choice, the Leadership Grid combined these two dimensions and recognized that each has needed positive elements. Organizations should strive for a high degree of the best that each provides, and neither can or should be sacrificed:

- **Concern for Production:** Focused on achieving high levels of productivity, results orientated, goal attainment, through structure.
- **Concern for People:** Emphasized the importance of addressing employees' needs and well-being.
- **Team Management Style:** Combined high concern for both production and people to achieve optimal results and high satisfaction.

Their model reinforced the idea that effective leadership involves balancing task and people-oriented approaches. How you interact with your people has everything to do with outcomes. You can achieve results and treat people in a manner that keeps them wanting to work with and for you. Most of us can relate to what sacrificing one for the other looks like, and it's never pretty. Most first-time supervisors who become responsible for leading their former teammates errantly think they will get the best outcomes by continuing to be a great friend to everyone in the same way they always had. This undermines structure, accountability, and results for relationships, becoming a lesson that is usually learned the hard way by all. Blake and Mouton's approach underscores that balance is necessary.

Hersey and Blanchard: Situational Leadership

Hersey and Blanchard's Situational Leadership model, introduced in 1969, emphasized adapting leadership styles to the situation and the readiness of followers:

- **Task Readiness:** Recognized that people need more structure when new to a task.
- **Adaptability:** Suggested that leadership style should evolve based on followers' experience and competence.
- **Encouraging Independence:** Advocated for encouraging experienced followers to take ownership and act independently.

This model highlighted the importance of situational awareness in leadership, encouraging leaders to be flexible and responsive to their team's needs.

Needs Theories of Motivation: Maslow, Herzberg, Alderfer, and McClelland

The needs theories of motivation provide insights into what drives people to take action They advocate the importance of creating conditions that allow employees to satisfy their needs, leading to increased motivation and engagement:

- **Maslow's Hierarchy of Needs:** Identified a progression of needs from physiological to self-actualization.
- **Herzberg's Two-Factor Theory:** Distinguished between hygiene factors that prevent dissatisfaction and motivators that promote satisfaction.
- **Alderfer's ERG Theory:** Simplified Maslow's hierarchy into three core needs: existence, relatedness, and growth.
- **McClelland's Acquired Needs Theory:** Focused on the need for achievement, affiliation, and power.

The importance of leaders creating opportunities for their people to feel good about themselves, be recognized by others, and achieve on the job cannot be overstated.

Process Theories of Motivation

Process theories of motivation, such as expectancy theory and equity theory, emphasize the role of perceptions in motivation. They acknowledge that people's perceptions of fairness and expected outcomes significantly impact their motivation and behavior:

- **Expectancy Theory:** Suggests that people are motivated when they believe their efforts will lead to desired outcomes.
- **Equity Theory:** Focuses on the importance of fairness and perceived equity in motivation.

We've all heard it – perception is reality. Think of the worker who winds up feeling that even if they try, it won't matter anyway. As a leader, do you help this worker see things from a different viewpoint or let them slowly fail and go away? Your role is to influence and clarify when perceptions are inaccurate.

CONTEMPORARY IDEAS IN MANAGEMENT AND LEADERSHIP

Today's thought leaders are contributing to management and leadership best practices for modern organizations. These practices are being utilized to drive organizational growth and development, foster stronger organizational culture, and get the best with and through people.

Amy C. Edmondson: Psychological Safety

Professor Amy C. Edmondson, Harvard Business School, introduced the concept of psychological safety in the 1990s: Her focus is on two areas:

- **Psychological Safety:** Ensures that employees feel safe to express themselves without fear of punishment.
- **Open Communication:** Encourages transparency and learning from front-line workers.

Edmondson's work emphasizes the importance of creating a safe environment where employees can openly share ideas and concerns, leading to innovation and continuous improvement. Otherwise, how will you get open communication about imminent hazards if people fear reprisal?

Change Management Theories

Change management theories, from Kurt Lewin to John Kotter, emphasize the importance of communication in successful change efforts. Make no mistake, implementing a new technology, especially a wearable, involves successfully managing people through change. Regardless of the model or theory that may be your favorite, all are consistent in the importance of:

- **Clear Communication:** Ensures that employees understand the reasons for change.
- **Inclusive Approach:** Involves employees in the change process to gain their commitment and support.

These theories highlight that effective communication and inclusion are critical at each stage of the process to overcoming resistance and achieving lasting organizational change.

Deming and Total Quality Management (TQM)

While the number of fundamental principles from W. Edwards Deming's Total Quality Management (TQM) varies depending on where you look, five that are most essential for leading companies:

- **Continuous Improvement:** This is an essential act of management that aims to optimize the company's overall performance and bring about lasting positive change. Encourages ongoing efforts to improve processes and quality.

- **Employee Involvement:** All employees should participate equally in achieving mutual goals and consider themselves an integral part of the organization. Stresses the importance of including workers in decision-making.
- **Fact-based Decision-Making:** Making decisions based on data-based knowledge makes it easier to satisfy customers.
- **Leadership:** Leadership provides direction, motivation, and guidance to the organization's quality improvement efforts.
- **Communication:** Effective communication of information helps ensure that everyone is working toward the same goals and following the same standards.

TQM and Deming's contributions underscore the value of a culture of improvement and the vital role of communication and engagement in achieving excellence. We need to view safety management not as a series of checklists and functions to be performed to achieve compliance, but rather as an ongoing process of improvement that uses data evidence to make decisions.

Understanding Human Factors

Human factors refer to the study of how people interact with their environments, tools, and systems. These include:

- **Ergonomics:** Designing workplaces that fit the needs and capabilities of workers.
- **Cognitive Load:** Managing the amount of mental effort required by tasks.
- **User-Centered Design:** Creating systems and processes that prioritize the user's experience.

Understanding human factors helps leaders create environments that enhance productivity, safety, and well-being. Data evidence can support decisions about work design to ensure safety. An understanding of how people process information and make decisions certainly can inform better safety practices.

Systems Thinking

Systems thinking involves understanding the organization as a complex system of interconnected parts:

- **Holistic View:** Seeing the organization as a whole, rather than a collection of separate parts.
- **Interdependencies:** Recognizing how different components of the organization influence one another.
- **Feedback Loops:** Identifying and understanding the cycles of feedback that affect system behavior.

Systems thinking enables leaders to identify underlying patterns and root causes of issues, leading to more effective problem solving and strategic decision-making.

It is incumbent upon today's safety and operations leaders to consider how all parts of the organization may be affected by changes to systems because this impacts safety and productivity.

Mental models and understanding how our brains work

Mental models are deeply ingrained assumptions and generalizations that influence how we understand the world:

- **Cognitive Biases:** Recognizing biases that affect decision-making and behavior.
- **Neuroscience:** Understanding how brain functions impact learning, memory, and emotions.
- **Adaptive Thinking:** Developing the ability to adjust mental models in response to new information and changing circumstances.

Work on the front lines is never as it is imagined or planned because employees are constantly adapting and almost always are successful. They are real-time risk managers, constantly making decisions on how to balance priorities and achieve expected outcomes safely. Yet we give them rules, policies, protocols, and standards to follow and seemingly never stop piling on. By understanding mental models and brain functions, leaders can better appreciate how workers go about processing information and making decisions.

KEY TAKEAWAYS

- Over the past century, we have learned a great deal about leading and motivating effectively through understanding human nature and psychology.
- By incorporating insights from these early theories as well as contemporary ideas, leaders can create environments where employees are motivated, engaged, and empowered to contribute to the organization's success.
- The evolution of management thinking teaches us that while tools and technologies may change, the fundamental principles of effective leadership remain constant.

BIBLIOGRAPHY

"Employee Engagement." *Gallup*, 2024, https://www.gallup.com/workplace/349484/state-of-the-global-workplace.aspx

Part IV

The Future of Work, Tom West

18 Factories of the Future – Integrating the Connected Worker in Industry 4.0

The concept of the factory has undergone significant transformations over the past century, evolving from the simple assembly lines of the Industrial Revolution to today's highly automated and digitized production environments. With the advent of Industry 4.0, the factory is once again poised for a revolution – one that integrates advanced technologies like the Internet of Things (IoT), artificial intelligence, and wearable devices to create smarter, safer, and more productive industrial environments.

Wearable technology, such as we've developed at MākuSafe, can play a crucial role in this transformation. By enabling the connected worker to interact seamlessly with the automated and connected systems around them, these devices ensure that the human element remains integral to the factory of the future. Our clients have led us to explore how wearable tech can go beyond safety to enhance productivity and efficiency in Industry 4.0, highlighting the interdependence of these factors and the challenges that lie ahead.

THE PROMISE OF INDUSTRY 4.0: SAFER, ADAPTIVE, AND MORE PRODUCTIVE FACTORIES

Industry 4.0 represents a paradigm shift in manufacturing, where technology is not just an aid but an inherent aspect of the production process. The factory of the future is envisioned as a highly adaptive and interconnected environment, where machines, systems, and humans work in harmony to achieve unprecedented levels of efficiency and safety.

Wearable technology like MākuSafe is perfectly suited for this vision by integrating workers into the digital fabric of the factory. In our case, devices can monitor environmental conditions, detect potential hazards, and even adapt the surrounding systems to ensure a safer work environment. For instance, environmental sensors in wearables can detect when temperatures rise beyond safe levels and, when integrated into facility systems, automatically activate air handlers to cool the area, thereby preventing heat-related incidents.

But the potential of wearable tech goes beyond just safety. By collecting data on worker interactions with machines and systems, these devices can help optimize

DOI: 10.1201/9781003432302-22

workflows, reduce downtime, and increase overall productivity. In an Industry 4.0 environment, the factory is no longer a static entity but a dynamic system that continuously adapts to the needs of both workers and machines.

INTERDEPENDENCE OF SAFETY, EFFICIENCY, AND PRODUCTIVITY

In the traditional view of manufacturing, safety, efficiency, and productivity are often seen as separate, sometimes competing, priorities. However, Industry 4.0 challenges this notion by demonstrating the interdependence of these factors. When safety is enhanced through technology, efficiency and productivity naturally follow.

For example, consider a scenario where wearable devices are used to monitor the presence and activity of workers on a production line. If a worker enters a hazardous area, the system can automatically slow down or stop nearby machines to prevent accidents. This not only protects the worker but also ensures that the production process is not disrupted by incidents, thereby maintaining efficiency and productivity.

Moreover, data collected from wearables can provide insights into how work is being performed, allowing for more accurate cost accounting and quality control. By understanding which teams or workers are more productive and why, managers can make informed decisions to improve overall factory performance. This interdependence highlights the importance of viewing safety, efficiency, and productivity as mutually reinforcing goals in the context of Industry 4.0.

If our goal is continuous learning in order to improve systems and create resilience in processes, and we understand that no one knows the realities of work as well as the people performing it every day, then enabling easy communication and creating this new habit has such tremendous potential. Forward-thinking organizations recognize that this is so valuable that they no longer want to wait for shift change meetings to hear improvement ideas about safety, quality, and production, but instead want that input during a shift in real time so they can make rapid changes immediately.

CHALLENGES OF INDUSTRIAL DIGITAL TRANSFORMATION

While the potential benefits of Industry 4.0 are clear, the path to achieving this vision is fraught with challenges. One of the biggest hurdles is the digital transformation of industrial operations, particularly the integration of data from various sources into a centralized, actionable system.

Many organizations have made significant investments in Industrial IoT (IIoT) devices, collecting vast amounts of data from machines and systems. However, this data is often fragmented, incomplete, or not timely enough to be useful. In contrast, data from wearable devices like MākuSafe can provide real-time, contextual information about workers' interactions with their environment. This data is not only easier to collect and centralize but also offers insights that might be more challenging to obtain from traditional IIoT efforts.

For instance, while some organizations have spent years trying to prove value in use cases and connect all their IIoT devices, MākuSafe can typically be deployed in a large facility and start collecting data within half a day. The data can be exchanged with other cloud platforms. This rapid deployment and ease of use make wearable technology a compelling option for companies looking to enhance their digital transformation efforts without the complexity and delays associated with IIoT integration.

A RESPONSIVE CHOCOLATE FACTORY

A classic episode from the 1950s television show *I Love Lucy* provides a humorous yet insightful illustration of the challenges that can arise in a production environment. In the episode, Lucy and her friend Ethel work on a chocolate factory production line, wrapping candies as the conveyor belt speeds up uncontrollably. Unable to keep pace, the duo resorts to stuffing chocolates into their mouths and hats, leading to chaos on the production floor.

While this scenario is played for laughs, it highlights a very real issue that persists in many factories today – the inability of production systems to adapt to the presence and pace of human workers. In the factory of the future, such situations could be avoided through the use of wearable technology. Imagine a production line that can detect the number of workers present and automatically adjust its speed and volume to match their capacity, ensuring both safety and productivity.

Moreover, wearable devices could track which workers are assigned to which lines, how long they have been working, and what tasks they are performing. This data could be used to conduct efficiency studies, optimize labor allocation, and ensure that workers are compensated accurately based on the complexity of their tasks. The insights gained from wearable tech could help managers understand why certain teams are more productive than others and identify areas for improvement.

ENHANCING SAFETY AND PRODUCTIVITY WITH WEARABLE TECH

The ability of wearable devices to detect worker presence has significant implications for safety and productivity. For example, if a worker approaches a machine too closely, the wearable could trigger the machine to slow down or stop, preventing accidents. Additionally, by integrating with cloud-based systems like HR databases,

FIGURE 18.1 Chapter 18 sidebar QR chaos-to-control – Sidebar Video – From Chaos to Control: How Wearable Tech Can Transform Production Efficiency https://makusafe.com/blog/chaos-to-control/

wearables could verify whether a worker's credentials, training, or licenses are up to date before allowing them to operate certain equipment or enter secure areas.

Traffic mapping is another area where wearables can have a substantial impact. By detecting the movement patterns of workers throughout the factory, lean process engineers can identify wasted time or unnecessary travel, leading to more efficient workflows. This data-driven approach to process optimization is a key component of Industry 4.0, where continuous improvement is achieved through the intelligent use of technology.

The next frontier will be harnessing the power of AI fueled by all of this data being gathered from devices and workers. By utilizing generative and predictive modeling, AI can provide insights for companies about where and how they can optimize their operations.

As factories evolve, so too do the expectations of the workers who operate within them. Today's employees are increasingly tech-savvy and expect the tools they use to make their jobs easier, safer, and more connected. Wearable technology meets these expectations by providing leaders with real-time information about their environment, helping them keep workers free from hazards and enabling them to communicate with their supervisors.

WEARABLES AS HUMAN-CENTRIC TECHNOLOGY

The University of Tennessee Reliability & Maintainability Center (RMC) installed and tested MākuSafe technology in their training factory for College of Engineering students as well as industry and government professionals. The implementation was overseen by Dr. Klaus M. Blache, RMC Director and Research Professor.

Dr. Blache has written extensively about various areas of manufacturing and continuous improvement for both processes and people. He is a champion of safety and change management, having worked for General Motors and served as chairman of the Society of Maintenance & Reliability Professionals. In his view, technology like MākuSafe is a significant contributor to workplace safety.

"The holistic well-being of employees will continue to play a greater role in employee safety, engagement, and productivity. A key attribute of Industry 4.0 is being more 'human-centric'. That refers to using digital technology to enhance and enable people to work at their best capability and contribute to human ingenuity."

As a seasoned professional who has authored two books, lectured globally, and helped numerous Fortune 500 companies with strategies and implementation, Klaus recognizes how wearables will help transition the current workforce skills and culture to Industry 4.0 and beyond.

"Based on my over 30 years of working across manufacturing, ergonomics/safety, maintenance & reliability, and now in academia helping industry and government agencies improve their operations, I view MākuSafe as a unique value-added choice for improving health, safety and productivity. MākuSafe data provides valuable real-time EHS insights in potential areas of concern to improve worker safety, absenteeism, turnover and overall well-being."

CONCLUSION

In the factory of the future, wearable tech is not just a tool for management but an empowering device for workers. By integrating the connected worker into the automated and digitized factory environment, wearable devices like MākuSafe help bridge the gap between human labor and machine efficiency. This symbiosis ensures that the factory of the future is not only more productive but also a safer and more satisfying place to work. Wearable tech can play a crucial role in creating a factory environment that is adaptive, responsive, and inherently safe – a true reflection of the promise of Industry 4.0.

KEY TAKEAWAYS

- The integration of wearable technology into Industry 4.0 represents a significant leap forward in the evolution of the factory.
- By connecting workers to the digital systems around them, these devices enhance safety, productivity, and efficiency.
- Achieving this vision requires overcoming challenges related to data integration and digital transformation.

19 The Future of EHS Embraces Technology and New Management Approaches

In the ever-evolving landscape of workplace safety, the future of Environment, Health, and Safety (EHS) management is upon us. It's time to shed outdated practices and embrace innovation to ensure the well-being of workers while enhancing operational efficiency. Our parting chapter explores the pivotal role of technology and new management approaches in shaping the future of EHS, highlighting the imperative for organizations to adapt or risk falling behind.

EMBRACING INNOVATION IN EHS

Historically, safety management has operated in isolation, disconnected from the technological advancements transforming other sectors. Today, the convergence of cutting-edge technology and progressive management philosophies is driving a paradigm shift in EHS practices. Organizations must recognize the interdependent nature of safety, productivity, and efficiency. It is crucial to move away from siloed objectives toward holistic approaches that prioritize worker well-being without compromising operational excellence.

Innovative solutions are now available that integrate safety management with broader operational goals. These solutions include sophisticated data analytics, real-time monitoring systems, and AI-driven insights that help organizations anticipate and mitigate risks before they materialize. This evolution is not merely about adopting new tools but about fostering a culture that values and applies technological advancements to continuous improvement.

HARNESSING THE POWER OF DATA

One of the most significant developments in modern EHS is the ability to gather, analyze, and utilize data to optimize operations and ensure worker safety. Imagine an advanced manufacturing facility where systems and equipment adapt in real time to the workers interacting with them, creating a symbiotic relationship between safety and productivity.

Real-time data collection and analysis empower organizations to identify patterns, predict potential hazards, and implement preventative measures. For instance, wearable technology can monitor individualized heat indexes, sound exposure, air quality,

DOI: 10.1201/9781003432302-23

and physical strain. This data-driven approach not only enhances safety but also improves efficiency and productivity by minimizing downtime and preventing accidents before they happen.

However, the integration of technology in EHS is not without its challenges. Common concerns regarding technology persist, including worker receptiveness, privacy issues, and potential impacts on organizational culture and morale. Addressing these concerns requires transparent communication and a commitment to using data responsibly and ethically.

HUMAN AND ORGANIZATIONAL PERFORMANCE

Central to the evolution of EHS is a shift toward human-centric approaches that focus on understanding the realities of work as it is performed. Safety luminaries such as Todd Conklin, Sydney Dekker, and Bob Edwards emphasize the importance of moving beyond blame and punishment toward cultivating a culture of learning and resilience. Safety is not merely the absence of incidents, but the presence of capacity in redesigned systems for people to fail safely, even on imperfect days.

Frameworks like Safety 2.0 and similar concepts underscore the need for organizations to build resilience and capacity into work processes, enabling workers to adapt and respond to unexpected challenges. This involves creating an environment where workers feel empowered to report hazards, share insights, and participate actively in safety initiatives. The goal is to learn from everyday work experiences and continuously improve safety practices with high engagement and participation.

DATA-DRIVEN DECISION-MAKING

The advent of wearable technology and advanced analytics empowers organizations to gain real-time awareness of risks and potential hazards. By collecting data associated with leading factors, organizations can proactively implement controls and improve mitigation efforts.

For example, at MākuSafe, our workforce wearable tech gathers data in real time about potential hazards. This includes individualized heat index, sound exposure, air quality, physical work effort, and ergonomic risks. Our clients routinely report significant improvements in safety metrics. Many have seen reductions in incident and claims frequency and severity exceeding 50–90%, or strain and exertion incidents dropping to zero within their first 12 months of use.

Data-enabled decision-making not only increases the effectiveness of management processes but also closes the continuous improvement loop, ensuring progress toward desired outcomes. By using real-time data, organizations can implement targeted interventions, monitor their effectiveness, and adjust strategies as needed. This iterative process supports a proactive safety culture where risks are managed dynamically and improvements are sustained over time.

ELEVATING FRONTLINE VOICES

Effective EHS management goes beyond compliance, amplifying the voices of workers. Technology can serve as a tool for empowerment, enabling employees to actively participate in safety initiatives and contribute to a culture of continuous improvement. Moreover, initiatives aimed at enhancing ESG (environmental, social, and governance) measures can bolster organizational performance, win new business, and attract and retain top talent in an increasingly competitive landscape.

By giving workers a platform to share their experiences and insights, organizations can gain a deeper understanding of the realities of work. This inclusive approach not only improves safety outcomes but also enhances worker engagement and morale. When workers feel heard and valued, they are more likely to buy into safety initiatives and contribute to a positive safety culture.

Furthermore, the integration of technology in EHS can support broader organizational goals, such as sustainability and social responsibility. By aligning safety initiatives with ESG objectives, organizations can demonstrate their commitment to ethical and responsible business practices, which can enhance their reputation and competitiveness in the market.

TOWARD STRATEGIC TRANSFORMATION

Ultimately, the goal of modern EHS management is to facilitate organizational transformation and sustainable competitiveness. By embracing technology and adopting progressive management approaches, organizations can move beyond traditional metrics of success toward achieving strategic objectives. From reducing incident rates to enhancing operational efficiency and productivity, the integration of technology and new management paradigms holds the key to a safer, stronger, and more resilient future.

Organizations that prioritize EHS as a strategic imperative can also achieve outcomes that extend beyond compliance. By integrating safety into their core business processes and applying technology to drive continuous improvement, they can create a culture of excellence that supports long-term success. This requires a commitment to innovation, collaboration, and a willingness to adapt to changing circumstances.

In conclusion, the future of EHS management is characterized by innovation, collaboration, and a commitment to continuous improvement. Organizations must embrace technology, leverage data-driven insights, and prioritize human well-being to thrive in an increasingly complex and dynamic world. By doing so, they can not only ensure the safety and well-being of their workforce but also position themselves for long-term success in a rapidly evolving business landscape.

As we move forward, let us remember that the journey toward a safer and more sustainable future is ongoing. It is through collective efforts and a shared commitment to excellence that we can truly realize the full potential of EHS in the 21st century. The future of EHS is not just about preventing incidents but about creating an environment where safety, productivity, and efficiency are mutually reinforcing. By embracing innovation and adopting a holistic approach to safety management, we can build resilient organizations that are equipped to navigate the challenges and opportunities of the future.

Part V

Appendix 1
Case Studies

In this appendix, we provide an overview of several case studies conducted recently using our wearable technology and have summarized key findings. These reports were all published by third parties or the users themselves. Links are included to access the full report for each.

The latest MākuSafe case studies, whitepapers, and reports will continue to be updated regularly and can be accessed here along with relevant content that may be of interest.

FIGURE A.1 QR Code 1 to https://makusafe.com/downloads/

NATIONAL SAFETY COUNCIL CASE STUDY OVERVIEW: NFI INDUSTRIES' USE OF MĀKUSAFE WEARABLE TECHNOLOGY

Introduction The National Safety Council's Work to Zero initiative, in partnership with SafetyTech Accelerator, conducted a case study highlighting NFI Industries' innovative approach to reducing musculoskeletal disorders (MSDs) in the workplace. MSDs are a significant concern, impacting nearly one quarter of the global workforce, with injuries such as strains, carpal tunnel syndrome, and tendonitis resulting from repetitive movements, awkward postures, and forceful exertions.

The Solution: MākuSafe Wearable Technology To address these challenges, NFI Industries piloted the MākuSafe system, a safety data and analytics solution. MākuSafe uses wearable technology to monitor and analyze worker movements, providing real-time insights into factors that can contribute to MSDs. The system collects data on physical exertion, movement types, incident timing, and location, which safety leaders use to better understand potential risks and support employees in preventing injury.

Pilot Program and Results NFI deployed 120 MākuSafe wearables across three locations, focusing on material-handling workers engaged in physically demanding tasks like loading and unloading. By combining the data provided by MākuSafe with targeted lifting training and behavior-based safety strategies, NFI achieved a remarkable 74% reduction in injuries at the pilot sites. Additionally, employee surveys showed a high level of favorability, with workers voluntarily embracing the wearables as part of the company's safety culture.

Key Lessons Learned: As NFI continues to expand the use of MākuSafe, several lessons emerged:

- Integrating behavior-based safety processes can enhance the organization's safety culture.
- Transparent communication is essential – employees should understand the technology's purpose and how data will be used.
- Regular feedback from both employees and managers is critical for continuous improvement.
- IT department engagement is crucial for addressing configuration, security, and compatibility concerns.
- It's important to fully understand the technology's capabilities and limitations to avoid being misled by unnecessary features.

Conclusion NFI's adoption of MākuSafe technology showcases the potential of wearable devices and data-driven insights to significantly reduce workplace injuries, creating a safer and more engaged workforce.

About NFI NFI Industries is a third-party logistics provider headquartered in Camden, New Jersey, employing over 16,800 associates nationwide. Their supply chain solutions span domestic and international operations.

For a deeper dive into this case study, download the full report here

FIGURE A.2 QR Code 2 to https://makusafe.com/wp-content/uploads/2023/06/WTZ-NSC-MakuSafe-Case-Study.pdf

The National Council on Compensation Insurance (NCCI) published research in 2024 exploring the impact of safety technologies on workplace environments, including wearables, AI, and computer vision systems. The report, part of a series, emphasized the rapid evolution of workplace safety and the importance of technological innovations in workers' compensation. NCCI conducted interviews with various stakeholders, including employers using innovative safety technology, to understand the implementation challenges and the benefits these technologies provide.

Key takeaways from the research include:

- Safety technology requires a strong partnership between the employer, technology provider, and insurer.
- The cost of implementing such technologies is not always a barrier, as the long-term savings in workers' compensation and workplace safety outweigh initial expenses.
- Employee buy-in is essential, achieved through transparency, education, and company leadership adopting the technology themselves.
- The technologies have resulted in both improved safety and increased productivity in industries like manufacturing and warehousing.

One prominent example highlighted was Life Line Emergency Vehicles, a company utilizing MākuSafe wearables. Initially, Life Line started with 40 devices, which expanded to 160 after seeing positive outcomes. Key successes include:

- **Data-Driven Insights:** Life Line began conducting daily reviews of wearable data, which led to monthly meetings with their technology provider and quarterly insurer reviews. These discussions identified ergonomic challenges, resulting in solutions like the installation of lift tables and changes to work processes.
- **"Good Catches":** Over 335 potential hazards, or "good catches," were reported, with 82% successfully addressed. This proactive hazard identification system prevented injuries and helped maintain a safer work environment.
- **Significant Cost Savings:** Workers' compensation costs were reduced by 60% between 2021 and 2023, highlighting the financial advantages of wearables.
- **Improved Workplace Safety:** Life Line saw fewer lost time hours, reduced trip hazards, and cleaner, more efficient workspaces.
- **Employee Engagement:** By promoting transparency and involving management in wearing the devices, Life Line addressed privacy concerns and fostered employee buy-in, making the wearables a standard part of their safety culture.

Overall, the Life Line experience underscores the transformative role that wearable technology can play in enhancing workplace safety, reducing costs, and improving operational efficiency.

For a deeper dive into this case study, download the full report here

FIGURE A.3 QR Code 3 to https://makusafe.com/blog/enhancing-workplace-safety-with-wearable-technology-a-case-study/

Workers Compensation Insurance Carrier Findings with Wearable Safety Technology Solution

A large worker's compensation insurance carrier piloted the MākuSafe safety technology with six of its policyholders having industrial operations over the course of two to six months, resulting in significant real reductions in claims. This findings report highlights some of the hazards identified and the overall results achieved during the test periods, including claims frequency reduced by 50% and claims severity reduced by 90%.

Company A (@ 6 months)

- 87% decrease of indicators per active device from first half (1,206) to second half (159) of the trial
- 90% reduction in work comp claim severity – $3,089before MākuSafe, $311 after
- Employee voice memo alerted management of equipment overload switch tripping, which likely prevented a $10,000 (average for their SIC) equipment breakdown claim or possible fire.

Company B (@ 6 months)

- 30% decrease of indicators per active device from first half (454) to second half (317) of the trial
- 30% reduction in work comp claim severity – $120,575 before MākuSafe and $84,091 after
- Indicator disparity of work cells conducting the same work gave management the insight to investigate and identify ergonomic hazards present in the high indicator cell. If the corrective action prevents just a single overexertion injury in the work cell, that would be an$18,000 loss avoided (the company's average overexertion).

Company C (@ 3 months)

- 39% decrease of indicators per active device from first half (53.2) to second half (32.5) of the trial
- 91% reduction in work comp claim severity – $10,051 before MākuSafe and $899 after

- Employee voice memos were used to report slip-fall hazards and other employees not wearing the required slip-resistant shoes. A single slip-fall injury averages $9,630 for this manufacturer.

Company D (@ 2.5 months)

- 3% decrease of indicators per active device from first half (100) to second half (96) of the trial
- 70% reduction in work comp claim severity – $24,320 before MākuSafe and $7,317 after
- Number of indicators leading management to completely overhaul their ergonomics program and collaborating with workers on task analysis to reduce unnecessary motions and force, which may result in significant operational improvements and lower workcomp mod (currently 1.16)
- High CO_2 levels help identify scrap wood burners not venting properly, where correcting chimney efficiency will have a positive impact on worker health and performance.

Company E (@ 2 months)

- 46% increase of indicators per active device from first half (57.5) to second half (84.2) of the trial
- 0 work comp claims since MākuSafe, but no claims six months prior either; they do average $9,000/year
- A device detected a trip. Management investigated and found the worker tripped on an air hose laying on the ground, so a hose reel was installed to eliminate the hazard. Preventing just a single trip-fall due to this improvement would be a savings of $10,000 (the average trip-fall for their SIC group).

Company F (@ 2 months)

- 8% decrease of indicators per active device from first half (27.6) to second half (25.5) of the trial
- Work comp claim analysis not available, not a policyholder

For a deeper dive into this case study, download the full report here

FIGURE A.4 QR Code 4 to https://makusafe.com/wp-content/uploads/2023/03/Work-Comp-Carrier-Findings-Program-Summary-1.pdf

CASE STUDY BY WORKERS COMPENSATION INSURER: THIRD PARTY LOGISTICS

OVERVIEW: WEARABLE TECHNOLOGY CASE STUDY WITH CONNER STRONG & BUCKELEW

This case study highlights a successful pilot program of MākuSafe wearable technology, led by Conner Strong & Buckelew, in collaboration with a leading third-party logistics provider. The aim was to enhance workplace safety by addressing known risks, such as lifting-related injuries, and identifying additional hazards like slips, repetitive motion, and heat exposure.

Key Findings:

- The logistics provider sought improved visibility into safety risks, beyond lifting and pushing, to enhance their employee safety protocols.
- Conner Strong & Buckelew recommended MākuSafe as a wearable technology solution to monitor workplace hazards, providing real-time data on environmental and ergonomic risks.
- By combining wearable technology with a lifting mechanics training program, the company improved employee behavior and training, resulting in a **56% reduction in lifting-related injuries**.
- The company saved **$450,000 in Workers' Compensation losses** in just one year.
- The pilot data also reinforced the company's behavior-based safety program, offering actionable insights into employee safety practices.

For a deeper dive into this case study, download the full report here

FIGURE A.5 QR Code 5 to https://makusafe.com/news/case-study-conner-strong-buckelew-finds-incredible-results-with-wearable-tech/

Case Study Summary: Data-Driven Worker Safety Improvement at Tom's Quality Millwork (TQM)

Background: This case study highlights the use of MākuSafe wearable technology at Tom's Quality Millwork, a major hardwood manufacturer, to improve workplace ergonomics and reduce worker injuries. A team of undergraduate industrial engineering students from the University of Wisconsin-Madison collaborated with TQM to analyze data collected from MākuSafe devices worn by shop floor workers. The aim was to reduce risks associated with high-force tasks like pushing, pulling, and repetitive movements, which are linked to long-term injuries and worker compensation claims.

Key Findings: The study focused on TQM's Line 3, which operates on two shifts. Workers rotate every 2.75 hours across four workstations, each handling tasks of varying difficulty depending on the product size. The MākuSafe data identified patterns in high-risk activities, including:

1. **Push/Pull Activities:** The data revealed frequent and elevated occurrences of push/pull movements on certain days, correlating with the production of larger and heavier boards.
2. **Time and Position Impact:** The frequency of push/pull movements decreased throughout the day, and positions 2 and 4 were found to have significantly higher occurrences of these activities. Certain workers were also identified as experiencing more push/pull hits, suggesting a need for further ergonomics adjustments.
3. **Job Rotation and Fatigue:** The analysis demonstrated that task frequency and job rotation influenced the strain on workers. Push/pull hits were notably higher during the morning rotations compared to the afternoon, indicating worker fatigue as a contributing factor to safety risks.

Recommendations: The student research team developed several practical recommendations to improve worker safety and reduce injury risk:

1. Introduce anti-fatigue mats at key positions.
2. Adjust the placement of scrap bins to reduce unnecessary movements.
3. Shorten the shift rotation length from 2.75 hours to 2 hours.
4. Decrease task frequency, especially for repetitive high-force tasks.
5. Ensure exclusive use of counterbalance carts for final product sorting to reduce manual handling.
6. Reorganize stacking carts to minimize strain.
7. Limit the number of boards handled simultaneously.

Outcomes: By following these data-driven and ergonomically informed recommendations, TQM is expected to lower the risk of worker injuries, reduce worker compensation claims, and maintain productivity on the shop floor. The implementation of these safety improvements offers a clear example of how wearable technology can drive meaningful workplace changes.

For a deeper dive into this case study, download the full report here

FIGURE A.6 QR Code 6 to https://makusafe.com/news/data-driven-worker-safety-improvement-at-toms-quality-millwork/

Construction General Contractor Safety Success with MākuSafe Wearable Technology

MākuSafe collaborated with the forward-thinking leadership team at The Weitz Company, on their Mission Critical projects. Through this partnership, Weitz achieved a remarkable milestone: a 0.0 Total Recordable Incident Rate (TRIR) over more than 31 months on sites where MākuSafe was used. This achievement is a testament to their dedication to worker safety, their embrace of innovative solutions, and their commitment to operationalizing technology beyond the pilot stage.

Weitz's Approach to Safety with MākuSafe The Weitz Company has a long-standing commitment to safety, and as their focus on Mission Critical and data center projects grew, they sought out new technologies to protect their workforce. MākuSafe's wearable devices collect and analyze data to help safety teams predict potential hazards before incidents occur. In a large multi-employer construction environment with many moving parts, small near-miss incidents often go unreported. MākuSafe's technology helps prevent these unreported near-misses from escalating into serious accidents by providing real-time data on conditions like environmental factors, motion, and worker location.

In a practical example shared by a Senior Project Manager, MākuSafe data revealed multiple slipping motions near a site entrance during a snowy day. Thanks to the alerts, Weitz's logistics team quickly addressed the hazard, preventing potential injuries before they could occur.

Meeting ESG Expectations The company's innovative approach to safety also aligns with the growing environmental, social, and governance (ESG) expectations from their clients. As more companies seek to ensure that their partners prioritize worker well-being and sustainable practices, proactive use of MākuSafe technology supports these objectives. By utilizing real-time safety data and implementing preventative measures, Weitz demonstrates their commitment not only to safety but also to the broader social responsibility of protecting workers and minimizing incidents – a key ESG requirement.

Key Benefits: Safety, Efficiency, and Health Weitz has deployed MākuSafe across multiple job sites, including their trade partners and subcontractors. This technology doesn't just protect workers from physical hazards but was also instrumental during the COVID-19 pandemic. With contact tracing capabilities and worker density mapping, MākuSafe provided critical data to reduce the risk of viral transmission on job sites, helping maintain health and productivity.

A Culture of Safety The Weitz leadership emphasizes the importance of embedding safety into the company's culture. By using MākuSafe technology, they've successfully engaged workers, improved safety performance, EMR and TRIR have improved significantly. As the first construction company to adopt MākuSafe, Weitz has continued to expand its use across various business units, setting an industry standard for safety.

The partnership between MākuSafe and The Weitz Company demonstrates the power of innovation in creating safer, more efficient work environments while also meeting the ESG expectations of clients and stakeholders.

For a deeper dive into this case study, download the full report here

FIGURE A.7 QR Code 7 to https://makusafe.com/news/makusafe.contributes-to-0-0-trir-on-data-center-construction-projects-the-weitz-company/

Appendix 2
Wearable Technology Endorsements from Industry Professionals

We are grateful that so many of our peers were willing to provide their thoughts on how MākuSafe is influencing the safety landscape globally. We were so overwhelmed by responses from our contacts and mentors in the industry and related fields that we wanted to include them, in their entirety, in this book.

Along the way, we have learned from each and every one of these folks, and our approach has been shaped by them. Our hope is that you take something from their comments here. We hope that you'll find it encouraging that so many thought leaders from the safety space and its fringes are seeing that safety has the opportunity for significant progress by embracing the principles and approaches we've suggested in this writing.

Gabe, Mark, and Tom

In today's rapidly evolving industrial landscape, safety management is undergoing a transformative shift – and *Safety and Wearable Technology* is the definitive guide to understanding how wearable technology is revolutionizing the field. As industries face growing pressures to improve safety outcomes, this book is a timely exploration of how data-driven insights from wearable devices can provide a new layer of protection for workers and companies alike.

This book, authored by experts Gabriel Glynn, Mark Frederick, and Tom West, is an indispensable resource for safety professionals, industry leaders, and anyone looking to understand the future of safety management. It breaks down complex topics like sophisticated sensors, IoT applications, and machine learning in a way that's accessible and actionable, making it clear why wearable safety technology is not just a trend – but a critical tool in reducing life-altering incidents and preventing fatalities.

Readers will appreciate how the authors go beyond technology itself to address the cultural and managerial shifts necessary to implement these innovations successfully. With chapters on green flags in safety culture, the imperative of preventing serious injuries, and the evolving role of safety leadership, the book provides a comprehensive roadmap for integrating wearable technology in ways that truly enhance safety outcomes.

In a world where predictive analytics and connected workers are becoming the norm, *Safety and Wearable Technology* underscores why embracing these advancements is essential – not only for compliance but also for creating safer, smarter, and more efficient workplaces.

Corrie Pitzer
CEO & Founder, Safemap International Pty. Ltd.

I worked for the Occupational Safety and Health Administration for many years, and it was our job to investigate workplace incidents and place blame on the employers. Yes, the employer was wrong (and cited) unless they could prove that the blame belonged to the employee! Todd Conklin convinced me, although after my retirement from OSHA, that "blame fixes nothing," "learning and improving is vital," and "learning is deliberate." But learning requires understanding and understanding requires data! Gabe, Mark, Tom and the entire MākuSafe team are committed to helping employers learn and improve through the deliberate acquisition and analysis of data. What a concept! As a 35+ year safety professional, I have recently seen the future and the future is MākuSafe!

Doug Fletcher, former OSHA Official
President, Fletcher Safety Consulting, Inc.
Host "What's The Hazard" Podcast

I highly recommend this book to anyone interested in the future of workplace safety and operational excellence. The authors brilliantly integrate Safety 2.0 concepts from thought leaders like Todd Conklin and Sidney Dekker with their own expertise in cutting-edge Industry 4.0 solutions. This book clearly demonstrates how these complementary systems can revolutionize safety management and business operations.

Chad Veach, CSP, ARM, CPCU
Assistant Vice President-Risk Improvement Innovation, EMC Insurance

Historically, safety tech has lagged behind, but this lack of "technical debt" offers a unique advantage. By embracing wearable technologies and predictive analytics, as outlined in *Safety and Wearable Technology*, safety leaders can bypass outdated systems and move directly into the future of proactive safety management. Claire Jackson's journey is a perfect example of how we can seize this moment to make workplaces safer. It's an essential read for anyone ready to embrace the next era of safety.

Scott Cuthbert
CEO & Co-Founder, Safeopedia Inc.
Canada

With over 35 years of experience leading safety and health programs, I've seen firsthand the complex challenges that safety practitioners face. As injury rates plateau and fatalities rise, professionals around the world find themselves at a critical juncture, asking: How can we break through the barriers to achieve the safety performance we desire?

Safety and Wearable Technology: Impact, Applications, and Implementation in Industry reveals the groundbreaking potential of wearable technology to predict and prevent incidents before they occur. By integrating real-time data, AI, and human insight, this book presents a proactive approach that's reshaping the future of safety. Technology is poised to take us closer to our goal of ensuring every worker returns home safely. Within these pages is a beacon of hope in reaching this lofty endeavor. Are you ready to embrace its potential? Your journey starts here.

Scott Gaddis
Vice President, Global Practice Leader, Safety and Health
Intelex Technologies

I've been tracking wearable tech for over a decade, starting from my involvement in an "emerging issues" discussion group at an insurance company. Since then, I've kept a close eye on the evolution of safety technologies and have spoken at numerous conferences on the subject. This book is an important addition to the conversation, providing clear insights into how wearable safety technology is changing the way we think about risk management.

Insurance entities are not just in the business of paying claims – they are in the business of protection. The ultimate goal is to predict and prevent losses, and wearable safety tech plays a critical role by providing the data we need to identify risks before they result in injuries. After all, you can't solve a problem you haven't identified and analyzed. This book equips readers with the tools to do exactly that.

Today's safety directors are more data-driven than ever, and wearable safety tech offers unprecedented insights. This tech allows us to passively gather metrics that would have taken days of observation in the past. By analyzing data on where risks occur, how often, and under what conditions, we can take action more quickly and effectively.

While this book will certainly be a go-to resource for safety professionals, it also offers significant value to anyone in operations. As I often say, safety, quality, and productivity are like a three-legged stool – when one improves, so do the others. The authors understand this, and their work reflects a broader perspective that touches on all aspects of business performance.

One of the key strengths of this book is its balanced approach. The authors not only explain the benefits of wearable tech but also address potential challenges and objections. For instance, I had always wondered why haptic feedback was left out of the MākuSafe device, but after reading the book, I understand the reasoning behind that decision. It forces readers to think more broadly and critically about wearables and how they can be best implemented in their own workplaces.

I first learned about the MākuSafe concept more than eight years ago, when it was still a sketch on a napkin. I knew it was a groundbreaking idea then, and this book reinforces that belief. The authors are among the smartest people I know, and every conversation with them leaves me more informed and more optimistic about solving workplace safety challenges. I highly recommend this book to anyone who wants to stay ahead of the curve in safety and risk management.

Chris Murphy, CSP, CRM, CPCU
VP Commercial Risk & Safety, Holmes Murphy

I have watched the MākuSafe technology platform grow and expand for more than four years. Every time I meet with the team, I marvel at the advancement in the technology and how they have really improved the platform over time. I am delighted that the company has made such a huge commitment to research and development, with nearly half of the workforce dedicated solely to product development. I think that is an amazing trait of a startup to continue to heavily invest in the product and not be lured solely by sales and marketing of the product.

Their technology platform has great architecture and can interface with many "dashboard" platforms to integrate with existing systems versus a standalone system. This equates to a better interface and higher likelihood of acceptance and sustainability in an organization.

Wearables are truly the future of health and safety. They provide the most proactive means to address hazards and exposures. Although near-miss reporting is a significant strength of my company, it requires a human act to complete, with many competing priorities. Wearables act independently and without prompting to provide nearly real-time evaluation of hazards. It is clearly the most proactive thing that can be done in the spectrum of injury prevention.

MākuSafe heavily invests in their customers as well, ensuring the product implementation and support teams are engaged in the launch and continued follow-up with sites. We have a site that is using the MākuSafe technology. Every aspect of the sales, installation, start-up, and user experience has been exceptional. The staff is top notch and really helpful. They are truly invested in the success of the product to work and make the user experience successful. Let's face it – workers' acceptance of new technology is mixed. For some, there are reservations on what is being monitored and how the data is being used. MākuSafe is very helpful at promoting worker acceptance through education, such that the worker embraces the use of the technology, knowing the nobler cause is their personal well-being/health and safety.

We have had some interesting revelations from our implementation and use of the MākuSafe. We had an operation that was seeing increased levels of VOCs in an area where no solvents are present. Through the proactive technology of MākuSafe, we were able to determine the source as workers using an air wand to remove some residual solvent from a degreasing operation. The air wand created the vapors that were detected by the wearable. Without this, we would have never known this exposure was occurring. Due to this proactive approach, we were able to look at the degreasing operation to determine options to reduce the residual solvent and eliminate the exposure to our employees.

In another operation, the wearable noted that one employee had higher energy expenditure and higher scores relative to ergonomic exposure. After reviewing the job, we noted the job was not set up well for workstation adjustability. The stature of the worker was not well suited to the height requirements since one of the job elements also included a hard-to-reach area. The local ergonomics team was able to review this and address the issue.

I think MākuSafe has become the market leader in wearable health and safety technology. Their platforms and user interface are excellent and only further evolve to become even better. Their commitment starts at the top, with the senior leadership of Gabe, Mark, and Tom. They are truly authentic in their leadership style and have built an excellent product and team. They are trustworthy and committed to excellence.

Mark Jones, PhD, MBA, CIH, CSP
EHS Director, Plastipak Packaging
Named one of America's Safest Companies by EHS Today
Past President and Senior Board Member, Michigan Safety Conference
30 years in Safety with Ford Motor Company

As a seasoned safety professional with over 40 years of experience, I have witnessed significant changes in the field. From my early days of using basic tools to measure flow rates, I now find myself immersed in cutting-edge concepts like the Fourth Industrial Revolution. While the methods may have evolved, the core purpose of safety remains unchanged – identifying and addressing the underlying causes of accidents.

Regardless of the specific exposure, be it chemical or physical, safety professionals strive to identify and mitigate the events at the bottom of the accident triangle. These events encompass the activities and actions that put employees at a higher risk of accidents. For instance, in the case of a chemical exposure, it could be the frequency of slips while carrying chemicals. By analyzing data, we can determine that employees carry chemicals 30 feet four times a day, resulting in 40,000 instances per year. Of these, there are 80 near-misses, 22 instances requiring medical attention, and 5 cases of missed work. As safety professionals, our goal is to reduce both the frequency of carrying chemicals and the likelihood of slipping with each occurrence.

Similarly, we address physical hazards, such as back injuries, by focusing on the frequency of lifting boxes. For example, individuals on a packaging line lift boxes 10 times per hour, resulting in 160,000 lifts annually. By examining the data, we uncover 140 near-misses, 18 instances requiring medical treatment, 9 cases of lost work days, and, unfortunately, 1 permanent disability. Our aim is to reduce the number of lifts and minimize actions that increase the risk of injury, such as twisting, lifting below the knees, and lifting above the shoulders.

Focusing on the bottom of the accident triangle has proven highly effective in reducing accidents and injuries. By addressing unsafe acts and conditions, we can prevent near misses, medically required injuries, time loss, and even fatal or permanent total disability incidents. The challenge lies in quantifying these elements accurately to determine where improvements should be made. Wearable technology presents a viable solution. For instance, MākuSafe's matchbox-sized wearable sensor platform allows us to collect data that helps identify problem areas and prioritize corrective actions. Unlike traditional safety methods that rely on observations, wearable technology provides reliable data to analyze critical behaviors, leading to informed decision-making. *Safety and Wearable Technology: Impact, Applications, and Implementation in Industry*, written by Gabriel Glynn, Mark Frederick, and Tom West, helps readers understand the importance and usefulness of wearable safety tech. It offers valuable insights for implementing this technology successfully in various industries.

This book helps bridge the gap between professionals who seek data-driven solutions and the potential of wearable technology. Divided into five parts, it covers the necessity of data in safety practices, explores the functionalities of wearables and machine learning, provides implementation insights, offers a glimpse into the future, and presents case studies demonstrating the diverse applications of wearable technology. The authors' credibility is unquestionable, with ample experience in implementing wearable technology solutions across various industries. By sharing their experiences with early adopters, they guide readers through successful implementation, helping them avoid costly mistakes.

However, the book goes beyond mere technological explanations. Rather than relying solely on beeping alarms to notify employees of wrongdoings, it focuses on utilizing wearable technology to eliminate exposure within the workplace. By targeting the unsafe actions and conditions that drive loss, this book equips safety professionals with the knowledge they need to reduce the likelihood of accident consequences.

Whether you are a safety professional seeking to enhance your understanding of wearable safety tech or an industry leader in need of practical implementation strategies, *Safety and Wearable Technology: Impact, Applications, and Implementation in Industry* is an essential resource. It empowers readers to comprehend the pros and cons of wearable safety tech and provides invaluable guidance on successfully integrating these technologies into their facilities. Prepare to embark on an enlightening journey that will revolutionize your approach to workplace safety.

Rick Fineman
Vice President of Loss Control
Berkshire Hathaway Homestate Companies
Workers Compensation Division

The secret to innovation isn't ideation – it's inspiration and iteration. The passionate team behind MākuSafe are pioneers behind the design and daily use of wearable technology that truly achieves their mission of making workplaces safer. The key to their success has been tight collaboration with a wide range of partners and the ability to translate learnings into product design over a number of years. I'm thrilled that all this hard-earned knowledge has been condensed and summarized into a highly readable book that will broaden the audience who may benefit from the experience and insights of the MākuSafe team. Highly recommend!

Rob Galbraith, CPCU, CLU, ChFC
Founder and CEO, Forestview Insights

Author of the bestseller *The End Of Insurance As We Know It: How Millennials, Insurtech, and Venture Capital Will Disrupt the Ecosystem*, Rob has been called "a global insurtech thought leader," "a mentor to many in the insurance industry," "the most interesting man in insurance," "the prophet," and "the WD40 of insurance."

As a 33-year safety professional who has been working with MākuSafe as a customer for seven years, I wholeheartedly endorse MākuSafe, a pioneer industrial safety technology. In an era where workplace safety is paramount, this startup stands out for its innovative approach to safeguarding employees and streamlining safety protocols.

MākuSafe has developed a cutting-edge solution that sets a new benchmark for safety technology. Their product leverages advanced sensors, real-time data analytics, and intuitive user interfaces to deliver unprecedented levels of employee protection and operational efficiency. With the ability to integrate these technologies into existing operational systems, they make it easier for organizations to proactively manage risks and ensure a safer working environment.

What truly sets MākuSafe apart is their commitment to both technological excellence and user-centric design. Their team of experts collaborates closely with industry leaders and customers to tailor solutions that address real-world challenges, ensuring their technology is practical and effective. This dedication to understanding and responding to the needs of their clients is a testament to their passion for enhancing workplace safety.

Using MākuSafe's safety technology is a commitment to creating safer, more productive workplaces. Their innovative solutions are transforming industrial safety. I strongly recommend MākuSafe to any organization looking to elevate their safety performance and protect their employees from the risk of injury. Their forward-thinking approach and unwavering dedication make them an invaluable partner in the quest for a safer, more secure working environment.

Bryon Snethen
(Retired) Vice President, Risk Improvement
EMC Insurance

As the Director of Client Risk Management at Henriott, an independent insurance agency and risk management consulting firm committed to helping clients mitigate and recover from critical incidents, I understand firsthand the importance of proactive safety measures. When it comes to workplace safety, especially in high-risk industries, the real priority is protecting your employees' lives and well-being. Ensuring a safer, healthier workplace is critical in today's tight labor market, and managing this risk – whether through self-insurance or traditional coverage – can make or break a company.

That's why *Safety and Wearable Technology: Impact, Applications, and Implementation in Industry* is a game changer for businesses striving for a strong safety culture and a healthy workforce. The rise of wearable safety technology is an essential evolution in risk management, providing real-time insights and predictive analytics that go beyond traditional safety measures.

The expertise of Gabriel Glynn, Mark Frederick, and Tom West is evident, having developed solutions over an eight-year period that position them as leaders in this space. Their approach demystifies the implementation of wearable technology and offers clear guidance on overcoming the cultural and operational challenges associated with these tools, ensuring that the transition to wearable safety tech isn't just a tech fad but a real strategy to protect employees and reduce costs, turning reactive strategies into proactive prevention.

For example, in "Chapter 6, Sophisticated Sensors," the authors explore how cutting-edge wearable devices equipped with environmental and biometric sensors are transforming workplace safety. These sensors monitor everything from air quality to worker movements, delivering real-time data that enables companies to react swiftly to potential hazards, reducing the risk of injuries and keeping employees safe. This proactive approach turns data into action, preventing incidents before they occur – a true revolution in safety management.

This book offers a thorough exploration of how industries can effectively integrate these innovations into their operations. It provides valuable, practical advice for organizations at every stage of their safety journey, whether they are early adopters or just beginning to explore the integration of Industry 4.0 principles.

At Henriott, we help companies both large and small manage their risks, and wearable technology is becoming one of the most effective tools for that. Whether you're a risk manager, safety professional, or a business leader looking to reduce claim frequency and severity, this book provides the roadmap you need. It's practical, actionable, and written by people who know how this tech works in the real world.

If you're serious about staying ahead of the curve in safety management and ready to take control of your insurance costs – whether through self-insurance or traditional means – *Safety and Wearable Technology: Impact, Applications, and Implementation in Industry* needs to be on your desk or iPad.

Zachary S. Finn MSM, AINS, ARM, HACCP
Director of Client Risk Management, Henriott

As a safety professional with over two decades of experience and a focus on wearable technologies, I have witnessed firsthand the evolution of safety wearables and their transformative impact on workplace safety. The history of wearables, as outlined in this book, highlights their deep-rooted presence in human society, dating back centuries to innovations like glasses and pocket watches. The authors' work in the wearable technology space and their commitment to injury prevention for almost a decade underscore the importance of the insights they share in this book.

Today, safety wearables represent the next frontier in leveraging technology to enhance workplace safety. The exponential growth in wearable technology adoption, as evidenced by the staggering market figures, underscores their increasing significance in various industries, including safety.

The historical narrative of wearables as cultural disruptors, life savers, personal entertainment devices, and always-connected tools resonates deeply with the evolution of safety wearables. Just as pocket watches revolutionized timekeeping and pacemakers extended lives, modern safety wearables hold the promise of revolutionizing workplace safety practices.

As the owner of a safety consulting company, I can foresee a future where wearable safety technology becomes ubiquitous on job sites, enhancing not only the well-being of workers but operational efficiency for safety professionals. This book serves as a valuable resource for both seasoned professionals and newcomers to the field, shedding light on the profound impact of wearables on safety practices and paving the way for a safer and more technologically advanced work environment.

I want to thank MākuSafe for allowing me to read this book and share in this experience. Here's to revolutionizing injury prevention!

Heather Chapman, MS, CSP, CHMM, CEAS III
Owner, Paradigm Safety
AIHA Total Worker Health Task Force Member
AIHA Ergonomics Committee Member
ISO 45001 Lead Auditor

I have worked in the insurance technology sector for the past 47 years, which included leading the development of two insurance agency management systems, with the last one for my own company, which was acquired by Vertafore in 1995. That system, Sagitta, is still running some of the largest insurance brokerage firms in the world. About 12 years ago, I pivoted my consulting practice from fee-for-service to equity-for-advice and, today, I am an equity adviser for nine insurtech startups.

MākuSafe is, sadly, not one of my equity insurtechs, but we became acquainted when I was running an insurtech competition for ACORD in 2019. MākuSafe won the competition that year, and I have followed their rise to prominence in the wearable safety technology space with a sort of secondhand pride ever since.

More than half the technology products that I review every year are risk-mitigating in nature, and I believe this will be the future of the insurance business. Selling a person or a business an insurance policy is necessary because we will always have unforeseen accidents to contend with, but making a person or a business safer creates a much more enduring business relationship.

While artificial intelligence is all the rage at the moment, I am a firm believer in the efficacy of machine-augmented human intelligence as a superior solution for many of our challenges. Strategically placed cameras with AI-enabled video analytics are an excellent addition to any workplace, but nothing beats a wearable device that can measure most of the environmental impacts for workers. And, perhaps the most important feature of MākuSafe wearables, the ability for workers to press a button and verbally report a safety threat or near accident situation is of paramount importance to safety-conscious employers and their insurers.

The story of MākuSafe is an invaluable one for our society!

Frank Sentner
Owner, Sentwood Consulting
Advisor & Board Member, InsurTech Hartford

Something is broken in the workplace, and there is a solution.

As the President and CEO of a portfolio of companies heavily concentrated in manufacturing, I saw firsthand that there was a problem in the workplace that was exposing employees to risk of injury, long-term bodily harm, or worse. Traditional responses to safety concerns were not resulting in substantive changes. In looking for innovative solutions, I found two important answers. The first was that meaningful actions to counter safety threats must rely on data that gives information specific to not only the work being done but the threats that are present, threats that are recognized, or the threats that are not acknowledged. The second is that utilizing that data requires a very different process of involving those closest to the work in designing and implementing systems to counter threats and to build organizational capacity. Lastly, when implemented successfully, these will lead not just to a safer workplace but to a stronger culture of engagement and can boost productivity and profitability.

Why should you read this book? You will learn about wearable safety technology that collects data to provide you with the information you need to make sound business decisions that protect your employees. You will also learn how to introduce and utilize wearable safety devices to efficiently engage those close to the work in the process of building organizational capacity. It is as important to have management read this as it is to have frontline workers read this. There must be alignment of interests, perspective, and approach to move the needle on safety and culture. The authors of the book bring years of pertinent direct involvement of developing the devices, introducing them into a variety of industrial workplaces, and working with teams at those companies on interpreting data and building solutions from the data.

Something is broken in the workplace, and there is a solution. *Safety and Wearable Technology: Impact, Applications and Implementation in Industry* will get you on the road to developing your solution.

Daniel Goldstein
CEO of Go ESOP LLC
Executive Fellow at the Rutgers Institute for the Study of Employee Ownership and Profit Sharing
Fellow at Ownership at Work
Former CEO, President, and Board Member at Folience

As a safety technologist with extensive experience designing wearable and connected work initiatives in high-risk environments, I've witnessed how wearable safety technology can enhance health and safety risk management. This book provides an essential and thorough exploration of the holistic integration of wearables into workplace systems, offering both practical insights and the foundational knowledge needed to leverage these tools effectively.

Drawing on years of hands-on experience, the authors leverage their diverse backgrounds to highlight the transformative potential of wearable technology. They provide invaluable guidance on how to implement wearables successfully, ensuring that readers are equipped to navigate the complexities of integrating this technology into risk management practices.

I highly encourage all health and safety professionals, along with their support teams, to read this book. It offers a comprehensive understanding of how wearables can enhance safety outcomes and improve worker well-being in today's connected world of work

Cam Stevens, MSc Human Factors
Founder & CEO, Pocketknife Group
Founder, Safety Innovation Academy
Trusted Guide & Coach for SafetyTech, Work Design & Digital Transformation
Australia

In today's fast-paced work environment, timely and reliable data is vital for an effective safety or continuous improvement (CI) program. Wearable technology provides these professionals with actionable insights, allowing them to proactively address root causes before they escalate. Companies using these wearables and the advanced AI analytic software are gaining significant benefits without being overwhelmed by data. For organizations with limited resources, focusing on real issues efficiently is crucial to success.

Chris H. Hill
Director, Engineering Services
Director, TAP Program
Iowa State University Center for Industrial Research and Service (CIRAS)

The future of insurance requires us to be proactive and provide solutions for risk management before claims even happen. But to provide the right solutions, we need the best data. Wearable safety products like MākuSafe are revolutionizing the industry by providing insurers and policyholders real-time information to stay ahead of the risks and protect workers.

As part of our robust loss control program at EMC Insurance, we've partnered with MākuSafe to offer policyholders this technology at preferred pricing and have seen significant improvements in workplace safety culture as a result – an incredible win for safety managers. Moreover, the data collected from these wearables has the potential to enhance our underwriting accuracy, allowing us to tailor policies more precisely to the risk profiles of our policyholders.

This book is a consequential read for anyone interested in reducing worker injury rates and is a tremendous example of how Insurtech companies are shaping the insurance industry of tomorrow.

Scott Jean
President and CEO
EMC Insurance

Wearable technology in the industrial safety space is still fairly new. While various offerings continue to better define their functionality and value, it is clear that MākuSafe has emerged as the leading platform to impact predictive outcomes.

By capturing data sets important to the health and safety of individual workers and overall facilities, and leveraging data analytics, MākuSafe empowers safety leaders to dramatically improve safety metrics and react in real time to needed modifications in job and workplace design.

Advancements in smart manufacturing and other forms of digital technology applications have impacted productivity and competitiveness in many ways. It is long overdue that a platform emerges to connect workers to the automated environments around them to greatly impact worker safety outcomes as well as productivity and efficiency. MākuSafe has found the perfect balance of data, analytics, and communication to leadership in real time, to truly impact safety in the workplace utilizing technology that is robust and cost effective.

Paul G. Jantsch
Founder and Principal, 3D Capital Investors LLC

In today's VUCA (Volatile, Uncertain, Complex, and Ambiguous) world, time has become a precious commodity for employees at all levels. The demands of the modern workplace often leave us with insufficient time to accomplish our tasks effectively. Therefore, it is essential for us to enhance our ability to identify priorities and focus areas more efficiently.

Fortunately, we now have access to innovative technologies that can significantly improve our time management. Tools such as artificial intelligence, automation, and wearables are reshaping how we work, allowing us to be both effective and efficient. At L'Oréal, we are embracing "new generation tools" that add value by conserving time and eliminating unnecessary distractions.

In the realm of health, safety, and wellbeing, these advancements have a vital role to play. Safety professionals are tasked with the important responsibility of selecting the right tools at the right time, tailoring their approach to align with the organization's cultural journey. By leveraging these technologies, we can create a safer and more productive work environment for everyone involved."

Malcolm Staves, FIChemE
Global Vice President Health & Safety
L'Oréal S.A.

Since first meeting Gabe Glynn in 2017, I have been impressed with his vision to empower workers with wearable technology. During my initial conversations on his nascent company (three employees at the time!), Gabe shared his vision: to inform workers to improve their decision-making capability; to companies to simplify through automation the collection of "actionable intelligence" (data trends); and to insurers to prospectively avoid injuries/illnesses in the workforce.

By now, that vision has matured to a successful mission. With an international impact, MākuSafe has promoted real-time data gathering that allows workers, companies, and their insurers to make data-driven decisions.

In this text, you will find the information to argue for wearable technology deployment addressing hazards of heat, forceful movement, noise, and some chemical compounds. You will also find that through the company's technology development, they have continually improved the utility of the technology and made a compelling cost/benefit analysis case for adopting the technology.

The book is a testament to the education and experiences of the company principals and the staff, as it offers practical promotion of employee engagement and metrics for transparency in exactly what is being collected and shared for data.

Read it, highlight it, then start your journey into wearable technology. There is no time to waste in pursuit of it.

Respectfully,
Steven D. Jahn, CIH, MBA, Fellow of AIHA
Past Chair, Real-Time Detection Systems Volunteer Group, American Industrial Hygiene Association

As a professional in the risk and occupational health and safety space for over 20 years, I am like many who have found myself hopeful in the benefits from wearable technology that align both strategically and tactically toward effective safeguards, communications, and controls for those that are closest to the risk: our workforce. This book is an important read for anyone wanting to learn about not only the benefits of today's wearable safety technology, but also strategic considerations for application specific to your workforce and successful implementation.

As today's world of work includes increasing complexity, length and depth within our supply chains, acknowledging how we are developing solid coupling-points between those performing the work, those impacted by potential risk exposure and driven from the desire to improve, achieving safe work in our shared world of work is taking on new meaning and urgency. Technology plays an increasingly important role in achieving this.

Better knowledge in health and safety, from Serious Incident/Fatality (SIF) intervention strategy to systems thinking and management is shedding new light on the importance of how quickly we recognize risk, determine if appropriate safeguards are in place, and enable end-to-end communication between the workers and those charged in resourcing them for safe work. I've asked these three questions below throughout my career, and I've found that leveraging the right technology, such as effective wearable solutions, enables stronger communications in each of these three areas:

1. Is work happening like we agreed/believe it's happening?
   How is the corporate end intentionally informed of how work is really happening at the working end? Has the scope of work changed? Do conflicting work instructions force employees to 'figure it out' and assume new risk?
2. What informs us/provides assurance that work is happening at acceptable levels of risk?
   What mechanism *exists for workers to easily communicate when experiencing different/new levels of risk? How are we prioritizing SIF risk related to lethal energy exposure and insufficient safeguards? How does our work system monitor/capture information for leadership to continually be informed as to the nature of work, safeguards, and resources?*
3. How are we improving?
   Do we have a systems approach that prompts continual monitoring and assessment of risk? Have we enabled those closest to the risk to capture improvement ideas and connect them to resources and leadership for improvement? Do our tools and methods for risk monitoring /assessment capture data in a manner that informs the organization (leadership) as to how work is being performed for safe outcomes, and informs leadership decisions for resources and planning?

As much as these questions have helped me in my career, what's important to recognize is the role of technology has enabled the continuity of communications within the complexity of today's work environment, while capturing data to inform the organization as to current levels of risk and continual improvement. Wearable technology plays a critical role within a healthy work system, as it connects risk information at the working end of the organization to those in a position to enable resources, change and support specific to the risk experienced by the worker. What's especially important to recognize is that the right wearable technology enables a strategic focus and intervention strategy on the type of risk that hurts people the most and creates tactical methods for monitoring and assessment at the working end, specific to the risk experienced by the worker and their job tasks.

As a safety professional who now works at a technology company focused on effective use of technology to manage risk across our supply chains, I'm grateful for the work MākuSafe is doing in the wearable technology space, as well as on the educational side toward prioritizing serious risk, strategy, and implementation specific to their clients work environments. I've coined the phrase "The right technology enables our organizations to manage risk at the speed of business", and it's refreshing to see others sharing the same drive and mission toward a better future.

Scott DeBow, CSP, ARM
Health, Safety & Environmental Director at Avetta
Author: Safety Management Systems in a Joint-Employer Environment
ASSP Board of Directors

The folks at MākuSafe haven't just penned a tutorial on wearable technology, they have introduced the interrelationship between human safety and operational health. The field of occupational safety and health is in desperate need of real-time risk data. Most workplaces are dynamic in nature and at present, safety and health risk data often comes from point sources such as assessments, inspections, and observations. This approach is analogous to checking automobile health 50+ years ago – when we had to pull over to check the oil, coolant, and air pressure. Unfortunately, there is a lot of unmanaged risk growth that happens in between these health checks. The technology outlined in the book will allow organizations to collect a real-time risk data dashboard and make educated decisions to create a safer and more successful work environment. The information herein is certain to initiate a step change in the way organizations manage their human and operational dynamic.

Peter Susca
Founder & Principal
OpX Safety / Operational Excellence LLC

Safety and Wearable Technology: Impact, Applications and Implementation in Industry is a timely resource for the occupational health and safety field. This provides an in-depth exploration into how wearable technology is actively reshaping safety management across diverse industries. With topics ranging from data-driven safety insights and IoT integration to the nuances of machine learning, the authors illuminate the powerful role technology can play in enhancing workplace safety. Beyond the technical insights, the authors explore the organizational culture and management practices essential for adopting these innovations effectively. It is a compelling read for safety professionals and industry leaders who are committed to building safer, smarter workplaces.

Nir Keren, Ph.D.
Associate Professor, A&BE, Iowa State University
Director, Occupational Safety Program, NIOSH Heartland Education and Research Center for Occupational Health and Safety
Director, VirtuTrace Laboratory for Applied Decision-Making Research in Virtual Reality

"Safety and Wearable Technology: Impact, Applications, and Implementation in Industry" is a must-read for anyone in safety and health dedicated to continuous improvement, considering the use of wearable technology as part of their safety improvement efforts, or looking for innovative ways to gain a better understanding of work as experienced by their employees. MākuSafe has been a leader in the wearable safety technology space for nearly a decade. They recognized early the transformative potential of wearable technology in safety and have been instrumental in normalizing the use of wearables to enhance workplace safety.

Wearable technology, spearheaded by organizations like MākuSafe, is here to stay and will redefine the landscape of workplace safety. The team at MākuSafe exemplifies entrepreneurship and a commitment to continuous learning. This book is a testament to their expertise and vision.

Lisa Brooks, CIE
Principal, Nexus HSE

Lisa Brooks is a certified industrial ergonomist and has been an ergonomics, S&H practitioner for over 30 years. She is known for her passion and energy and has proven leadership abilities, and distinctive technical expertise in ergonomics, human and organizational performance (HOP), and fatal and serious injury & illness prevention. She led global HSE networks at ORCHSE Strategies and The National Safety Council for 8 years after spending 23 years in industry in various global safety and ergonomics leadership roles for GE, International Paper, and Alcoa.

Glossary

Advanced manufacturing: Manufacturing that uses innovative technologies and processes to improve the design, production, and performance of products. It incorporates techniques like automation, robotics, and additive manufacturing (3D printing).

Artificial intelligence (AI): The field of computer science dedicated to creating systems that can perform tasks requiring human-like intelligence, such as learning, problem-solving, and understanding language.

ASTM E2920: A metric measuring the severity and nature of a workplace injury, used mainly to identify cases with meaningful connections to work and potential consequences that would have value for future prevention.

BBS (Behavior-Based Safety): An approach to improving workplace safety by focusing on the behaviors of employees through observing and analyzing behaviors, providing feedback, and encouraging safety practices.

Connected worker: An employee who uses digital tools and technologies, such as wearable devices, mobile apps, or IoT-enabled equipment, to access information, communicate, and receive real-time updates while on the job in order to enhance efficiency and safety.

DART (days away, restricted, or transferred): The metric used to track workplace injuries and illnesses that result in an employee being unable to work (days away), assigned to light duty (restricted), or transferred to another job as a result of the injury or illness. DART rates are used to assess the severity and impact of workplace incidents on employee productivity and can indicate a company's overall safety performance.

EHS (Environmental Health & Safety): The management and regulatory practices aimed at protecting the work environment, ensuring the health and safety of employees while complying with legal and industry standards.

Hazard: A situation or occurrence that poses a potential danger or harm to employees, property, or operations.

HOP (Human and Organizational Performance): A framework used to analyze and improve the ways people work within an organization; the focus on optimizing performance by understanding human behavior, system interactions, and organizational processes.

Human-centric technology: Technology that prioritizes user needs and experiences, ensuring that devices are intuitive, ethical, and improve quality of life.

Industry 4.0 –Digital Transformation: The ongoing Fourth Industrial Revolution, which is characterized by the integration of technologies like IoT, AI, and advanced robotics into manufacturing processes.

IoT (Internet of Things): A network of connected devices that collect and share data over the internet, enabling automation and smarter decision-making.

IIoT (Industrial Internet of Things): The application of IoT technologies specifically in industrial settings, such as manufacturing and transportation. Unlike general IoT, which connects consumer devices, IIoT connects industrial machines, sensors, and systems to monitor and optimize industrial processes, improve safety, and increase efficiency through real-time data analytics and automation.

Machine learning (ML): A subset of artificial intelligence focused on developing algorithms that enable computers to learn from and make predictions or decisions based on data without being explicitly programmed for each task.

PPE (personal protective equipment): Gear and clothing designed to protect individuals from injuries or illnesses while working in hazardous environments.

Risk: The likelihood that a hazard will result in harm as well as the severity of that harm if it occurs.

SIFs (serious injuries and fatalities) or P-SIF (potential SIFs): Incidents that result in, or have the potential to result in, severe harm or death to workers. The concept is used to prioritize and address the most critical safety risks, with the goal of preventing the most serious types of workplace accidents through targeted interventions and safety measures.

Smart factory: An interconnected and automated production environment where efficiency is enhanced through flexibility, monitoring, and adaptive control of manufacturing processes.

Index

O

P

R

S

T

V

W

Z

Kolibri 360 GmbH
Ferdinand-Jühlke-Straße 7
99095 Erfurt, Deutschland
produktsicherheit@kolibri360.de